Implausibles

A compendium of conceptual physics

in 2 parts :

basic review of mathematics

exposition on conjectures of physics
(Mathematical Archaeology)

David L. Birdsall, Ph.D.

BZB

Acknowledgments:

Computer Programming was done using GNU **gfortran Fortran95 compiler** from the gcc-9.2.0-32 package of programs, https://gcc.gnu.org, and Arduino Language **C/C++ compiler**, https://www.arduino.cc .

Amber Energy Minimization and Molecular Graphics were rendered under **Chimera** software from UCSF, http://www.cgl.ucsf.edu/chimera,
and **Xfit** from the **XtalView** suite of CCMS programs by Duncan McRee, email dem@scripps.edu .

Some plotting graphics were prepared through GNU **Graph** program, http://www.forum.padowan.dk
and **Microsoft Mathematics v4.0.110.8.0000**, Microsoft Corporation, 2010 ;
otherwise all figures were designed with Microsoft **Paint** and **Paint 3D** applications.

Graphic Interpolations were conducted using applications from

Graphing Calculator v1.11	email uclahlaw@gmail.com
TechCalc v1.8.1	http://www.roamingsquirrel.com

The mathematical review (Implausibles, part 1) was aided by much study and many references.
Three very useful texts were:

Calculus and Analytical Geometry, George B. Thomas, Jr., Addison-Wesley Publishing Company, Inc., 1972
Physics The Easy Way, Robert L. Lehrman, Barron's Educational Series, Inc., 1998
Chemistry, Martin S. Silberberg, McGraw-Hill, 2003

Word processing was completed with GNU **OpenOffice** program, http://www.openoffice.org .

The book cover was designed and created using GNU **Gimp 2.10** image editing software, https://www.gimp.org .

PDF files for printing were created under **OpenOffice** and collated with **PDF reDirect v2**, http://www.exp-systems.com .

Cover art : molecular surfaces of calculated $(GXPGXP)_4$ structures

 ISBN: 978-0-9786979-7-6
Printed in The United States of America

Implausibles

part 1

basic review of mathematics

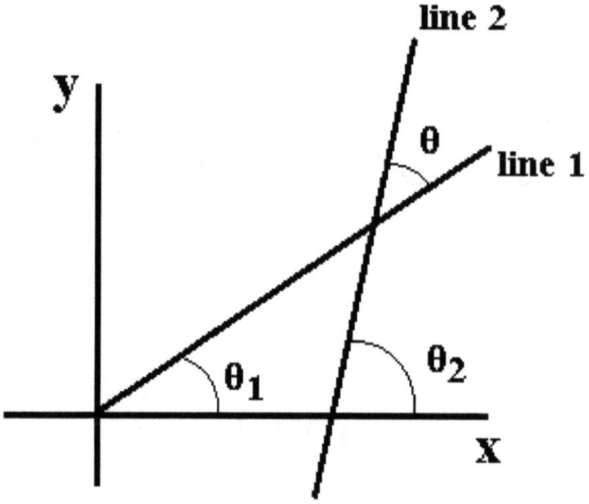

$$\tan\theta = \left| \frac{m_2 - m_1}{1 + m_2 m_1} \right|$$

Implausibles

part 1 : Basic Review of Mathematics

Listings

Review of Mathematical Physics

Before we delve into the implausible of our possible contentions and theories, first let us review some aspects of mathematics used to elucidate basic physical principles and observations.

Algebra

Graph Characteristics

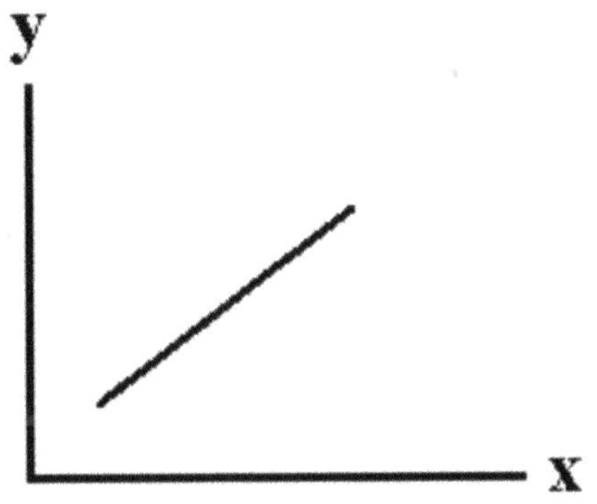

linear : $y = mx + b$

$y - y_1 = m(x - x_1)$

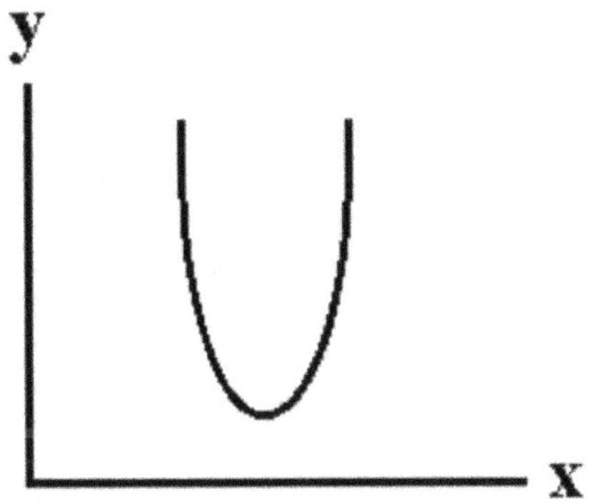

quadratic : $y = ax^2 + bx + c$

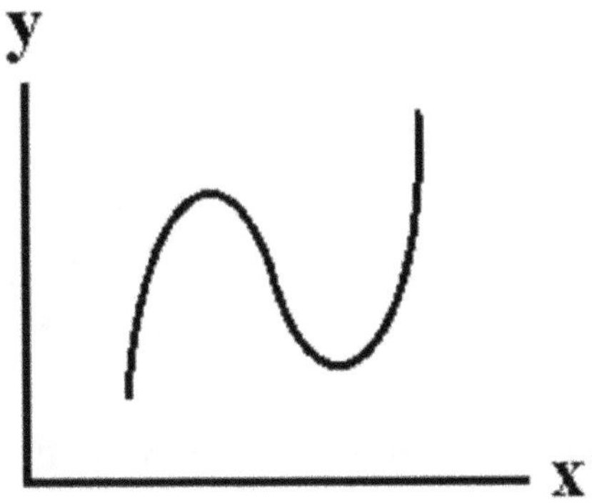

cubic : $y = ax^3 + bx^2 + cx + d$

Parametric Equations

A third variable is used to relate functions of the other two.

e.g., let $t ==$ parameter for $f(t)$ and $g(t)$; $\therefore$ $x = f(t)$ and $y = g(t)$

This shows how x and y vary with t ; one notes the domain and range of (x, y) according to the variable t; e.g., (x, y) == position while t == speed and direction (sign of t) . One goes back to rectangular, e.g. (x, y), by eliminating the parameter (t): solve for t in terms of x, or y, then substitute in the 2nd equation to get $y = f(x)$, or $x = f(y)$.

Quadratic Functions

Standard form : $y = f(x) = ax^2 + bx + c$

c = y-intercept (value of y when $x = 0$) ;

at most there are two x-intercepts (found when $y = f(x) = 0$).

When a is positive : the curve opens up (on the graph).

When a is negative : the curve opens down.

Another form (standard for parabolas) is : $f(x) = a(x - h)^2 + k$, where vertex is : (h, k) ; (x – h) moves the vertex positively h units along, and the axis of symmetry is : x = h ; k == units up on y axis.

vertex of quadratic function curve is at : $\left(\frac{-b}{2a} , \frac{4ac - b^2}{4a} \right)$

axis of symmetry is (the line) : $x = \left(\frac{-b}{2a} \right)$

The Quadratic Formula (for calculating x) is : $\frac{\sqrt{-b \pm b^2 - 4ac}}{2a}$

(If the radical is negative, then x is undefined and does not occur.)

The quadratic equation is always of a parabola.
The vertex is at [-b / 2a , f(-b / 2a)].
The axis line is : x = -b / 2a , or x = h .
For graph transformations, (x – h) moves the vertex positively h units (i.e., to the right).

When $a > 0$, the parabola opens upward.
When $a < 0$, the parabola opens downward.
When $f(x) = ax^2$, the vertex is at (0, 0).
As $| a | \uparrow$, y increases and the parabola becomes steeper (more narrow).

When in the $ax^2 + bx + c$ form, one can complete the squares to find the vertex, this being the maximum or minimum of the quadratic function:

If a is not "1" then factor to get the x^2 term alone (i.e., without a coefficient).
$ax^2 + bx + c = (ax^2 + bx + d) - d + c$, where $d = (b / 2)^2$.
$\therefore = (x + d^{1/2})^2 - d + c$, where the vertex is $[-d^{1/2} , (-d + c)]$.
Here, $(-d + c) = k$, in the $f(x) = a(x - h)^2 + k$ form.

e.g., $f(x) = 2x^2 + 8x + 7$
$= 2(x^2 + 4x) + 7$
$= 2(x^2 + 4x + 4 - 4) + 7$, as $(4 / 2)^2 = 4 = d$
$\therefore 2(x^2 + 4x + 4) - 8 + 7 = 2(x + 2)^2 - 1$
and the vertex is at (-2, -1);
$-d^{1/2} = -4^{1/2} = -2$, $-2d + c = -(2 \bullet 4) + 7 = 8 + 7 = -1$

Astronomy:

Lunar Day (== the rotation of the moon around the earth) == Synodic Period,
the average of which is 29 days + 12 hours + 44 minutes
$\therefore$ Daylight on a point of the moon lasts around 2 weeks (or half the synodic period).

Biochemistry

Bilateral Animal

dorsal (top; back)
anterior (head) ← → posterior (tail)
ventral (bottom)

Enteric Cells == cells of the intestinal lining

Enzyme Catalysis

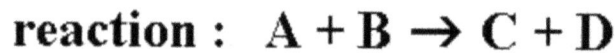

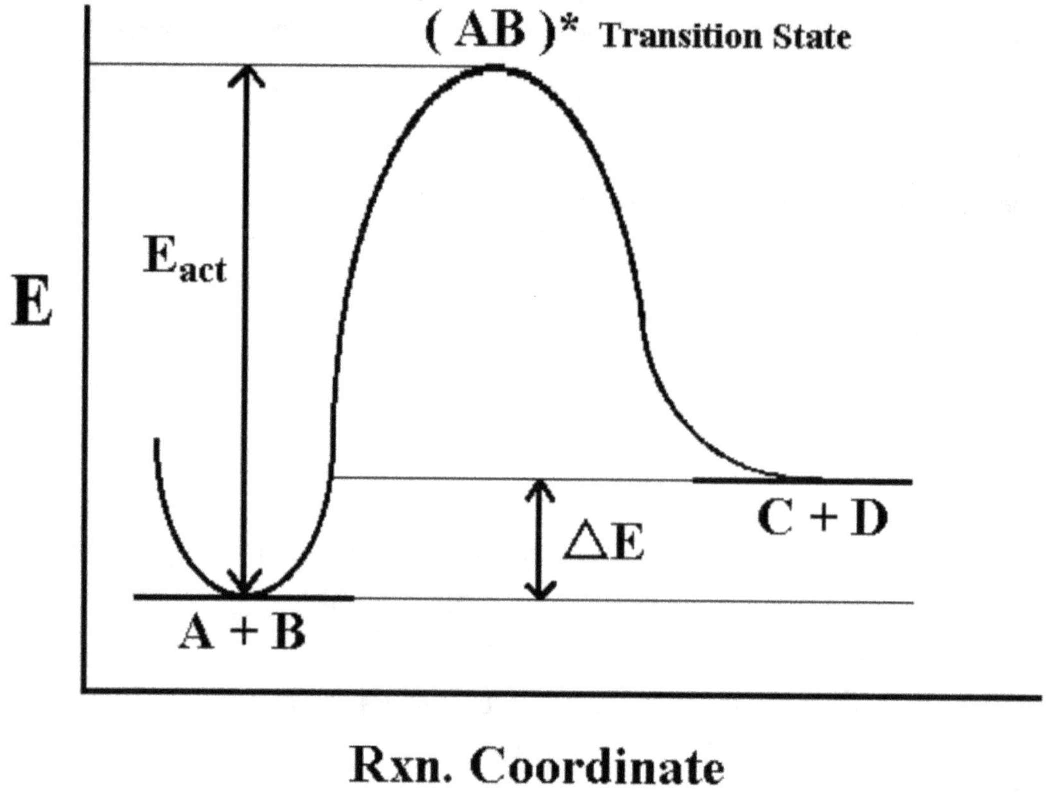

reaction : A + B → C + D

E_{act} is lowered (shortened) by the catalysis of enzymes.

Genetic Code

The genetic code consists of 4 different DNA bases (A, T, G, C) with a 3 base codon coding for each amino acid.

Therefore, there are $4^3 = 64$ possible 3 base codons (for any combination of 4 bases).

Glycolysis

Glycolysis is the anerobic (without use of oxygen) breakdown of Glucose to 2 molecules of Pyruvate, to make 2 molecules of ATP.
It is considered the most ancient form of biological energy production, and is therefore found in most organisms.

Photosynthesis

Photosynthesis is an anabolic (biosynthetic) process to create carbohydrates.
It therefore uses light energy for biosynthetic processes.
The Calvin cycle (C3 metabolism) in plants uses

$$3CO_2 \rightarrow \text{Glyceraldehyde-3-phosphate.}$$

This occurs in chloroplasts, the light being collected by chlorophylls (a, b) in Thylakoid Stacks, utilizing Photosystems I and II (light and dark reactions).
For green plants: $6CO_2 + 6H_2O\ (+ 6H_2O) \rightarrow C_6H_{12}O_6 + 6O_2\ (+ 6H_2O)$

Reactions:

$$CO_2 + 2\ H_2O + \text{light energy} \rightarrow (\ CH_2O\ )_{carbohydrate} + H_2O + O_2$$

When glucose is produced:

$$6\ CO_2 + 12\ H_2O + \text{light energy} \rightarrow C_6H_{12}O_6 + 6\ H_2O + 6\ O_2$$

Plants are autotrophs since they produce their own foods (carbohydrates) through photosynthesis.

For Purple Sulfur Bacteria:

$$CO_2 + 2\ H_2S + \text{light energy} \rightarrow (\ CH_2O\ ) + S$$

∴ the O_2 in regular photosynthesis does not come from CO_2.
The incorporation of carbon from CO_2 into CH_2O can occur in the dark, and therefore, as found by Calvin, does not directly depend on sunlight.
This light-independent component of photosynthesis is called the Calvin Cycle.
The reaction that produces O_2 : $2\ H_2O + \text{sunlight} \rightarrow 4\ H^+ + 4\ e^- + O_2$

is the light-dependent component of photosynthesis.

photosystem II

light $\overset{\text{photon}}{\rightarrow}$ **chlorophyll** $\overset{e^-}{\rightarrow}$ **pheophytin** $\overset{e^-}{\rightarrow}$ **electron tramsport chain**

(ETC)

using plastoquinne

↓

starts protpn pump that makes ATP using ATP Synthase

ETC → **cytochrome complex**

↓

plastocyanin

cyclic photophosphorylation

plastoquine

photosystem I

light → **chlorophyll** $\overset{\rightarrow}{e^-}$ **ferrodoxin** ⟶ **Calvin Cycle**

photon

(energy transduction photon → e-)

(brought back to high energy by proton)

NADP   **NADPH**

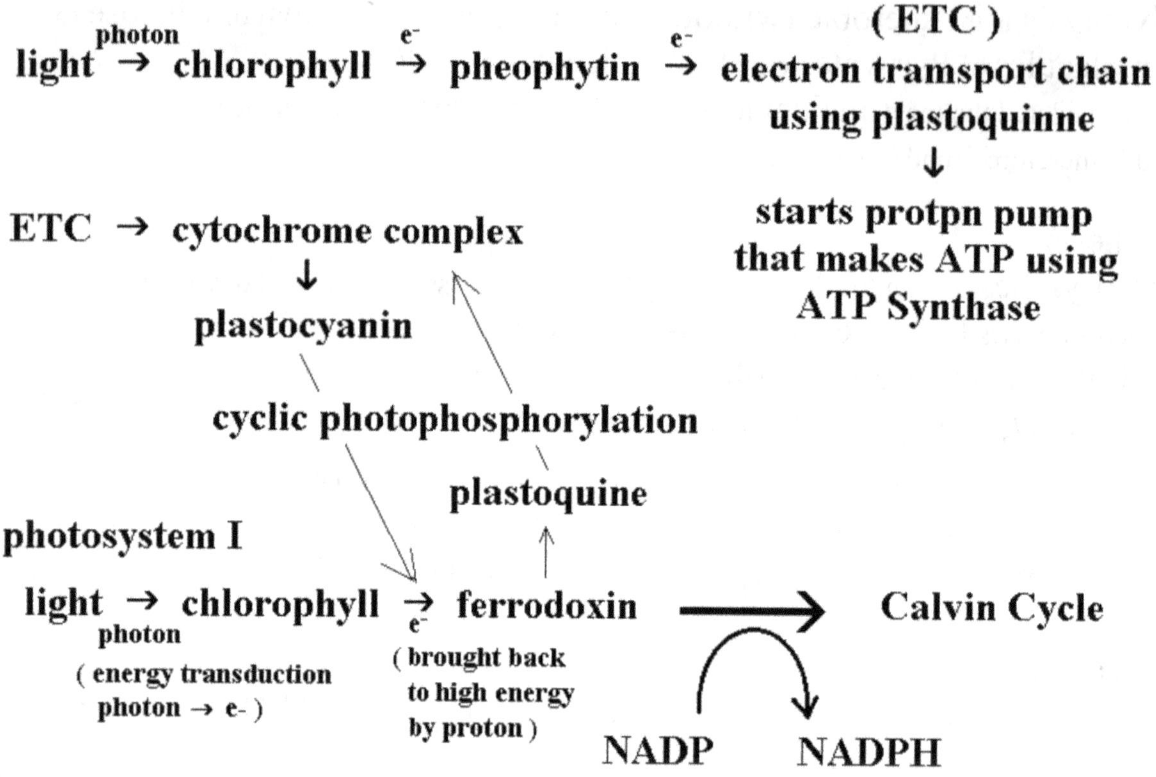

Calvin Cycle :

Rubisco

CO_2 **+ Ribulose BisPhosphate** → **(2) 3-phosphoglycerate**

(CO_2 fixation)

ATP, NADPH

⬇

ADP, NADP

glyceraldehyde 3-phosphate

↓

glucose → **sucrose, starch**

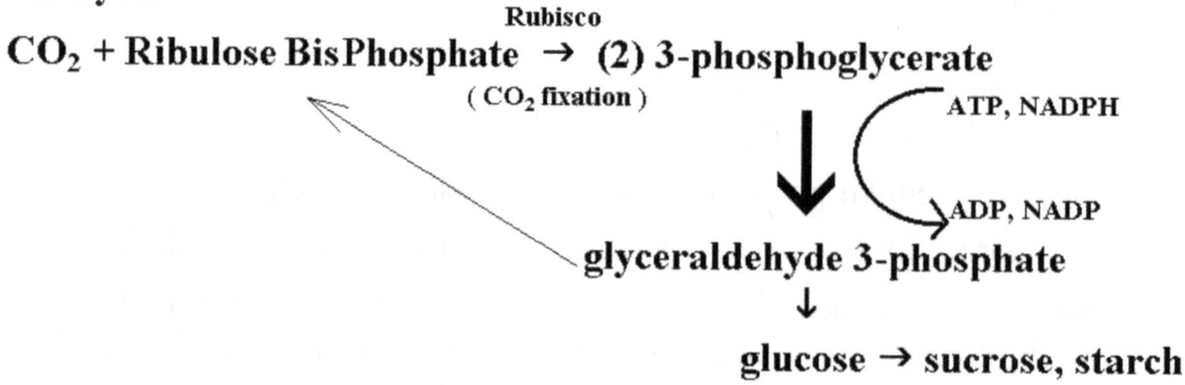

Rubisco : ribulose – 1, 5 – bisphosphate carboxylate/oxygenase
          fixes CO_2 (to Ribulose BisPhosphate) and is

the most common enzyme on earth (though it is slow for an enzyme).

It also fixes O_2 to Ribulose BisPhosphate, when CO_2 is low, making a toxic compound that must be broken down (releasing CO_2). This is called photorespiration and consumes ATP.

Chromophores of the Cytochrome Complex:

P680 light of 680nm (red-ish) also absorbs in the blue (~500nm).

P700 light of 700nm (red-ish) likewise absorbs in the blue (~500nm).

Reduced molecules

Reduced molecules (high hydrogen atom content, like sugars) have higher potential energy (can be oxidized for chemical energy) since their electrons are not as closely held by their atoms as for oxidized molecules (e.g., CO_2).

Respiration

Respiration causes the production of ATP (an energy carrier due to unstable phosphate bonds).

Glycolysis Pathway

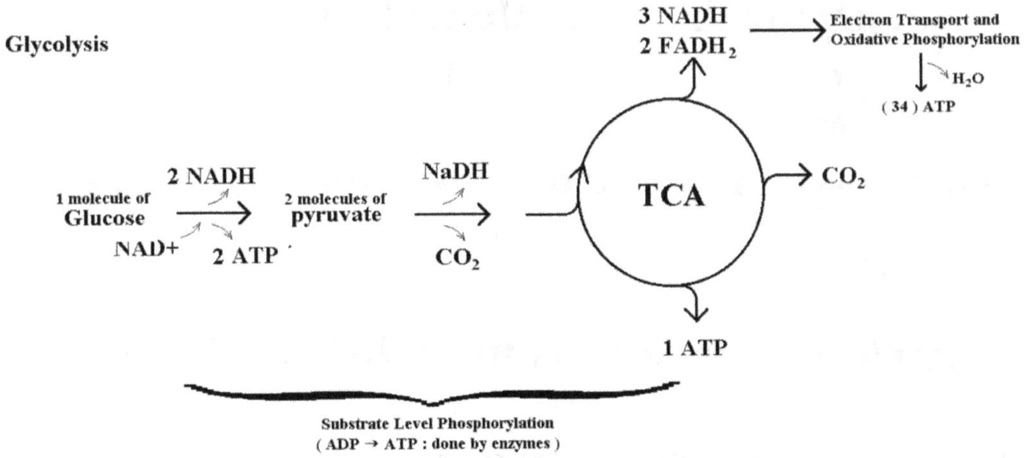

$$C_6H_{12}O_6 \underset{\text{oxidation}}{\overset{\text{reduction}}{\rightleftarrows}} 6CO_2 + 6H_2O + \text{energy (in ATP)}$$

Cellular Respiration is in the direction of reduction.

Oxidative Phosphorylation of O_2 :

$O_2 \rightarrow H_2O$ (reduced oxygen) , as ADP→ATP (phosphorylation)

Electron Transport : the electrons of reduced molecules are step-wise transferred to more oxidized molecules (using up their potential energy). Protons (H^+) are produced and pumped across cellular membranes (inner membrane of mitochondria for eucaryotes, cell membranes for procaryotes). This sets up a proton gradient for protons to come back into the cell via (the enzyme) ATP synthetase, which allows ADP → ATP to occur.

NADH, $FADH_2$ are reduced electron carriers. In electron transport and oxidative phosphoylation: they give electrons to O_2 to make H_2O, along with the protons available.

Other electron donors can be : H_2, H_2S, CH_4.
Two electron acceptors are NO_3^- and SO_4^{2-}.
When O_2 (as electron acceptor) is not present or available,
then fermentation occurs with intermediates as electron acceptors.
This leads to production of ethanol or lactic acid (→ 2 ATP).
Fermentation allows for NADH → NAD^+ , to continue glycolysis.

The TCA cycle initiates as:

coenzyme A (↓) **oxaloacetate** (↓)

pyruvate → Acetyl CoA → citrate (to start the cycle)

pyruvate → 3 CO_2 after full TCA cycle

Krebs Cycle == TCA or Citrate Acid Cycle , where small carboxylic acids like citrate, malate, succinate, oxaloacetate are oxidized to CO_2
(but the carboxylic acids are remade in the cycle).

Calculus

Absolute Maximun or Minimum

To determine the absolute max. or min. of f(x) , on a closed interval of x :

1) determine critical points using $f'(x) = 0$

2) determine f(x) for the lower and upper limits (endpoints) of the interval

∴ The Abs. Max. will have the largest y ; the Abs. Min. will have the smallest y .

(They need not be at critical points, which is why those points must be checked.)

The calculus here usually involves f(x) or $f'(x)$ == a quadratic (function), so that you have a parabola. This guarantees the function's vertex is at an Abs. Min. or Max.

The vertex is where (for x) $f'(x) = 0$.

If, for the quadratic term ax^2, a is positive, the parabola opens up (∪) and the vertex (critical point) is an Abs. Min.

If a is negative, the parabola opens down (∩) and the vertex is an Abs. Max.

If the term is cubic : ax^3 ,

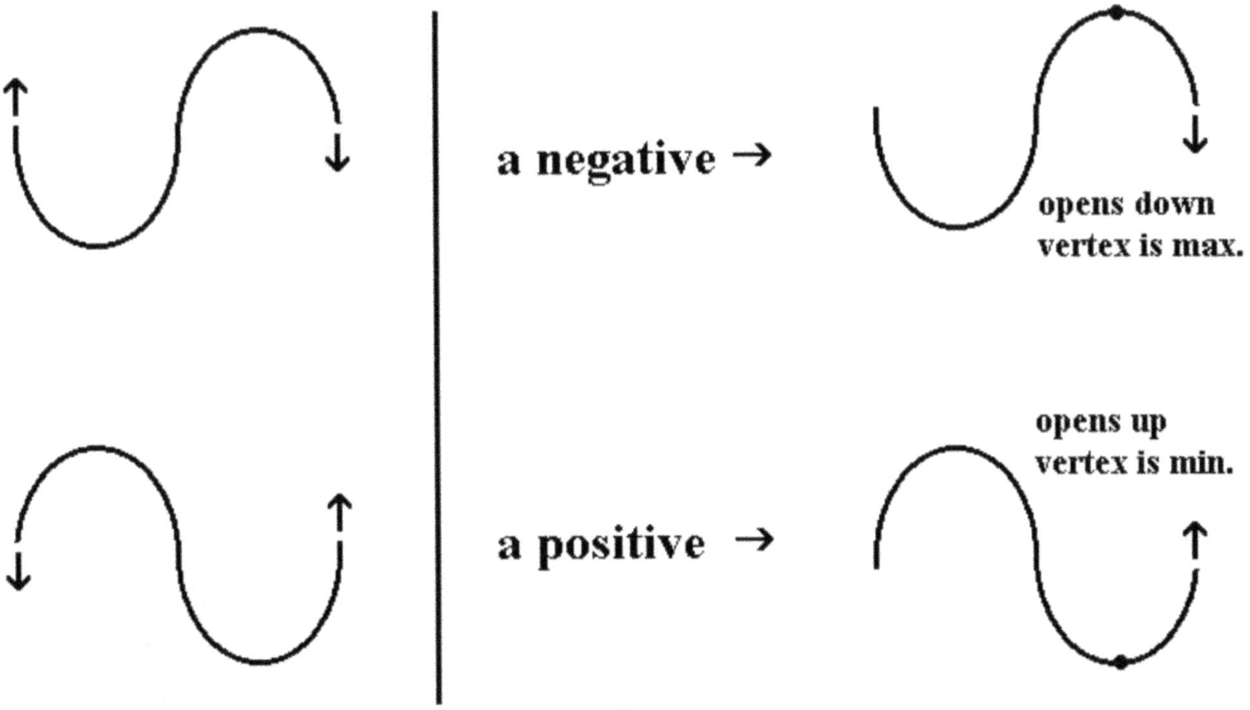

Area between two curves

If f(x) is above g(x) for the interval a <= x <= b , then

$\int_a^b [\, f(x) - g(x) \,]dx$ is the area between them.

To find the points where f(x) and g(x) intersect, set f(x) = g(x) and solve for x .

Chain Rule (or Power of a Function) for Derivatives

If $f(x) = [\, u(x) \,]^n$, then $f'(x) = n[\, u(x) \,]^{n-1} \bullet u'(x)$, where n is a real number.

The Power Rule for Derivatives says: if $f(x) = x^n$, then $f'(x) = nx^{n-1}$.

Note: one should always express final answers with positive exponents (for clarity).

For negative exponents: $b^{-n} = 1 / b^n$, $1 / b^{-n} = b^n$.

For fractional exponents (exponents and roots):

$$b^{n/d} = b^{\frac{n}{d}}$$

n ← exponent
d ← root

$$b^{n/d} = \left(\sqrt[d]{b^n}\right) = \left(\sqrt[d]{b}\right)^n$$

algebraic — **numerical (used in calculations as more convenient)**

e.g.

If $f(x) = 5x^{1/3}$

$$f'(x) = (\,5 \bullet 1/3\,)x^{\frac{1}{3}-\frac{3}{3}} = \frac{5}{3}x^{-2/3} = \frac{5}{3}\left(\frac{1}{\sqrt[3]{x^2}}\right) = \frac{5}{3\sqrt[3]{x^2}}$$

If $f(x) = 7\sqrt{x} = 7x^{1/2}$

$$f'(x) = (\,7 \bullet 1/2)x^{(1/2)-1} = \frac{7}{2}x^{-1/2} = \left(\frac{7}{2}\right)\left(\frac{1}{x^2}\right) = \frac{7}{2\sqrt{x}}$$

If $f(x) = \frac{1}{\sqrt{x}} = 1 \bullet x^{-1/2}$

$$f'(x) = 1 \bullet (\,-1/2\,)x^{(-1/2)-1} = -(1/2)x^{-3/2} = \frac{-1}{2\sqrt[2]{x^3}} = \frac{-1}{2\sqrt{x^3}}$$

$$= \frac{-1}{2\sqrt{x \bullet x \bullet x}} = \frac{-1}{2\sqrt{x^2 \bullet x}}$$

$$= \frac{-1}{2x\sqrt{x}}$$

Chain Rule for Parametric Equations

If $x = f(t)$ and $y = g(t)$, then

$$\frac{dy}{dx} = \frac{dy}{dx} \bullet \frac{dt}{dx} = \frac{dy / dt}{dx / dt}$$

This is equivalent to the Chain Rule when replacing an expression of x by u :
$nu^{n-1}\, du = f'y = dy / dx$.

$dy / dx = f'(x)$	derivative of f(x)
$dy = f'(x)\, dx$	
$dy = f'(x)$	differential of f(x)

Definite Integral Characteristics

They never integrate to a constant C, since they have numerical limits.

$\int_a^b f(x)dx = -\int_b^a f(x)dx$

$\int_a^c f(x)dx = \int_a^b f(x)dx + \int_b^c f(x)dx$ for a =< b =< c

$\int_a^a f(x) = 0$

Explicit Equations for y = f(x)

e.g., $y = 5 - x^2 - x \therefore f'(x) = -2x - 1$

Fundamental Theorem of Calculus

The definite integral is the area under the curve,
while f(x) is (and must be) continuous and non-negative over the interval a =< x =< b;
$\int_a^b f(x)dx = f(b) - f(a)$

Implicit Equations (implicit for y) : How do you differentiate?

e.g., $x^2 + x + y = 5$
Differentiate equation term by term, regarding y as a function of x .

$$f'(x) \equiv \frac{d(x^2)}{dx} + \frac{d(x)}{dx} + \frac{d(y)}{dx} = \frac{d(5)}{dx}$$

$$= 2x + 1 + \frac{dy}{dx} = 0$$

$$\therefore \frac{dy}{dx} = f'(x) = -2x - 1$$

Note: $\frac{d(3y)}{dx} = 3\frac{dy}{dx}$

Using Chain Rule with implicit differentiation

let $f(x) = [\ u(x)\ ]^3$ $\therefore f'(x) = 3[\ u(x)\ ]^2 \bullet u'(x)$

let $f(x) = y^3$ (i.e., $u(x) = y$) $\therefore f'(x) = 3y^2 \bullet (\ dy / dx\ )$

For Product Rule:

e.g., $xy = 5$

$[\ x \bullet d(y) / dx\ ] + [\ y \bullet d(x) / dx\ ] = 0$

$x(\ dy / dx\ ) + y = 0$, $dy / dx = -y / x$

e.g., find derivative of $y^3 = 2x^2 + 1$ using implicit differentiation :

$d(\ y^3\ ) / dx = [\ d(2x^2\ ) / dx\ ] + [\ d(\ 1\ ) / dx\ ]$

$3y^2(\ dy / dx\ ) = 4x + 0 \ \therefore dy / dx = 4x / 3y^2$

e.g., differentiate $y^3 + y^2 = 3x - x^7$

$[\ d(\ y^3\ ) / dx\ ] + [\ d(\ y^2\ ) / dx\ ] = [\ d(\ 3x\ ) / dx\ ] - [\ d(\ x^7\ ) / dx\ ]$

$3y^2(\ dy / dx\ ) + 2y(\ dy / dx\ ) = 3 - 7x^6$

$(\ dy / dx\ )(\ 3y^2 + 2y\ ) = 3 - 7x^6 \ \therefore dy / dx = (\ 3 - 7x^6\ ) / (\ 3y^2 + 2y\ )$

Integration by Substitution

e.g., integrate $\int (\ 3x - 5\ )^5 dx$

1) let $u = 3x - 5$ [if there are more than one expression, let u be the most complex one]

2) take the derivative du / dx ; here, du / dx = 3
3) solve for dx : dx = du / 3
4) Substitute u, du in original problem

$$\int(3x - 5)^5 dx = \int u^5(du / 3) = (1 / 3)\int u^5 du$$

Integrate (if possible) for u

$$(1 / 3)\int u^5 du = (1 / 3)(u^6 / 6) + C = (u^6 / 18) + C$$

Change back to original variables

$$\int(3x - 5)^5 dx = (u^6 / 18) + C = [(3x - 5)^6 / 18] + C$$

e.g., integrate $\int(2x + 3)\exp(x^2 + 3x + 5)dx$

let $u = x^2 + 3x + 5 \therefore du / dx = 2x + 3 , dx = du / (2x + 3)$

$\therefore \int(2x + 3)e^u du / (2x + 3) = \int e^u du = e^u + C$

$\therefore e^u + c = \exp(x^2 + 3x + 5) + C$

Notations for Derivatives

If, for example, $y = f(x) = x^2 - 4x$

Derivative Notations can be :

$$f'(x) , \frac{dy}{dx} , \frac{df}{dx} , \frac{d(x^2 - 4x)}{dx} , D_x(x^2 - 4x)$$

$$\text{and also } y' , f' , \frac{df(x)}{dx} ;$$

$$\text{for 2nd derivative : } f''(x) , y'' , \frac{d^2(x)}{dx^2}$$

Partial Differentiation

With respect to one variable, all of the other variables are treated as constants.

Notations:

$\partial f / \partial x , f_x(x, y) , f_x$

$\partial f / \partial y , f_y(x, y) , f_y$ etc.

e.g., for $z = x^2 - y^3 ; \partial z / \partial y = -3y^2$

Power Rule for Derivatives

If $f(x) = cx^n$, then $f'(x) = cnx^{n-1}$

If $f(x) = u(x) + v(x)$, then $f'(x) = u'(x) + v'(x)$

If $f(x) = u(x) - v(x)$, then $f'(x) = u'(x) - v'(x)$

Product Rule for Derivatives

If $f(x) = F(x) \bullet S(x)$, then $f'(x) = F(x) \bullet S'(x) + F'(x) \bullet S(x)$

e.g., for $f(x) = x^2(x^3 - 1)$:

$$\therefore \quad f'(x) = x^2 \frac{d(x^3 - 1)}{dx} + (x^3 - 1) \frac{d(x^2)}{dx}$$

$$= x^2(3x^2) + (x^3 - 1)2x \quad \text{etc.}$$

Quotient Rule for Derivatives

$$\text{If} \quad f(x) = \frac{N(x)}{D(x)} ,$$

$$\text{then} \quad f'(x) = \frac{D(x) \bullet N'(x) - N(x) \bullet D'(x)}{[D(x)]^2}$$

Rate of Change

For the function $y = f(x)$:

the average rate of change = $\triangle y / \triangle x$

Instantaneous rate of change is :

$$\mathbf{f'(x)} = \lim_{\Delta x \to 0} \frac{(\, f(x - \Delta x) - f(x)\,)}{\Delta x} = \textbf{derivative}$$

$$\textbf{The point tangent to the curve} = \frac{dy}{dx}$$

The derivative is the slope of the tangent line to the curve f(x) at a (definite) point.
Where the slope is horizontal (to x-axis), slope = 0 ;
where it is vertical the slope is undefined.
These two conditions yield what are called 'critical points' on the curve.

Relative Minimums, Maximums, and Stationary Inflection Points

Set $f'(x) = 0$ and then solve for x → critical points, after solving f(x) for y .

If $f(x-1) >$ or $< f(x) >$ or $< f(x+1)$:	x is at a stationary inflection
If $f(x-1) < f(x)$ and $f(x+1) < f(x)$:	x is at a relative maximum
If $f(x-1) > f(x)$ and $f(x+1) > f(x)$:	x is at a relative minimum

----

If $f'(x-1) < f'(x) < f'(x+1)$:	x is at a relative minimum
If $f'(x-1) > f'(x) > f'(x+1)$:	x is at a relative maximum
But, when $f'(x) = 0$, if neither is true :	x is at a stationary inflection

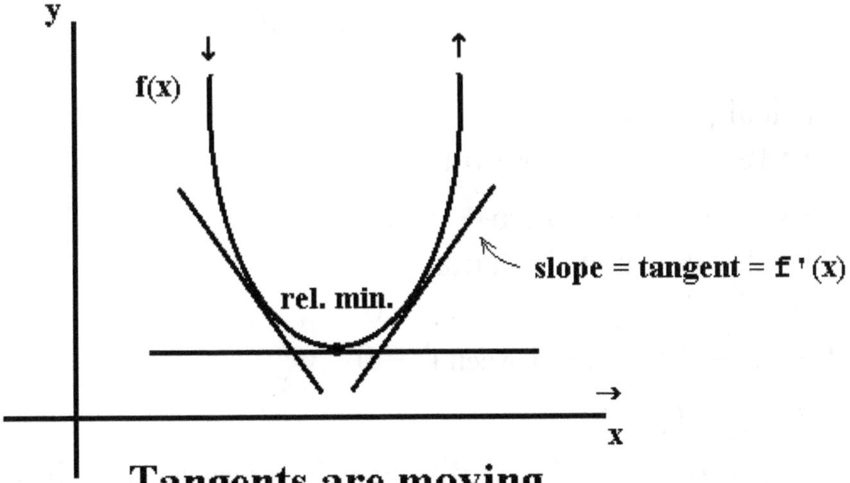

Tangents are moving concave-up along x

Second Derivative tests for Rel. Max. and Rel. Min.

$f''(x)$ → rate of change of slope ($= f'(x)$) of the tangent

(∴ going from left to right, along x):

If x is a critical point, ∴ $f'(x) = 0$,

If $f''(x) > 0$, f(x) is moving concave-up → critical point is rel. min.

If $f''(x) < 0$, f(x) is moving concave-down → critical point is rel. max.

If $f''(x) = 0$, the test fails;

∴ use f(x) and $f'(x)$ to determine the critical points of (x, f(x)).

Second Derivative (partial derivative) tests for Critical Points of Bivariate Functions

If $f_x = 0$ and $f_y = 0$ (1st derivatives),

let $D(x, y) = f_{xx}(x, y)f_{yy}(x, y) - [f_{xy}(x, y)]^2$.

When you evaluate D at the critical point:

If D is negative, the critical point is a saddle point.

If D is zero, the test is inconclusive.

If D is positive:

If f_{xx} at critical point is negative, critical point is a maximum.

If f_{xx} at critical point is positive , critical point is a minimum.

To find the critical points:

1) take f_x, f_y of the function
2) set them equal to zero
3) solve them simultaneously

e.g., $f(x, y) = x^2 - 2x + y^3 - 3y^2 - 9y + 6$

$f_x = 2x - 2$; ∴ x = 1 when $f_x = 0$

$f_y = 3y^2 - 6y - 9$

set $3y^2 - 6y - 9 = 0$ and divide by 3 (reduce the equation coefficients)

$y^2 - 2y - 3 = 0$

$(y - 3)(y + 1) = 0$ → $y = 3$, $y = -1$

If $x = 1, y = 3$:

$f(x, y) = f(1, 3) = -22$

$\therefore [x, y, f(x, y)] == (1, 3, -22)$ is a critical point

If $x = 1, y = -1$

$f(x, y) = f(1, -1) = 10$

$\therefore (1, -1, 10)$ is another critical point

To classify the critical points:

$f_{xx}(x, y) = \delta(2x - 2) / \delta x = 2$

$f_{yy}(x, y) = \delta(3y^2 - 6y - 9) / \delta y = 6y - 9$

$f_{xy}(x, y) = \delta(3y^2 - 6y - 9) / \delta x = 0$

$D(1, 3) = 24$, $\therefore$ positive ;

since f_{xx} is also positive, $(1, 3, -22)$ is a minimum.

$D(1, -1) = -24$, $\therefore$ negative → $(1, -1, 10)$ is a saddle point

Second Partial Derivatives (f_x , f_y are 1st partial derivatives)

e.g., for $z = f_x(x, y) = 3xy - x^2 + 5y^3$

$\partial z / \partial x = f_x(x, y) = \partial(3xy)/ \partial x \ + \ \partial(-x^2)/ \partial x \ + \ \partial(5y^3)/ \partial x$

$= 3y + (-2x) + 0 = 3y - 2x = f_x$

$\partial z / \partial y = f_y(x, y) = \partial(3xy)/\partial y \ + \ \partial(-x^2)/ \partial y \ + \ \partial(5y^3)/\partial y$

$= 3x + 0 + 15y^2 = 3x + 15y^2 = f_y$

There are (in this case) four 2nd order partial derivatives,
since $f_x = 3y - 2x = \partial z / \partial x$:

1) $\partial^2 z / \partial x^2 = f_{xx}(x, y) = \partial(3y)/\partial x - 2\partial(x)/\partial x = 0 - 2 = -2$

y is constant ; we have the derivative of f_x with respect to x .

$\partial z / \partial x = f_x = 3y - 2x$

2) $\partial^2 z / \partial y^2 = f_{yy}(x, y) = \partial(3x)/ \partial y + 15\partial(y^2)/\partial y = 0 + 30y = 30y$

x is constant ; we have the derivative of f_y with respect to y .

$\partial z / \partial y = f_y = 3x + 15y^2$

3) $\partial^2 z / \partial y\partial x = f_{xy}(x, y) = 3\partial(y)/\partial y - \partial(2x)/\partial y = 3 - 0 = 3$

x is constant ; we have the derivative of f_x with respect to y .

$\partial z / \partial x = f_x = 3y - 2x$

4) $\partial^2 z / \partial x\partial y = f_{yx}(x, y) = 3\partial(x)/\partial x + \partial(15y^2)/\partial x = 3 + 0 = 3$

y is constant ; we have the derivative of f_y with respect to x .

$\partial z / \partial y = f_y = 3x + 15y^2$

The mixed partials (cross partials), f_{xy} and f_{yx} , are always equal.

A point (a, b) for which both $f_x(a, b) = 0$ and $f_y(a, b) = 0$ is a critical point (on a 3D surface it is a rel. min., rel. max., or saddle point for stationary inflection). $z = f(x, y)$ makes a 3D surface : Bivariate Function.

Summary of Operations in the Calculus :

operation	**Differentiation Rules** $f(x)$	$f'(x)$
power	$f(x) = x^n$	$f'(x) = nx^{-1}$
constant	$f(x) = c$	$f'(x) = 0$
const. function	$c \bullet u(x)$	$f'(x) = c \bullet u'(x)$
special case	$c \bullet x^n$	cnx^{n-1}
sum	$u(x) + v(x)$	$u'(x) + v'(x)$
difference	$u(x) - v(x)$	$u'(x) - v'(x)$
Product Rule	$F(x) \bullet S(x)$	$F(x)S'(x) + S(x)F'(x)$
Quotient Rule	$\dfrac{N(x)}{D(x)}$	$\dfrac{D(x)N'(x) - N(x)D'(x)}{[\, D(x)\,]^2}$
Chain Rule	$[\, u(x)\,]^n$	$n[\, u(x)\,]^{n-1} \bullet u'(x)$

e exponential $e^{u(x)}$ $u'(x) \bullet e^{u(x)}$

e^x is always positive; it is asymptotic to x-axis (i.e., $y = 0$).

$$\text{If } f(x) = e^x \text{, then } f'(x) = \frac{d(e^x)}{dx} \bullet e^x = 1 \bullet e^x = e^x$$

Logarithmic Rule $\ln u(x)$ $\dfrac{u'(x)}{u(x)}$

$\log_{(base)}$, $\ln_{(e)}$; base > 0

u(x) must always be positive since logarithms only work on positive numbers.

$\ln x = y$ **logarithmic form**
$e^y = x$ **exponential form**
} **they are equivalent statements**

$$\text{Note: } f'(\ln x) = \frac{d(x)/dx}{x} = \frac{1}{x}$$

operation	**Integration Rules : $\int f'(x)\,dx = f(x) + C$**	
	f'(x)	**f(x)**
power functions	$\int x^n dx$	$\frac{x^{n+1}}{n+1} + C$ ($n \neq -1$)
logarithmic	$\int e^x\,dx$	$e^x + C$
	$\int 1/x\,dx$	$\ln\lvert x \rvert + C$
	$\int x^{-1}\,dx$	

---

sum of integrals	$\int [\, f(x) + g(x) \,]dx = \int f(x)dx + \int g(x)dx$
constant	$\int kf(x)dx = k\int f(x)dx$

Note: $\frac{d(x)}{dx} = 1 \quad \therefore \quad \int dx = x + C$

(derivative of x is 1 , so integral of 1 is x)

$$\int \ln x\,dx = (x\ln x) - x + C$$

$$\int e^{ax}\,dx = (\,1 / a\,)e^{ax} + C$$

$$f'[\,(\,x\ln x\,) - x\,] = x(\,1 / x\,) + \ln x\,(\,1\,) - 1 = 1 + (\,\ln x\,) - 1 = \ln x$$

$$f(x) = a^u \quad \rightarrow \quad f'(x) = a^u \bullet (\,\ln a\,)\;\; du\;\; dx$$

(u is differentiable function of x)

Trapezoid Rule

For function f(x), its integration (area under the curve) can be approximated from f(a) to f(b) by the trapezoid rule:

$$\int_{x=a}^{x=b} f(x)\,dx = (b - a)\left[\frac{f(a) + f(b)}{2}\right]$$

Chemistry

Acids and Bases

Definition types:

Arrehnius (must occur in water)

H^+ donor == acid

OH^- donor == base

Bronsfed-Lowry (protron transfer)

H^+ donor == acid

H^+ acceptor == base

Lewis (electron pair transfer to make adducts, covalent bonds)

electron pair acceptor == acid

electron pair donor == base

Atomic Orbitals

Quantum Numbers for Energy Levels

n == Principal Quantum Number → size of orbital

l == Angular Momentum Quantum Number → shape of orbital

m == Magnetic Quantum Number → orientation of orbital

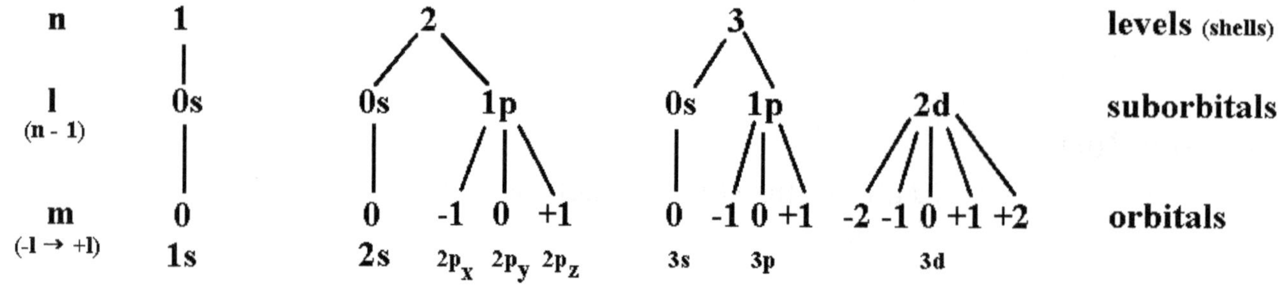

Atomic Radii

Generally (for covalent bonds)

across period (of Periodic Table) → radius decreases, as electrons are more tightly held

down group (column of Table) → radius increases, as atoms get larger with more electrons

Ionic bonds

metalic: loss of electrons ∴ positive → smaller radius than (corresponding) neutral atom

non-metal: gain of electrons ∴ negative → larger radius than (corresponding) neutral atom

For either type

smaller radius across Table, larger down Table;
but positive ions are smaller than negative ions

Concentration of pure water

18g → 1 mole H_20 ∴ 18g / mole

1 liter water == 1000g

1000g / (18g / mole) = 55.5 moles (per liter) == 55.5 M

Concentrations of solutions

molarity = moles / liters == moles / volume

molality = moles / kilograms == moles / mass

Molality is often used in physical chemistry because it is (generally) independent of temperature.

Center of Mass

For center of mass (CM) coordinates from particles

m_1 at (x_1, y_1, z_1) and m_2 at (x_2, y_2, z_2) :

The reduced mass $= m_1 m_2 / (m_1 + m_2)$

$$CM_x = (x_1 m_1 + x_2 m_2) / (m_1 + m_2)$$

$$CM_y = (y_1 m_1 + y_2 m_2) / (m_1 + m_2)$$

$$CM_z = (z_1 m_1 + z_2 m_2) / (m_1 + m_2)$$

Covalent and Ionic Bonding Diagrams (conceptual) :

e.g., Florine : (9 protons),

9 electrons - 2 = 7 valence electrons in outer shell

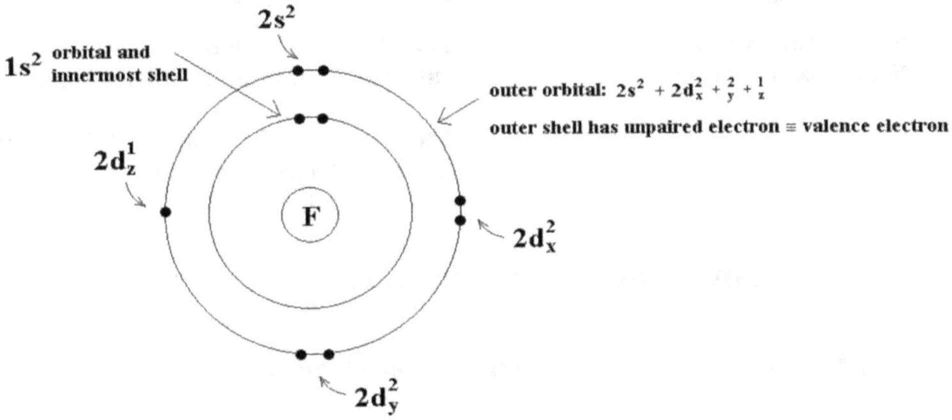

e.g., Sodium : (11 protons) ,

11 electrons - 2 - 8 = 1 in outer shell (principal energy level: 3)

$1s^2$ ↑ $3s^1$

($2s^2 + 2d^6$)

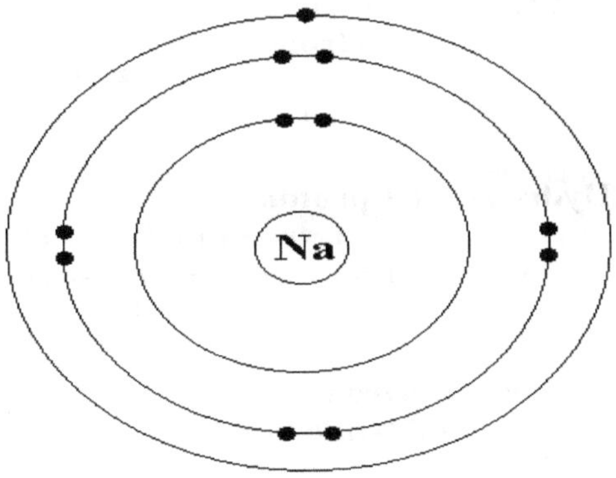

NaF would be ionic:
F would take the outermost Na electron.

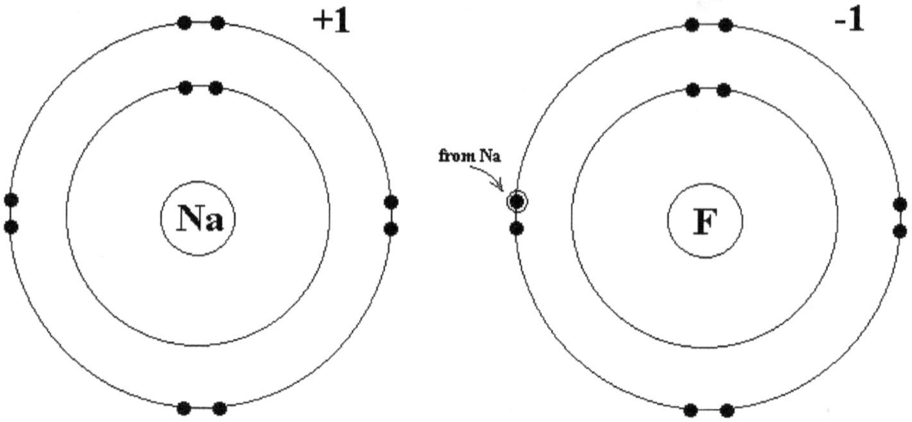

Effective ionic radius is smaller than (neutral) Na due to loss of the outer shell.

Effective ionic radius is slightly greater than (neutral) F since outer shell has gained an electron.

More electrons in an outer shell means less shielding of outer electrons by inner electrons, so the outer shell increases slightly in radius.

Both atomic outer shells are complete for ionic molecule NaF.

H_2O has covalent bonding (sharing of electrons).

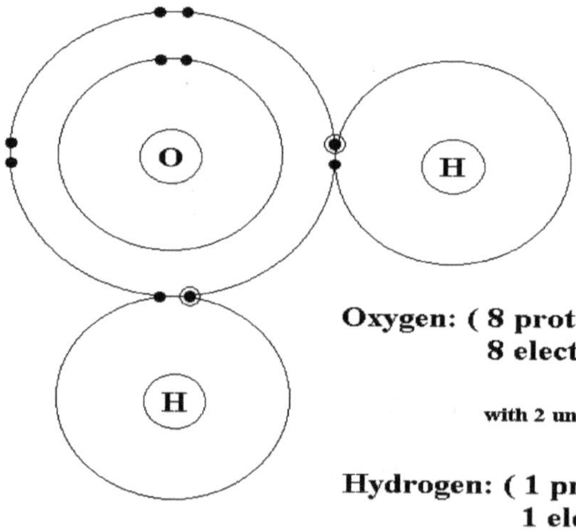

Oxygen: (8 protons),
8 electrons - $\underset{1s^2}{2}$ = 6 in outer shell
$2s^2 + 2d_x^1 + 2d_y^1$
with 2 unpaired electrons → 2 valence electrons

Hydrogen: (1 proton),
1 electron in outer shell ($1s^1$)
since it is unpaired it is a valence electron

Unpaired outer shell electrons ≡ valence electrons which can form covalent single, double, and triple bonds.

$O_2 \equiv O = O$ Double Bond

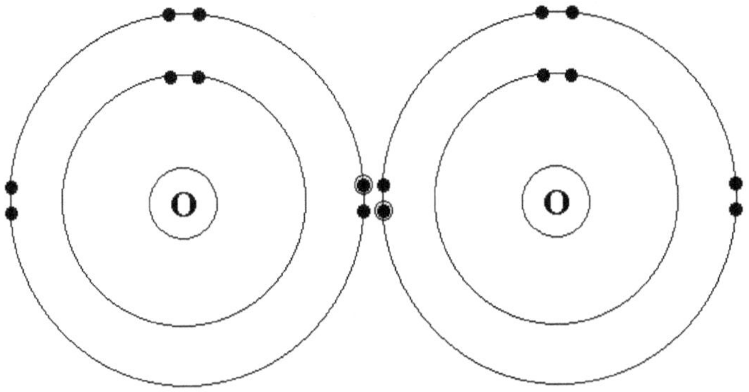

Nitrogen: 7 electrons (5 outer shell),
3 unpaired outer shell electrons
→ 3 valence electrons

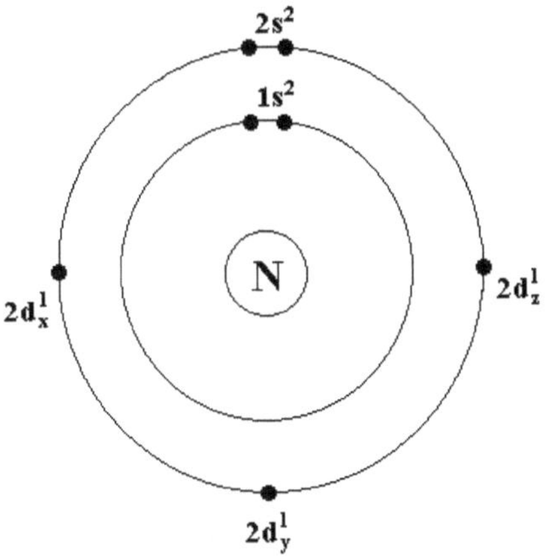

N_2 Triple Bond

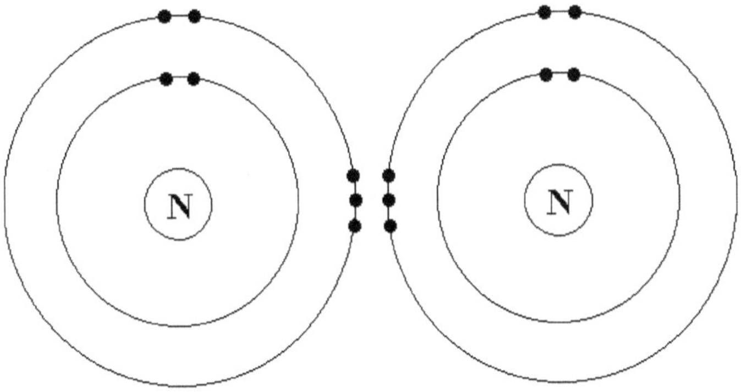

Definitions

Energy == the capacity to do work (or create heat)

A change in potential energy equals (or leads to) kinetic energy.

Entropy == an amount of thermo-energy transformed or transferred into energy (in a closed system) not useful for work or not available for useful work

Force == a propensity or tendency of a mass (for motion)

Inertia == the property of a mass to resist an acceleration

Energy Levels

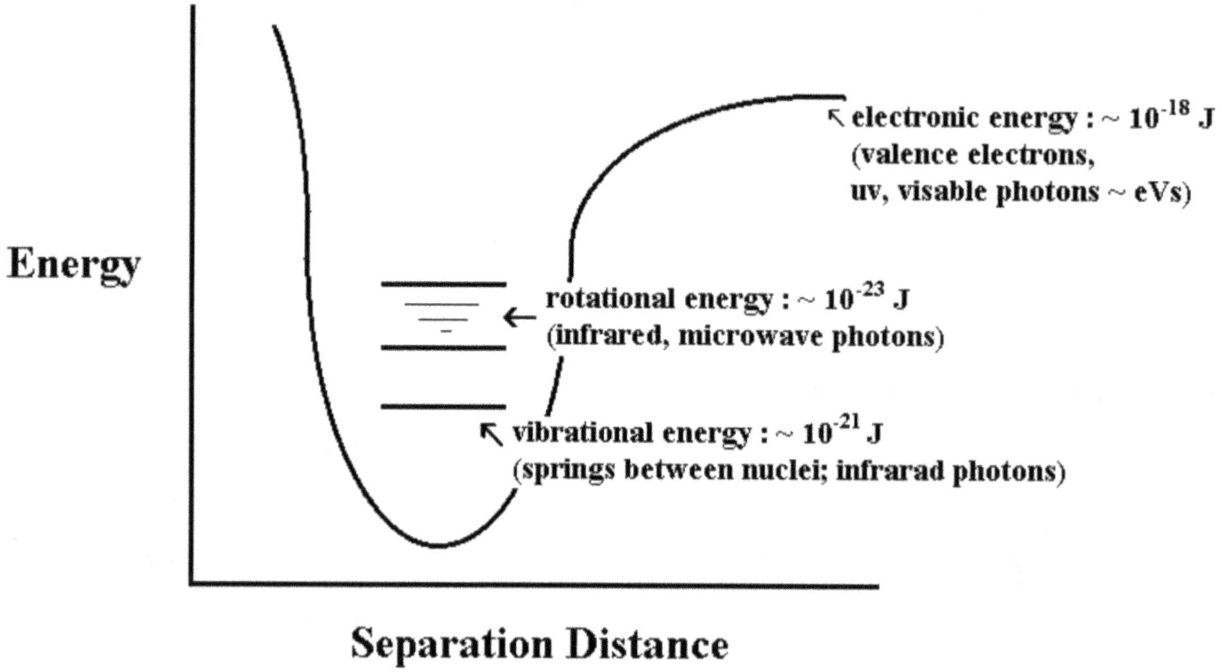

$E_{total} = E_{elec} + E_{vib} + E_{rot}$.

Nuclear energy levels are ~ 10^{-12} J .

1 mole == 6.022 x 10^{23} (of anything or unit considered) == Avogadro's Number
The mass of 1 mole = the molecular weight (of that considered), usually in grams.
One mole of a substance is the same # of atoms or molecules of material as there are atoms in 12 grams of carbon-12.

Valence electrons are electrons in the outer principal energy level, and are used for making chemical reactions

e.g.: energy levels $(1, 2, 3, 4, \ldots\ldots)_{s,\ p,\ d}$ == n

e.g.

$$4s^2p_x^1\ p_y^1\ p_z^1$$

electron ↙ (points to the superscript 1)

↑ n ; ↖↗ *l*

(x, y, z) → m combination

$l = 0 \rightarrow s$

$l = 1 \rightarrow p$

1 anstrom == 1Å = 1 x 10^{-10} meters = 0.1 x 10^{-9}m = 0.1 nm
∴ 10Å = 1 nanometer == 1 nm

1 Dalton (D) = 1 atomic mass unit (amu) = 1.66 x 10^{-27}kg ,
where u(nit) == (1 / 12) the mass of ^{12}C

Weight
Weight = force of gravity ∴ w = mg == Newtons (in SI units)

Equilibria
given : aA + bB ⇆ cC + dD
$k = [\ C\ ]^c[\ D\ ]^d / [\ A\ ]^a[\ B\ ]^b$ == equilibrium constant

$\triangle G^0 = -RT \bullet \ln k == -2.3RT \bullet \log k$
k == e^[$-\triangle G^0/RT$] = 10^[$-\triangle G^0/2.3RT$]
Gibb's Free Energy ($\triangle G^0$) for substances in their standard states always have a concentration [ ] = 1M
At 25^0 c (room temperature), 2.3RT = 1.364 kcal / mol .

$\triangle G^0 = \triangle H^0 - T\triangle S^0$ (at conditions of constant pressure) ;
△H == enthalpy, the change in bond energies and strengths
△S == entropy, the change in freedom of motion during the reaction
exothermic rxn. : △H < 0 ; endothermic rnx. : △H > 0
exergonic rxn. : △G < 0 ; endergonic rxn. : △G > 0

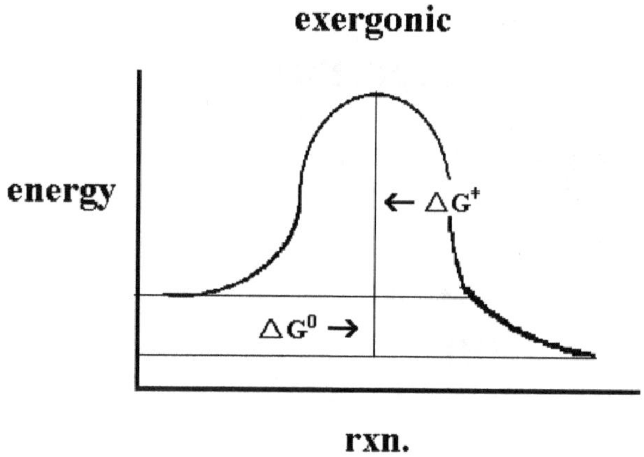

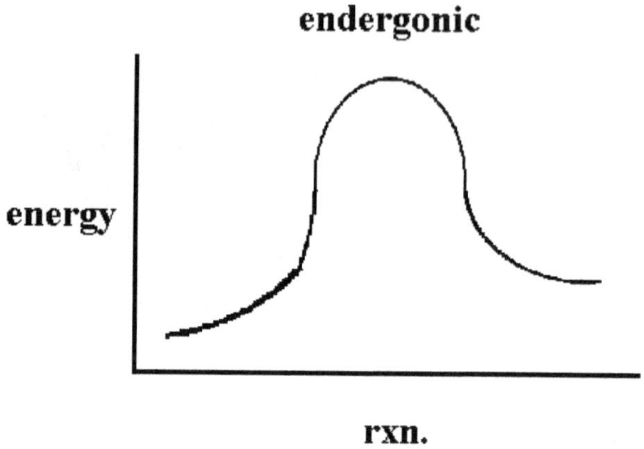

The rate of reaction = $\nu = \kappa[\,A\,]^a[\,B\,]^b$, 2nd order rxn.,
where κ == the rate constant of the reaction.
$\kappa = \nu^{\ddagger} \bullet e\hat{}[\,\Delta G / RT\,] = \nu^{\ddagger} \bullet e\hat{}[\,-\Delta H^{\ddagger} / RT\,] \bullet e\hat{}[\,\Delta S^{\ddagger} / R\,]$

The rate of reaction is related to activation energies
(when going from reactants to transition states).
$\Delta G^{\ddagger}$ == activation free energy
$\Delta H^{\ddagger}$ == activation enthalpy (an energy)
$e\hat{}[-\Delta G^{\ddagger} / RT]$ == fraction of molecules reaching the transition state
$\gamma^{\ddagger}$ == frequency of molecules reaching the transition state

Van't Hoff Equation

$\ln(\,k_2 / k_1\,) = (\,-\Delta H_{rxn.}^{0} / R\,)[\,(\,1 / T_2\,) - (\,1 / T_1\,)\,]$,
where k == equilibrium constants, T == Kelvin temperatures,
R == the gas constant = 8.314 Joules/(mol•∓kelvin temp.)

Henderson-Hasselbalch Equation

$pH = pK_a + \log(\,[\,A^-\,] / [\,HA\,]\,)$
where HA is acid (e.g., Acetic Acid) and A^- is conjugate base or salt (e.g., acetate)

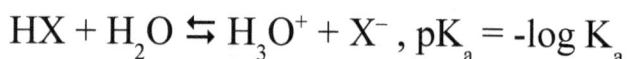

$HX + H_2O \leftrightharpoons H_3O^+ + X^-$, $pK_a = -\log K_a$
$K_a = [\,H_3O^+\,][\,X^-\,] / [\,HX\,][\,H_2O\,] == [\,H_3O^+\,][\,X^-\,] / [\,HX\,]$
when water is the solvent (and therefore is in great excess).

As $pK_a \downarrow \rightarrow$ stronger acid (especially when $pK_a < 0$) ; when $K_a < 1$, log is negative.

pH

$pH = -\log[\ H_3O^+\ ] == -\log[\ H^+\ ]$

$k_a = [\ H_3O^+\ ][\ A^-\ ] / [\ HA\ ]$

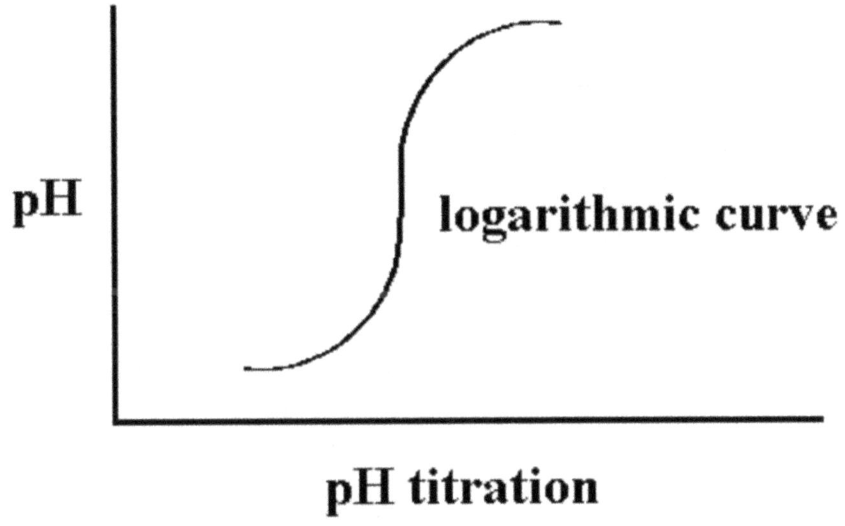

Reaction Types

Single Replacement Rxn. : $Ax + B \rightarrow Bx + A$ (exchange of cations)

Double Replacement Rxn. : $NaOH + HCl \rightarrow NaCl + HOH$ (example)

Reaction Rates

(at constant temperature)

rate = $k[\ A\ ][\ B\ ]^2$, where the reaction is 2nd order in B and 1st order in A

(1 + 2 =) 3rd order overall , with k == rate constant

First Order Reaction: for one reactant

rate = $k[\ A\ ] = -\Delta[\ A\ ] / \Delta t$

Integration yields:

$$\ln([A]_0 / [A]_t) = kt = \ln[A]_0 - \ln[A]_t$$

Rearranging in y = mx + b form (straight line) yields (for plots):

$$\ln[A]_t = -kt + \ln[A]_0$$

First Order Reaction

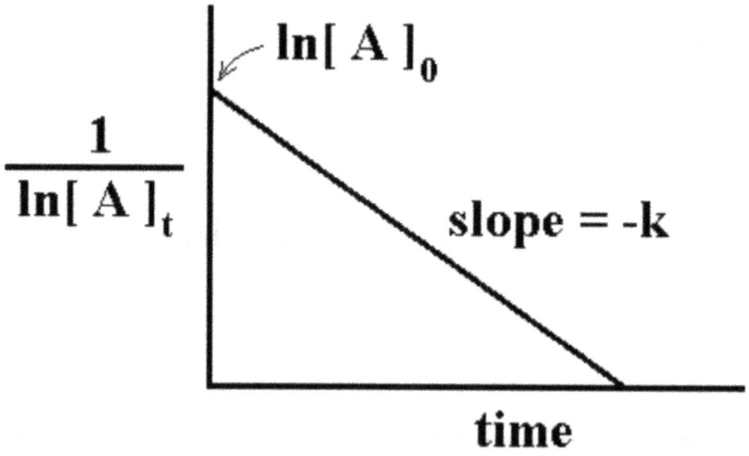

For a 1st order reaction, k is in units of seconds^{-1} ($k == s^{-1}$)

2nd Order Reaction: for one reactant

rate = $-\triangle[A\} / \triangle t = k[A]^2$

Integration yields:

$$(1 / [A]_t) - (1 / [A]_0) = kt$$

Second Order Reaction

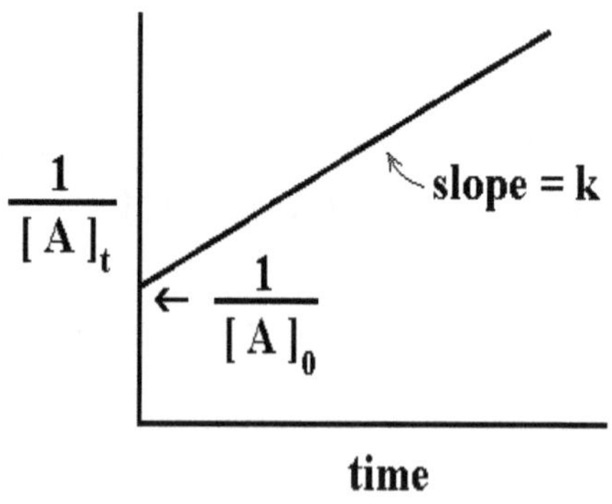

For a 2nd order reaction, k is in units of liters / (moles•seconds)

Zero order reaction: for one reactant

rate = - r[A] / r[A]$_0$ = k[A]$_0$ = k

Integration yields:

$[A]_t - [A]_0 = -kt$

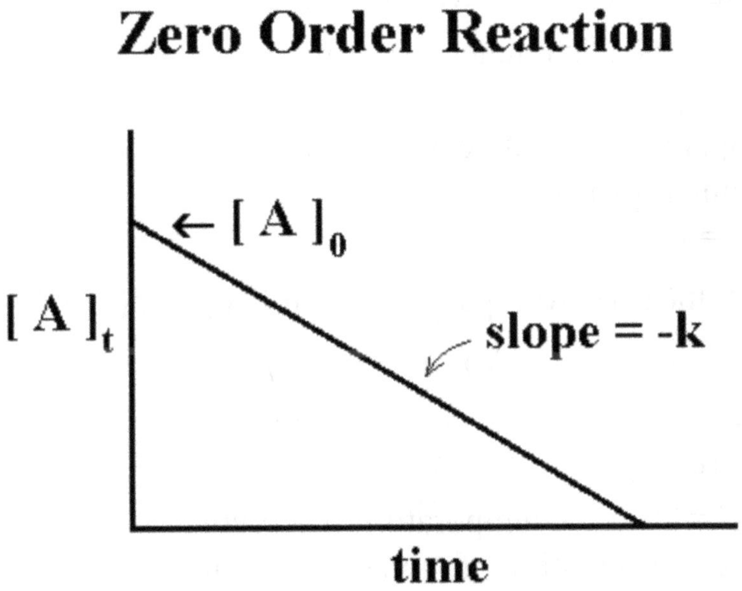

For a zero order reaction, k is in units of moles / (liters•seconds)

The half-life of a 1st order reaction is a constant independent of reactant concentration:

$\ln([A]_0 / [A]_t) = kt$;

at $t_{1/2}$, $[A]_t = \frac{1}{2}[A]_0$ (with a rate = k[A])

$\therefore \ln([A]_0 / \frac{1}{2}[A]_0) = \ln(2) = kt_{1/2}$,

$t_{1/2} = \ln(2) / k == 0.693 / k$

Concentration Decrease for 1st order reaction (rate = k[A]):

initial conc. $[A]_0$ x 1st half-life ½ x 2nd half-life½ x 3rd half-life ½ x ……

(1 half-life) (2 half-lives) (3 half-lives)

→ final conc. = $[A]_0 \bullet (\frac{1}{2})^{\#\ \text{half-lives spent}}$

Concentration Decrease for zero order reaction:

Half-life = $[\,A\,]_0 / 2k$; rate = k

($t_{1/2}$ contracts with time)

Concentration Decrease for (simple) 2nd order reaction:

Half-life = $1 / k[\,A\,]_0$; rate = $k[\,A\,]^2$

($t_{1/2}$ increases or "stretches" with time)

Effect of Temperature on Rate Constant, k

Arrhenius Equation:

$k = A \bullet e^{\wedge}[-E_a / RT]$;

$\therefore \ln(k) = A \bullet e^{\wedge}[-E_a / RT] = \ln(A) + \ln(e^{\wedge}[-E_a / RT])$

$\therefore \ln(k) = \ln(A) - E_a / RT$

(in y = b + mx form),

where E_a = activation energy ,

T = absolute temperature, in Kelvin ,

A == orientation of molecules colliding ,

R == universal gas constant = 8.314 J/(mol•Kelvin)

= 0.0821 atm•Liters / (mol•Kelvin)

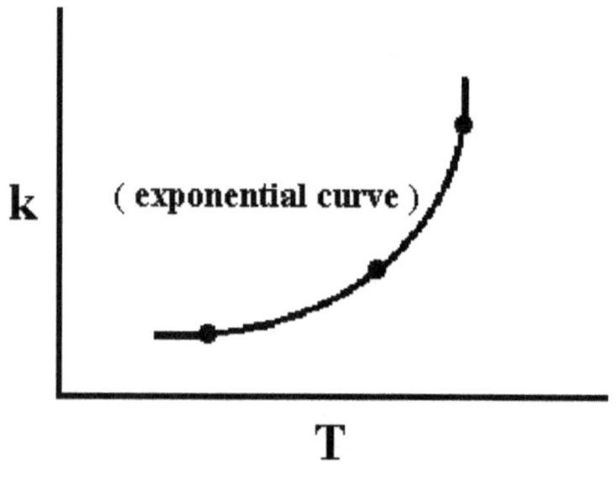

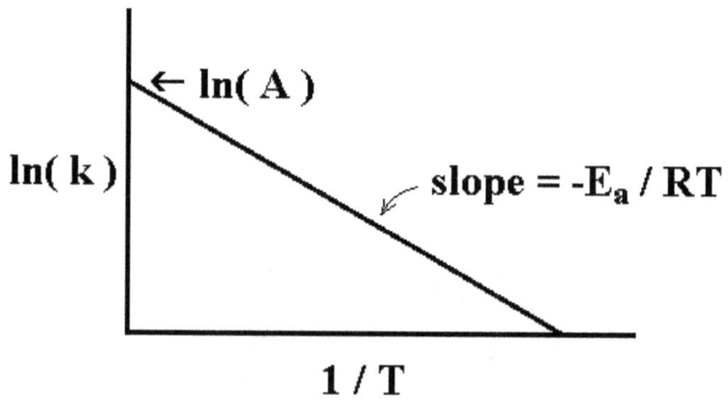

$\ln(k_2 / k_1) = (-E_a / R)[\,(1 / T_1) - (1 / T_2)\,]$

Solute Effects on Boiling and Freezing Points

Liquid (solvent) boils at a certain vapor pressure and temperature.

Adding a solute lowers the # of liquid molecules available to vaporize (at a surface)

→ boiling point increases (boiling point elevation)

Adding a solute lowers the vapor pressure

→ solution must be cooled lower to freeze ∴ freezing point depression

At the freezing point, the vapor pressure of ice and water are equal.

Specific Heat Capacity

SpHeat == $c = q / (m \bullet \Delta T)$, where c = heat capacity in joules/(grams•degree), q = heat in joules, m = mass in grams, ΔT = temperature change in celsius

e.g., if it takes 125 joules of heat to raise 111g of iron by 2.5 degrees celsius, then the specific heat capacity of iron is:

$c = 125\ j / (111\ g \bullet 2.5°c) = 0.45\ j/(g \bullet °c)$

Electricity

Conservation of electric charge:

The net electric charge in an isolated system remains constant.

$$\vec{F}_{1\ on\ 2} = k \cdot \frac{q_1 q_2}{r^2} \cdot \hat{r}$$

$$= \frac{1}{4\pi\epsilon_0} \cdot \frac{q_1 q_2}{r^2} \cdot \hat{r}$$

$\hat{r}$ = unit vector 1 → 2

$k = 9.0 \times 10^9\ N \bullet m^2/c^2$ **c = vacuum speed of light**

$\epsilon_0 = 8.85 \times 10^{-12}\ c^2 / N \bullet m^2$ == permittivity

E = F•d == force times distance

∴ use vector addition of forces

Coulomb's Law :

$$\vec{E} = \frac{\vec{F}}{q^*}$$

q* = positive test charge in electrical field
charge e = 1.6 x 10^{-19} coulombs
proton: $^{+}$e electron: $^{-}$e

$$\underset{\text{electric field}}{\vec{E}} = \frac{q}{4\pi\epsilon_0 r^2}\hat{r}$$

A point charge q produces an electric field at a distance r.

Gauss' Law (one of Maxwell's Equations) :

$$\Phi_E = \frac{Q_{\text{net, enclosed}}}{\epsilon_0}$$

Q = surface clarge

Gauss' Law is for the electric flux over a closed surface (a volume) to an enclosed surface.

$$\Phi_E = E_{\perp}A = EA_{\perp} = EA\bullet\cos\theta \qquad A = \text{area}$$

Electric and Magnetic Interactions

The two fundamental parameters that determine electric and magnetic interactions are:

electric → ε_0 == permittivity of the vacuum

magnetic → μ_0 == permeability of the vacuum

The dielectric constant == $\varepsilon / \varepsilon_0 = \kappa$ (ε not of a vacuum), a measure of e^- (charge) polarizability (dipoles δ^+, δ^-).

The speed of light in a medium (which is transparent to the light) is:

$$v = 1 / (\varepsilon \bullet \mu)^{1/2} ; c = 1 / (\varepsilon_0 \bullet \mu_0)^{1/2} .$$

In general : $\varepsilon >= \varepsilon_0$, $\mu >= \mu_0$; for most materials μ is very close to μ_0 .

∴ ε mostly determines the speed of light through the material.

The (material's) index of refraction, $n = (\varepsilon \bullet \mu / \varepsilon_0 \bullet \mu_0)^{1/2} \sim\sim \kappa^{1/2}$.

∴ The speed (v) of light in a medium is: $v = c / n$

or more generally, $\lambda = v / f = c / nf$.

v and λ change in a medium due to the material's n, but not the light's f.

Unlike through a vacuum, interactions of electromagnetic (EM) radiation (with matter) are dependent on the frequency (or wave-length) of the light. This dependence is called dispersion (of the light).
For example, one plots: n vs λ.

When the material is not transparent to the EM, near wave-lengths at which the EM is absorbed (by the material and ∴ is not transparent to the material), the index of refraction (n) rises with increasing λ in an atypical behavior called anomalous dispersion due to resonance, where the radiation frequency matches a natural absorption frequency of the material and can be readily absorbed (i.e., absorption is due to resonance).
Wherever there is an absorption peak,
there is a corresponding anomalous dispersion.

Electric Currents

Capacitance, C

Capacitance is the capacity of an electric field to store energy between parallel conducting plates.

$Q = CV$, ∴ $C = Q / V$ = (by convention) coulombs / volts = farads

The electric field E stores energy / volume between the capacitor (or condenser) plates.

R•C == time

Current, I

$I = \triangle Q / \triangle t$ == (by convention) coulombs / seconds = amperes

Ohm's Law

$V = I \bullet R$, where R resistance is in ohms

Power, P

$P = IV = I^2R$ == rate of loss of electrical energy (i.e., heat of the wire).

Resistance, R

$R = \rho L / A = 1 / G$, where ρ is an Intrinsic Property of the material called the resistivity and $1 / \rho = \sigma$ = the conductivity,
and G = the conductance (measured in siemens),
L = length of the resistor.

Electric Circuit Diagram

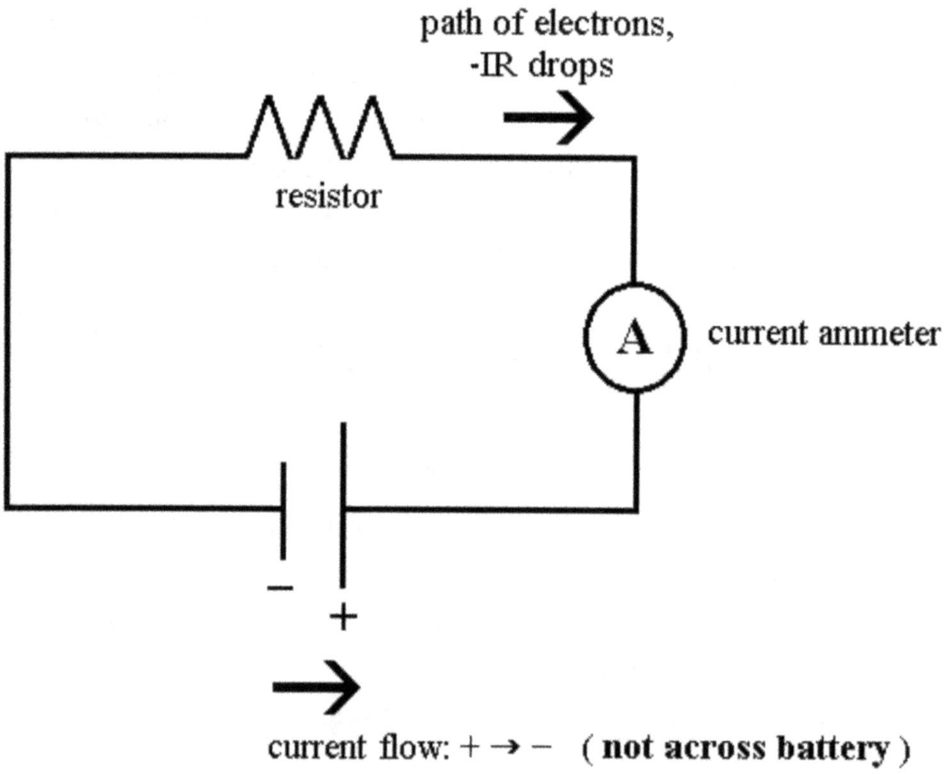

For a series circuit:

$R_T = R_1 + R_2$

$1 / C_T = (1 / C_1) + (1 / C_2)$, C == capacitance

RC == time

$RC = \tau$ as the circuit time constant.

Electric Dipoles

For electric diploes: $V_{dipole} = p \bullet \cos\theta / 4\pi\epsilon_0 r^2$,
where p == electric dipole moment = $q \bullet d$
(i.e., the distance between poles times either charge),
r = distance from center of dipole to an observation point,
$\theta = \angle$ between dipole axis and line to observation point.

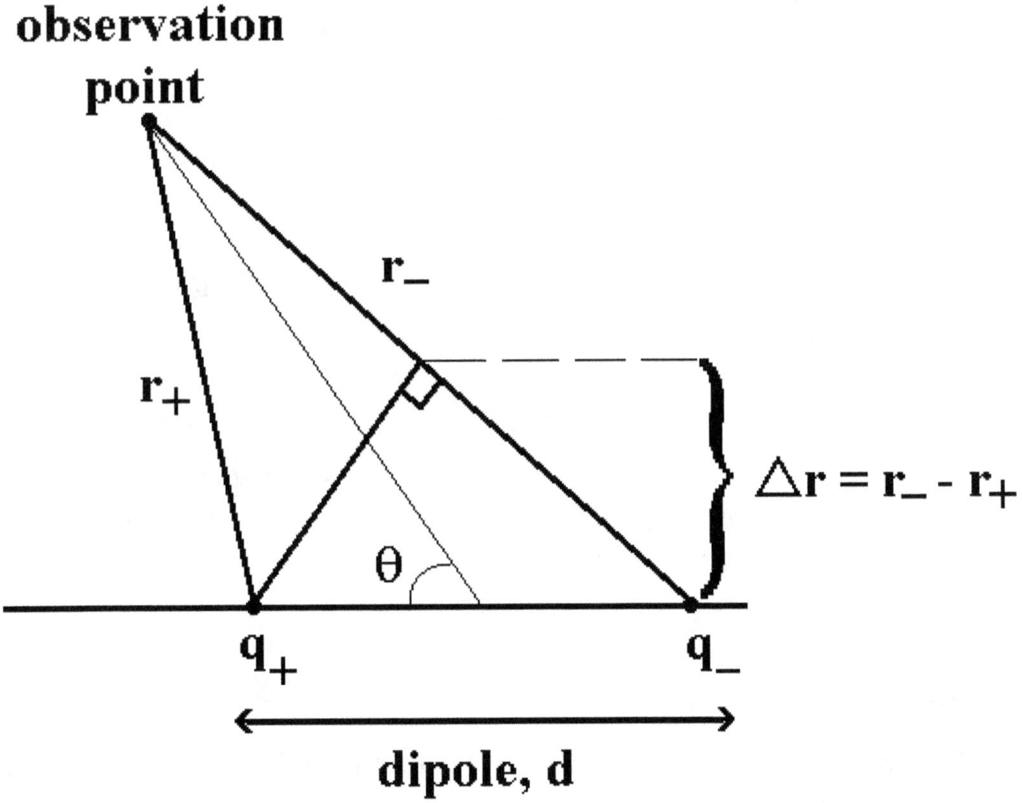

$V_{diople} = (1 / 4\pi\epsilon_0)[(q / r_+) - (q / r_-)]$ dipole voltage

$E_{dipole} = (1 / 4\pi\epsilon_0)[(q / r_+^2) - (q / r_-^2)]$ dipole electric field

Electric Flux

Electric Flux is the number of electric field lines hitting an area or surface.

Electric Force

Electric force is a conservative force: the work done by an electric force in moving a charged particle (between two points) is independent of path and depends only on starting and ending locations.

Electric Potential

Electric Potential == voltage, V

$V_{(r)} = PE_{E,(r)} / q_0$, $PE = q_0 V_{(r)}$

$\therefore \Delta V = \Delta PE_E / q_0$; $1\ eV = (1.6 \times 10^{-19} C)(1V) = 1.6 \times 10^{-19} J$

Electric potential of a point charge:

$V_{(r)} = q / 4\pi\epsilon_0 r$; $1\ J / C = 1$ volt (a scalar function)

The electric potential at a point is the external work needed to move a unit positive charge from infinity (or far away) to that point along any path.

The change in electric potential equals the negative of the work done by the electric forces (== equal and opposite to the work done by the external forces):

e.g., For a flashlight using two 1.5V batteries (= 3V), each unit of charge (1C) that moves through the light bulb from one side of a battery to the other side has used 3J worth of battery energy.

Electric Potential Energy (due to force of q_1 on q_2 by distance r):

$PE_{E,(r)} = q_1 q_2 / 4\pi\epsilon_0 r$

The electrical work done in moving a particle between 2 points is:

$-\Delta PE_E = w$

Electrode Half-reactions:

Anode → oxidation
Cathode → reduction

Electrophoretic Mobility

$U = v / E = q / f$, where v = velocity, E = applied electric field, q = charge, f = friction factor

$$\vec{v} = \frac{q\vec{E}}{f}$$

Energy

$E = E_0 / \kappa$, where κ is the dielectric constant

$PE = (1/2)CV^2 = (1/2)(\epsilon_0 A / d)(Ed)^2 = (1/2)\epsilon_0 E^2 A \bullet d$,

$PE / volume = (1/2)\epsilon_0 E^2$

$V = V_0 / \kappa$ and $C = C_0 \bullet \kappa$, where V_0 volts and C_0 capacitance are for vacuum conditions with no dielectric.

Force on a point charge (q_0) in an electric field, E

$$\vec{F} = q_0 \vec{E}$$

$w = F\triangle x = q_0 E_x \triangle x$, $\triangle V = rPE_E / q_0 = -E_x \triangle x$, $E_x = -\triangle V / \triangle x$

$\therefore F_x = -\triangle PE_E / \triangle x$

On equipotential surfaces (or lines) no work is required to move a charge because there is zero potential difference;
the surface or lines are ⊥ to the E electric field lines.

Light Energy

Non-ionizing radiation can be absorbed according to the Beer-Lambert Law: $A = \log (I_0 / I)$;

$I = I_0 \bullet e^{-\varepsilon xc}$ or, converting to base 10:

$\log(I / I_0) = -\varepsilon xc \bullet \log(e) = -\varepsilon xc / 2.3$;

$\therefore A_{OD} = \varepsilon xc / 2.3 = \log(I_0 / I)$,

where A_{OD} is now called the optical density (e.g., an OD of 2 means only 10^{-2} == 1% of light is transmitted).

ε = the molar extinction coefficient, x = distance through the sample, c = the molar concentration of the (liquid) material that absorbs the light.

For the elastic scattering of light:

the scattered light is of the same λ (wave-length) as the incident light.

Fluorescence (time scale of ~10^{-9} seconds) has longer λ than the incident λ.

One can use fluorescence to measure local conformations of molecules, ligand binding, and molecule rotations if polarized fluorescence is used (causing rotation of emitted polarized light).

Phosphorescence (time scale of ~10^{-3} seconds) has a longer time delay than fluorescence.

Magnetism

$F_M = q \bullet v \bullet B$, where q == charge, v = velocity (of charge), and B = magnetic Field.

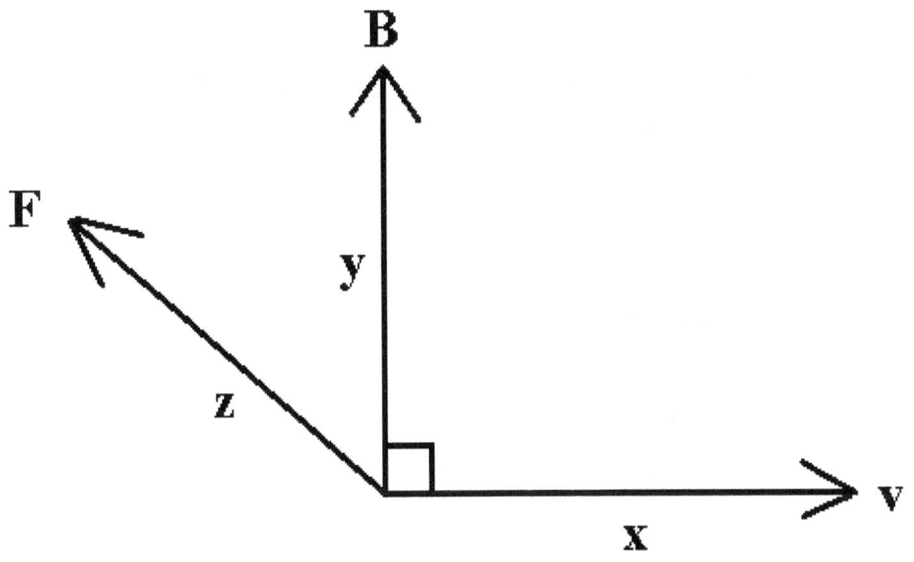

Right-Hand Rule : (x) (y) (z)
curl fingers from v to B ; thumb points as F_M

$F_M = q \cdot v \cdot B = m \cdot a = (1/2)(mv^2/r)$, where m = mass,
a = acceleration, and r = radius
(for the analog of F being a centripetal force):

q circles ⊥ to B, steered by F_M, with a speed (or velocity) proportional to radius r.

∴ $q/m = v/(r \cdot B)$, used for the mass spectrometer as a charge to mass ratio.

$KE = (1/2)mv^2 == eV$ (= electron Volt): e = electron charge, V = volt

∴ $v = (2eV/m)^{1/2}$, $m = (e \cdot r^2/2V)B^2$.

In general, $F_M = q \cdot v \cdot B \sin\theta$, θ = angle between v and B ;
only the velocity component ⊥ to B contributes to F_M.
The charged particle moves in a helical path around the B field.
The axial velocity along the B field is constant.

Since the magnetic force on a charged particle is always $\perp$ to the particle's velocity, the velocity is always directed along the instantaneous displacement of charge;

$\therefore$, magnetic forces can never do any work :

$w = F\triangle s$, s being distance;

only the component of force directed along the displacement $\triangle s$ can do work.

$\therefore$ Magnetic forces only steer a charge's velocity vector, but do not change its magnitude. Thus, the kinetic energy does not change → no work is done (on the charged particle).

With a wire carrying current, in a uniform B field, the electrons will feel a force transverse (i.e., $\perp$) to the wire as well as to B. The B field can not do work, but the the electrons pulling on the protons (+ charges) via the electrons' E field will provide the kinetic energy to pull (or push) the wire up or down, while in B.

$\therefore$ The E field does this work.

Current $I = \triangle Q / \triangle t$; $\therefore$ $Q = I\triangle t$

For a wire of length L, with current I:

$L = v_{drift}\triangle t$, where v_{drift} = time for a charge to flow the distance L.

$\therefore$ $F_M = Qv_{drift}B = Q(L / \triangle t)B = I\triangle t(L / \triangle t)B = I \bullet L \bullet B$,

with $L \perp B$.

Using the right-hand rule: curl the fingers from L to B, allowing the thumb to give the F_M direction.

The torque on a closed loop (carrying a current) in a uniform B Field is:

$\tau = I \bullet A \bullet B$, with A == area of loop

$\mu = I \bullet A$ = magnetic dipole moment; $\therefore$ (in general)

$\tau = \mu \bullet B \bullet \sin\theta$, with θ = angle between B and μ .

$\triangle w = \tau\triangle\theta$, $PE_\mu = -\mu \bullet B \bullet \cos\theta$ (always E doing the work)

The magnetic field of a wire's current is, then:

$B = \mu_0 \bullet I / 2\pi \bullet r$ (measured in teslas, T),

where μ_0 = the permeability of the vacuum = $4\pi \times 10^{-7}$ T•m/A

and r = the radius of the wire's cross-section, m = meters.

Solenoids

The solenoid, a tight helical coil of wire, can be used to provide a uniform magnetic field throughout its interior, similar to the way by which a capacitor produces a uniform electrical field between its plates.

For a closed Amperian Loop

Ampere's Law of circulation is : $\Sigma B_{\|} \bullet \triangle l = \mu_0 \bullet I_{enclosed}$

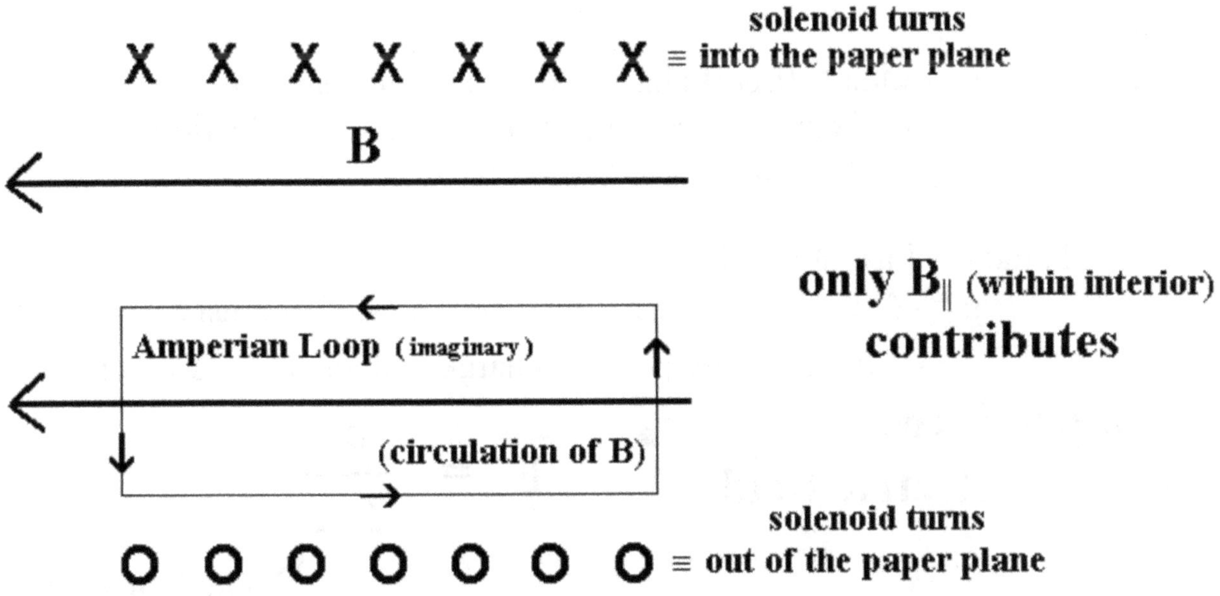

The magnetic flux is defined as:

Magnetic Flux

$\Phi_B = \bar{B}_\perp A = \bar{B} \cdot A\cos\theta$

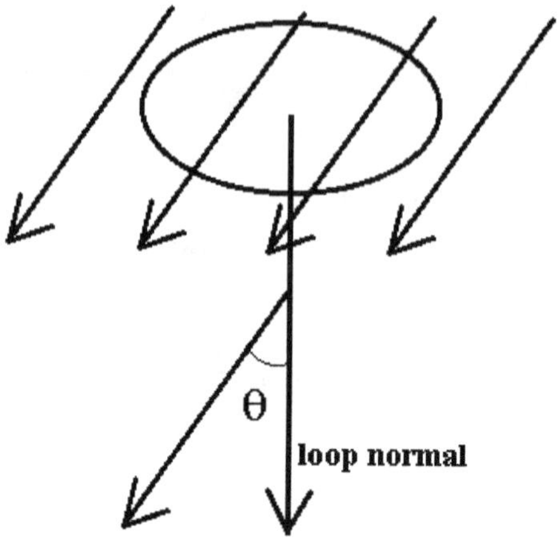

$\bar{B}_\perp$ ≡ **component of the average B ⊥ to the face of a wire loop**

Φ_B ≡ **a measure of the number of magnetic field lines that cross the area of the loop**

θ = **angle between B and loop normal**

A = **area of loop**

Only when A, θ, or B change at the loop will there be an induced current in the loop: electromagnetic induction (due to time changes in A, θ, or B).

Faraday's Law (for a coil):

$\varepsilon = -\Delta\Phi_B / \Delta t$, where ε = averaged induced emf (an energy rate) == driving energy per unit charge (with same units as volts).

ε produces an average electric field : $\vec{E} = \dfrac{\varepsilon}{\Delta x}$

↖ **length of wire**

∴ when there is a changing Φ_B, the coil acts as a battery, producing an induced emf and generating an E field within the wire that will produce a charge flow (of electrons).

Accelerating electric charges produce time varying magnetic fields and also generate electric fields (and ∴ produce electromagnetic radiation).

The induced emf is always of a polarity such as to oppose the charge of magnetic flux that created it (Lenz's Law).
That is, the induced B, of the induced emf, will have an opposite direction to the B of the current that caused the induction (current of the other coil).

AC Generator:

An AC generator is a coil of N turns of area A, made to rotate at a uniform angular velocity ω in a region of uniform B.

$\Phi_B = N{\bullet}B{\bullet}A{\bullet}\cos(\omega t)$

$\therefore\ \varepsilon = -\triangle\Phi_B / \triangle t = -N{\bullet}B{\bullet}A{\bullet}(\triangle\cos(wt) / \triangle t)$

$=$ *via calculus* $N{\bullet}B{\bullet}A{\bullet}\omega{\bullet}\sin(\omega t) = \varepsilon_{max}{\bullet}\sin(\omega t)$.

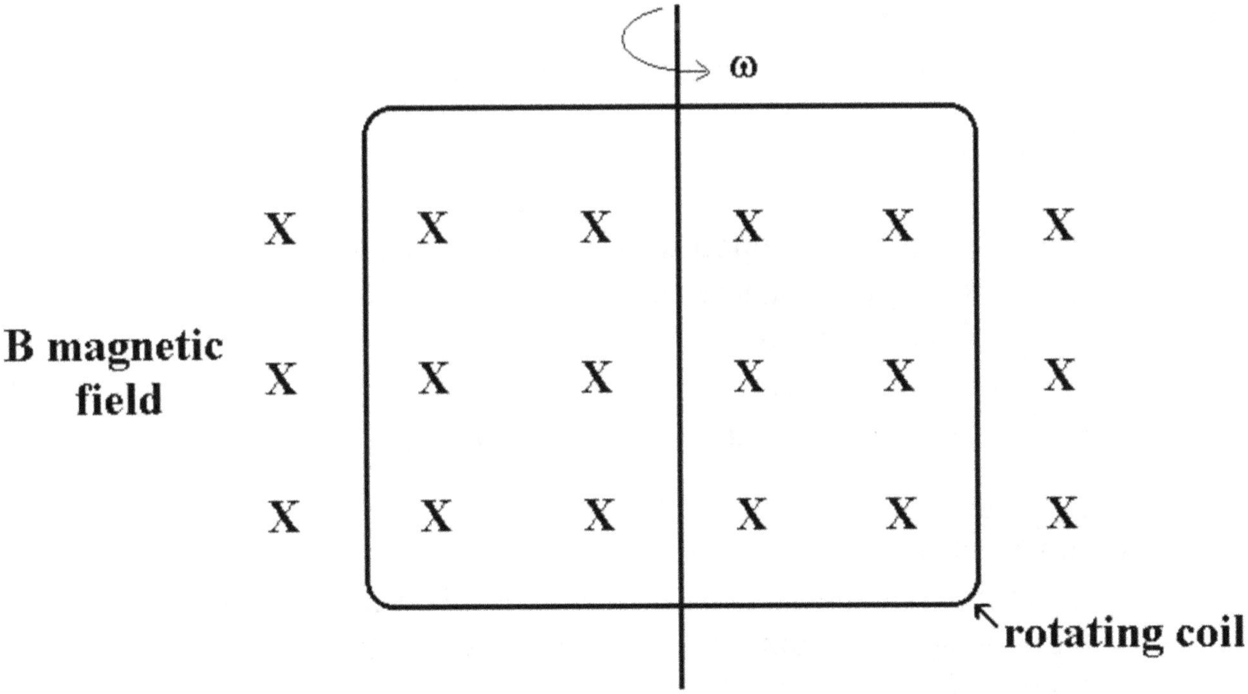

As the coil rotates the induced emf varies sinusoidally and is called an AC voltage.

Maxwell's 4 Equations (qualitatively given):

They are a summary of previous results for Electromagnetic Radiation, along with Maxwell's conjecture that a magnetic field induces an electric field; e.g., there is no current between the plates of a capacitor, but the currents on either side of the plates, that charge the plates, induce a B that in turn induces an E between the plates.

1) Gauss' Law for electric fields connects E to electric charges and holds under electrostatics (Coulomb's Law and also as generally).
2) For a magnetic field, there is no magnetic 'charge' (or monopole); ∴ all B field lines will form closed loops.
3) Changing magnetic flux is connected with an induced E (Faraday's Law).
4) A current or changing electric flux is connected with an induced B (Maxwell).

Maxwell condensed mathematically (via calculus) these laws to a minimal set of equations which seem inviolable even today (as classical physics).

Relativity theory is supported by the laws in that:
If an observer experiences (and measures) an E, a second observer (of the 1st) will measure (for the 1st) a B (and visa versa).
∴ Electric phenomena and magnetic phenomena are manifestations of the same thing: electromagnetic radiation.

Quantum Mechanics

Quantum Mechanics incorporates particle-like properties of electromagnetic radiation (along with wave-like particles), where the fundamental quantum of radiation is the photon (introduced by Einstein): Photons carry discreet amounts of energy and momentum that can be localized in space (just like a particle).
$E = hf$, where h == Plank's constant = 6.63 x 10-34 J•s ,
and f == frequency of the photon = c / λ ,
λ being the wave-length of the photon and c the speed of light in a vacuum.

$\hbar = h / 2\pi$ (note: momentum, $p = h / \lambda$)
angular momentum, $L = \hbar / 2\pi$

It is surprising that a mass-less photon may have momentum, but equations for the energy and momentum of a photon can explain a number of phenomena that can not be explained by a theory based solely on classical electromagnetic waves.

A photon can behave as a wave packet, having a limited extent in space, which can change its spatial dimensions in response to interactions with the external world.
The greater the spatial extent (to the limit of an infinite sine wave) the closer is the frequency content to a pure frequency.
∴ A single photon can act more like an extended wave or like a particle depending on its spatial extent.

The intensity of a classical wave
(which is equal to the energy per unit time per unit area
and ∴ is proportional to the square of a wave's amplitude)
corresponds not to a single photon
but to the # of photons per second (each with a particular energy, E; therefore, a wave packet density).

The average value of the Poynting vector,
$S = c \bullet (P \bullet E / V) = c \bullet \varepsilon_0 \bullet E^2$,
gives this intensity as: $I = (1 / 2) c \bullet \varepsilon_0 \bullet E_{max}^2$,
where power in watts == $P = E / t$,
with t in seconds, E in joules and V in volume.

Schrodinger Equation (qualitatively) says: an electron can act wave-like as (can) a photon; e.g., with a large enough de Broglie wave-length one can have electron diffraction.

When a photon acts as a wave packet of spacial extent, this implies that the square of the electric field must be a measure of where the photon is located.

The Schrodinger Equation plays (for quantum mechanics) the same role as Maxwell's equations for electromagnetic radiation. The Schrodinger Equation allows one to compute the space- and time-dependences of the wave function (Ψ) for any quantum system.

In the simplified, time-independent and one dimensional form of a one dimensional box of length L, with boundary conditions == nodes at $\Psi_{(0,t)} = \Psi_{(L,t)} = 0$, the particle moving back and forth from $x = 0$ to $x = L$, the possible standing waves are:
$\Psi_{n(x,t)} = A \bullet n \bullet \sin(n\pi\lambda / L)$.

Using the relationship between de Broglie wave-length and (non-relativistic) momentum:

$p = h / \lambda$, $\lambda_n = 2L / n$;

$\therefore$ $p \bullet n = nh / 2L$, $E_n = p \bullet n^2 / 2m$.

$\therefore$ $E_n = n^2 \bullet (h^2 / 8mL^2)$.

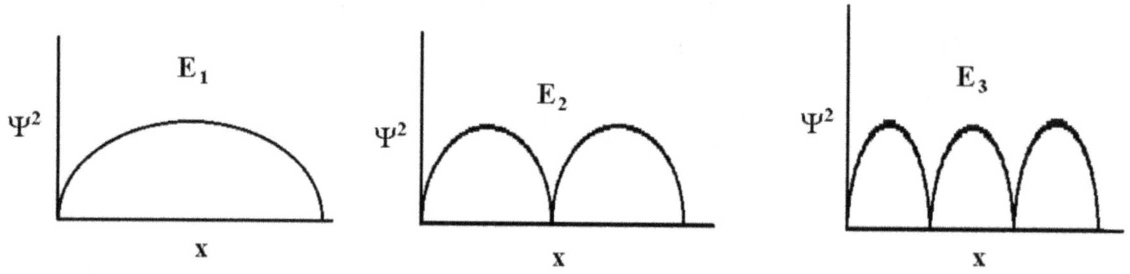

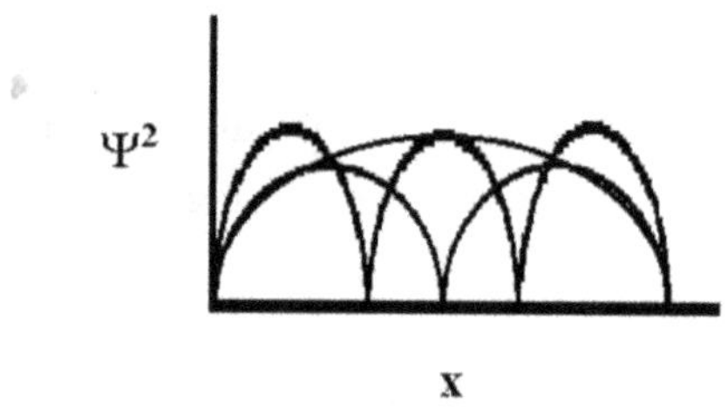

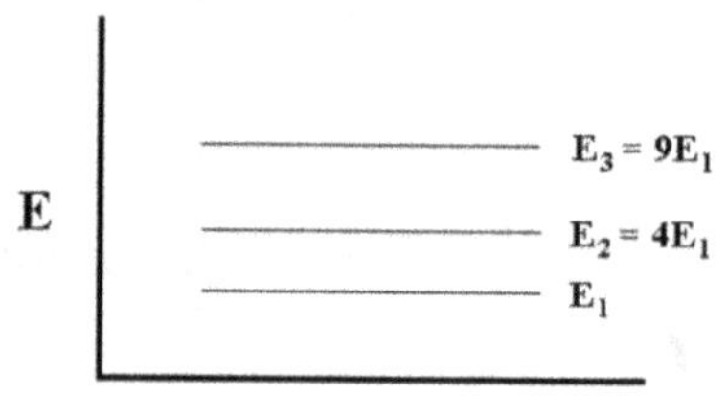

Longer wave-lengths correspond to lower energy states.

The simplified Schrodinger Equation
(time independent and one dimensional):
$(-h^2 / 8\pi^2 m)(d^2\Psi_{(x)} / dx^2) + PE_{(x)}\Psi_{(x)} = E\Psi_{(x)}$.

The particle can not have zero kinetic energy (KE),
but rather must have a minimum energy (= E_1).
∴ It has a zero point energy (even at temperature of absolute zero).

Heisenberg Uncertainty Principle

The resolution of uncertainty is comparable to the wave-length of a photon: $\triangle x \approx \lambda$.

If a photon is used to determine the location of an electron, some fraction of the photon's momentum is imparted to the electron:
∴ $p \approx h / \lambda$;
∴ The principle is: $\triangle x \triangle p \approx h$
(i.e., uncertainty in momentum along the x-direction).
This is the minimum uncertainty product possible
(the uncertainty relation).

If either $\triangle x$ or $\triangle p$ is known exactly, then $\triangle p$ or $\triangle x = 0$, so we can't know the other ($\triangle x$ or $\triangle p$) since it goes to infinity.

Another 'Heisenberg Pair' is: $\triangle E \triangle t \approx h$.
$\triangle E$ and $\triangle t$ are conjugate variables (as are $\triangle x$ and $\triangle p$) due to the uncertainties linking them.

The observed phenomenon of 'tunneling' is available:
e.g., if $\triangle t$ is short enough, $\triangle E$ may be large enough to "tunnel" through energy barriers.

Quantum Mechanics provides a wave function Ψ dependent on both time and position: The square of the wave function, $\Psi_{(x, y, z, t)}^2$, when multiplied by the volume of a small region $\triangle v$ located at (x, y, z), represents the probability that the electron will be found within that volume at that position at the specified time.

∴ $\Psi_{(x, y, z, t)}^{2} \bullet \triangle v$ = probability to find an electron within $\triangle v$ at (x, y, z) at time t.
∴ Ψ^2 is a probability density == probability per unit volume.

Ψ for matter waves is analogous to E for photons: intensity, I α E^2.
Ψ^2 α probability of finding an electron in $\triangle v$.

A Normalization Condition (scale for quantifying Ψ) is:

$$\Sigma\, \Psi^2 \bullet \triangle v = 1\,,$$

where the summation is over all the volume of the system.

Special Relativity (Lorentz - Einstein)

1) the principal of relativity is: all laws of physics are the same in all inertial frames of reference.
2) The speed of light in a vacuum (c) has the same value in all inertial reference frames.

Classical momentum: $p = mv = m \bullet (\triangle x / \triangle t)$, v = velocity and m = mass
Relativistic momentum: $p = mv / [\,1 - (v^2 / c^2)\,]^{1/2} = \gamma \bullet m \bullet v$,
where γ = the Lorentz Factor; $\gamma = 1 / [\,1 - (v^2 / c^2)\,]^{1/2}$.

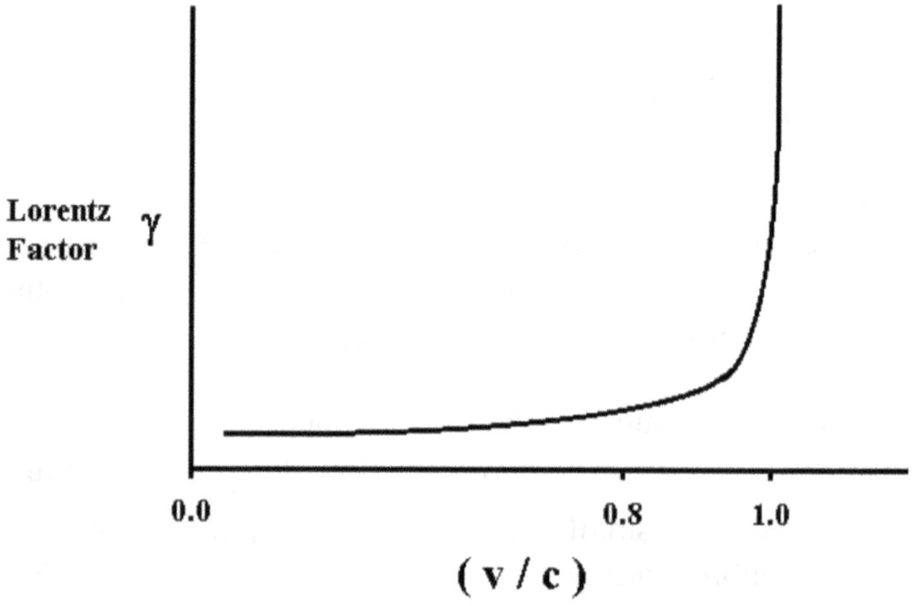

Force == rate of change of momentum = mv / t = ma , where a = acceleration ;
∴ As v approaches c, the force needed for that change,
as well as the momentum needed, increases drastically.
Since the force needed for a particle of any mass to approach [v / c = 1]
is infinite, a particle with any mass can never reach v = c .

Kinetic Energy, KE = (1 / 2)mv^2 →
KE = (mc^2 / [1 - (v^2 / c^2)]$^{1/2}$) – mc^2 = γmc^2 – mc^2 ,
where γmc^2 == the relativistic energy and mc^2 == the rest energy.

Note: by binomial theorem, $(1 - x^2)^{-1/2} = 1 + (x^2 / 2) \ldots$.
∴ KE = $mc^2 \bullet [1 + (v^2 / 2c^2)] - mc^2 = (1 / 2)mv^2$ as (v / c) << 1
∴ KE = $\gamma mc^2 - mc^2 = E - mc^2$
∴ E = KE + mc^2 (i.e., kinetic energy + rest mass energy)

$E = mc^2$ is also of classical physics, but deals with rest mass and rest energy.

Relativistic Mass : $m / [1 - (v^2 / c^2)]^{1/2} = \gamma \bullet m$, where m = rest mass.
$\gamma \bullet m$ increases drastically as the particle's speed approaches c;
the relativistic mass would become infinite at c.

Kinetic energy and momentum:
non-relativistic KE = $u^2 / 2m$ → $E^2 = p^2c^2 + m^2c^4$
== (relativistic momentum)^2 + (rest energy)^2 →
(for a mass-less particle at m = 0 or γ >> 1) E = pc .

Photoelectric Effect
E = pc , E = h*f* = hc / λ
∴ momentum of a photon, p = E / c (which is particle-like)
= h / λ (which is wave-like), where λ is the de Broglie particle wave-length.

If λ is too long (too low of energy), it will not have (by the lone photon itself) enough energy to eject an electron from a metal surface.
The minimum energy needed to eject an electron is symbolized as
ϕ == work function (often in eVs).

An ejected electron has a maximum KE.
The photoelectric effect says: $KE_{max} = hf - \phi$, and
(since KE is always positive) the minimum frequency of light needed is $f_{min} = \phi / h$.

f_{min} is independent of the intensity of the light and only depends on KE_{max} , where light intensity == # electrons / second / cross-sectional area.

Force and Energy

Definitions for heat, work, energy (in calories, joules, British Thermo-Units)

1 cal = 4.186 J
1 kcal = 4186 J = 1 Cal
1 BTU = 1055 J

Energy is the source of force (enegry == work)

work = force time distance , $w = f \cdot d$
kinetic energy == $KE = (1/2)(mv^2)$
1 joule (of energy) = $1 kg \cdot m^2 / s^2$
The force of gravity == $f_{gravity} = mg$
The gravitational potential energy at height $h = mgh$ ($= -w_{gravity}$)

w = KE + PE = a constant , where the total is a constant for conservative forces (i.e., not dependent on path or velocity; ∴ potential energy, PE, is allowed here).

Power = $P = \Delta w / \Delta t = f \Delta x / \Delta t = f \Delta v$; $P = w / t = f \cdot v$
[1 Joule/second == 1 watt]

Energy is a scalar quantity, while force is a vector quantity (i.e., has direction).

For centripetal forces:

$f_{net} = ma_{cent} = m(v^2 / r)$,

where f_{net} is in the direction of the centripetal acceleration.

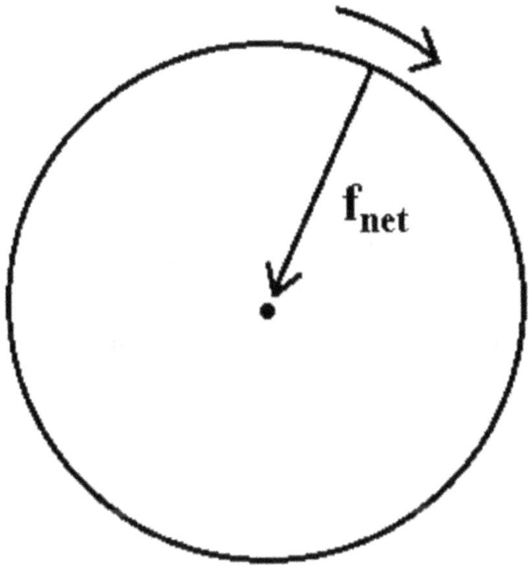

$f = ma = m(v / t) = mv / t = p / t$, where p == momentum

(allows for problems with $\triangle m$)

$KE = (1 / 2)mv^2$ and $p = mv$; $\therefore$ $KE = p^2 / 2m$

Impulse = $f\triangle t = \triangle p = p_{final} - p_{initial}$

Force, F == a physical interaction (a push or pull)

$F = m \bullet a = (m \bullet v) / t$, where m = mass and a = acceleration ;

mv = momentum (v being a velocity) ;

mass is inertia (subject to measurement) or gravitational (subject to being)

For the force of gravity (due to the mass of the earth):

$F = m \bullet g$,

$g = 9.8$ meters/seconds^2 = 9.8 Newtons/kilogram

== earth's gravitational field .

Force = mass times acceleration == (for example) kg•(meters/seconds^2)

Horizontal component of force = $force_{total}$•cos(0°) ; cos(0°) = 1
Vertical component of force = $force_{total}$•sin(90°) ; sin(90°) = 1

e.g., A skier of 80 kg pushes at constant velocity on a horizontal surface.
What is the horizontal component of the force (assume the skier pushes to the horizontal at an angle of 60°)?

$f_x = f_{total}\cos(0°)$, but the skier is weighted down by 80 kg.

At an angle of 60°, cos(60°) = 0.5

$\therefore$ $force_{along\ x} = f_{total}$•(0.5) = f_{total} / 2

The total force is 80 newtons (i.e., weight of the skier);

(one Newton = 1N == 1kg•meters/seconds^2)

$\therefore$ $force_{along\ x}$ = (80 / 2) newtons = 40 newtons

Geography

earth's diameter = 7917.5 miles

Geometry

Areas and Volumes

Area of a Circle : $A = \pi r^2$ (where r == a radius)
Surface Area of a Sphere : $A = 4\pi r^2$
Volume of a Sphere : $V = (4 / 3)\pi r^3$

Area of △ = (1 / 2)•base•height
Area of any △ = (1 / 2)(product of 2 sides)(sin of ∠ between the sides)
e.g., this is true for oblique triangles.

For a △ thus:

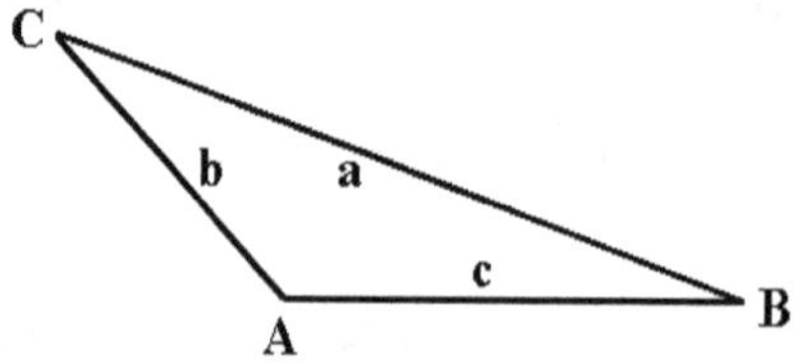

angles : A, B, C
sides : a, b, c

Area of $\triangle = [(s - a)(s - b)(s - c)]^{1/2}$, where $s = (a + b + c) / 2$

Inclination of a line (to x-axis)

slope == $m = \tan \theta = \triangle y / \triangle x$

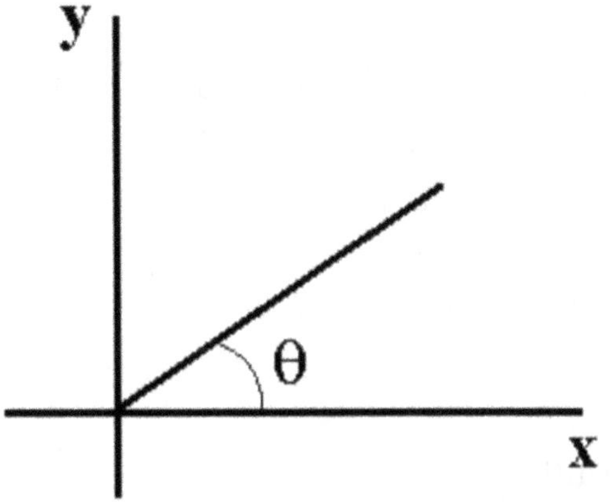

for 2 lines: $\theta = \theta_2 - \theta_1$ == ∠ between 2 lines ($\theta_2 > \theta_1$)

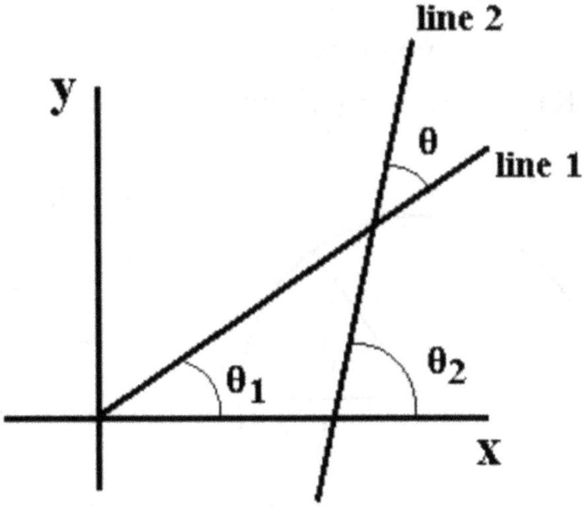

$$\tan \theta = \left| \frac{m_2 - m_1}{1 + m_2 m_1} \right|$$

Perpendicular ($\perp$) distance

between a point (x, y) and a line (of general form) $Ax + By + C = 0$:

$d = | Ax + By + C | / (A^2 + B^2)^{1/2}$

Linear Systems

The system (of equations) : $\{ax + by = c , dx + ey = f \}$ yields

$x = (ce - bf) / (ae - db)$

$y = (af - cd) / (ae - db)$

Motion, Trajectories and Gravity

Acceleration due to gravity, g

$g = 9.8$ meters / second^2

distance traveled == $s = (1 / 2)gt^2$ or $s = v_{initial}(t) + (1 / 2)gt^2$

Angular Momentum, L

$L = I\varpi$, where $\varpi = \Delta\theta / \Delta t$ in 2π radians / revolution

I = moment of inertia : rotational analog of mass = mr^2, r == radius

τ = torque : rotational analog of force = $Force_{perpindicular} \bullet radius$

$KE = (\frac{1}{2})mv^2$, v == velocity = $r\varpi$;

$\therefore$ rotational KE = $KE_{rot} == (\frac{1}{2})m(r\varpi)^2 = (\frac{1}{2})I\varpi^2$, where $I = mr^2$

distance
$s = r\theta$, θ in rads

velocity
$v = r\varpi$
($\varpi \equiv$ angular velocity)

acceleration
$a_{tangental} = r\alpha$
($\alpha \equiv$ angular acceleration)

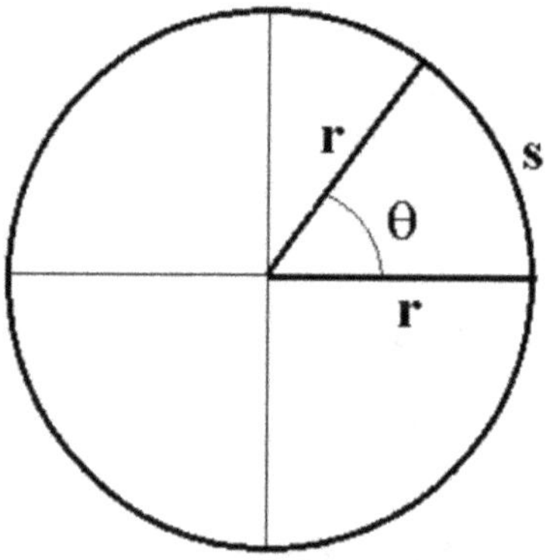

$\tau = I\alpha$

$\tau = F_{\perp} r$ **(in Newton-meters, though not an energy)**

$\tau = r_{\perp} F$
↑
(moment or level arm)

work **(for pure rotational motion)**

$\Delta w = F_{\perp} r \Delta\theta = \tau\Delta\theta$, $s = r\Delta\theta$

Free Fall Equations:

	physical variables
$s = (1/2)at^2$	s, a, t
$v(t) = v_o + at$	v, a, t
$x(t) = x_o + v_o t + (1/2)at^2$	x, a, t
$v^2 = v_o^2 + 2a(x - x_o)$	v, a, x

Forces and Weights

The force against gravity for a mass, m, is: $mg = T \bullet \sin\theta$, where T == tension

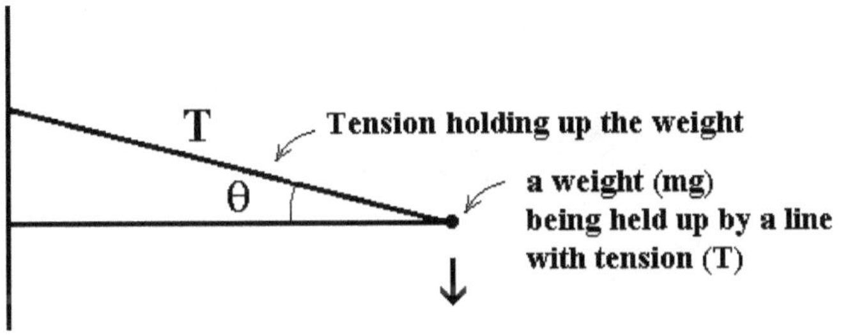

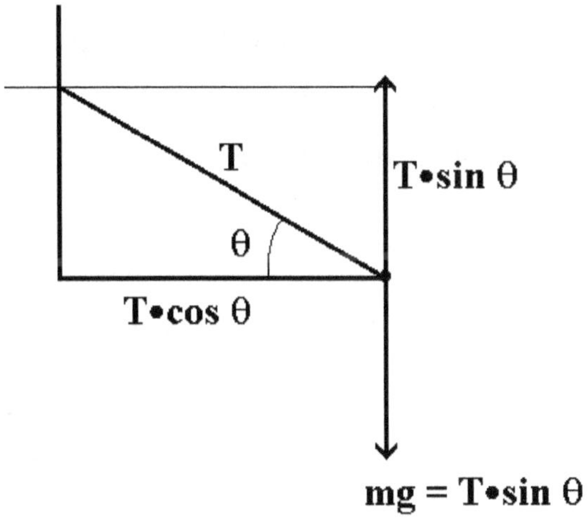

impulse= $F\triangle t = m\triangle v$ == change in momentum
mgh = gravitational potential energy (h = height, for a fall via gravity)
power (= rate of work) = $\triangle E / \triangle t$

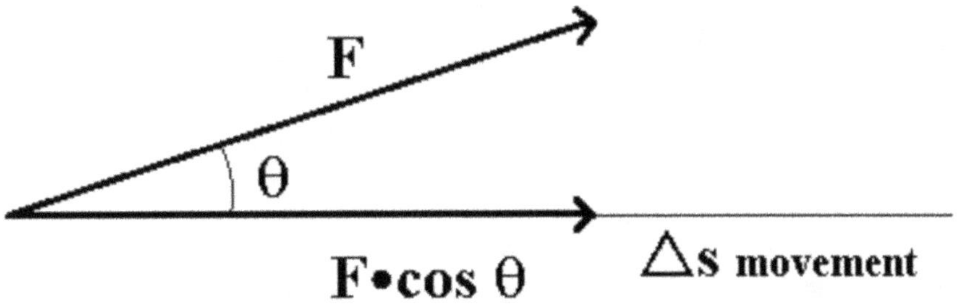

$$w = F \bullet \triangle s \bullet \cos\theta$$

Gravitational Attraction (gravitational force)

$f_{gravity} = G(m_1 m_2 / r^2)$, where r == distance between the masses m_1 and m_2

$G = 6.67 \times 10\text{-}11\ N{\bullet}m^2/kg^2$

Newton's 3 Laws of Motion

1) objects in constant velocity retain that velocity until stopped by a force
2) $F = m{\bullet}a$
3) equal and opposite forces are exerted by objects and masses (upon each other)

Newton's Law of Gravitational Attraction (Gravitational Force)

$\mathbf{F_{(M\ on\ m)} = G(M{\bullet}m / r^2)}$

Divide by m to get the gravitational field produced by M : $g_M = G(M / r^2)$

Polar Coordinates in Rotational Motion

$\theta = s / r$, with θ in radians : $\therefore\ s = r{\bullet}\theta$; when $s = r$, $\theta = 1$ radian

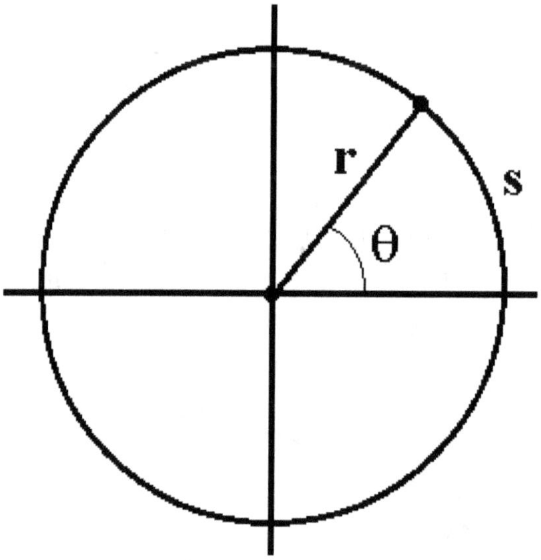

$s = 2\pi r$, in radians

$2\pi == 360^\circ = 1$ revolution

$\therefore$ 1 radian $= 360^\circ / 2\pi = 57.3^\circ$

A radian is unit-less since s and r use the same units

1 rad $= 180^\circ / \pi$, 1 degree $= \pi \bullet$ rad $/ 180^\circ$

angular velocity = $\omega = \triangle\theta / \triangle t$
instantaneous angular velocity (i.e., as $r \bullet t \rightarrow 0$) =

$$\omega = \lim_{(\triangle t \rightarrow 0)} \triangle\theta / \triangle t$$

linear velocity, $\triangle s / \triangle t = r\triangle\theta / \triangle t$, $\therefore$ linear velocity $v = r \bullet \omega$

average acceleration, $\overline{\alpha} = \triangle\omega / \triangle t$
instantaneous acceleration, $\alpha = \lim_{(\triangle t \rightarrow 0)} \triangle\omega / \triangle t$
tangential or linear acceleration, $a_{tang} = r \bullet \alpha$
centripetal acceleration, $a_{cent} = v^2 / r$

$a_{tang} = 0$ in uniform circular motion, but not a_{cent}

Since the velocity changes with radius, $a_{cent} = v^2 / r = \omega^2 r$;
a point closer to the center of a wheel rotates more slowly than a point farther from the wheel's center.

Constant angular acceleration == uniform circular motion.

Divide linear variables by r to get angular variables:
e.g., $\theta = s / r$, $\omega = v / r$, i.e. for free fall equations ($\triangle s = \triangle x$)

Rotational Kinetic Energy:
$KE = (1/2)mv^2 == (1/2)m(\omega r)^2 = (1/2)m\omega^2 r^2$
One can define I == moment of inertia : $I = mr^2$; $\therefore KE = (1/2)I\omega^2$
There are different formulas for I for different shapes,
due to internal forces' maintaining shape and rigid body symmetries.

Torques, τ
have units of energy, normally Nm or Newton•meters
They are a form of work: $\tau = I\alpha = mr^2\alpha$ (for simple cases) $= rF_{\perp}$;
$\therefore rF_{\perp} = mr^2\alpha$

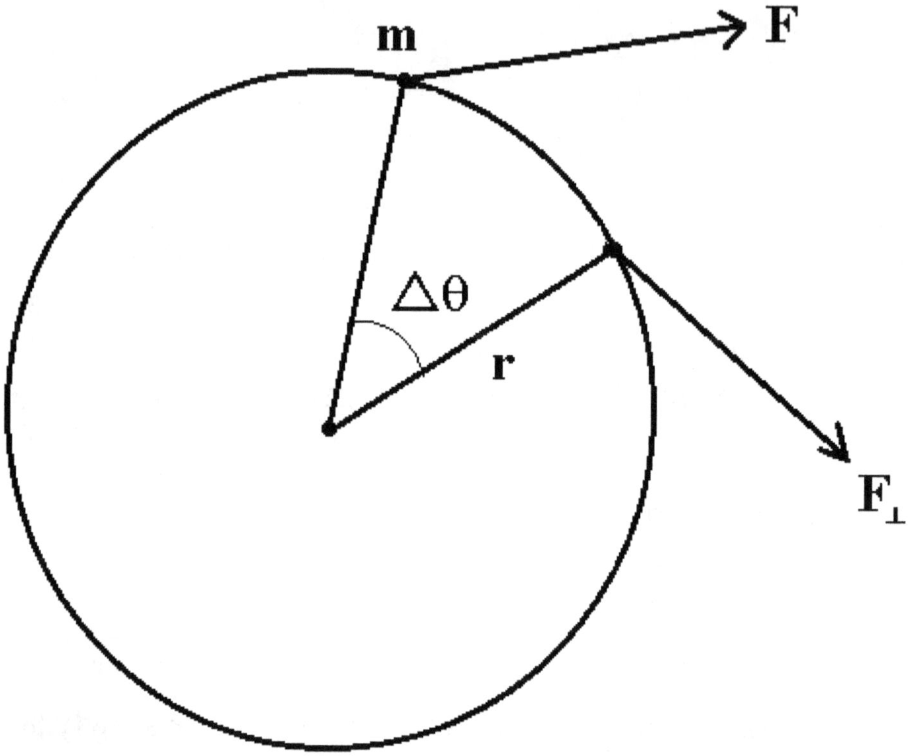

For physical comparisons (of rotational to linear analogues):

$$(\tau , I , \alpha) == (\text{Force} , \text{mass} , \text{acceleration}_{\text{linear}})$$

angular momentum, $L = I\omega == p = mv$ (as an analogue) .

$$\tau = rF_{\perp} = r\bullet(F\bullet\sin\theta) = (r\bullet\sin\theta)\bullet F = r_{\perp}F$$

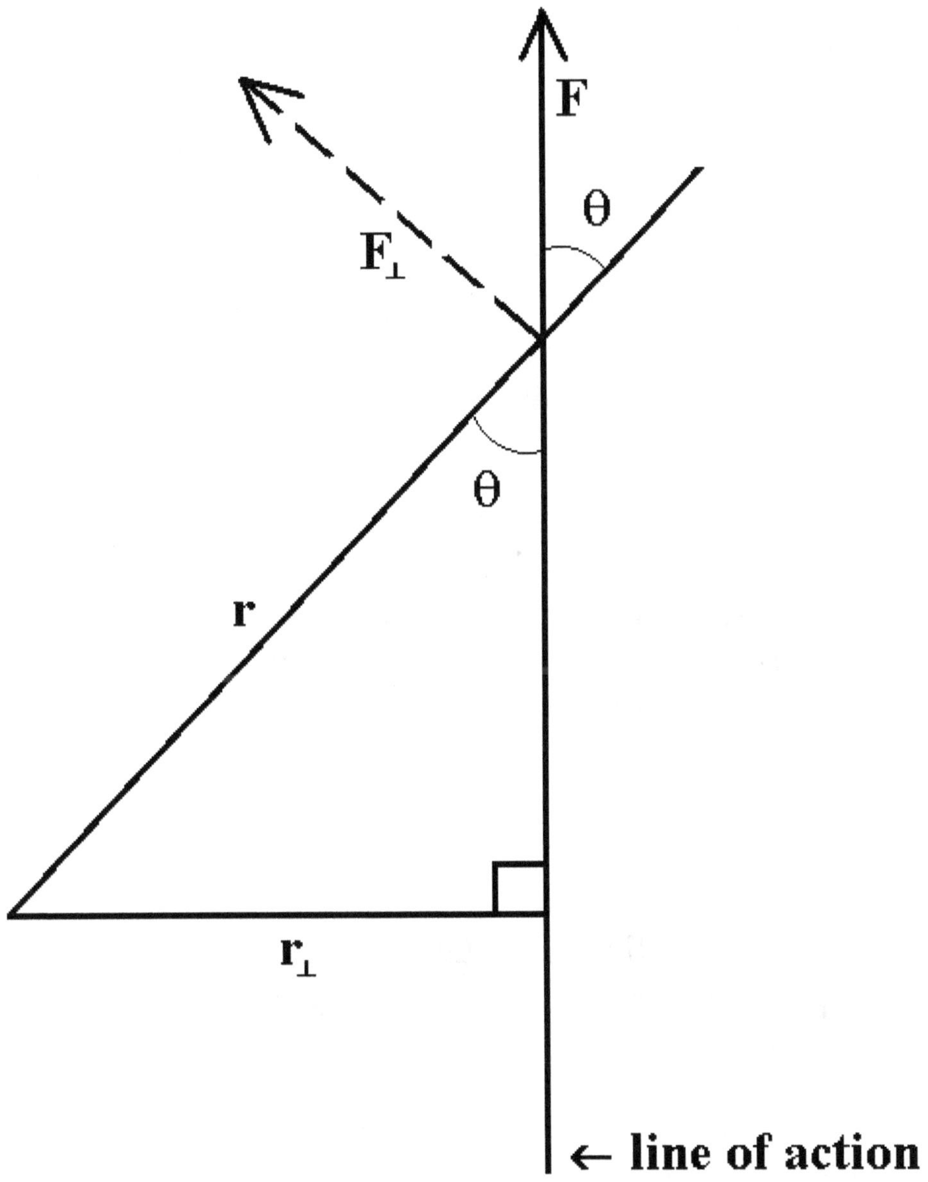

$\sin\theta = r_\perp / r$

$r_\perp$ is perpendicular to F (and line of action).

$F_\perp$ is perpendicular to r ; $r_\perp$ is called the moment (or lever) arm.

Projectile Motion

$s = \frac{1}{2} gt^2$, where s = projected distance, g = acceleration due to gravity, and t = time ; the constant g = 9.8 meters / second^2 ;

e.g. , If a rock is thrown (or projected) off a 50 meter cliff and lands 90 meters away on the ground, how fast was its speed when it hit the ground?
The vertical (projected) distance of the fall = 50 meters == s .
When the rock comes to rest on the ground (neglecting any rolling, which does not contribute to the 90 meters landing site), the weight of its mass due to the earth's gravity is empirically (i.e., measured as) independent of time;
thus, we must establish the time duration of the fall.

$\therefore$ $s = \frac{1}{2} gt^2$ == 50 meters = $\frac{1}{2} gt^2$; solve for t .
$t = (2 \cdot 50 \text{ meters} / 9.8 \text{ metes/second}^2)^{1/2} = 3.19$ seconds
The speed (velocity, v = distance / time) is
the horizontal distance traveled in this 3.19 seconds.
v = (horizontal distance traveled) / t = 90 meters / 3.19 seconds = 28.2 m/s
= 28 m/s to 2 significant figures (since 90 meters is at 2 significant figures).

Initial (v_i) and Final (v_f) Velocities

$$v_f^2 = v_i^2 + 2a(x - x_o) ,$$

where a == acceleration, x_o == starting distance , x = horizontal distance

e.g., A ball is thrown off an 8 ft ledge with a velocity of 22 ft/s and at an angle of 28° .
Find the ball's final velocity :

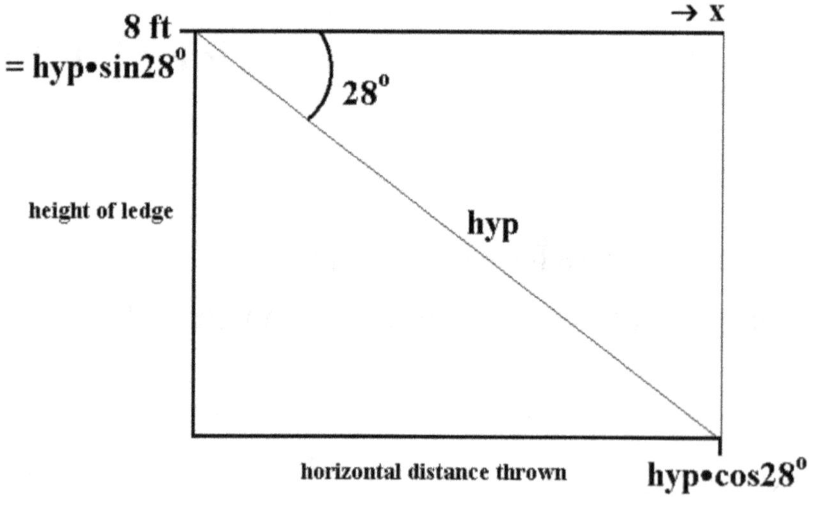

The hypotenuse (hyp) for the projection
(needed to find the horizontal distance thrown) is hyp•sin28°,
which is equal to the height of the ledge.

$\therefore$ hyp•sin28° = 8 ft , hyp = 17.04 ft.

The horizontal distance thrown, then, is

hyp•cos28° = (17.04 ft)•cos28° =15.05 ft .

$v_f^2 = v_i^2 + 2a(x - x_0)$, with $x_o = 0$ ft and $x = (x - x_o) = 15.05$ ft

a = acceleration due to gravity = g = 9.8 ft/s²

$\therefore v_f^2 = (22 \text{ ft/s})^2 + 2(9.8 \text{ ft/s}^2)(15.05 \text{ ft}) = 778.9 (\text{ft/s})^2$

$v_f = (778.9 (\text{ft/s})^2)^{1/2} = 27.91 \text{ ft/s}$

e.g., A ball is thrown straight up and 2.2 seconds later is caught
(at the same height it was initially thrown).
What was its initial velocity (v_i) and how high did it rise?

The ball is thrown against gravity.
The amount of speed it loses as it rises is: $\triangle v = g\triangle t$;

$\therefore \triangle v = (9.8 \text{ meters/second}^2)(2.2 \text{ seconds} / 2) = 10.78 \text{ m/s}^2$

(assuming it reaches its peak height in ½ the time).
But, the final speed (v_f) comes from throughout the entire motion:

$\therefore v_f = g\triangle t = (9.8 \text{ m/s}^2)(2.2 \text{ s}) = 21.56 \text{ m/s}$

$v_f = v_i + a\triangle t == v_i + g\triangle t$

$\therefore v_i = v_f - g\triangle t = 0.0 \text{ m/s}$, since $g\triangle t = v_f$

The speed up is +10.78 m/s and the speed down is -10.78 m/s .
The height the ball reaches is:

$s = (½)gt^2 = (½)(9.8 \text{ m/s}^2)(2.2 \text{ s} / 2)^2 = 5.39 \text{ meters} = 5.4 \text{ m}$

e.g., A ball is thrown horizontally from the roof (i.e., top)of a building,
at a speed of 5 meters / second.
It hits the ground 20 meters from the building. How high is the building?

The ball is essentially falling vertically from the building: the only force on it is gravity since a constant velocity gives it no additional force.

At 5 meters / second if it travels 20 meters (horizontally to the ground), then it has fallen 20 meters / (5 meters / second) = 4 seconds .

It falls due to the acceleration of gravity.

$V_{final} = V_{initial} + a \cdot t$, where a == acceleration = g (= acceleration due to gravity).
But the vertical distance traveled = S , $S = [(V_i + V_f) / 2] \cdot t$;
∴ with $V_f = V_i + g \cdot t$ and $V_i = 0$ (i.e., initial vertical velocity) ,

$S = V_i \cdot (t) + (1 / 2)gt^2 = (1 / 2)gt^2$;

$2S / g = t^2$, $t = (2S / g)^{1/2}$

$g = 9.8$ meters / second^2

∴ $S = gt^2 / 2 = [(9.8$ meters / second$^2) \cdot 4$ seconds] / 2

= 78.4 meters ~ = 78 meters

e.g., A rock is thrown up in the air and comes down.
All the time (going up, going down) it is in free fall, affected by the acceleration due to gravity (9.8 m/s^2).

If its velocity is 20 m/s , after one second its velocity is 20 m/s – 9.8 m/s : every second it looses 9.8 m/s of velocity;
9.8 m/s^2 == (9.8 m/s) per second .

When it reaches the top of its rise, its velocity is zero, and it starts to fall down: its velocity is now negative, till it reaches the ground (going -20 m/s).
Its acceleration (during the transit or travel) is always g.
Free fall == vertical velocity.

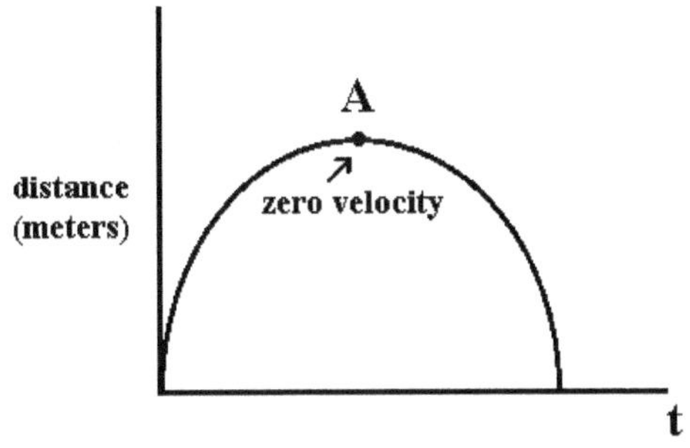

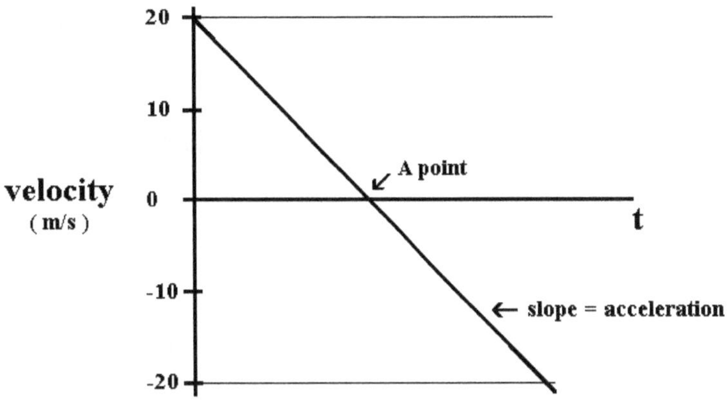

Velocity

Average Velocity, $\triangle v$

$\triangle v = [\, f(t + \triangle t) - f(t)\,] / \triangle t$, where $\triangle s = f(t + \triangle t) - f(t)$

Numbers and Statistics

Absolute Value problems

e.g., $|\, x - 2\,| = 3$ implies both $x - 2 = 3$ and $-(x - 2) = 3$
(two different equations) ; $\therefore$ $x = 5$ and $x = -1$;
but check for extraneous solutions
(solutions which don't work out after | | operation,
or variable = 0 == undefined and thus no solution).

Arithmetic Progression

$a_n = a_1 + (n - 1)d$,
where n == n^{th} term, a_1 == initial term, and d == common difference between terms.

In general : $a_n = a_m + (n - m)d$

e.g., If $a_1 = -12$ and $a_{24} = 66$, what does a_{42} equal?

First find d (the common difference of the arithmetic progression) from

$a_n = a_1 + (n - 1)d$

$66 = -12 + (24 - 1)d$

Now use this d with n = 42 to calculate a_{42} :

$$a_{42} = -12 + (42 - 1)d$$

Binomial Special Products

$(x + y)(x - y) = [\, x(x - y) + y(x - y) = x^2 - xy + xy - y^2 \,] = x^2 - y^2$

$(x + y)^2 = x^2 + 2xy + y^2$

$(x - y)^2 = x^2 - 2xy + y^2$

$(x + y)^3 = x^3 + 3x^2y + 2xy^2 + y^3$

$(x - y)^3 = x^3 - 3x^2y + 3xy^2 - y^3$

$u^3 + v^3 = (u + v)(u^2 - uv + v^2)$

$u^3 - v^3 = (u - v)(u^2 + uv + v^2)$

u, x and/or v, y can be real numbers or expressions or variables.

Binomial Theorem

Expansion of $(x + y)^n$ for n objects picked r at a time is:

$(x + y)^n = x^n + nx^{n-1}y + \ldots. + nC_r x^{n-r}y^r + \ldots. + nxy^{n-1} + y^n$,

where

$$nC_r = \binom{n}{r} = \frac{n!}{(n - r)!r!}$$

The binomial coefficient is a combination.

$$\text{e.g.,}\quad 8C_2 = \binom{8}{2} = \frac{8!}{(8 - 2)!2!} = \frac{8!}{6! \bullet 2!} = \frac{8 \bullet 7 \bullet 6!}{6! \bullet 2!} = \frac{8 \bullet 7}{2 \bullet 1}$$

When r ≠ 0 and r ≠ n , one can use simple factors:
(same # of factors above and below)

$$8C_2 = \binom{8}{2} = \frac{\overbrace{8 \bullet 7}^{\text{2 factors}}}{\underbrace{2 \bullet 1}_{\text{2 factors}}} = 28$$

Note: $0! = 1$, $1! = 1$

Note: the **Combination** $nC_r = nC_{n-r}$

e.g. ,

$$\binom{8}{2} = \binom{8}{8-2} = \binom{8}{6} = \frac{8\cdot7\cdot6\cdot5\cdot4\cdot3}{6\cdot5\cdot4\cdot3\cdot2\cdot1} = \frac{8\cdot7}{2\cdot1}$$

Note: a **Permutation**, where order is important, is:

$nP_r = n! \,/\, (n - r)!$

Box and Whisker Plots, and Quantiles

For a given set of data: $a_1, a_2, a_3, \ldots\, a_n$

1) put them in numerical order (e.g., increasing order)
2) find the median of the data; e.g., if there are 17 points, the median is the 9th (odd) point. If there are 18 points, the median is (even → odd) the average of the 9th and 10th points → Q_2 quantile
3) find the median for the 1st half (upper half) → Q_1 quantile
4) find the median for the lower half → Q_3 quantile
5) the plot is then (on some linear scale for the a_s):

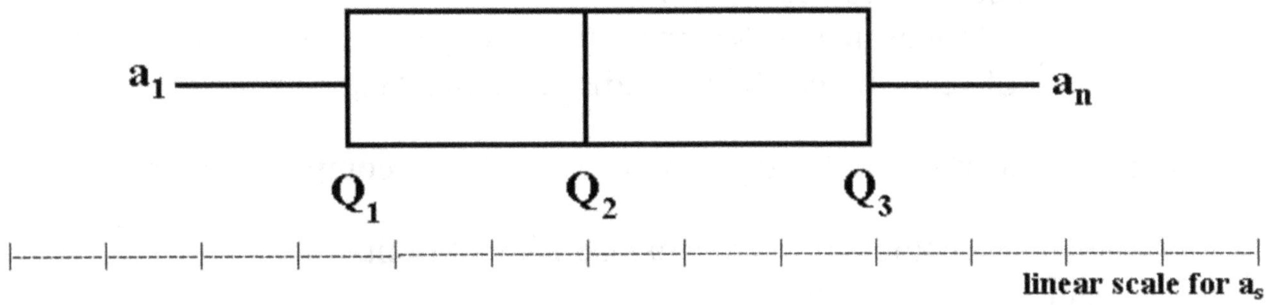

Completing the Square (to solve quadratic equations)

e.g., for $x^2 + 6x - 7 = 0$

1) get x^2 term without coefficient, and put loose term to other side

   $x^2 + 6x = 7$

2) divide coefficient of x by 2 , square this number , add to both sides , simplfy

   $6x / 2 = 3x \;;\; 3^2 = 9$

$(x^2 + 3x + 9) = 7 + 9$
$(x + 3)^2 = 7 + 9 = 16$

3) take the square root
$[(x + 3)^2]^{1/2} = (16)^{1/2}$; $x + 3 = (16)^{1/2} = +,- 4$

4) solve for x
$x = -3 +,-4$

e.g, for $4x^2 -2x -5 = 0$

1) $4x^2 - 2x = 5$
$x^2 - (2 / 4)x = 5 / 4$

2) $x^2 - (1 / 2)x + (1 / 4)^2 = (5 / 4) + (1 / 4)^2$
$[x - (1 / 4)]^2 = (5 / 4) + (1 / 16)$

3) $x - (1 / 4)^2 = [(5 / 4) + (1 / 16)]^{1/2} = (21 / 16)^{1/2} = +,-(21)^{1/2} / 4$

4) $x = (1 / 4) +,-(21)^{1/2} / 4$

Compound Interest

The formula for Annual Compound Interest is: $A = P(1 + r/n)^{nt}$, where

A = future value with interest
P = principal (initial) amount
r = annual interest rate (a fraction)
n = # of times interest is compounded per year (usually 12, for 12 months)
t = # of years (money is invested); per annum means $t = 1$

Simple Interest is, then, $A = P(1 + rt)$; ie., $n = 1$ (only compounded once a year).

When interest is compounding to increase without limit:
$\lim_{(n \to \infty)} (1 + 1/n)^n = e$,
where n == # of times interest is compounded in a year

Simple Interest here is: $I_{simple} = r \bullet B_0 \bullet m_t$, where

r = interest rate == (percentage, % / 12) per annum
B_0 = initial balance
m_t = # of time periods (i.e., months)

e.g., at a 12.99% interest rate: $r = 0.1299 / 12$;
and after 3 months $I_{simple} = (0.1299 / 12) \bullet B_0 \bullet 3$

Note that: $e = 1 + (1/1!) + (1/2!) + (1/3!) + \ldots\ldots \sim= 2.718281828$

For compounding interest:

Future Value = (Present Value)$(1 + r/n)^n$, where

r = (fractional) interest rate

n = # of periods (months)

$\therefore$ at infinity of n : $(1 + r/n)^n \rightarrow (1 + 1/n)^n = e$

Degrees

Degree → DMS (Degrees, Minutes, Seconds) conventions

60 minutes to a degree

60 seconds to a minute

$\therefore$ x.y degrees → x degrees + 60(.y) minutes

If 60(.y) minutes → y.z , then this is 60(.z) seconds

DMS → Degree

D + (M / 60) degrees + (S / 3600) degrees = total degrees

Determinants

Given $a_1x + b_1y = c_1$

$a_2x + b_2y = c_2$

one can use determinants to solve for x and y :

The common denominator is

$$\begin{vmatrix} a_1 & b_1 \\ a_2 & b_2 \end{vmatrix} = (a_1 \bullet b_2) - (a_2 \bullet b_1)$$

If denominator = 0,
then the lines (for the two equations) are coincident or parallel.

The x numerator is $\begin{vmatrix} c_1 & b_1 \\ c_2 & b_2 \end{vmatrix}$ The y numerator is $\begin{vmatrix} a_1 & c_1 \\ a_2 & c_2 \end{vmatrix}$

$$\therefore \quad x = \frac{\begin{vmatrix} c_1 & b_1 \\ c_2 & b_2 \end{vmatrix}}{\begin{vmatrix} a_1 & b_1 \\ a_2 & b_2 \end{vmatrix}} \qquad y = \frac{\begin{vmatrix} a_1 & c_1 \\ a_2 & c_2 \end{vmatrix}}{\begin{vmatrix} a_1 & b_1 \\ a_2 & b_2 \end{vmatrix}}$$

$$\begin{vmatrix} a_1 & b_1 & c_1 \\ a_2 & b_2 & c_2 \\ a_3 & b_3 & c_3 \end{vmatrix} \equiv \begin{vmatrix} + & - & + \\ - & + & - \\ + & - & + \end{vmatrix}$$

$$= a_1\begin{vmatrix} b_2 & c_2 \\ b_3 & c_3 \end{vmatrix} - b_1\begin{vmatrix} a_2 & c_2 \\ a_3 & c_3 \end{vmatrix} + c_1\begin{vmatrix} a_2 & b_2 \\ a_3 & b_3 \end{vmatrix}$$

Exponential and Logarithmic Functions

for $a > 0$ and $a =/= 1$

$a^0 = 1$, $a^1 = a$, $a^m \bullet a^n = a^{m+n}$,

$a^m / a^n = a^{m-n}$, $(a^m)^n = a^{mn}$

$a^{-m} = 1 / a^m$

If $f(x) = a^x$ ($= y$) , $f^{-1}(x) = \log_a x$ (when $a^y = x$)

$\log_a 1 = 0$, $\log_a a = 1$

$\log_a mn = \log_a m + \log_a n$

$\log_a (m / n) = \log_a m - \log_a n$

$\log_a x^m = m\log_a x$

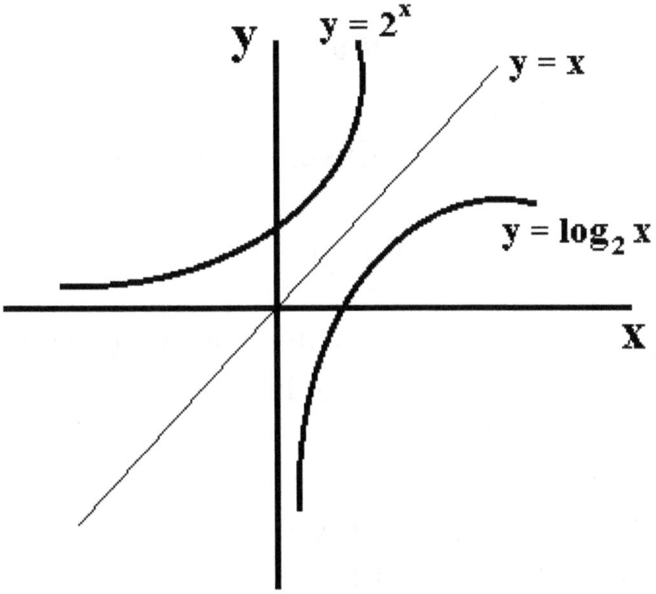

Given $a = e^b$; $\therefore$ $\ln(a) = b$

$\ln(a^u) = u \bullet \ln(a)$

$\ln(a \bullet x) = \ln(a) + \ln(x)$

$\ln(x / a) = \ln(x) - \ln(a)$

$\ln(x^n) = n \bullet \ln(x)$

$a^u = e^{u \bullet \ln(a)}$

$\ln(1) = 0$, $\ln(0) = -\infty$

$\log(1) = 0$, $\log(0) = -\infty$

$\log(10^2) = 2$

$2 \bullet \log(10) = 2$, $\log 10 = 1$

$a^u \bullet a^v = a^{u+v}$

$a^{-u} = 1 / a^u$

$(a^u)^v = a^{uv}$

$(a \bullet b)^u = a^u \bullet a^v$

e.g., solve for x : $4^{2x} = 7^{(x-1)}$

convert from exponents to logarithms

$2x \bullet \log(4) = (x - 1) \bullet \log(7)$

solve for x (as normal) → $x = -2.373....$

Given $y = a^x$, then $\log_{(a)} y = x = \log(a^x)$

$a^{-x} = (1/a)^x = 1/a^x$

$\log_a a^x = x$; $\therefore$ $a^\wedge\log_a x = x$ (inverse functions)

Given $a = e^\wedge\log_{(e)} a$, then $a^x = (e^\wedge\log_{(e)} a)^x = e^\wedge x\bullet\log_{(e)} a$

e == natural base , ln == natural logarithm , log == logarithm

$\ln(e^r) = r\bullet\ln(e)$; $\log(10^r) = r\bullet\log(10)$

To change base of logarithms

changing log of x from base a → base b :

$\log_a(x) = \log_b(x) / \log_b(a)$

e.g., Find x if $20^x = 10^{130}$ (i.e., changing base from 20 to 10) :

$20^x = 10^{130}$

$\log_{20}(20^x) = \log_{10}(10^{130}) = 130$

$\therefore \log(20^x) = \log(10^{130}) = 130$

$x\bullet\log(20) = 130$

$x = 130 / \log(20) = 99.9 \approx 100$

$e = 2.71828....$, $e^{-1} \sim= 0.3679 == 36.8\%$

$e == (1/0!) + (1/1!) + (1/2!) + (1/3!) + ...$

$e^r == (r^0/0!) + (r^1/1!) + (r^2/2!) + (r^3/3!) + ...$

$e^{\ln(A)} = A$; $e^{m\bullet\ln(A)} = A^m$; $e^{n + m\bullet\ln(A)} = A^m\bullet e^n$

$0! = 1$ and $1! = 1$ (by definition) ; $e^{i\pi} = -1$; $\therefore$

Euler's Rule: $e^{i\pi} + 1 = 0$ (this relates exponential to trigonometric functions)

Exponential Graphs

of $f(x) = ax^n$

Graphs for ax^n

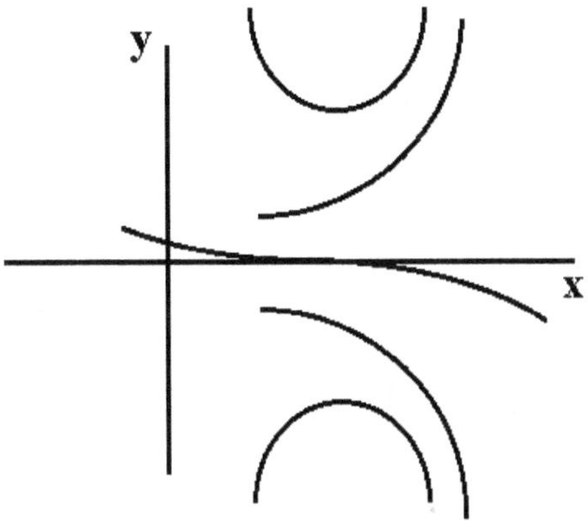

$a > 0$
concave away from x-axis

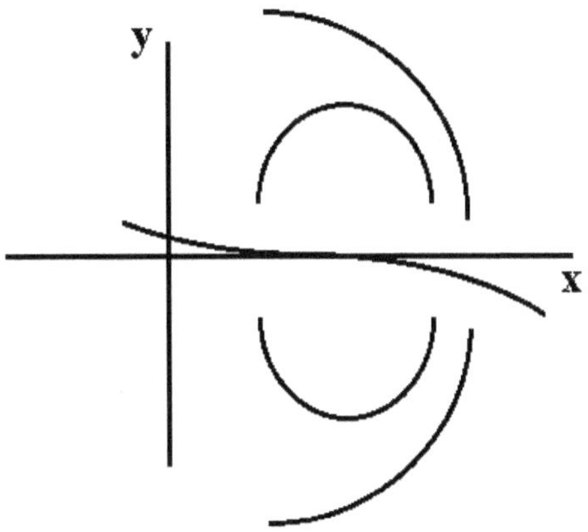

$a < 0$
concave towards x-axis

n is even
a is positive

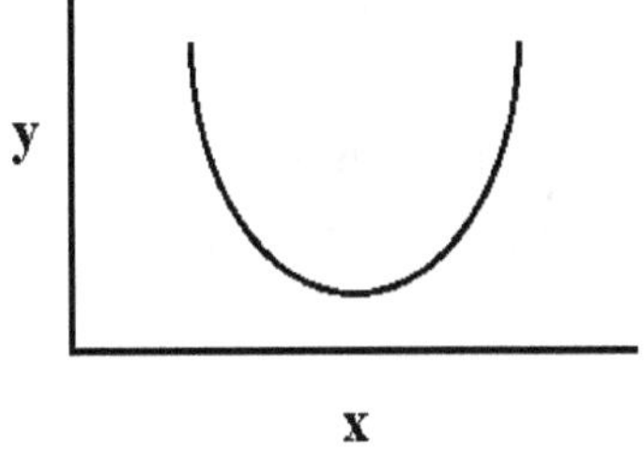

n is even
a is negative

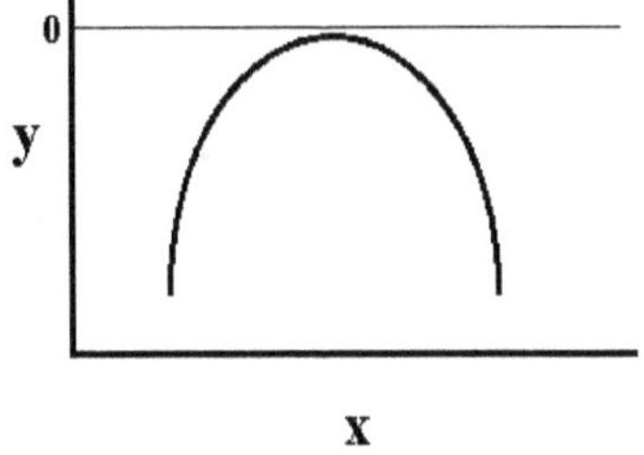

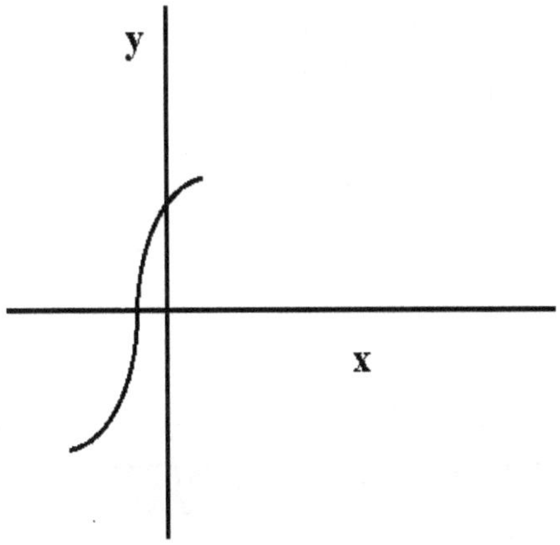

n is odd
a is positive

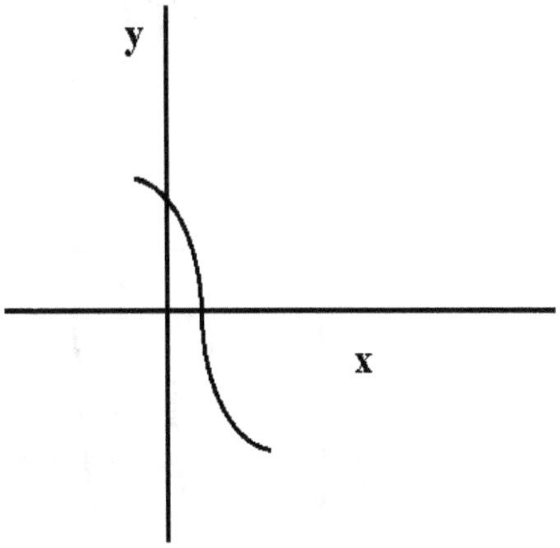

n is odd
a is negative

Extremas

When graphing a function a local extremum (a maximum or minimum) may occur in a region which is not overall maximum or minimum. It may also be global. Functions with many extremas can be very difficult to graph.

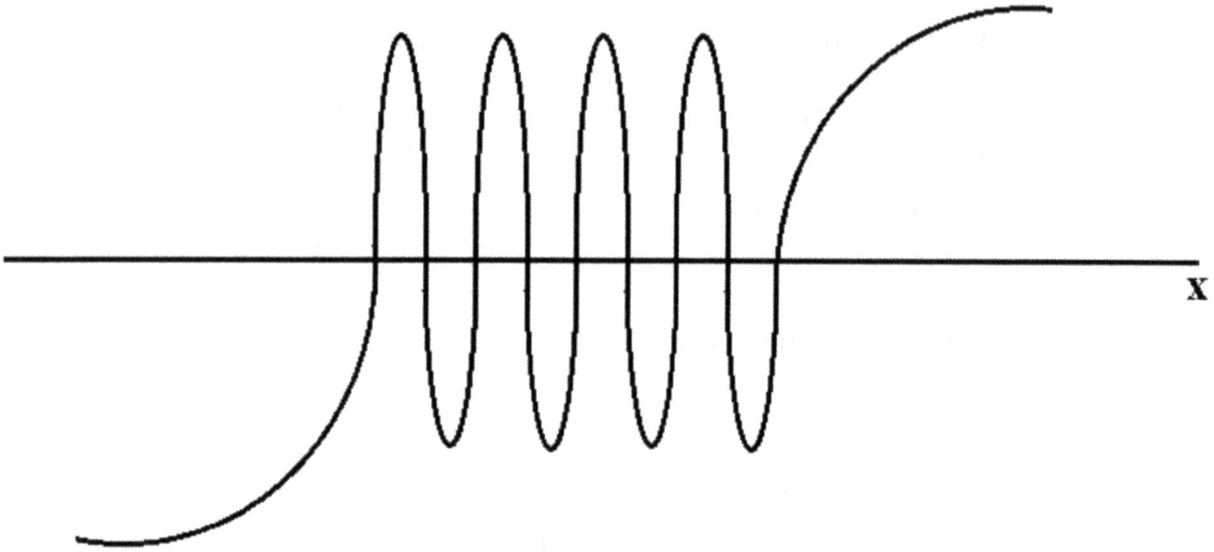

Examples: cos (1 / x) and sin (1 / x) both near zero

Another example is $\sin(e^{2x+9})$ near 0 and 1 , which can have (at x = 0 or 1):

$$[(e^{11} / \pi) - (1 / 2)] - [(e^{9} / \pi) - (1 / 2)] + 1 = 19058 - 2579 + 1 = 16480$$

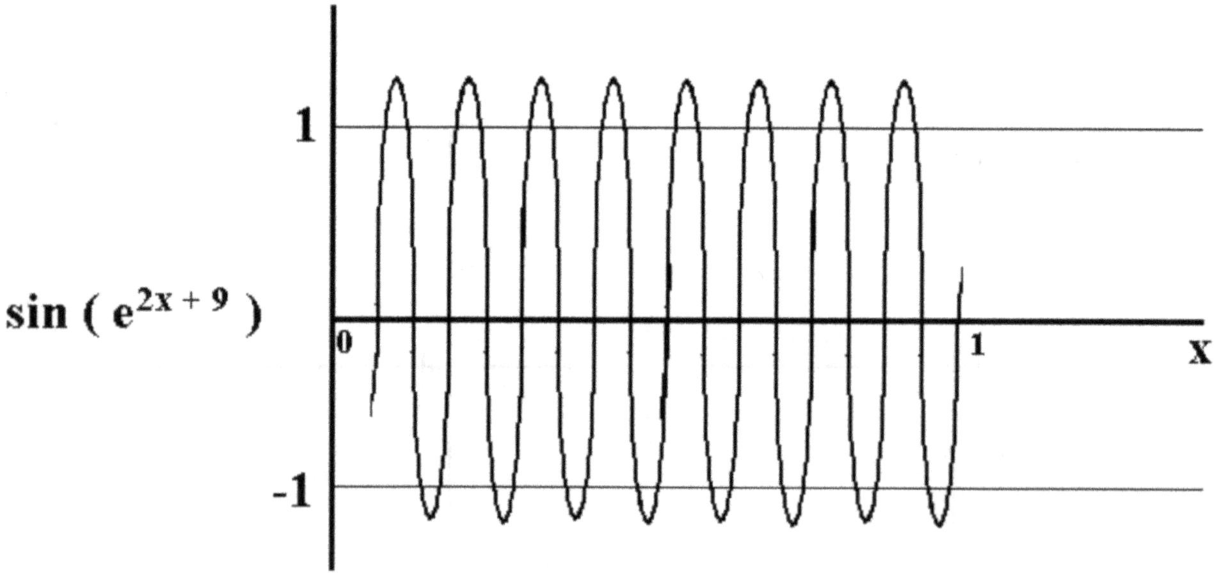

i.e., an extrema is in the closed interval [0, 1] .

Factoring Polynomials

e.g., factor $x^3 + x^2 - x - 1$

1) $x^2 - 1 = (x + 1)(x - 1)$ from intuition or experience

$\therefore\ x^3 + x^2 - x - 1 = x^3 - x + x^2 - 1$ rearrange terms

2) $x^3 - x + x^2 - 1 = x^3 - x + (x + 1)(x - 1)$

3) $x^3 - x + (x + 1)(x - 1) = x(x^2 - 1) + (x + 1)(x - 1)$

$= x(x^2 - 1) + 1(x^2 - 1)$ a common factor here is $(x^2 - 1)$

$= x(x^2 - 1) + (x^2 - 1) == (x + 1)(x^2 - 1)$

$= (x^2 - 1)(x + 1)$ standard factoring form

Factoring a polynomial by grouping

Generally, group terms to try and find a common factor.

e.g., $x^3 - x^2 + 4x - 4$

$= (x^3 - x^2) + (4x - 4)$

$= x^2(x - 1) + 4(x - 1)$

$= (x - 1)(x^2 + 4)$

e.g., $6x^3 - 7x^2 - x + 2$

$= x(6x^2 - 7x - 1) + 2$

After exploring possibilities, we find that :

$6x^2 - 7x + 1 = (x - 1)(6x - 1)$

$\therefore$ We change -1 to +1 (in $6x^2 - 7x - 1$) by adding +2 and subtracting 2x

$x(6x^2 - 7x + 1) \quad + 2 - 2x$

$= x(x - 1)(6x - 1) + 2(1 - x)$

Now, we need a common factor.

$\therefore$ We can have $(1 - x)\bullet(-1) = (-1 + x) = (x - 1)$.

$(-1 / -1)\bullet[x(x - 1)(6x - 1) + 2(1 - x)]$

$= [-x(x - 1)(6x - 1) + 2\bullet(-1)(1 - x)] / -1$

$= [-x(x - 1)(6x - 1) / -1] + [2(x - 1) / -1]$

$= x(x - 1)(6x - 1) - 2(x - 1)$

$= (x - 1)[x(6x - 1) - 2]$

Factoring a trinomial by grouping

for form $ax^2 + bx + c$:

Look for ac such that $\boldsymbol{a} + \boldsymbol{c} = b$ to allow $bx = \boldsymbol{a}x + \boldsymbol{c}x$

e.g., $2x^2 + 5x - 3$

$ac = (2)(-3) = -6$

$b = 5$; $(+6)(-1) = -6$ and $(+6) + (-1) = 6 - 1 = 5$

$\therefore\ 5x = 6x - x$

$\therefore\ 2x^2 + 6x - x - 3 = (2x^2 + 6x) - (x + 3)$

$= 2x(x + 3) - 1(x + 3) = (x + 3)(2x - 1)$

Functions

Odd function : $f(-x) = -f(x)$

e.g., $\sin(-x) = -\sin(x)$

Odd functions are symmetric about the origin:

$f(x) = (\frac{1}{2})x^3$, $f(-x) = -f(x)$

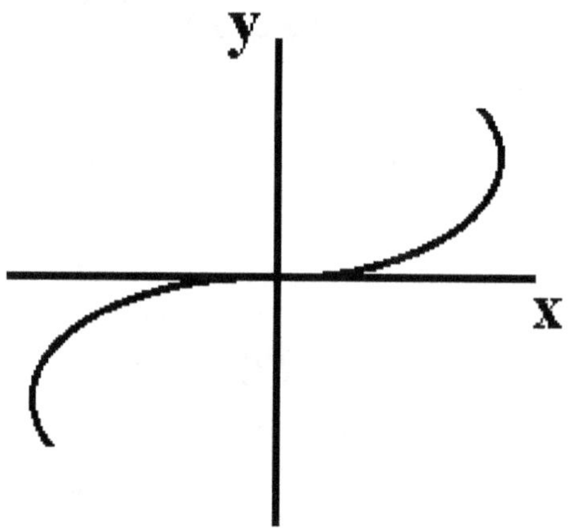

Even function : $f(-x) = f(x)$

e.g., $\cos(-x) = \cos(x)$

Even functions are symmetric about the y axis:

$f(x) = x^2 - 2$, $f(-x) = f(x)$

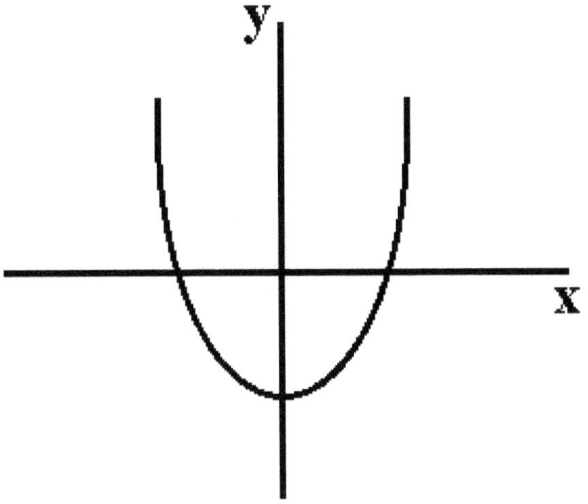

For a one-to-one function, y is a function of x only if each x has only one y , and ∴ cuts the graph vertically at only one x value.

If f yields a single output for each input and yields a single input for each output, then f is one-to-one: a horizontal line cuts the graph for x (value) at only one point. A function that is increasing or decreasing on an interval is one-to-one on that interval.

If f is one-to-one with domain x and range y , then there is a function f^{-1} with domain y and range x .
f^{-1} is the inverse of f . They are symmetric along the $y = x$ line.
To find the inverse of $y = f(x)$, interchange x and y and then solve for y .
$f^{-1}(y_0) = x_0$ if and only if $f(x_0) = y_0$.

Given $y = f(x)$: $y = -f(x)$ is a reflection of $y = f(x)$ on the x axis.
Likewise, $y = f(-x)$ is a reflection of $y = f(x)$ on the y axis.

$y = |f(x)|$ makes y always positive.
$y = f(|x|)$ makes the graph symmetric on the y axis.

Geometric Series

A geometric series is : ($a_1 = a$) + ($a_2 = ar$) + ($a_3 = ar^2$) + ($a_4 = ar^3$) + …. ,
where a == the starting term and r == a common ratio .

e.g., If $a_1 = 8$ and $a_6 = 8192$, what is a_{42} equal to ?

$a_6 == a \bullet r^5 = 8192$; $\therefore\ r^5 = 8192 / 8 = 1024$

$\therefore\ r = (1024)^{1/5} = 4$; i.e., $4^5 = 1024$

[$\log(r^5) = \log(1024) = 3$; $5 \bullet \log(r) = 3$, $\log(r) = 3 / 5 = 0.6$]
[$\text{antilog}(\log r) = \text{antilog}(0.6) = r = 4$; i.e., $10^{0.6} = 4 = r$]

$a_{42} = a \bullet r^{41} = 8 \bullet 4^{41} = 3.87 \times 10^{25}$

Hyperbolic Functions

e.g., cosh, sinh

Just as cos(t) , sin(t) form a circle with a unit radius,
cosh(t), sinh(t) form the right half of an equilateral hyperbola
e.g., as for : $x^2 - y^2 = 1$

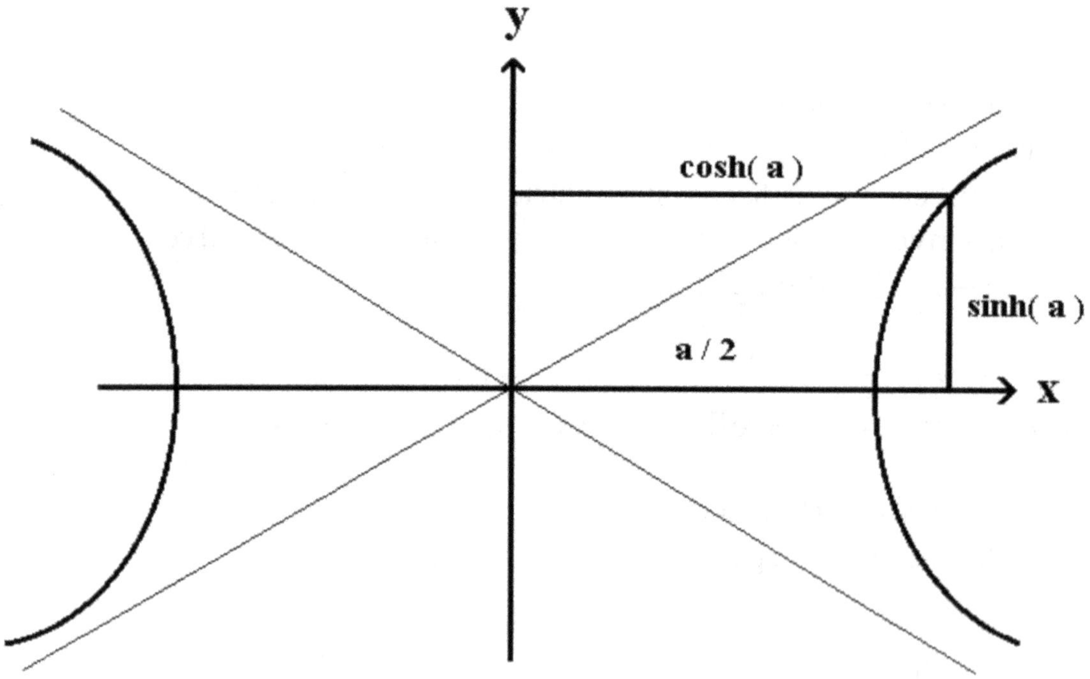

The hyperbolic angle is the size of the hyperbolic area (hyperbolic sector; e.g., a / 2).
The hyperbolic functions may be defined in terms of the legs of a right triangle covering this sector.

The hyperbolic functions occur in solutions of some important differential equations.

Inequalities

Flipping symbols with signs : $a < b$, but $(-1)a > (-1)b$

Imaginary Numbers

The imaginary unit, $i = (-1)^{1/2}$, is used in order for oscillatory solutions to exist:

e.g., $e^{ix} = \cos(x) + i \bullet \sin(x)$

Inverse Functions

These must be one to one functions; ∴ only one x per y == horizontal test.

e.g., $f(x) = (2x + 8)^3$

$y = (2x + 8)^3$ or exchange x for y, and solve for y == $f^{-1}(x)$.

$(y)^{1/3} = 2x + 8$

$(y)^{1/3} - 8 = 2x$

$[(y)^{1/3} - 8] / 2 = x == f^{-1}(y)$

Inverse Relationship

e.g., find the inverse relation of $x^{3y} = -5$

The inverse to an exponential is a logarithm.

$\log x^{3y} = \log(-5)$

$3\log x^y = \log(-5)$

$\log x^y = \log(-5) / 3 = y\log x$

∴ $y = [\log(-5) / 3] / \log x = \log(-5) / 3\log x$

Long Division of polynomials

e.g., $\dfrac{x^2 - 2}{x + 1}$

Fill in empty terms: $x^2 - 2 = x^2 + 0x - 2$

$$\begin{array}{r|l} & x - 2 \\ x + 1 & \overline{x^2 + 0x - 2} \\ & x^2 + x \\ & \overline{-x - 2} \\ & -2x - 2 \\ & \overline{+x - 0} \end{array}$$

$x(x + 1) = x^2 + x$ This looks like $x^2 + 0x$ → subtract

$-x - 2$ → **leaves a remainder**

$-2(x + 1) = -2x - 2$ → $-2x - 2$

x term is left as a remainder → $+x - 0$

$$\therefore\ x + \frac{-x - 2}{x + 1} = x + \frac{-(x + 2)}{x + 1}$$

$$= x - \frac{x + 2}{x + 1} \quad \text{(answer)}$$

$$\therefore\ \frac{x^2 - 2}{x + 1} = x - 2 + \frac{x}{x + 1} = x - 1 - 1 + \frac{x}{x + 1}$$

$$= x - 1 - \frac{x + 1}{x + 1} + \frac{x}{x + 1} = x - 1 + \frac{-(x + 1) + x}{x + 1}$$

$$= x - 1 - \frac{1}{x + 1} \quad \text{(answer)}$$

Try

$$\begin{array}{r|l} & x - 1 \\ x + 1 & \overline{x^2 + 0x - 2} \\ & x^2 + x \\ & \overline{-x - 2} \\ & -x - 1 \\ & \overline{-1} \end{array}$$

$-1(x + 1) = -x - 1$ → $-x - 1$

-1 → **leaves a real-number remainder** $\equiv \dfrac{-1}{x + 1}$

$$\therefore\ \frac{x^2 - 2}{x + 1} = x - 1 - \frac{1}{x + 1} \quad \textbf{(answer, in most reduced form)}$$

Note: $x - 1 - [1/(x+1)] = x - 1[(x+1)/(x+1) - [1/(x+1)]$
$= x - [(x+1)/(x+1)] - [(1/(x+1)]$
$= x - [[(x+1)+1]/(x+1)]$
$= x - [(x+2)/(x+1)]$ as shown above.

Matrices

addition

$$\begin{bmatrix} a & b \\ c & d \end{bmatrix} + \begin{bmatrix} e & f \\ g & h \end{bmatrix} = \begin{bmatrix} a+e & b+f \\ c+g & d+h \end{bmatrix}$$

scalar multiplication

$$2\begin{bmatrix} a & b \\ c & d \end{bmatrix} = \begin{bmatrix} 2a & 2b \\ 2c & 2d \end{bmatrix}$$

transposition

$$\begin{bmatrix} a & b \\ c & d \end{bmatrix}^T = \begin{bmatrix} a & c \\ b & d \end{bmatrix}$$

matrix multiplication

$$\begin{bmatrix} a & b \\ c & d \end{bmatrix} \bullet \begin{bmatrix} e & f \\ g & h \end{bmatrix} = \begin{bmatrix} ae+bg & af+bh \\ ce+dg & cf+dh \end{bmatrix}$$

dot products

determinants

$$\det\begin{bmatrix} a & b \\ c & d \end{bmatrix} = \begin{vmatrix} a & b \\ c & d \end{vmatrix} = ad - bc$$

$$\det(AB) = \det(A) \bullet \det(B)$$

↖ **product of square matrices**

complex numbers

$$a + ib \longleftrightarrow \begin{bmatrix} a & -b \\ c & d \end{bmatrix}$$

Linear Equations

A == m x n matrix

x = column vector (i.e., n x 1 matrix)

$$Ax = b$$

$$\underset{A}{\begin{bmatrix} a & b \\ c & d \end{bmatrix}} \underset{x}{\begin{bmatrix} e \\ f \end{bmatrix}} = \underset{b}{\begin{bmatrix} ae + bf \\ ce + df \end{bmatrix}}$$

$\therefore\ Ax = b == A_{m,1}x_1 + A_{m,2}x_2 + \ldots. + A_{m,n}x_n = b_n$

e.g., $A_{1,1}x_1 + A_{1,2}x_2 + \ldots. + A_{1,n}x_n = b_1$

Mod Division

Mod division yields the integer remainder after integer division:

e.g., 34 mod 3 = 1

34 / 3 == 33 / 3 = (11) + 1 (integer remainder)

Partial Fraction Decomposition

(of rational polynomial fractions):

e.g.,

$$\frac{x+7}{x^2 - x - 6} = \frac{A}{(x-3)} + \frac{B}{(x+2)}$$

$^{a}x^2 \; {}^{b}{-x} \; \overset{c}{-6} = (x-3)(x+2)$

$ac = -6$

$a + c = b = -3 + 2 = -1$

$$\therefore \quad \frac{A \bullet (x+2)}{(x-3)(x+2)} + \frac{B \bullet (x-3)}{(x+2)(x-3)} = \frac{x+7}{(x^2 - x - 6)} = \frac{x+7}{(x-3)(x+2)}$$

$$\therefore \quad A(x+2) + B(x-3) = x+7$$

Choose any convenient values of x to get A, B

e.g., let x = -2

A(-2 + 2) + B(-2 – 3) = -2 + 7

↓ ↓ ↓

0 + -5B = 5 ∴ B = -1

let x = +3

A(+3 + 2) + B(+3 – 3) = +3 + 7

↓ ↓ ↓

A(+5) + 0 = 10

5A = 10 ; ∴ A = 2

$$\therefore \quad \frac{x+7}{x^2 - x - 6} = \frac{2}{(x-3)} + \frac{-1}{(x+2)} = \frac{2}{x-3} - \frac{1}{x+2}$$

Pascal's △ :

$(x+y)^0 = 1$	→	1	0th row
$(x+y)^1 = 1x + 1y$	→	1 1	1st row
$(x+y)^2 = 1x^2 + 2xy + 1y^2$	→	1 2 1	2nd row
$(x+y)^3 = 1x^3 + 3x^2y + 3xy^2 + 1y^3$	→	1 3 3 3	3rd row

etc.

e.g., $(x+1)^3$ has (1 3 3 1) form : $(1)x^3 + 3x^2(1) + 3x(1)^2 + 1(1)^3$

$\therefore (x+1)^3 = x^3 + 3x^2 + 3x + 1$

e.g., for $(a + 2b)^8$: to get coefficients use n, r, and ${}_nC_r$ for terms

1st term = ${}_8C_0 a^{8-0}(2b)^0$ $\quad$ n = 8, r = 0

2nd term = ${}_8C_1 a^{8-1}(2b)^1$ $\quad$ n = 8, r = 1

6th term = ${}_8C_5 a^{8-5}(2b)^5$ $\quad$ n = 8, r = 5

${}_8C_5 \rightarrow 56$, $a^{8-5} \rightarrow a^3$, $(2b)^5 \rightarrow 2^5b^5$

$56 \cdot 2^5 \rightarrow 1792$

$\therefore$ 6th term = $1792a^3b^5$ $\quad$ etc.

For $(x - y)^n$, use alternate signs: e.g., $(x - y)^2 = x^2 - 2xy + y^2$ $\quad$ etc.

Periodic (Trigonometric) Functions

$f(x + p) = f(x)$, where p is a positive number.

The smallest p is the period of f ; ∴ the graph represents every p units along x axis.

sin x, cos x, csc x, sec x have period of 2π .

tan x, cot x have period of π .

For $f(x) = A \cdot \sin(bx)$:

Amplitude = A , period = $2\pi / b$

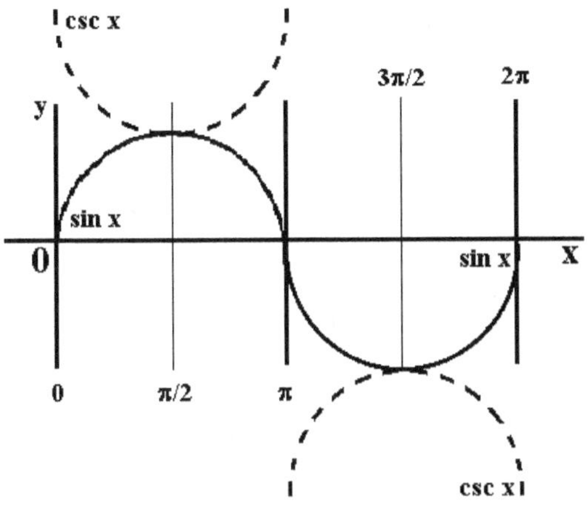

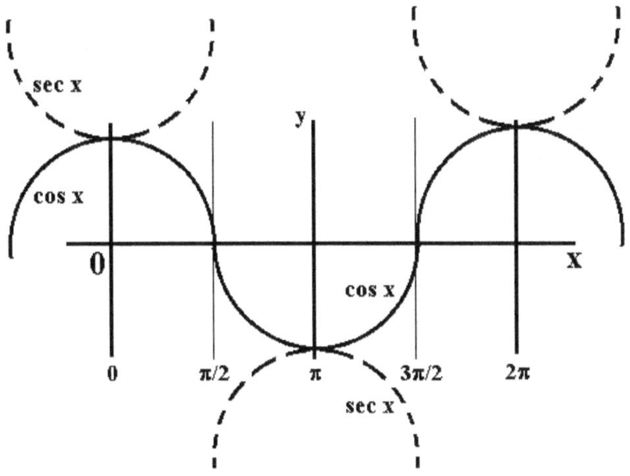

For g(x) = tan(cx) : period = π / c , the amplitude is infinite.

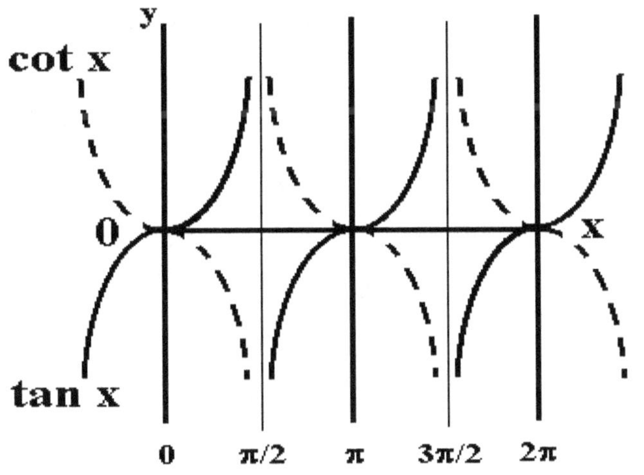

For inverses to trigonometric functions:

restrict the domain for one-to-one function.

sin x : -π / 2 =< x =< π / 2

cos x : 0 =< x =< π

tan x : -π / 2 < x < π / 2

e.g., $\sin^{-1}$ x == angle whose sine is x == arcsin x

(as one-to-one functions, they are symmetric about the y = x line)

$\sec^{-1}(x) = \cos^{-1}(1 / x)$, $\csc^{-1}(x) = \sin^{-1}(1 / x)$,

$\cot^{-1}(x) = (\pi / 2) - \tan^{-1}(x)$

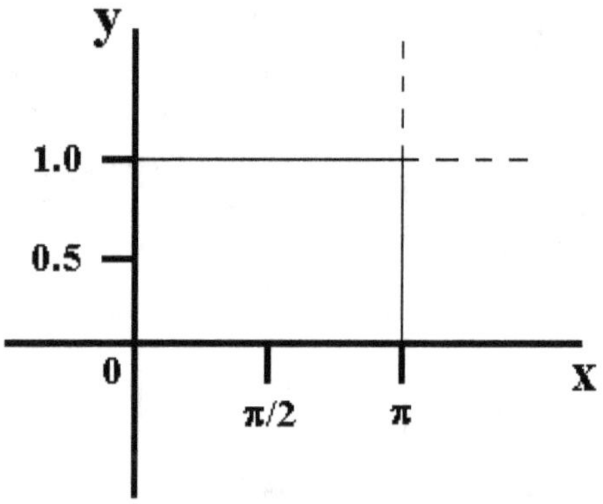

function

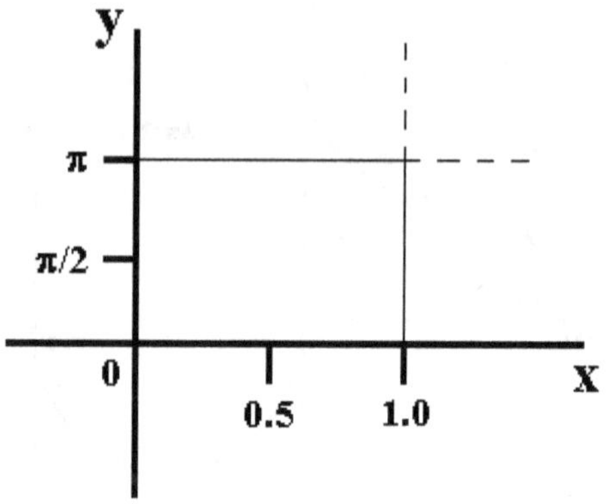

inverse or arc function

Permutations

Order of elements or members is important for calculating permutations.

e.g., How many permutations (n) occur for 2 odd and 5 even numbers?

Let n = O + E ∴ 7 = 2 + 5

The # of permutations of 7 numbers with 2 odd and 5 even is:

$$P = \frac{n!}{O!E!} = \frac{7!}{2!5!} = \frac{7 \cdot 6 \cdot 5 \cdot 4 \cdot 3 \cdot 2 \cdot 1}{(2 \cdot 1)(5 \cdot 4 \cdot 3 \cdot 2 \cdot 1)} = \frac{7 \cdot 6}{2} = 21$$

How many of these permutations have O starting?

∴ n = 7 – 1 = 6 , O = 2 – 1 = 1

$$P_O = \frac{6 \cdot 5 \cdot 4 \cdot 3 \cdot 2 \cdot 1}{(1)(5 \cdot 4 \cdot 3 \cdot 2 \cdot 1)} = 6$$

How many of these permutations have E starting?

∴ n = 7 – 1 = 6 , E = 5 – 1 = 4

$$P_E = \frac{6 \cdot 5 \cdot 4 \cdot 3 \cdot 2 \cdot 1}{(2 \cdot 1)(4 \cdot 3 \cdot 2 \cdot 1)} = \frac{6 \cdot 5}{2} = 15$$

Check : 15 + 6 = 21

Point-slope equation of a line:

The equation of a line with slope m and running through point (x_1, y_1) is
$y - y_1 = m(x - x_1)$; e.g., for $y - 3 = -2(x - 2)$
point (x1, y1) = (2, 3) and slope m = -2

The slope-intercept form of a line is $y = mx + b$, with slope m and y-intercept b .

Probability Distribution:

The probability distribution for a normal (Gausian) curve is $p = (\chi - \mu) / \sigma$,
where μ = mean, σ = standard distribution.
They are used to find χ, the population from the mean.

Quadratic Equation Solution by Inspection

e.g., find values of x for $x^2 = 3x + 4$

$x \bullet x = 3x + 4$

$x (x - 3) = 4$

From inspection we can see that two values of x are : x = 4 , x = -1

since $4(4 - 3) = 4 \bullet 1 = 4$ and $-1(-1 - 3) = -1 \bullet -4 = 4$

$\therefore x^2 = 3x + 4$; $x^2 - 3x - 4 = 0 \rightarrow (x - 4)(x + 1) = 0$

i.e., $x - 4 = 0 \rightarrow x = 4$; $x + 1 = 0 \rightarrow x = -1$

Quadratic Formula

For the quadratic equation $ax^2 + bx + c = 0$,

$x = [-b +,-(b^2 -4ac)^{1/2}] / 2a$

Radicals

$$\sqrt[n]{a^m} = \left(\sqrt[n]{a}\right)^m$$

$$\sqrt[n]{a} \cdot \sqrt[n]{b} = \sqrt[n]{a \cdot b}$$

$$\sqrt[m]{\sqrt[n]{a}} = \sqrt[m \cdot n]{a}$$

$$a^{1/n} = \sqrt[n]{a}$$

$$a^{m/n} = \sqrt[n]{a^m}$$

$$a \cdot \sqrt[n]{a^m} = a^1 \cdot a^{m/n} = a^{(1+m/n)}$$

Rates (r)

Relationship between Finite Rate (a) and Instantaneous Rate (b) : $a = e^b$

Finite Rate == definite or discreet interval

Instantaneous Rate == continuously

If $N_t = N_0 a^t$, then $N_t = N_0 e^{bt}$ ($\therefore\ a = e^b$)

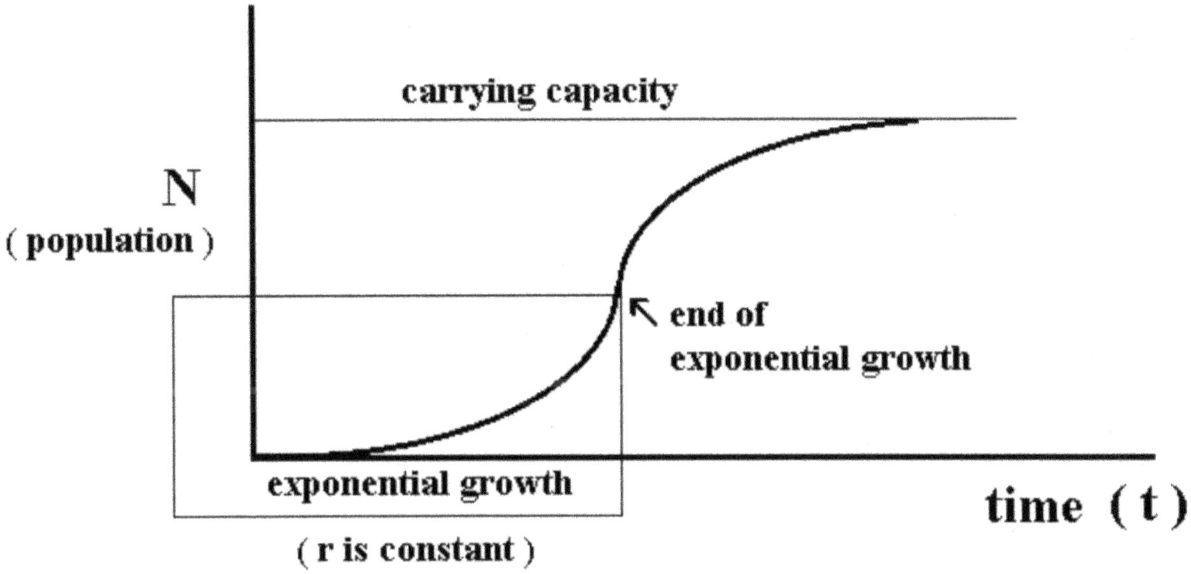

Sequences

Arithmetic Sequence

sum of sequence = $s_n = (n / 2)(a_1 + a_n)$

$a_n = d(a_1) + c$, $c = a_1 - d$; $\therefore\ a_n = a_1 + (n - 1)d$

where d == common difference between terms .

Geometric Sequence

based on ratio between consecutive terms

$a_n = a_1 r^{n-1}$, $a_{n+1} = r \bullet a_n$, where r == the ratio ; (e.g,. $a_{10} = a_4 r^{10-4}$)

sum = $\Sigma_{(i=1 \to n)} a_1 r^{i-1}$

for finite geometric sequence sum up $a_1[(1 - r^n) / (1 - r)]$

(for infinite sum, $r^n \to 0$, $| r | < 1$)

Sigma Notation, Σ

e.g., Express in sigma notation the series: 11 + 18 + 37 + ... + 102

The answer is $\Sigma_{n = 1, 8}\ n^2 + 4n + 6$

If the form (of the answer) is: $ax^2 + bx + c$, then you can take the first 3 values as forming 3 equations with 3 unknowns and solve for (a, b, c) with n (= x) = 1 to 3 :

$$a(1)^2 + b(1) + c = 11$$
$$a(2)^2 + b(2) + c = 18$$
$$a(3)^2 + b(3) + c = 37$$
$$\rightarrow a = 1, b = 4, c = 6 \ \therefore\ 1n^2 + 4n + 6$$

∴ If the form of the answer is: $ax^2 + bx + c$,
then that's (a) times as many terms to test.

Standard Deviation (σ)

$\sigma = (\text{variance} == \nu)^{1/2}$; thus $\sigma^2 = \nu$.

$\nu = (1 / N)\bullet\Sigma_{(i = 1 \rightarrow N)}(x_i - \mu)^2$,

where μ = mean (i.e. average x), x_i = sample values, N = # of values

(when all of the x values are used; when only a subset of x values are needed, use $1/(N - 1)$ in the radical).

68% == 1σ from the mean : ∴ +,- $1\sigma + \mu$

95% == 2σ from the mean : ∴ +,- $2\sigma + \mu$.

Statistical Definitions

mean	==	average value (of a set of values)
median	==	½ of values are smaller than or equal to the median, and ½ of values are larger than or equal to the median
mode	==	most common value in a set

Trigonometric Functions

Basic Formulas

for sine → sin, cosine → cos, secant → sec,
cosecant → csc, tangent → tan, cotagent → cot

$\sin\theta = 1 / \csc\theta$; $\cos\theta = 1 / \sec\theta$; $\tan\theta = 1 / \cot\theta$
$\tan\theta = \sin\theta / \cos\theta$; $\cot\theta = \cos\theta / \sin\theta$

θ deg.	0^0	30^0	45^0	60^0	90^0	180^0	270^0	
θ radians	0	$\pi / 6$	$\pi / 4$	$\pi / 3$	$\pi / 2$	π	$3\pi / 2$	**$0^\circ \rightarrow 360^\circ$**
sin θ	0	1 / 2	$\sqrt{2} / 2$ $= 1/\sqrt{2}$	$\sqrt{3} / 2$	1	0	-1	**$0 \rightarrow 2\pi$**
cos θ	1	$\sqrt{3} / 2$	$\sqrt{2} / 2$ $= 1/\sqrt{2}$	1 / 2	0	-1	0	**$-1 \rightarrow 1$**
tan θ	0	$\sqrt{3} / 3$	1	$\sqrt{3}$	–	0	–	($0 \rightarrow \pi$) **$0 \rightarrow 1$**

$\sin^2\theta + \cos^2\theta = 1$
$1 + \tan^2\theta = \sec^2\theta$
$1 + \cot^2\theta = \csc^2\theta$

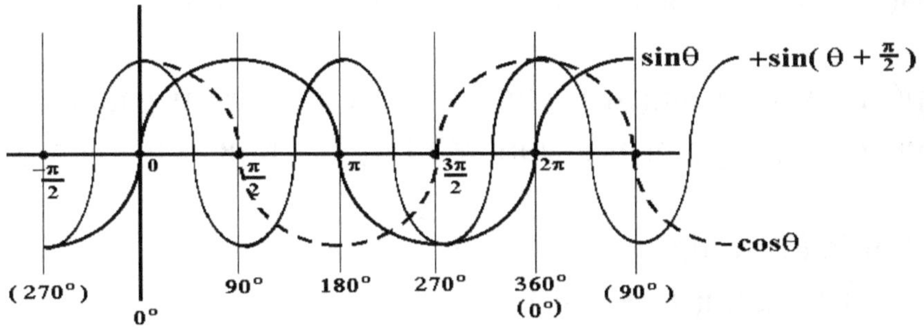

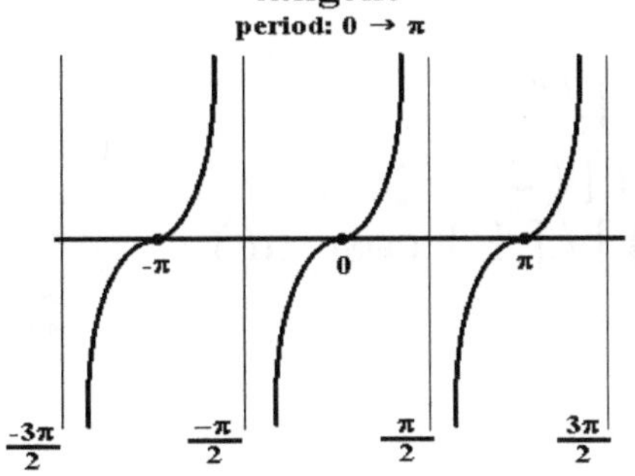

Sum to Product Formulas

$\sin(u) \cdot \sin(v) = \frac{1}{2} [\cos(u - v) - \cos(u + v)]$
$\cos(u) \cdot \cos(v) = \frac{1}{2} [\cos(u - v) + \cos(u + v)]$
$\sin(u) \cdot \cos(v) = \frac{1}{2} [\sin(u + v) + \sin(u - v)]$
$\cos(u) \cdot \sin(v) = \frac{1}{2} [\sin(u + v) - \sin(u - v)]$

$\sin(x) + \sin(y) = 2\sin[(x + y) / 2] \cdot \cos[(x - y) / 2]$
$\sin(x) - \sin(y) = 2\cos[(x + y) / 2] \cdot \sin[(x - y) / 2]$
$\cos(x) + \cos(y) = 2\cos[(x + y) / 2] \cdot \cos[(x - y) / 2]$
$\cos(x) - \cos(y) = -2\sin[(x + y) / 2] \cdot \sin[(x - y) / 2]$

Difference Angle Formulas

$\sin[(\pi / 2) - u] = \cos(u)$ $\cos[(\pi / 2) - u] = \sin(u)$
$\tan[(\pi / 2) - u] = \cot(u)$ $\cot[(\pi / 2) - u] = \tan(u)$
$\sec[(\pi / 2) - u] = \csc(u)$ $\csc[(\pi / 2) - u] = \sec(u)$

$\sin(u + v) = \sin(u) \cdot \cos(v) + \cos(u) C \sin(v)$
$\sin(u - v) = \sin(u) \cdot \cos(v) - \cos(u) \cdot \sin(v)$
$\cos(u + v) = \cos(u) \cdot \cos(v) - \sin(u) \cdot \sin(v)$
$\cos(u - v) = \cos(u) \cdot \cos(v) + \sin(u) \cdot \sin(v)$

$\tan(u + v) = [\tan(u) + \tan(v)] / [1 - \tan(u) \cdot \tan(v)]$
$\tan(u - v) = [\tan(u) - \tan(v)] / [1 + \tan(u) \cdot \tan(v)]$

Double Angle Formulas

$\sin(2u) = 2\sin(u) \cdot \cos(u)$
$\cos(2u) = \cos^2(u) - \sin^2(u) = 2\cos^2(u) - 1 = 1 - 2\sin^2(u)$
$\tan(2u) = 2\tan(u) / [1 - \tan^2(u)]$

$\sin^2(u) = [1 - \cos(2u)] / 2$
$\cos^2(u) = [1 + \cos(2u)] / 2$
$\tan^2(u) = [1 - \cos(2u)] / [1 + \cos(2u)]$

Even / Odd Identities

sin(-u) = -sin(u)	cos(-u) = cos(u)	tan(-u) = -tan(u)
csc(-u) = -csc(u)	sec(-u) = sec(u)	cot(-u) = -cot(u)
(odd identity)	(even identity)	(odd identity)

Physical Properties

Extensive == depends on the amount (of matter)

e.g., volume, length, mass

Intensive == independent of amount

e.g., density

The division of extensive properties yields intensive properties

e.g., mass / volume = density

For a solid, uniform material of one density: $m = \rho \bullet t \bullet A$,
where m = mass, ρ = density, t = thickness, and A = area ;
∴ t•A == volume

Plot Types:

Log vs Log

For **Log vs Log** plots: $S = A^z$, where z == slope of line (S, A in log)

Quantum Mechanics

Analogy for Quantum Mechanics

Say a coin-flip experiment (for heads or tails) involves 4 coins.
Each coin has 1 head and 1 tail.
The # of combinations of heads and tails
are distinctly: $4^2 = 16$ == 16 microstates.
The number of types of combinations is always N + 1 ; ∴ 4 + 1 = 5 types:
(all heads), (all tails), (1 head + 3 tails), (1 tail + 3 heads),
(2 heads + 2 tails) == 5 macrostates as the quantum number.

Each macrostate contains microstates.
Each macrostate can be described by occupation numbers for its microstates:

e.g., let head = 1, tail = 0 ; these are occupation numbers.
Each # is like an energy level for the number of atoms in it.

A set of energy levels is a macrostate:

macrostate (1 tail + 3 heads) has
microstates (0, 1, 1, 1), (1, 1, 0, 1), (1, 0, 1, 1), (1, 1, 1, 0).
These are the atoms in each energy level.

Each macrostate has a probability of occurring that can be different from the others, while each microstate has a probability, here, of 1 / 16 :

e.g., macrostate (all heads) has only one microstate, (1, 1, 1, 1)
macrostate (all tails) has only one microstate, (0, 0, 0, 0)
macrostate (1 head + 3 tails) has 4 microstates
macrostate (1 tail + 3 heads) has 4 microstates
macrostate (2 heads + 2 tails) has 6 microstates

The # of different microstates is related to the "randomness" of their macrostate.

If you have N objects, h of one type and t of another type, then the # of ways to combine them is N! / (h!t!) .
∴ for (2 heads + 2 tails) this number is 4! / (2!2!) = 6 .

for (2 heads + 2 tails) :

$$\frac{4!}{2!2!} = \frac{4!}{(2\bullet 1)(2\bullet 1)} = \frac{4\bullet 3\bullet \cancel{2}\bullet \cancel{1}}{\cancel{2}\bullet \cancel{1}\bullet 2\bullet 1} = \frac{4\bullet 3}{2\bullet 1} = \frac{12}{2} = 6$$

∴ macrostate (2 heads + 2 tails) is most probable :

$$\frac{6}{(1 + 1 + 4 + 4 + 6)} = \frac{6}{16} = \frac{3}{8}$$

Note that as the number of coins (= N) increases, the probability of all heads or all tails becomes increasingly smaller.

If the coins are atoms, and if heads comes up (for a coin)
== the coin is energized by a quantum of energy,
then this is analogous to distributing energy quanta to the coins.

The macrostate with the most microstates distributes the energy most randomly and ∴ has higher entropy == heat flow.
Thus, heat flows from hotter to colder bodies (preferentially).
Microscopically, entropy is a measure of disorder (randomness).

A microstate is a particular order (of constituents) :
e.g., (2 , 1 , 0) → (2 + 1 , 0)
A macrostate contains all (relevant) microstates for a generalized order :
e.g., [2 + 1 , 0] == (2 , 1 , 0) , (1 , 2 , 0) .
This macrostate has 2 microstates.

The information for the # of microstates in a macrostate is contained in a function Ω , which is related to entropy:
Entropy $S = k_B \ln \Omega$, where k_B is the Boltzmann Constant.
Ω == the statistical weight of a system,
the macrostate being its occupation number.

Entropy == s = k•ln(w) , where k = Boltzmann Constant
= 1.38 x 10-23 Joules/Kelvin temp. = R / N_A ,
where R = universal gas constant, N_A = Avogadro's Number,
and w = # of ways of arranging components of a system.

Entropy is a statistical function dependent on occupation and Quantum numbers but indirectly on state variables (P, T, V) :
It is a measure of the likelihood of a particular macrostate (occuring), when given total values for energy and other conserved quantities.

For example, when flipping 100 coins the likelihood is ~90% that there will be 45 tails to 55 heads (i.e, the probability peaks there).
With much larger numbers of microstates (in thermodynamic systems) the range of parameters for the final macrostate becomes extremely sharply peaked.

The actual statistics used are:
N_E = energy quanta to be distributed (numbers of quanta)
N_A = # of atoms to receive the energy quanta

A = the # of arrangements of microstates
$A = (N_E + N_A - 1)! / (N_E!(N_A - 1)!)$
when calculated for closed systems;
∴ no transfer of mass,
only energy (thermal photons) between systems.

If the energy quanta to be used (N_E) is held constant,
and only the energy levels (ε) available to the atoms have their spacings changed,
this is the quantum mechanical equivalent to work.

$$\underset{\text{entropy}}{\Delta S} \geq \frac{Q^{\text{ input heat flow}}}{T_{\text{absolute temperature}}} \qquad \textbf{at constant P and T}$$

2nd Law of Thermodynamics: $\Delta S \geq 0$ (for the universe)

∴ For PV work (→ $P\Delta V$) V must increase (with incoming Q).
As V↑, movement allowed for each atom increases;
∴ ε decreases ($\varepsilon \propto 1 / L^2$ for a volume of L^3).

Changing the energy level spacing ∈ of the atoms in a system == work.
Changing the number of energy quanta a system has == heat flow.

Energy of a system is: $\in N_E / N_A$, with $\in \propto 1 / L^2$; L = atom confined to length L .
In a quasi-static system of constant T and very slowly increasing volume:
L↑ ∴ ∈↓ ∴ N_E↑ , N_A constant for closed system.

An increase in energy quanta (N_E↑) → increase in # of ways of
dividing up quanta among atoms == more disorder,
the microscopic reason for why entropy change is Q / T .

$T\Delta S$ == (loosely speaking)
a measure of the energy content of the "order" in a system.

For a closed quasi-static system, if heat flows into a system at constant temperature, the system's volume has to increase; otherwise, the internal energy would increase and the temperature would also increase.

Closed system: can have exchange of heat with other systems (i.e., its surroundings), but no exchange of mass (atoms).
Open system: can have exchange of heat with other systems as well as exchange of mass.

Open systems can appear to violate the 2nd law and have decreasing entropy. Life is fundamentally a process that reduces entropy in a series of self-organizing processes. There is no actual violation of the 2nd law because life can not occur as a closed system.

For an open system, the entropy (of the system and its surroundings) will be maximized. The free energy (energy "free" to do useful work) tends to decrease for an open system and events (e.g., chemical reactions) will proceed spontaneously so long as the free energy decreases.

Gibbs Free Energy: $G = H - TS = U + PV - TS$
Under conditions of constant T and P, the only energy changes that can occur within an open system are $P\triangle V$ work, heat flow (to or from the surroundings), and other useful work (e.g., chemical, electrical).

G must always decrease as a system approaches equilibrium, and must remain at that minimum value during equilibrium.

For an isothermal process: $\triangle G = \triangle H - T\triangle S$. The process will spontaneously occur when $\triangle G$ is negative (negative free energy).

$\triangle G = \Sigma (\triangle G_i) = \Sigma (\mu_i \bullet \triangle n_i)$, where n_i = # of moles of species i, and μ_i = Gibbs Free Energy per mole == chemical potential of species i.

For the chemical reaction: $n_A A + n_B B \leftrightarrows n_C C + n_D D$, with $\mu = \mu_i^0 + RT \bullet \ln (c_i)$

where μ_i^0 is a chemical potential at some standard condition and
c_i is the molar concentration (for each chemical participant),
then $\triangle G_{total}^0 = -RT \bullet \ln K_{eq}$.

$K_{eq} = e$^-($\triangle G_{total}^0 / RT$), $e^{-\triangle G / RT}$ == e^-$\triangle E / k_B T$ = Boltzmann Factor,
which gives relative populations of the two states separated by energy $\triangle E$.

$k_B T$ is an energy which (at room temperature of 20°c) = 4×10^{-21} J
= (1 / 40)eV , the thermal energy of a gas molecule ; 1 eV = 1.6×10^{-19} J

$\triangle E = E_2 - E_1$, N_2 / N_1 = e^-$\triangle E / k_B T$,
where N_1 and N_2 are populations considered (for their energies).

Internal energy (U) starts out as εN_E ;but as $\varepsilon\downarrow$ and $\triangle U$ (at constant T) remains constant, then N_E must ↑ → A↑ ∴ more disorder (randomness).

For a reversible process (∴ $Q = T\triangle S$, Q being heat) :

$\triangle S = 0$ when the volume change is so slow that T is constant and the system is in thermal equilibrium as a "quasistatic" process.

$S = k_B \ln\Omega$, where Ω contains information on the # of microstates in a given macrostate (i.e., occupation numbers), and k_B is the Boltzmann constant.
∴ Ω is a statistical weight of the system.

1st Law of Thermodynamics is Conservation of Energy,
where (for a closed system) $\triangle U = Q - W$; U == internal energy,
Q = heat added to system, W = work done by system on the surroundings.
A negative Q == heat leaving system; a negative W == work done on system.

2nd Law of Thermodynamics:

The total entropy of a closed system always increases, $\triangle S >= 0$.
$\triangle S = 0$ only in the special case of a reversible process, where the process is performed slowly enough so that the system remains in equilibrium throughout (∴ a quasi-static process).

For a quasi-static system there is a very slow change in volume (V).

$\triangle S >= Q / T$,
where Q = heat input and T = absolute (Kelvin) temperature.

Given a variety of different events that can occur (satisfying energy conservation and other conserved quantities), the one having the most possible (number of) microstates will be the one that occurs. This number of microstates for a macrostate is intrinsically related to the macrostate's increased "randomness."

Frictional forces are non-conservative because the thermal energy they produce can not be reversibly transformed back to mechanical energy. Mechanical energies are more "organized" and much less "random" in nature than thermal energies.

The arrangements of energy in a closed system are:

$A = (N_E + NA - 1)! / [N_E! \bullet (N_A - 1)!]$,
where N_E == # energy quanta and N_A == # atoms involved.

$PV = nRT = Nk_BT$, where N = # of molecules,
k_B = Boltzmann Constant = 1.38×10^{-25} Joules/Kelvin temp.,
n = # of moles = N / N_A,
N_A = Avogardro's number = 6.02×10^{-23} molecules / mole,
and R = the gas constant = $N_A \bullet k_B$ = 8.31 Joules / (mol•Kelvin temp.).

A state variable is dependent on the system itself.
$\triangle U$ (internal energy) is a state variable.

Heat, or a flow of thermal energy from a hotter body to a cooler one, is not a state variable. Neither is work a state variable.
∴ We do not write $\triangle Q$ or $\triangle W$, since $\triangle$ implies a state variable.

Macrostates for overall translational kinetic energy have much fewer microstates and ∴ much lower entropy than thermal macrostates (random diffuse motions). In time the translational K.E. macrostates "thermolize" to randomize its atoms'

motions, ∴ over time increasing its entropy for the 2nd law:
total entropy of a closed system always increases ($\Delta S >= 0$).

Comparison of Kinetics and Thermodynamics

Kinetics == how fast (is the time needed to reach equilibrium)

∴ involves equilibrium constants:

concentrations, rates of reaction

Thermodynamics == stabilities of products (done at equilibrium)

∴ involves energies: enthalpies (constant pressure), temperatures, position of equilibria (ΔG^0), transition states ($\Delta G^{\ddagger}$)

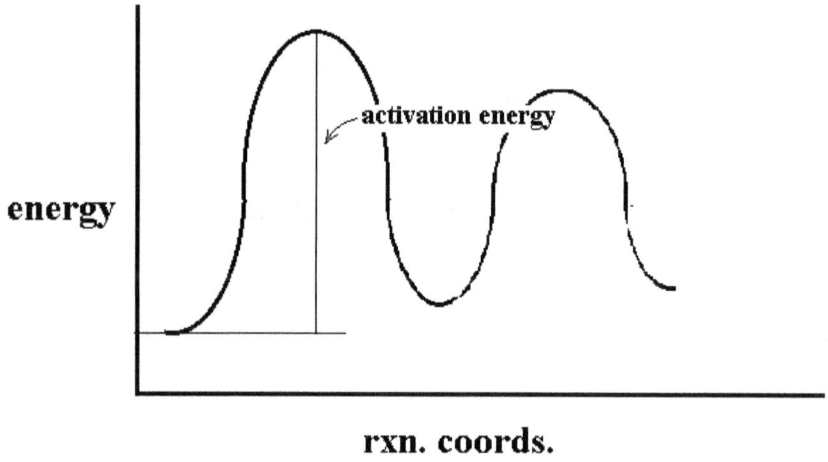

peaks ≡ transition states (states of equilibrium)

Rates are determined by activation energies.

$$\Delta G = -RT\ln(k)$$

Relativity

Special Relativity

$$\Delta t = \Delta t_0 / [1 - (v^2 / c^2)] , \Delta t > \Delta t_0$$

$$\Delta s = \Delta s_0 \bullet [1 - (v^2 / c^2)] , \Delta s < \Delta s_0$$

General Relativity

In space-time coordinates (x,y,z ; t) : gravity and acceleration are indistinguishable. This provides for a description of the attractions between masses, but does not explain the ***cause*** of their attractions.

Velocity Transformation

$v' = (v + u) / [1 + (u \cdot v) / c^2)]$

Thermodynamics and Kinetics

Thermodynamics defines:

1) feasibility of reaction
2) position of equlibrium
3) stability of a compound

Kinetics defines rate of reaction

(activation energy, energy barrier for compound formation)

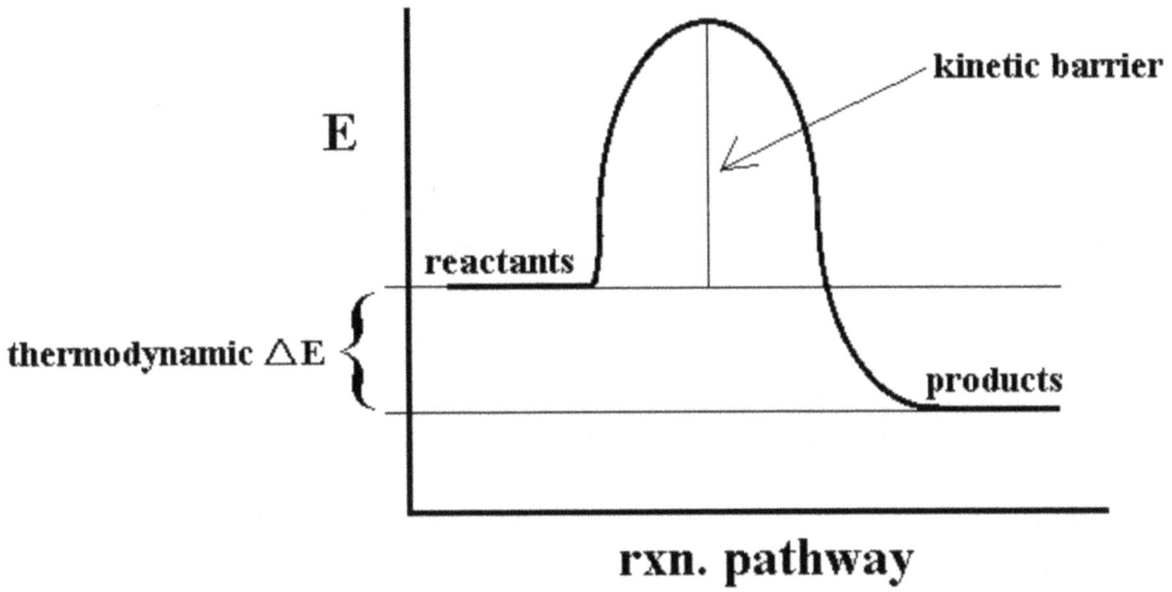

$\triangle G$ as free energy = $\triangle H - T\triangle S$

$\triangle G^o_{\text{standard state}} = \triangle H^o - T\triangle S^o$

$\triangle G^o = -RT \cdot \ln K_{eq}$

$\triangle G = \triangle G^o + RT \cdot \ln \boldsymbol{Q}$, $\boldsymbol{Q}$ = [products / [reactants]

== reaction quotient at starting position

Ideal Gas Law : PV = nRT , with n = # moles of gas

For an ideal gas (with Q == heat)

1) if P is constant : $W = P \cdot (V_{final} - V_{initial})$ → isobaric process
2) if T is constant : $W = nRT\ln(V_f / V_i)$, $\triangle U = 0$ → isothermal

process, PV is constant

3) if V is constant : no work is done ∴ $\triangle U = Q \rightarrow$ isochoric process
4) there is no exchange in heat (P, V, T remaining variable) : $Q = 0$
∴ $\triangle U = -W \rightarrow$ adiabatic process

For small amounts of work done (i.e. δW) :

$$\delta W = F\triangle x = PA\triangle x = P\triangle V$$

$Q = c \bullet m \bullet \triangle T$, where c = specific heat, m = mass, Q = heat
Specific Heat (c) == a measure of the heat absorption or release of a material as the temperature changes.

At a constant temperature,
$Q_{transformation} = L \bullet m$, where L == latent heat of fusion or vaporization.

For open systems, such as in biology, which are often at constant temperature or pressure, it is useful to consider Gibb's Free Energy changes:
$G = H - TS = U + PV - TS$, where H == enthalpy, V == volume, T == temperature, P == pressure, U == (internal) energy (of system).

$\triangle G < 0$ for spontaneous reactions,
$\triangle G > 0$ where reaction needs input of energy.

Enthalpy → chemical bond energies and heats of chemical reactions.

$H = U + PV$; for constant P, isobaric $\rightarrow \triangle H = Q$ ∴ heat flow .
$\triangle H$ is a state variable, while Q is not; ∴ Q is not a property of the system and depends on the system and its surroundings.
Isobaric ($P_{constant}$) conditions are fairly common and therefore tell us about heat flow using enthalpy.

For chemical reactions such as: reactants A + B → products C + D ,

$$\triangle H = H_C + H_D - H_A - H_B$$

When heat flows into the system, $\triangle H$ is positive == endothermic rxn.
(rxn. requires input of energy).
When heat flows out of system, $\triangle H$ is negative == exothermic rxn.
(rxn. may occur spontaneously).

Bond energies have $\triangle$Hs : bond disassociation energies required to break bonds.

Gibb's Free Energy (a state variable), $\triangle G$: it is related to chemical potential; it is a measure of the energy available to do useful work at constant pressure and temperature (typical for biology).

Useful Energy (able to do work) = sum of kinetic and potential energies
== mechanical energy;
i.e., energy not transformed into the internal energy of the material.

For chemical reactions:

$$\triangle G_{total}{}^{0} = -RT\ln K_{eq}$$

$$\ln K_{eq} = -\triangle G^{0} / RT \rightarrow \text{ratio of energies}$$

$$K_{eq} = e\hat{}-(\ \triangle G / RT\ ) \rightarrow \text{for moles}$$

$$K_{eq} = e\hat{}-(\ \triangle E / k_{B}T\ ) \rightarrow \text{for particles} = N_2 / N_1$$

$$= e\hat{}-(\ E_2 - E_1\ ) / k_{B}T,$$

where N_2 = population at E_2 and N_1 = population at E_1.

The Nerst Equation says: $\triangle G^{0} = -nFE^{0}$,
where F = Faraday constant, n = moles of electrons,
G^{0} = free energy of reduction, E^{0} = (standard) reduction potential.

For an electrolytic cell:

$E_{cell}{}^{o} = (\ RT / nF\ )\ln(\ K\ )$, where n = moles of e^{-},
R == gas constant = 8.314 J/(mol rnx•Kelvin temp.),
F == Faraday's Constant = 9.65×10^{4} J/(volts•moles e^{-})

$E_{cell} = E_{cell}{}^{o} - (\ RT / nF\ )\ln(\ Q\ )$ (== Nernst Equation),
where Q == rxn. quotient (= K at equilibrium)

At 25° c (= 298 K):

$E_{cell}{}^{o} = (\ 0.0257\text{•volts} / n\ )\ln(\ K\ )$, or in logarithms
$2.303\text{•}(\ 0.0257\text{•volts} / n\ )\ln(\ K\ ) = (\ 0.0592 / n\ )\log(\ K\ )$

Temperature Scales

(Fahrenheit, Celsius, Kelvin):

$T_f = (9/5)T_c + 32°$ f

$T_c = T_K - 273.15°$ c

0° c = 273K = 32° f

100° c = 373K = 212° f

Trigonometry

(△ of sides a, b, c and vertex angles A, B, C)

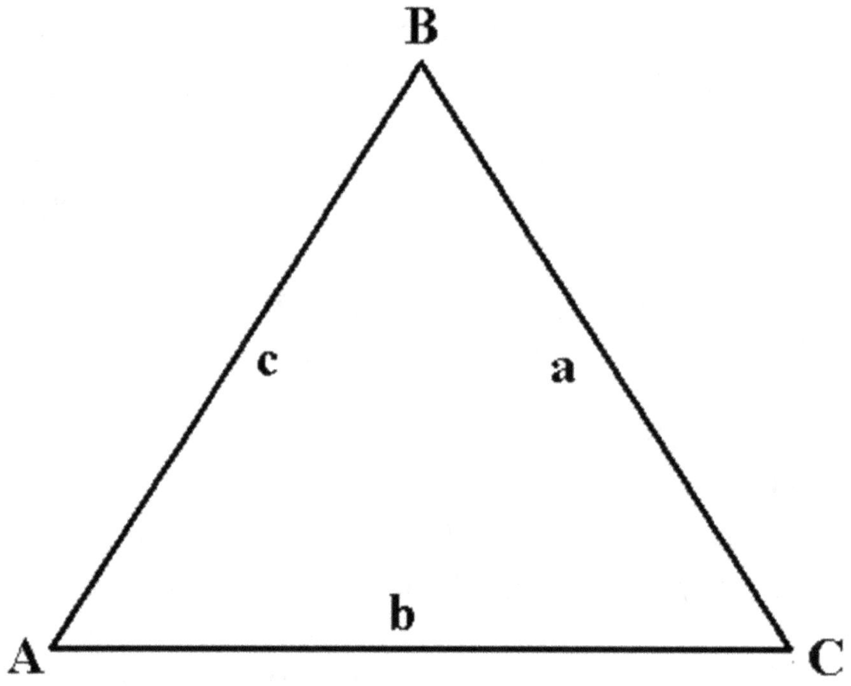

Law of Sines

$(a / \sin A) = (b / \sin B) = (c / \sin C)$

Law of Cosines

$a^2 = b^2 + c^2 - 2bc \bullet \cos A$

$b^2 = a^2 + c^2 - 2ac \bullet \cos B$

$c^2 = a^2 + b^2 - 2ab \bullet \cos C$

or more generally:

$$\begin{aligned} c^2 &= (\ a\cos\theta - b\ )^2 + (\ a\sin\theta - 0\ )^2 \\ &= a^2\cos^2\theta - ba\cos\theta - ba\cos\theta + b^2 + a^2\sin^2\theta \\ &= a^2\cos^2\theta + a^2\sin^2\theta - 2ab\cos\theta + b^2 \\ &= a^2(\ \cos^2\theta + \sin^2\theta\ ) + b^2 - 2ab\cos\theta \\ &= a^2 + b^2 - 2ab\cos\theta \end{aligned}$$

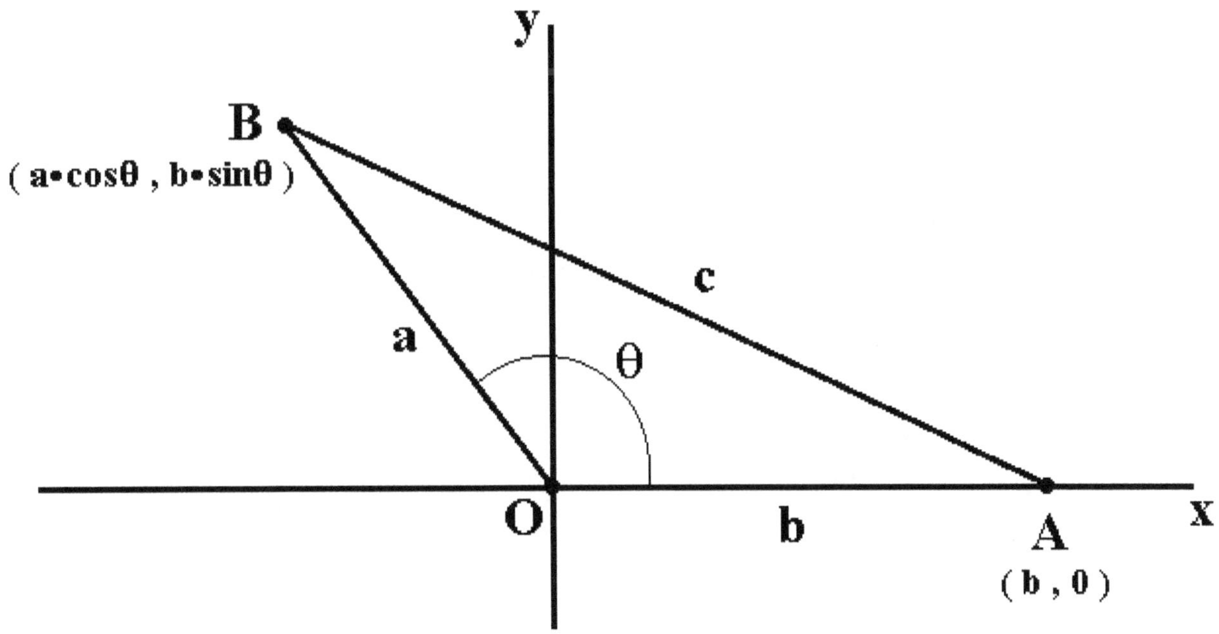

for △ with known sides but unknown angles

Unit Circle

$x = r \cdot \cos\theta$, $y = r \cdot \sin\theta$

$\cos^2\theta + \sin^2\theta = 1$

$d = [\,(x_2 - x_1)^2 + (y_2 - y_1)^2\,]^{1/2}$

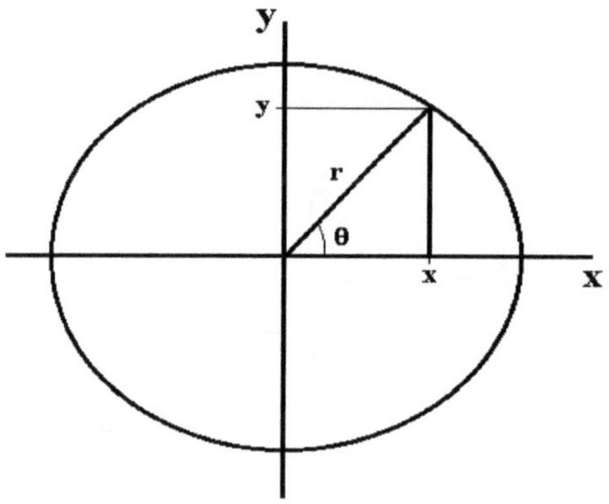

Formulas

tan(u) = sin(u) / cos(u)

cot(u) = cos(u) / sin(u) = 1 / tan(u)

sec(u) = 1 / cos(u)

csc(u) = 1 / sin(u)

For an acute angle θ :

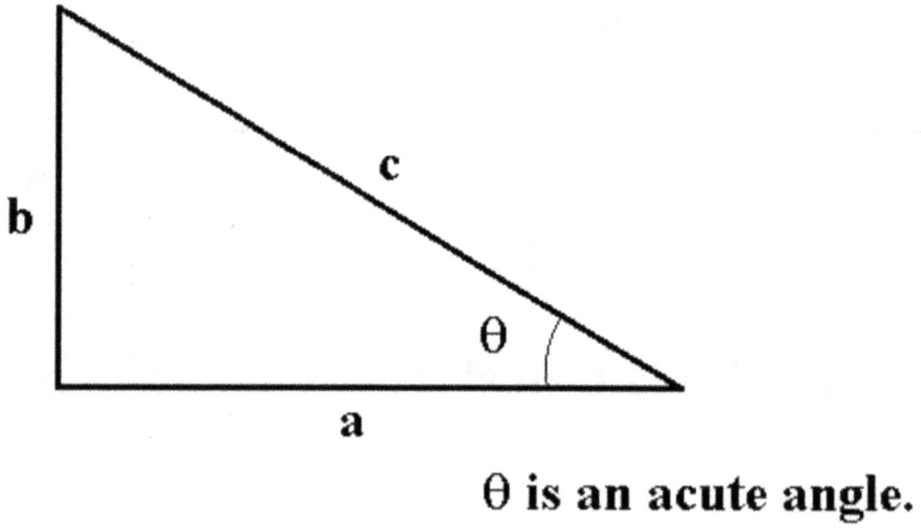

$\sin\theta = b / c$ $\cos\theta = a / c$ $\tan\theta = b / a$

For hypotenuse r and angle θ' :

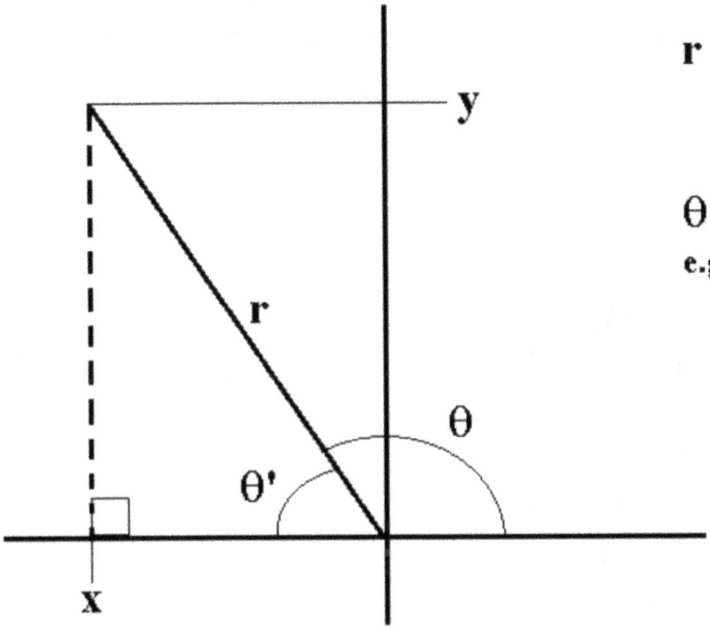

$$r = (x^2 + y^2)^{1/2} \neq 0$$

θ' ≡ reference angle

e.g. sin θ' = sin θ , except possibly in sign

For any θ :

sin θ' = y / r , cos θ' = x / r , tan θ' = y / x

csc θ' = r / y , sec θ' = x / r , cot θ' = x / y

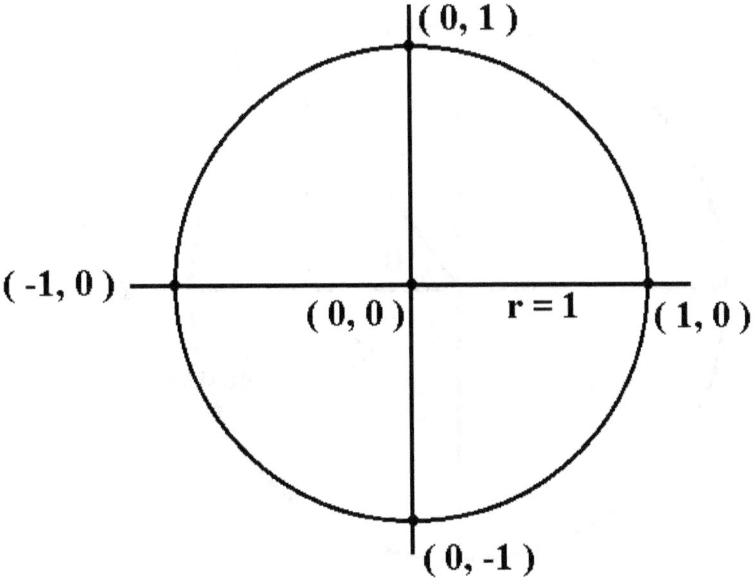

θ degrees	0°	30°	45°	60°	90°	180°	270°
θ radians	0	π/6	π/4	π/3	π/2	π	3π/2
sin θ radians	$(0)^{1/2}/2$	$(1)^{1/2}/2$	$(2)^{1/2}/2$	$(3)^{1/2}/2$	$(4)^{1/2}/2$	0	-1
cos θ radians	1	$(3)^{1/2}/2$	$(2)^{1/2}/2$	$(1)^{1/2}/2$	$(0)^{1/2}/2$	-1	0
tan θ radians	0	$(3)^{1/2}/3$	1	$(3)^{1/2}$	---	0	---

Vectors:

given point P = (1, 2) and point Q = (4, 5)

The vector from P to Q == **v** .

The components of **v** are < (4 – 1) , (5 – 2) > = (3, 3) .

The distance from P to Q = $[(4 – 2)^2 + (5 – 2)^2]^{1/2} = [3^2 + 3^2]^{1/2} = [18]^{1/2}$.

$[18]^{1/2}$ == the magnitude of **v** == || **v** || .

The unit vector **u** = **v** / || **v** || = (1 / || **v** ||)**v** .

i = < 1, 0 > unit vector , **j** = < 0, 1 > unit vector.

v can be stated as a linear combination of **i** and **j** :

v = < 3, 3 > = 3< 1, 0 > + 3< 0, 1 > = 3**i** + 3**j**

(3 and 3 are the horizontal and vertical components of **v**)

v + **v** = (3 + 3)**i** + (3 + 3)**j** = 6**i** + 6**j** = 2**v**

For a unit circle:

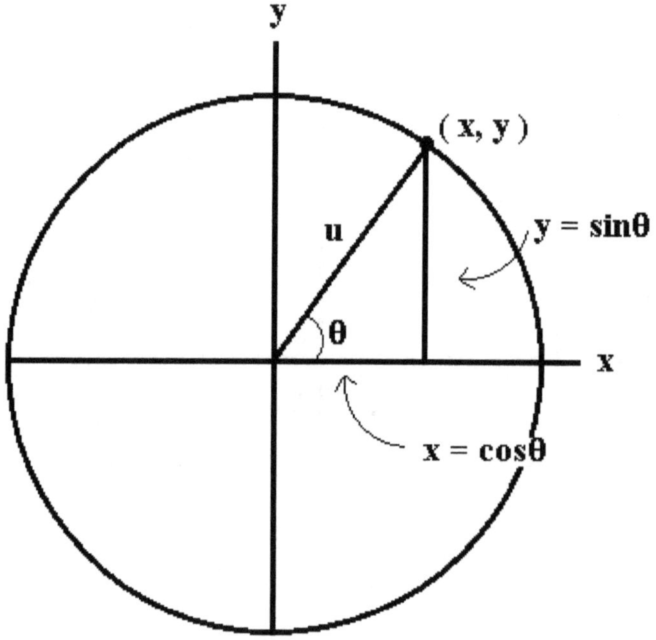

$\mathbf{u} = < x, y > = < \cos\theta, \sin\theta > = (\cos\theta)\mathbf{i} + (\sin\theta)\mathbf{j}$
θ = direction angle
$\therefore$ for $\mathbf{v}$, $\mathbf{v} = \| \mathbf{v} \| < \cos\theta, \sin\theta > = \| \mathbf{v} \|(\cos\theta)\mathbf{i} + \| \mathbf{v} \|(\sin\theta)\mathbf{j}$
$\mathbf{v} = a\mathbf{i} + b\mathbf{j} = \| \mathbf{v} \|(\cos\theta)\mathbf{i} + \| \mathbf{v} \|(\sin\theta)\mathbf{j}$
$\therefore \tan\theta = \sin\theta / \cos\theta = \| \mathbf{v} \|(\sin\theta) / \| \mathbf{v} \|(\cos\theta) = b / a$

Dot Product for $\mathbf{u} = < \mathbf{u}_1, \mathbf{u}_2 >$, $\mathbf{v} = < \mathbf{v}_1, \mathbf{v}_2 >$:
$\mathbf{u} \bullet \mathbf{v} = \mathbf{u}_1\mathbf{v}_1 + \mathbf{u}_2\mathbf{v}_2$ == a scalar (factorial and non-directional) value
$\mathbf{u} \bullet \mathbf{u} = \| \mathbf{u} \|^2$, $c(\mathbf{u} \bullet \mathbf{v}) = c\mathbf{u} \bullet \mathbf{v} = \mathbf{u} \bullet c\mathbf{v}$ (c == a scalar)
$\mathbf{u} \bullet \mathbf{v} = \| \mathbf{u} \| \, \| \mathbf{v} \| \cos\theta$, where θ is the angle (from $0 \rightarrow \pi$) between the vectors
$\therefore \cos\theta = \mathbf{u} \bullet \mathbf{v} / (\| \mathbf{u} \| \, \| \mathbf{v} \|)$; the vectors $\mathbf{u}$, $\mathbf{v}$ must be non-zero
$\| \mathbf{u} \| = [(\mathbf{u}_1 - 0)^2 + (\mathbf{u}_2 - 0)^2]^{1/2}$
$\| \mathbf{v} \| = [(\mathbf{v}_1 - 0)^2 + (\mathbf{v}_2 - 0)^2]^{1/2}$
If $\mathbf{u} \bullet \mathbf{v} = 0$, then $\mathbf{u}$ and $\mathbf{v}$ are orthogonal to each other ($\theta = 90^\circ == \pi / 2$).

If $\mathbf{u} = \mathbf{w}_1 + \mathbf{w}_2$ (by vector addition) and $\mathbf{w}_1$ is orthogonal to $\mathbf{w}_2$ and $\mathbf{w}_1$ is parallel or a scalar multiple of $\mathbf{v}$ (say c), then $\mathbf{w}_1$ is the projection of $\mathbf{u}$ onto $\mathbf{v}$ ($\mathbf{w}_1 = c\mathbf{v}$).
$\therefore \mathbf{u} = \mathbf{w}_1 + \mathbf{w}_2 = c\mathbf{v} + \mathbf{w}_2$
$\mathbf{u} \bullet \mathbf{v} = (c\mathbf{v} + \mathbf{w}_2) \bullet \mathbf{v} = c\mathbf{v} \bullet \mathbf{v} + \mathbf{w}_2 \bullet \mathbf{v} = c\| \mathbf{v} \|^2 + 0$ ($\mathbf{w}_2$ is $\perp$ $\mathbf{v}$)
$\therefore c = \mathbf{u} \bullet \mathbf{v} / \| \mathbf{v} \|^2$, and $\mathbf{w}_1 = \text{proj}_{\mathbf{v}}\mathbf{u}$ (i.e. projection of $\mathbf{u}$ onto $\mathbf{v}$) =
$c\mathbf{v} = (\mathbf{u} \bullet \mathbf{v} / \| \mathbf{v} \|^2)\mathbf{v}$.

For the x and y components of a vector, the vector's magnitude is:

$$|\vec{A}| = (A_x^2 + A_y^2)^{1/2}$$ **(Pythagoras' theorem)**

$$\cos\theta = A_x / |\vec{A}| \quad , \quad \sin\theta = A_y / |\vec{A}|$$

You can add vectors via their x and y components.

Wave Mechanics

ε_0 == permitivity in vacuum (electric interaction)

μ_0 == permeability in vacuum (magnetic interaction)

$v = c / n = 1 / (\varepsilon\mu)^{1/2}$, where v = speed of light in a medium,
c = speed of light in a vacuum, and n = index of refraction (of a medium)

$n = (\varepsilon\mu / \varepsilon_0\mu_0)^{1/2} \sim = (\kappa)^{1/2}$, where k = dielectric constant
(of a medium or material)

For an electric field (of energy E) and magnetic field (of strength B):

$E_{max} / B_{max} = c$

$E = hf = hc / \lambda$, momentum of photon == $P = h / \lambda$, $\lambda = v / f = c / nf$,
where f = frequency of (photon's) wave, λ = wavelength (of photon)

$v = 1 / T = \lambda f$, where frequency (in Hertz) == $f = 1 / T$,
T = time period (in seconds), v = wave velocity, and λ = wave length.

Harmonic waves:

$y(x) = A\bullet\sin(kx)$, where A = amplitude and k = wave number = $2\pi / \lambda$.

For a traveling harmonic wave:

$y(x, t) = A\bullet\sin(kx - \omega t)$, where ω = angular freq. of oscillation
$\omega = 2\pi f$

When the waveform is constant: $v = x / t = \omega / k = (2\pi f) / (2\pi / \lambda) = \lambda f$

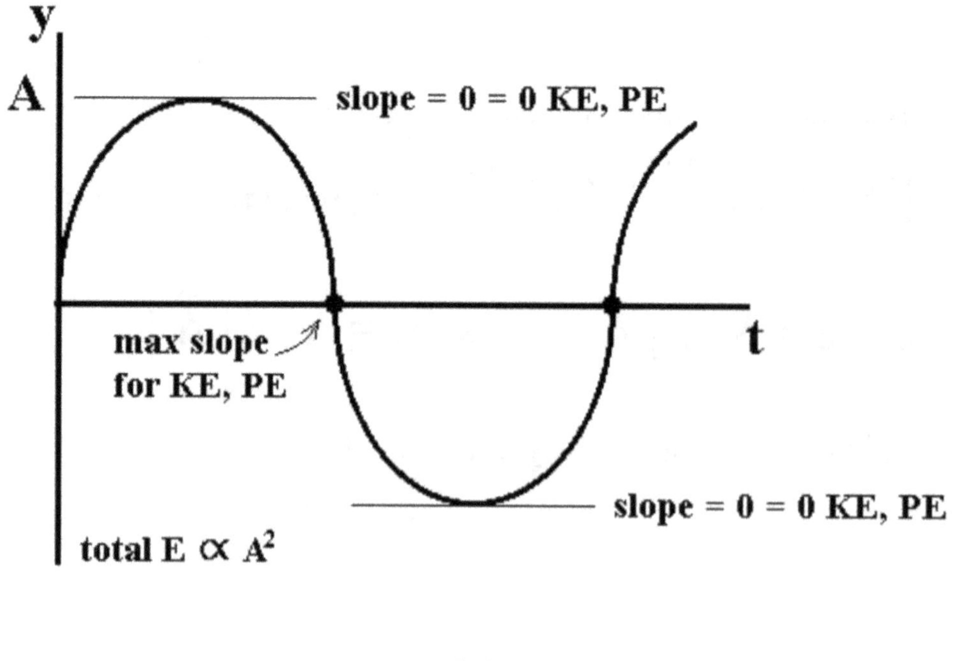

Fixed String Waves

velocity of wave = v_{wave} = $[\, F_T / (m / L) \,]^{1/2}$, where F_T = tension of string, (m / L) = mass density (== mass per unit length)

For two waves of same amplitude A, but one wave with a phase shift of ψ :

$y_1 = A \bullet \sin(kx - \omega t)$, $y_2 = A \bullet \sin(kx - \omega t + \psi)$

The superposition principle allows $y = y_1 + y_2$:

$$y = [\, 2A \bullet \cos((½) \bullet \psi) \,] \sin(kx - \omega t + (½) \bullet \psi)$$

new amplitude phase shift of $\psi / 2$

∴ a wave of same λ and f, with new A and phase shift

when $\psi = 0$ → total constructive interference, amplitude = 2A

when $\psi = \pi$ → total destructive interference, amplitude = 0

If the two waves are of same A but in opposite directions:

$$y = y_1 + y_2 = A \bullet \sin(kx + \omega t) + A \bullet \sin(kx - \omega t) = 2A \bullet \sin(kx) \bullet \cos(\omega t)$$

[opposite directions]

This is not a traveling wave, but a standing wave.

Standing waves only occur at resonant frequencies, at particular phase relationships between the two waves, giving rise to harmonics:

1st harmonic (= fundamental harmonic) $\lambda = 2L$, L = length of string (half of a wavelength fits on a string)

∴ fundamental frequency = $v_{wave} / 2L$ (i.e. inverse of round trip)

For nodes, here the amplitude is always zero when kx = 0 or some multiple of π.

Wavelength Harmonics

$\lambda_n = 2L / n$, with n = 1, 2, 3, …

harmonic frequencies == $v_{wave} / \lambda_n = nf_1 = f_n$,

where f_1 is the fundamental frequency.

Harmonics higher than n = 1 are called overtones.

Wave Motion

period == T = amount of time for one oscillation
frequency == (1 / T) = # of oscillations per second == Hertz = f = 1 / T
angular frequency = angular velocity == $\varpi = 2\pi f$ (in radians / second)
$\therefore\ \varpi = 2\pi f = 2\pi / T$

Energy Relationships:

$E = hc / \lambda$, where h = Plank's constant, λ = wavelength,
c = (vacuum) speed of light
kinetic energy = KE = $(1 / 2)mv^2 = (mv)v / 2$
$= pv / 2 = (mv)^2 / 2m = p^2 / 2m = m^2v^2 / 2m = mv^2 / 2$,
where m = mass, p = momentum, v = velocity

Wave Natures

Electromagnetic waves are transverse (oscillate vertically).
Sound waves in a fluid are longitudinal (oscillate horizontally),
and require pressure in a medium (i.e., matter).
Water waves are both transverse and longitudinal.

Harmonic or Sinusoidal Wave: $y(x) = A\sin(kx)$,
where wave number == $k = 2\pi / \lambda$.

Traveling Harmonic Wave : $y(x, t) = A\sin(kx - \varpi t)$, where
angular velocity == $\varpi = 2\pi f$ in radians / second ,
frequency == f = 1 / T == Hertz , period == $T = 2\pi / \varpi$, and $\nu = 1 / T = \lambda f$.

Diffraction, Reflection and Refraction

law of refraction:

$$\sin \theta_{incident} / \sin \theta_{refracted} = \nu_{incident} / \nu_{refracted}$$

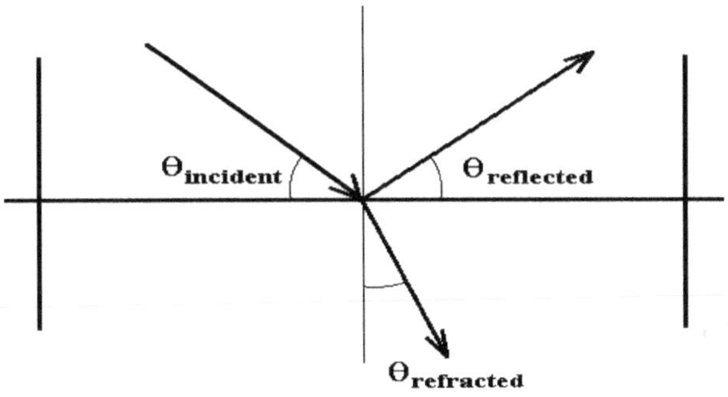

or

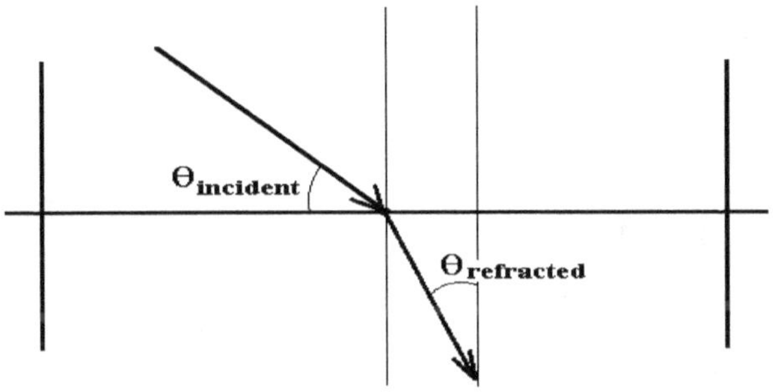

Diffraction

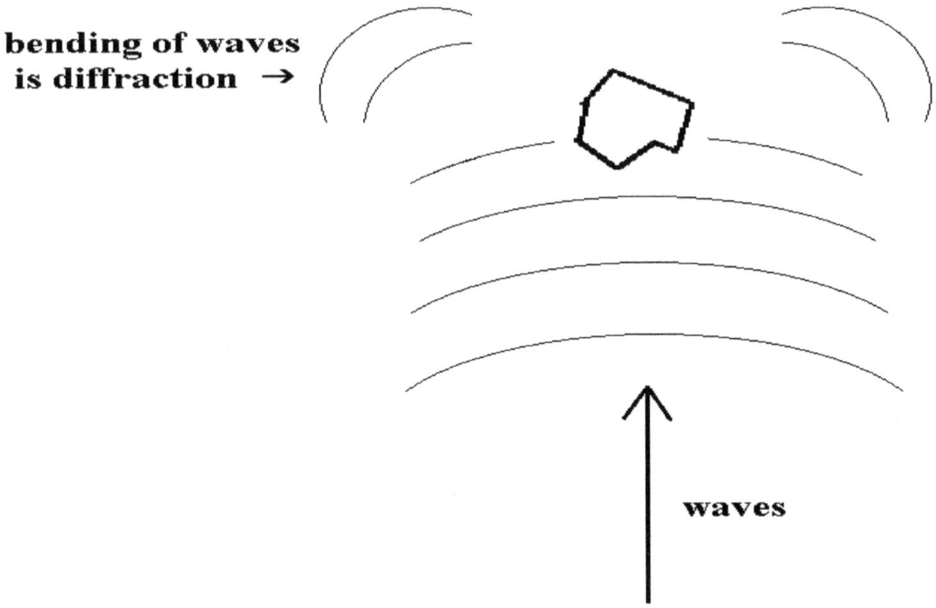

If the object is much larger than the wavelength hitting it, little diffraction occurs.
If the object is smaller or around the size of the wavelength, much diffraction (bending of the waves) occurs.

Doppler Effect

Doppler Effect

$$f' = f\left(\frac{1 \pm v_D / v}{1 \mp v_S / v}\right)$$

$\pm$: S and D are getting closer.
$\mp$: S and D are getting farther.

v_D = **Detector velocity**
v_S = **Source velocity**
v = **sound wave's velocity**

f = **sound wave's frequency**

f' = **sound wave's *apparent* frequency**

Implausibles

part 2

exposition on conjectures of physics
(Mathematical Archaeology)

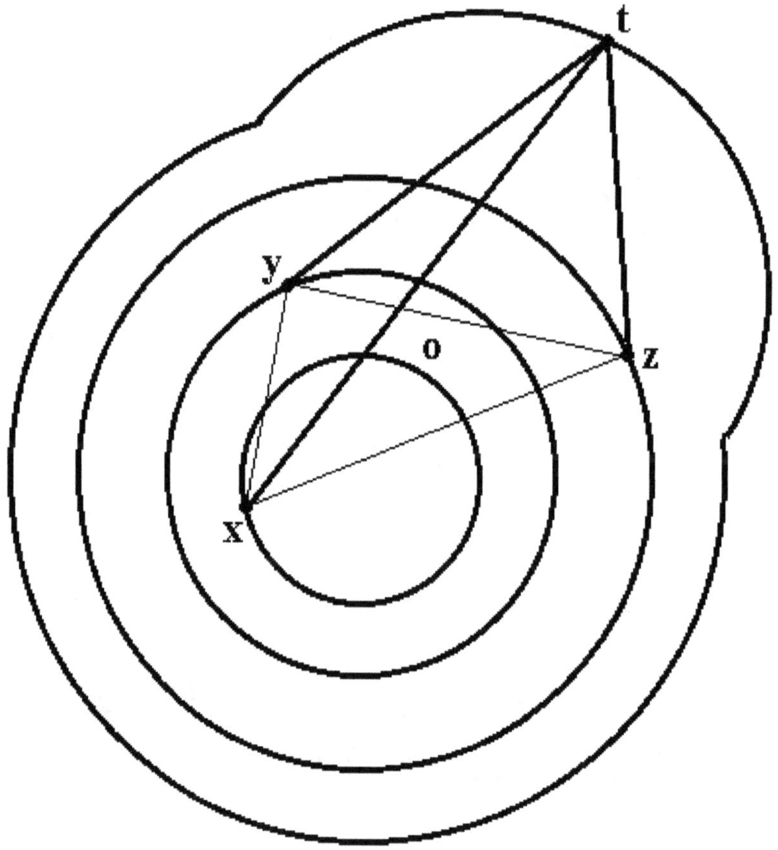
t
y
o
z
x

Implausibles

Now that we have reviewed some elements of employable mathematical physics, let us start our perusal of the implausible by listing a few curious postulates.

Suggestion of some Initial Plausible Contentions

1) inertia : a body in motion stays in motion unless acted upon by a force

2) size of an atom is ~ 10^{-8} cm (diameter); that of a nucleus ~ 10^{-13} cm; yet almost all of the weight of the atom is confined to the nucleus (electrons are much lighter than protons and neutrons)

3) amount of energy (E) *achievable* from a (rest) mass is $E = mc^2$

4) an electric field is (analogously) like a body of water. If two objects are "floating" in it, when they are close to each other pushing one (in the water) may disturb (the position of) the other. If the objects are very distant from each other, pushing one might hardly affect the position of the other; but if you "jiggle" the one object, setting up "waves" in the water, the position of the other object (distantly away) may be affected. The "jiggled" water acts as "particles" affecting the 2nd (distant) object. The smaller the "wave" (i.e. the higher its frequency) the more particle-like it behaves (in the field).

Let's assume that all (observable) phenomena are determined by measurements. Measurement is not equivalent to phenomena, but is (only) descriptive of such. Then energy can be descriptive of mass, without equivalency to mass. Then mass can be considered a field for/of measurement.

If the scale of measurement is much larger than the scale of what is being measured, details of what is being measured are uncertain: e.g., the position of a speck of matter in a drop of water, trying to be determined by a (displacement of) plank of wood much larger than the drop (in dimensions).

The equivalent reciprocal model of this would have the plank of similar dimensions as the speck, but the drop huge compared to the size of the plank or speck or both. In both models the speck is in (continuous) motion. In the first model its position (in the drop) is in question (i.e. uncertain), while in the second its motion (say, velocity) is uncertain (in speed and direction). Yet, for both illustrations these quantities are definite.

A physical phenomenon is not the same as its measurement. So to say mass is equal to its inertia is wrong. To say otherwise is to ascribe a philosophical or literary (rationality) law, such as to attribute infinity to physical nature as; for example, a corpse remains dead for eternity (as an indisputable fact).

From these implausibilities, one may compose the following postulates (of contradiction):

--- It is impracticable that a mass is ever at rest in an indefinite and (so called) infinite space, since in such a condition its position can not be (fully) defined.

--- If such a mass at presumed rest is given a force, so that it appears (sans frictional or direction-altering insults) it travels indefinitely (i.e. forever) at a constant velocity (direction and constant speed), its actual travel (of great distance) approximates that of a light beam (of no mass but only energy).

--- This transit is not actually straight but must be altered (or curved) due to gravitational influences of the (mass of) container of the "infinite" space (i.e. the universe, therefore the "walls" of the universe). This is, of course, destructive to the conjecture that the traveling mass (converted to light during the vastness of its travel) is ever so restricted to a finite (unchanging) velocity.

--- Presumably, the alternation of the transit (of the defined mass) may reveal characteristics of the universal walls and the gravitational effects of the mass(es) of these boundaries, such that assignment of a physical nature to time may be abrogated (to time as defined, rather, by these walls or boundaries).

--- Therefore, when in a local (reference) system, light is induced, this is caused by a mass (once at a local system resting) given a force of movement that is directly affected (by the universal boundaries) of such (system relative) vastness of transit as to convert its energy (of internal capacity) into observable energy (or "light"). Thus, the space which is subjected is equivalent to a mass ("jelly") and its physical limitations (e.g., of light's travel, speed and velocities through it).

--- The proposition that Brownium Motion is entropically inspired and (therefore) can not produce useful work is immediately destroyed by the fact of its observance and contemplation (since thinking is an application of useful work).

--- One, then, may propose that all probability is philosophy (rather than science), so that it may be described (or titled) as "mathematical philosophy." This characteristic, in the mathematical and thus relations of physical calculations (as manipulations of equations and relationships of quantities and qualities) can be demarcated (e.g., with "stars" or other consistent or constantly used symbols) such that, if the demarcations remain (or persist) after the mathematical developments are concluded, the results can be claimed to be of (at least a taint of) philosophical content.

The characteristic of a philosophy is demonstrated (illustrated) by the model of a decomposing corpse which, by thoroughly physical (or natural) means recomposes and (if "life" be an energy or energetic process) readopts animation (or life). This is a philosophical (or of literature) possibility from probabilities, which may be so slight of extent as to be less likely to occur within a multitude of times of the existence of the universe and so is unlikely ever to be measurable. Again, one might distinguish a measurement from its physics, for to fail to do so shows a sloppiness of mentoring (or thinking). That a measurement is possible does not command that it may occur. "Time" is a measurement, just as "temperature" is a measurement of an intensive physical property from "heat." Therefore, measurement is an exercise in establishing probabilities, and is distinct from physical events (i.e., measurements are probabilities). A scale (e.g., an "inch" or a ruler) is not the same as the distance it is used to measure (or quantify). **Nul/los** maths might be useful to illustrate such distinctions or distinguishing, with probability being "nulled" to zero (i.e., taking the place of zero).

--- What if we had a geometrical object, the discreet qualities (observable) of which depended on the exactness and unchanging geometry (say, opposing polarized planes remaining ordered by not only those planes but the others of the object, to retain the polarized opposition)? Then, an alteration of the geometry, as like a Lorentz contraction, should cause a change in (observable) quality of the object. Within the inertial reference frame of the object, its observer should notice (or measure) no change (in quality), but an observer outside of this reference frame should be able to detect the object's change of quality. The outside observer, in back of the reference frame and directly watching it traveling away, might notice the (transparent) object darkening (of light running from the front opposing plane to the back one), while the observer within the frame (while also observing the object similarly of direction as the outside observer) would not experience this darkening of the object. The opposing planes of the object would have to rotate (of one to the other) for the outside observer, as the object's physical exactness is altered. The observers, of course, could be detectors of film.

No two observers (of different reference frames) are ever moving uniformly (i.e., in straight lines, constant of velocities) with respect to each other if the universe is turning (or is in turmoil of motion). Therefore, while the two opposing planes can each undergo equivalent (Lorentz contractions) alteration, their spacial relationship to each other (to maintain the exactness of geometry that allows their consistent opposition) also changes for the external observer.

Because of the gravitational influences of all masses on each other, there is no "empty" space: any mass within a space moving along with that space, as that space itself is moving. (It's as if space itself were part of the "jelly" that is moving.) Thus, in measuring that space it seems to be irretrievably related to time (due to its motion). Since a universe is bounded by walls of mass (presumably), its entire space is so subjected (to time). But, as for conjecture, this allows for a "wall-less" universe where some of its space is not so subjected. There would, of course, be no mass in such space (as argued through induction). Then, one could just define (or suspect) that there are several subjected spaces in the universe whose masses do not relate to each other (physically, thus by gravitation), each with their own physical constants and mores. [Definitions, though, do not constitute proofs.] This is the "non-physical" supposition (for no physics between the "jellies").

Measurement of Space-masses

But of this supposition if non-physicality is adopted, where there are no physical relationships between "space-masses" (or masses of space), then this leads to the following: 1) true space (non-subjected space) allows for no physicality within it (and therefore none between space-masses); 2) the "universe" (of collected masses, space-masses, and non-subjected spaces) may be bounded by walls or (be) wall-less (as far as the space-masses are concerned). Therefore, space-masses (as describable) suspiciously take on the character (or description of qualities) of what have been termed "black holes," and the space of space-masses as that of "dark matter" and "dark energy."

So, in this view, let us say the universe is walled (i.e., "contained") but is subjected to the gravitational influences (or physical properties) only of itself (its size and distances between component parts or regions). The wall is then buffered (from other gravitational or physical effects or influences) by the non-subjected spaces to disallow the wall's influences on the space-masses (and non-subjected spaces) it contains. The non-subjected spaces (or "true" empty space) obstructs physical influences across and (therefore) between them. The wall architecture (shape or folds geometrically) is, then, allowed to be from simple (regular geometric shape) to very complicated or intricate (e.g., compartmentalized, and of varying densities of formation or location). Needless to say, since it (the universe) contains "everything," at some points or regions it must be rigid.

So, the measurement of time is restricted to the space-masses ("jellies"). What is the nature of this measurement? If any physical property changes, through a movement of position (be it energy or mass) within the space-mass, this movement must obviously be compensated for within the space-mass, since the compensation is prohibited within or at any adjoining non-subjected space (and, thus, not at the universe "walls"). This intra-compensation is a definition for time. The space-mass can not move within the non-subjected space nor abut (and disturb) the universe "wall," so the compensation must be internal.

But how was the space-mass created or grown within the non-subjected space, if it can have no movement within that space? If we assume (that) of universal walls, perhaps the space-masses derive from it as like due to invaginations (through the non-subjected space) connected to the walls by (relatively) thin tethers of mass. As such, the space-masses were not grown but are constants of projection from the walls. This would cause the walls to influence the space-masses (and visa versa), unless the tethers are so thin as to (physically) prevent this, the thinness of the wall constrictions, of course, preventing compensations between the given space-mass and other space-masses or portions of wall connected to it (by the constrictions of universe wall).

Energy Relationships and Time

Let, then, as for example,

t (as time) == change in quality ,

therefore, a measurement, allowing for non-trivial relationships

$d ==$ distance , speed $= d / t$, volume $= l \cdot h \cdot w$

(i.e., a box's space; if $l = h = w$, then it is of a cube) ;

For l, $d_1 = (x_2 - x_1)$, a change

$\therefore\ d_1 / t_x == d_1 / (x_2 - x_1)$ [Note the *analogy* (==) of change (in distance) to time.]

acceleration $= a = d / t^2 == d_1 / [(x_2 - x_1)(y_2 - y_1)] = a_1$

$\therefore\ s = s_{(l\,h\,w)},\ d_1 / [(x_2 - x_1)(y_2 - y_1)] = d_1 / [(x_2 - x_1)(z_2 - z_1)]$

$= d_1 / [(y_2 - y_1)(z_2 - z_1)]$ (for a cube)

$\therefore\ 1 / (y_2 - y_1) = 1 / (z_2 - z_1) = (x_2 - x_1) / [(y_2 - y_1)(z_2 - z_1)]$

These are accelerations along l (or from l's perspective).

$\therefore$ if d_1 / t_y is a speed, then $1 / t_y$ is an acceleration (for l).

Now, we allow for the suppositions:

1) maximum speed is for largest d or smallest t

2) minimum density is for smallest mass or largest space

3) (therefore) the density of the box is minimum when the speed is maximum (through the box)

density = ρ = m(ass) / volume = $m / s == m / t^3 = m / [t_x \cdot t_y \cdot t_z]$

For l, $a_1 = d_1 / [(x_2 - x_1)(y_2 - y_1)] = (x_2 - x_1) / [(y_2 - y_1)(z_2 - z_1)] = t_x / (t_y \cdot t_z)$

$\therefore (t_y \cdot t_z) = t_x / a_1$

$$\rho = m / [t_x \cdot (t_y \cdot t_z)] = (a_1 \cdot m) / [t_x \cdot t_x] = (a_1 \cdot m) / t_x^{\ 2} == (a_h \cdot m) / t_y^{\ 2} == (a_w \cdot m) / t_z^{\ 2}$$

Therefore, for any m: minimum a $\rightarrow$ min ρ $\rightarrow$ max speed (since $\rho \propto a$).

For example, the speed of light is at a maximum when the density is at a minimum; or, conceptually, light achieves its constant (and only) speed as its possible acceleration (along a length that is straight) is minimized.

Therefore, the concept of time devolves to that of a measurement, rather than a physical attribute. This would be so for each considered space-mass. Density causes changes in direction (of travel).

--- Energy is defined as being only potential or kinetic, and is generally described as an ability to do (useful) work. A reasonable proposal for it is: 'force x distance', where force is (or imparts) a propensity for motion. But energy may have a third form, to be called "transferable." To transfer (from one substance to another) is a propensity also, and thus is directly to be related to a change of condition (or quality) and therefore to time (as for a measurement, as we have correlated here).

$\therefore$, as an example, the propensity for a change in position $(x_2 - x_1)$ yields (or is equated to) a distance; and this change can not be instantaneous (or idealized), and so yields to a time (of measurement).

$\therefore$ the energy (of a substance or material or observation of phenomena) is:

$$E = P.E. + T.E. + K.E.$$

A mass stationary at a certain height, impacted by a gravitational field, is said to contain a potential energy (for falling), but what work is it doing? Apparently it is the work to maintain its stationary positioning. When the mass falls, its movement is caused by an acceleration due to gravity, and its potential energy is converted to kinetic energy. Its work is its movement (or fall, which might be useful in some way, while its P.E. may have been useful in some way when it was stationary). If the mass moves at a constant velocity perpendicular to the gravitational field (i.e., horizontally, thus not falling), it only has a momentum (m • v) and does no work. Even though it is moving (and indefinitely, as an idealization) its momentum is not impacted by any acceleration (gravity) and so it is said (mechanically) not to have P.E. nor K.E. (for a change in motion). Yet, when a falling mass reaches (hits) ground and stops moving, the impact with the ground disturbs the substance of the ground (causing displacements and heat),and this impact is a manifestation of work (as the K.E. of the mass is transferred to heat and movement energies for the ground).

But we can re-describe both processes (falling and momentum) in terms of T.E.: the stationary mass about to fall (vertically) has (apparent) P.E. sans (T.E. + K.E.). The falling mass has K.E. (sans T.E. + P.E.). The fallen mass hitting the ground and again stationary has T.E. (sans P.E. + K.E.) if its gravitational impact is exhausted, or has T.E. + P.E. (sans K.E.) if it remains under gravitational influence. The momentum movement (no longer under gravitational influence) of the mass clearly retains an energy of a (being as) stationary mass, but without the ability to do work. Apparently, it has T.E. lost and P.E. lost, with yet a kinetic component (for indefinite motion). If we say this is still a K.E., but of an non-transferable energy, then we must reformulate the energy components of the process (in terms of *observational* properties):

1) mass stationary but influenced by gravity has **P.E. + T.E.** | work : no time (no change)

2) mass falling under acceleration of gravity has **T.E. + K.E.** | work : time

3) mass hitting ground and stopping (still under gravity) has **T.E. + P.E.** | work : no time, | yet change (has | time of change)

4) stationary mass not impacted by gravity has **P.E.** | no work (no velocity, | no time)

5) moving mass not impacted by gravity has **K.E.** | no work (only momentum); | time but no change, | therefore $\mathbf{t \rightarrow \infty}$

Perhaps

6) (stationary) mass under conversion (to energy or a "different" mass), not impacted by gravity has **T.E.** | | work (of conversion): time, | but no movement

In reality, all masses are under gravitational influences (by their own beings and that of all others). Therefore, the ability of an energy to do work requires T.E., irrespective of an acceleration (or even any movement).

Of course, when a mass is given a momentum, it presumably retains the energy of that work of conference (as an internal energy) indefinitely (until released or transferred by friction or stoppage of movement). All resting masses must be accelerated (by a force) for movement $(f = m \bullet a)$. Since a distance must be associated with this force applied to a mass (during the acceleration into motion), work energy $(w = f \bullet d)$ is contributed to initiate the momentum, the acceleration of the mass decreasing down to a constant velocity (in an idealistic sense).

Since we assume instantaneous time does not occur (i.e., is not physically real), this leads to the interesting corollary that all change occurs during the time of a process. Therefore, for example, process **(3)** has a time duration of occurrence. Since we assume there is no physical infinity of movement, process **(5)**, which has time, must actually have (a definable) change (here of distance traveled). The mass does not change, but its position does (for a momentum). A conversion of mass must involve movement (within the mass), so process **(6)** probably has T.E. + P.E. + K.E. (but with observation it seems or looks as only T.E.).

Traditional (conventional) physical theory says that one can not distinguish between processes **(4)** and **(5)** within a closed inertial reference frame, since within the frame one could not notice the momentum of the mass (without comparing it to an outside reference frame); so it would seem (to an internal observer) to be undergoing process **(4)**. Yet here, both processes can be distinguished by conferring to either (and both) some T.E. For **(5)** it is in the form of friction (to retard and end the motion). For **(4)**, presumably some property of a stationary mass may be measured to change (or to be changing). In reality, of course, all masses are changing and therefore conduce to (6), and all are under gravitational attractions and so may conduce to $(1 \rightarrow 3)$. So, all (real) masses have T.E. + P.E. + K.E. and are undergoing some form of work (as well as heat transfer). But experimentally (i.e., of experimental observations) they are more distinguished (restrictive) of their types of energies.

Distinction between Real and Mathematical Models

Now we must come to a distinction for a model (to explain phenomena) being physical (real) or mathematical (idealistic), since a merely idealistic model may predict (correctly) observations but only for a range of experimental tactics, beyond which it can be (or should be) discarded, while a real model is correct for all perceivable ranges of experiment.

In particular we can take the photon (of light energy). It has a measurable momentum (since some of this may be transferred to the targets it hits; e.g., the Compton experiment of directing x-ray photons at graphite and finding their wavelengths increased, due to a transfer of momentum from the photons to the graphite). Yet, all photons are considered (as well as can be measured) to be mass-less.

If we consider the de Broglie wavelength as usually given:

$\lambda = h / mu$, where $E = hv = hc / \lambda$,

u = speed of mass m, $v_{freq.} = c / \lambda$, c = (vacuum) speed of light

and take momentum = $mu == p$; the speed of all photons = c ($== u$) ;

therefore, $\lambda = h / mc = h / p$, thus $p = h / \lambda$.

Therefore, p is measurable; yet $m = 0$ (which would lead to an infinite λ). So there is a (real or physical) contradiction of mathematical terms. This also brings to mind the famous "paradox of Zeno," where one can never travel through a distance since for the (duration of the) progress of the travel the remaining distance to travel may be halved to the infinitesimal. Of course, in reality the distance is really traveled (and even passed), for the infinite divisions of the remaining distances are only a mathematical idealization. For the photon, too, its measurement of wavelength may be an idealization (as against its momentum).

So, how can we distinguish the physical (real) from the mathematical (idealistic)? It certainly seems as if the proportionality constant h ($= E / v == E\lambda / c$) is suspect here (as Plank's constant), both momentum (p) and wavelength (λ) being experimentally determinable. (As well the distances in Zeno's paradox are determinable, until too small to measure.) One can easily conjecture that whenever a division is part of a mathematical term, the possibility of an idealization (or non-physical) behavior is introduced, since while addition and subtraction are complementary composites (each can be restricted to whole integers or whole "types" of numbers or physical terms), multiplication and division are not (actually) so. They (multiplication and division) are only treated as composite operations as according to convenience (of expression). Multiplication is directly a form of addition, which itself is (directly) a form of subtraction. But division is only similar to subtraction and is not directly related. (One could say division and subtraction are members of a family of reduction, with subtraction directly related to addition and thus to multiplication as members of a family of increment. The two families are related since aspects of either occur in each.)

It is notable that division is a fundamental characteristic or property of construction for the (mathematical) equation (and its inequality relations). Perhaps a different (mathematical) construction (to relate quantities and variables) is necessary (or at least useful) for describing (and predicting) real (physical) components of phenomena. Division itself, as a member of the reduction(ist) family, might be as indirectly related to multiplication as, say (for example), calculating a square (or n-) root of a number. So let us propose an avoidance of division (for expressions of "reals").

Taking the density of a space-mass, as constructed earlier:

$\rho_{(l\,h\,w)} = (a_l \cdot m) / t_x^{\,2}$, which is already modified by our consideration of time as a *change of distance* (i.e., Δx), we would first insist that the expression (for physical "reals") be restricted to: $\rho t_x^{\,2} = a_l m$ or $\rho t_{(x,y,z)}^{\,2} = a_{(l,h,w)} m$, where ( ) denotes corresponding to either(s).

An advantage here, for example, is that we can enforce (if desired) that the variables ρ, t, a,

m all remain integers (whole numbers), or any combination of them. We can replace the "=" sign with another that denotes (or helps to enforce) the restrictions (from division) that we impose: say, "***S***" (for *simultaneous*).

$$\therefore\ \rho t_{(x,y,z)}^{\ 2}\ \boldsymbol{S}\ a_{(l,h,w)}m$$

Obviously, the expression can be modified by additions, subtractions, and multiplications (while other forms of reduction, like square roots and percentages, may be suspect).

Therefore, an enhancement of the expression with a constant (k) could be (as an example):

$$(\ \rho t_{(x,y,z)}^{\ 2} + k\ )\ \boldsymbol{S}\ (\ a_{(l,h,w)}m + k\ )$$

Thus, (for our space-mass model under our conception of time) k is equivalent to an acceleration of a mass, which is equivalent to an amplification (t^2) of a density.

But what, then, is the new relationship of (x,y,z) for t^2 ?

For example: $t_y t_z = t_x / a_1 \ \rightarrow\ a_1 t_y t_z\ \boldsymbol{S}\ t_x$

$$t^2 == t_x^{\ 2} == t_x(a_1 t_y t_z) = a_1 t_x t_y t_z$$

$$k == \rho a_1 t_x t_y t_z = a_1(t_x t_y t_z \rho) == a_1 m$$

$$m == \rho(t_x t_y t_z)$$

Note, if a mass represents an energy, then $(t_x t_y t_z)$ or $(\ t_{(x,y,z)}\ )^2$ "energizes" the density (ρ).

So, energizing the density of a mass or accelerating it (the mass) are equivalent processes.

$\therefore$ it is the density of a mass that may be energized to allow the mass to be accelerated.

Note that we are using three types of expressive relationships here:

equivalence (==), equality (=), *simultaneous*-ness (***S***).

[defined] [idealistic] [physical]

This distinction is important for delimiting divisions, since this relates to the weaknesses and incorrectness of current physical theories: e.g., for the relativity theories there is a dependence on the constancy of the speed of light, but this is only produced by a (somehow) instantaneous velocity upon creation (of the light). The instantaneous is not physical (or real, although one might argue that the properties of the medium used change upon initiation of the light's creation and travel).

A creation (of energy) implies some component of force to be employed, such that a mass is accelerated to a given velocity and speed ($f = m \bullet a$), this acceleration producing an energy within some distance ($E = f \bullet d$) of motion. But to convert an acceleration to a velocity requires (mathematically) a division of quantities (as the velocity terms remain due to divisions):

$$a = d / t^2 \; ; \; v = at = (d / t^2)t = d / t$$

i.e., the required acceleration of (energy) production necessitates a division of (at least a) velocity:

$$a = v(1 / t) = (d / t)(1 / t) = d / t^2$$

That is, the manipulation of a mass to produce or create an energy (light) requires an acceleration of that mass, which involves a division of quantities. Only when the t (of 1/t) goes towards infinity can v become (and remain) constant:

$$a = v(1 / [t \rightarrow \infty]) == \text{“v”} \text{ (with no a)}$$

But infinity is not physical, and the (or this) division violates our restriction (to allow for the simultaneous).

Likewise, for quantum mechanical processes, quantum jumps in energies (to various energy levels) are considered instantaneous and so suffer (conceptually) similar objections for their production: a mass must be manipulated to provide them (energy jumps), and so an acceleration to the speed of energy quanta is treated as being instantaneous and therefore not physical.

--- How, then, can we “make simultaneous” a physical argument mathematically?

1) put the relationship (of expressions) in a form without divisions (to not idealize any quantities)

e.g., $a = b / c \;\rightarrow\; ac \,\boldsymbol{S}\, b$

If an expression has divisions which can be isolated, convert each to new (non-divisional) terms

e.g., $a + c = (b / d) + e$; therefore, create new term $f == b / d$, where $f \bullet d$ represents b ; let $f' == d$;

$$\therefore\; f' (a + c) = f' ((b / d) + e) = f'a + f'c \,\boldsymbol{S}\, b + f'e$$

2) modify (or enhance) expressions only with additions, subtractions, and/or multiplications of/to terms

e.g., $(ac \,\boldsymbol{S}\, b) + g \rightarrow g + ac \,\boldsymbol{S}\, g + b$

3) The simultaneous-ness (simultaneity) is not equality, in a relationship. There is a considerable difference (in **1**) between f and f ', but as long a b / d is treated as not legitimate (a term) the f and f ' terms (even by mistake of notice) can possibly be treated as interchangeable. Only the relationship between them must be maintained (or acknowledged).

Therefore, one can actually have (construct or consider):

f 'a + f 'c ***S*** b + f 'e , fa + fc ***S*** b + fc (Here is an attempt to run the illegitimate f through a simultaneous relation.) , or

fa + fc ***S*** fd + fe (another such attempt, as above)

3D structures from Amino Acid Sequences and associated Water Molecules

For example, (conceptually) the amino acid sequence of a protein, under certain conditions and as reproducible, determines the conversion of the sequence into a definite 3 dimensional (organic) structure, and allows for:

a.a. (for amino acid) Sequence ***S*** 3D structure

(While the *process* of a.a. seq. → 3D str. is not simultaneous, the definition of seq. is simultaneous with the 3D str. Protein translation and peptide folding through expulsion from the ribosome seem somewhat similar to plastic filament extrusion from a 3D printer's nozzle when the filament is not properly tethered to the forming print and/or build plate's heated bed.)

Note that a.a. seq. is not equivalent to 3D str. : a.a. seq. ==/== 3D str. ,

and a.a. seq. is not equal to 3D str. : a.a. seq. =/= 3D str.

But, a definition implies an equivalence.

Therefore: the

ordering of a.a.s along a polypeptide chain == a.a. sequence ∫3D protein structure

for a real physical process, where a **sum** of terms and expressions (w/o divisions) establishes the equivalence.

Characteristics are:

– each a.a. has a propensity to go to a (relative) position →

a unique (definite) 3D structure

- each (non-ending) a.a. is bounded by a previous and next a.a. (seq. ordering maintained) →

each atom (mass) has a unique (x,y,z); no overlapping of positions: that would be an idealization mathematically

- summation of (equal of kind) terms yields a definite value;

molecular interactions between a.a.s (attractions, charges, etc.) influence ordering of polypeptide → secondary (2nd) str.s promoted (adopted for 3D str.)

- each a.a. movement (to relative position) disrupts the greater mass of solvent (water):

- molecular ordering → 3D str. of protein causes (via entropy) greater disordering of water/solvent molecules than (a) disordered polypeptide (This argues that the process is entropically driven.)

There may be other characteristics usable, but each (relative) a.a. positioning requires at least 3 (per term) for a final (relative) 3D (or x,y,z) position.

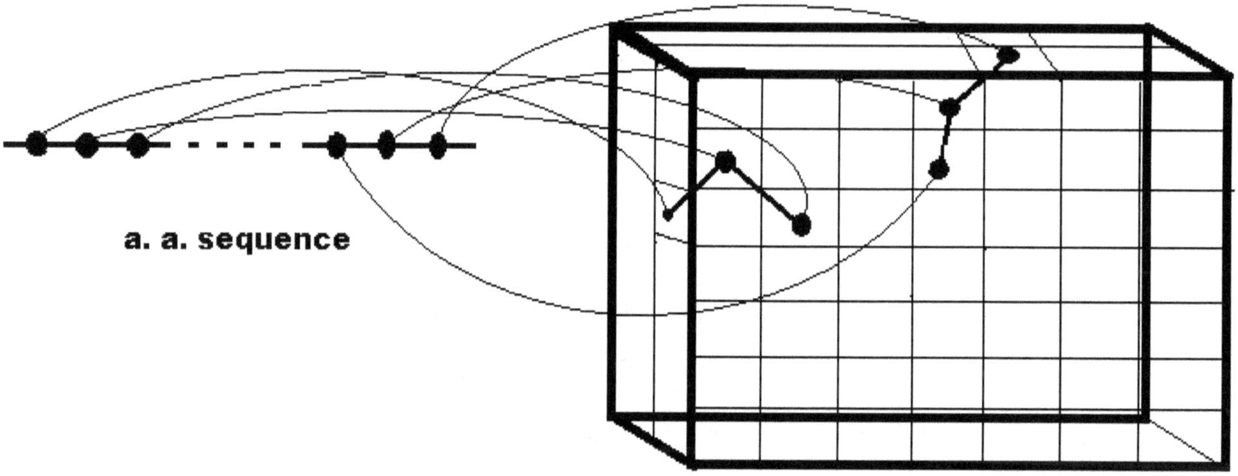

3D Structure
(with surface areas as simplification for cubes of volume)

Let's take 4 characteristics (per a.a. term):

1) water molecule disordering

2) a.a. (seq. order) bonding

3) molecular interactions (a.a.)

4) summation of terms

1) If we assume all peptide bonds are the same (of structure), to a first approximation, and so each has the same number of ordered water molecules for a disordered (or even straight) polypeptide chain, then we may disregard these waters and concentrate on the ordered waters around the a.a. side-chains.

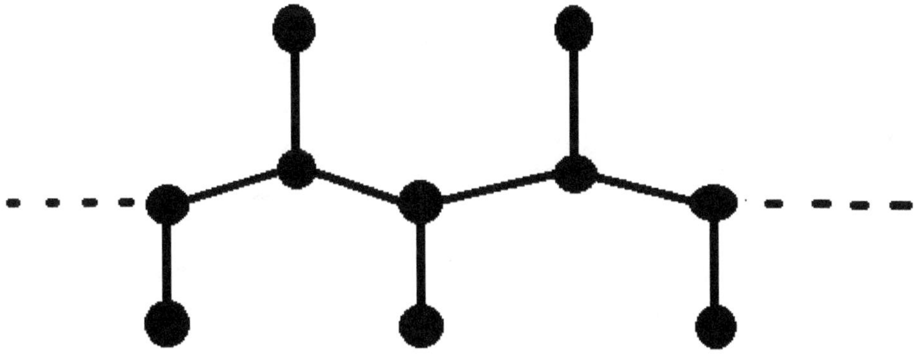

Roughly, the larger the side-chain mass, the larger the number of ordered water molecules around it (due to the side-chain surface area).

Therefore, mass of a.a. == # ordered waters

If the a.a. ends up on the surface of the 3D str., its # of ordered waters (for its side-chain) does not change (from that from it for a straight/disordered polypeptide). Let the smallest side-chain (glycine's) have a relative # of ordered waters be equal to 3, one for each (covalent) H-bond to its methyl carbon. And for the other side-chains, let this number of ordered waters be increased by 1 for each H (of the side-chain) and every other atom (e.g., O, N, S) that can conceivably form a (transient) hydrogen bond to a water molecule (since the partial charges of a water molecule include both δ+ and δ- as available for influences).

Therefore, contribution to term (for summation of terms) is : mass (of a.a.) → # waters

2) For a given a.a., the positional compatibility with the previous a.a. must be the same as that for the next a.a., but each a.a. position must be different from the two others (and all others). The post-compatibilities (for a given a.a.) are highest for the previous and next a.a., and include peptide bond linkages (whereas cystine linkages are only between cysteine side-chains). But if we give any (entire) a.a. a (relatively) equal volume of (positional) space (no matter the structural nature of the peptide), such that adjoining (adjacent) a.a. volumes allow for peptide bonding, then if we (ill-structurally) consider the peptide chain to be a linear line of equal boxes, this line of volume can be maximally condensed to a linear extent (i.e., sides of final 3D str.).

For example,

boxes of volume

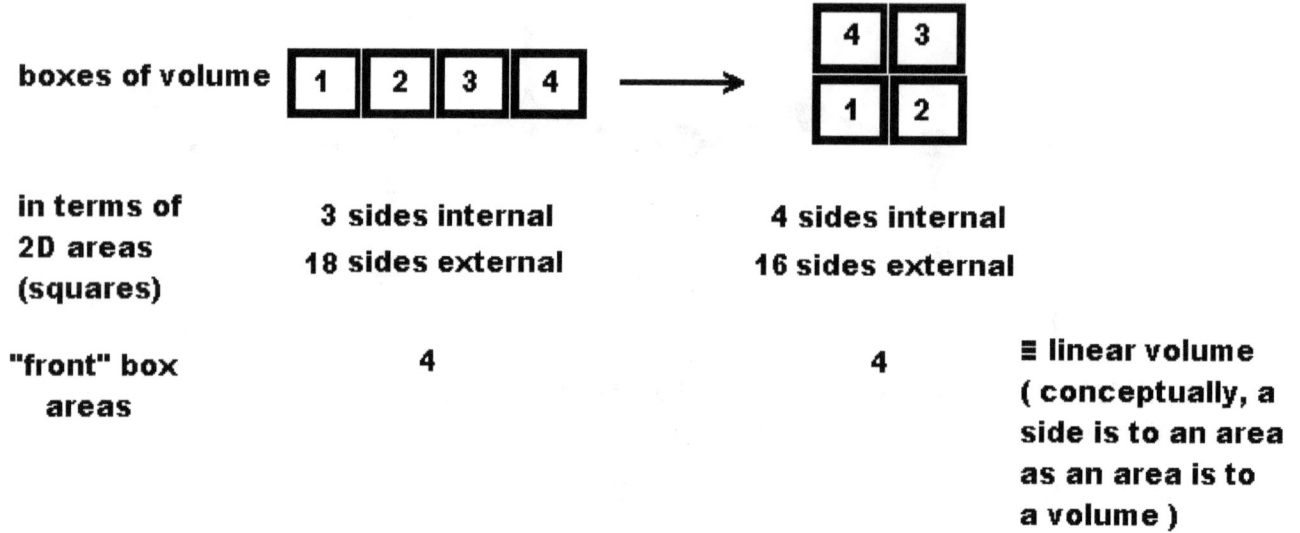

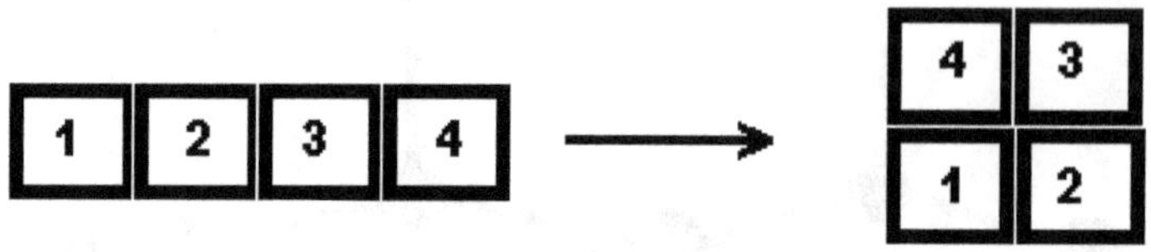

in terms of 2D areas (squares)	**3 sides internal** **18 sides external**	**4 sides internal** **16 sides external**	
"front" box areas	**4**	**4**	**≡ linear volume (conceptually, a side is to an area as an area is to a volume)**

This linear contraction reduces the # of (as per volume) surface sides. The linear volume (of the polypeptide) remains the same.

The linear contraction's extent would probably depend on how much disruption of water molecule ordering can be tolerated (by the conversion of the disordered polypeptide to a 3D str.).

If we give an arbitrary a.a.'s positioning a (relative) coordinate of (x,y,z), that of the previous a.a. must be (+- x, +- y,+- z) sufficient to maintain the α-carbon distances between the two a.a.s. Likewise, the coordinate of the next a.a. is (-+ x, -+ y, -+ z). These relative coordinates can be for any stage of the (linear → 3D final str.) process. But each coordinate must be unique (per stage), i.e., the only one available.

The α-carbon to α-carbon distance between proximal (ordered) a.a.s is, of course, a (constant) radius:

$$\sqrt[2]{\pm\Delta x^2 \underset{+}{,} \pm\Delta y^2 \underset{+}{,} \pm\Delta z^2}$$

or

$$\sqrt[2]{\mp\Delta x^2 \underset{+}{,} \mp\Delta y^2 \underset{+}{,} \mp\Delta z^2}$$

But $\sqrt[2]{} = (\ )^{1/2}$ **is *idealistic* .**

centered on (x,y,z). But this radial relationship must be constantly maintained for each of the (major) atoms of the (structurally unchanging) peptide bond:

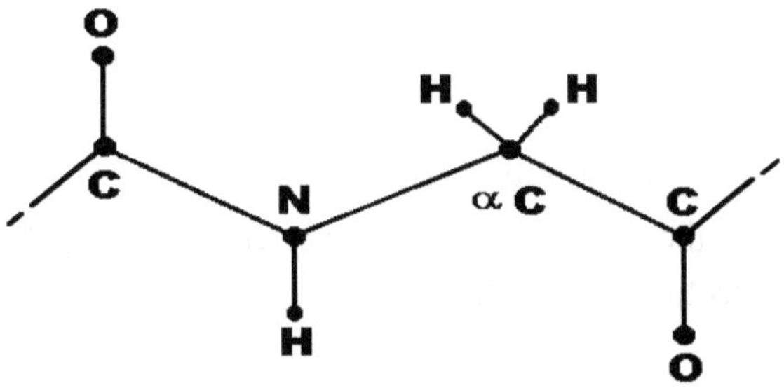

(i.e., also for the carbonyl oxygens, carbonyl carbons, amide nitrogen, nitrogen hydrogen, and each α-carbon hydrogen).

Conjecture:

Two of these atom-to-atom consistent (in length) radii are enough to establish a plane of direction, and a third radius produces another plane. These planes are actually circles. Two circles intersect at two points (unless by coincidence, for all points) at most, when they have equal radii. A third circle intersects at a unique point not of one of the previous two, establishing the 2D location (for it) that maintains its (distal) relationship (at its new position) with the new positions of the other two points.

Actually, these circle intersections promote regions of solution (each circle having a solution for each atom type) for the shifted a.a. position. And the radii are of spheres with solutions confined to the surfaces (circumferences).

The following figure:

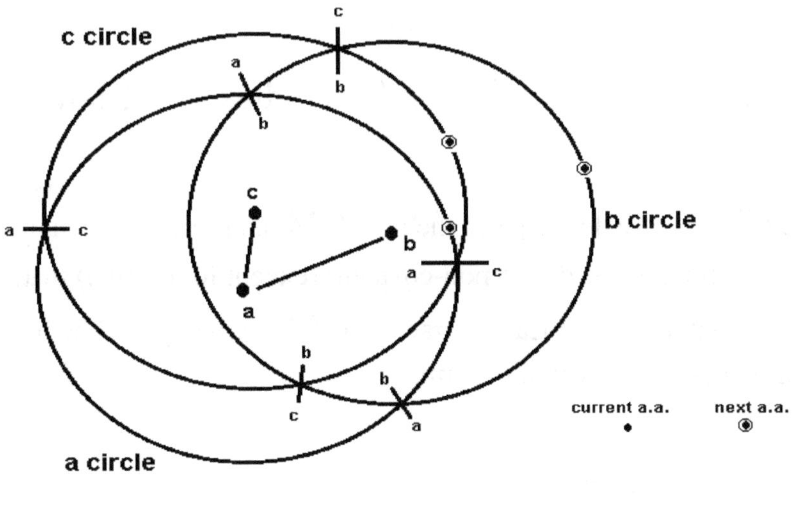

is an example solution for one a.a. and the next a.a. Therefore the intersects don't seem to directly predict the locations of possible solutions (for the next a.a. position) in this 2D simplification. But anywhere on circle a (for atom a) allows positioning of c on circle c and b on circle b, etc.

But if we connect the common intersect(ion) types (ab, ac, bc) by lines, the three lines apparently intersect at the center of mass of the abc figure. (For spheres the lines would be planes, and the intersection of the three planes would form a line for the center of mass, CM.)

Extending a radius from the (current) center of mass establishes another sphere as a solution (surface) for the next (peptide bond atoms) center of mass, although the (abc) atoms can be in any orientation about the CM. Since the CM is due to the (abc) positions, it is independent of the extent (length) of any radius (from it). The CM solution (surface) sphere helps to avoid having coincidental atomic positions between the a.a.s (peptide bonds nascent and extended for 3D str.).

Therefore, based on peptide bond (linking) positioning:

(a.a.)-1 →	a.a. →	(a.a.)+1
x -+ dx, y -+ dy, z -+ dz	x, y, z	x +- dx, y +- dy, z +- dz
where: **dist. (dx, dy, dz) = radius**		**dist. (dx, dy, dz) = radius**

Contribution to term : CM solution surface of radius = dist. between α-carbons, etc.

3) Where on the CM surface do we place the next CM, and in what orientation of peptide bond atoms? Both answers must be due to non-covalent (except for cystine) a.a. interactions.

If conditions are such that the linear contraction (of the polypeptide chain) is strong, then a globular (protein) str. is favored to form.

We can first divide the CM sphere solution surface into two parts:

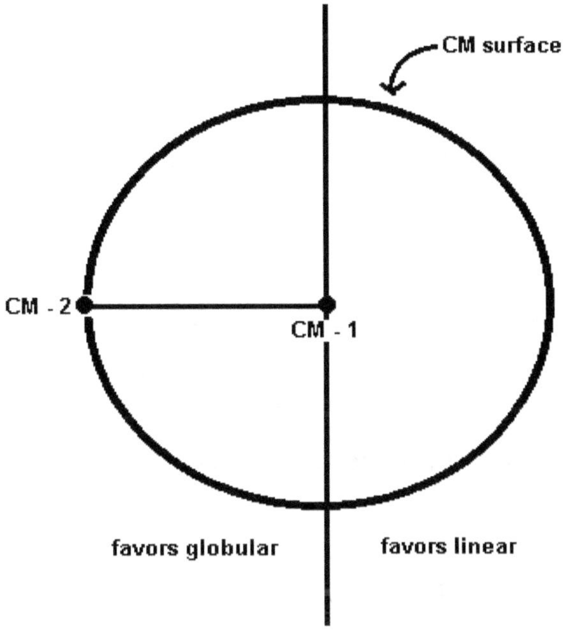

If globular shape is favored, CM must avoid CM-2; therefore, the half hemisphere has a "hole" around CM-2. But this is idealistic (e.g., ½ hole). Actually, the whole sphere (surface) is available to CM. (Even the CM-2 position might be accommodated, subject to the orientations of the a.a.s using it.)

So, again, we do not divide (or partition) the sphere's surface. For a position to be directed to a point on the (3D) surface, it must have (at least) 3 uncorrelated conditions of influence: e.g., bulk size (as the a.a. size increases, less linear contraction is favored), water disordering (as this increases, more linear contraction is favored), side-chain charge: to be accommodated most proximal for CM-2 (or +2) and CM-1 (or +1). But these conditions are somewhat correlated: larger size favors more water disordering; side-chain charge favors water ordering, so neutralization (effective) favors water disordering. And, the effect of mass size to peptide bond atoms should be rigid for their configuration for a CM, to maintain same atom distances to previous and next a.a.s.

The more or less spontaneous adoption of a disordered polypeptide chain for a definite (unique or defined) final 3D str. might suggest that this promotion is produced and predicted by the behaviors of the solvent molecules (e.g., waters). It is notable to observe that a disordered polypeptide has (i.e., starts with) many varied possible orientations (of its constituent a.a.s) as it progresses to a defined final 3D str.

Implausibles : part 2

Now, suppose we have a polypeptide of the following limited extent, where the bounded waters (to be considered) are restricted to the side-chains:

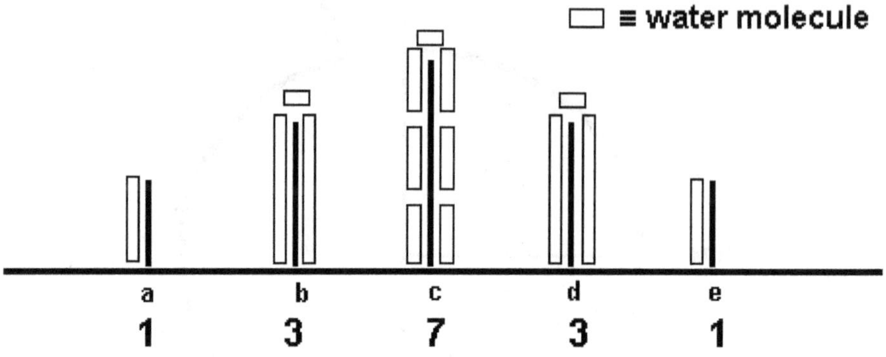

as a "disordered" conformation

The (molecular) goal would be to disorder as many waters (per given side-chain "density" of waters) as structurally possible:

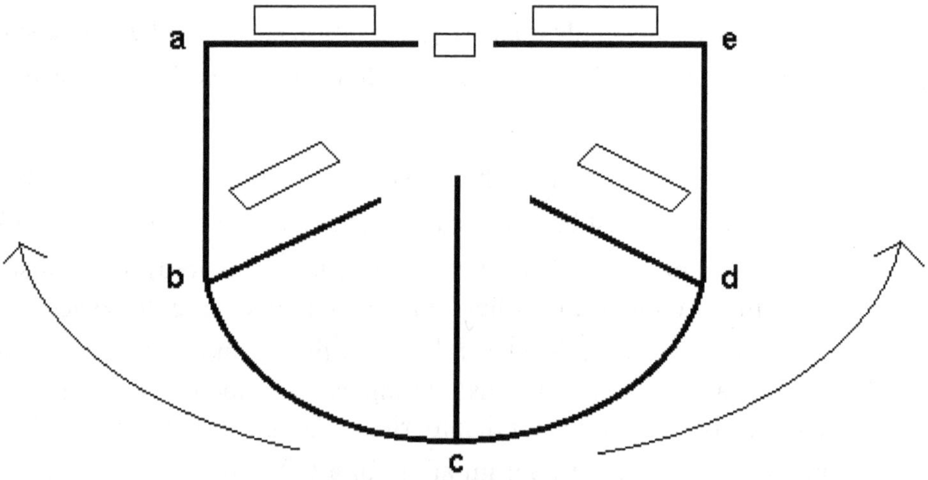

∴ a.a. (sidechain) c looses 7 - 1 = 6 waters
a.a.s b and d each loose 3 - 1 = 2 waters
a.a.s a and e each loose 1 - 1 = 0 waters

Therefore, adoption of this (peptide chain) conformation disorders $(6 + 2 + 2 =)10$ waters. Thus, all of the "disordered" conformations (each) start out with the same # (and positions) of ordered waters and result with the same # of disordered waters for the (final) 3D str. adopted (i.e., consistency of process): the selection of the position (of the a.a.) on the CM solution surface is guided by the (side-chain) water disordering.

If we, to a first approximation, consider each side-chain to be aliphatic, with (only) size determining the # of waters ordered to it (via dispersion forces), and then subtract from this # the waters to remain ordered (due to charges, polarities, etc.), the # remaining are to be lost (disordered) and a (linear) map of disordering is designed (or revealed):

$$\text{a1 - b3 - c7 - d3 - e1} \longrightarrow \text{a1 - b1 - c1 - d1 - e1}$$
$$\text{-0 -0 -0 -0 -0} \text{-0 -2 -6 -2 -0}$$

or
$$\binom{1}{0}_a \binom{3}{0}_b \binom{7}{0}_c \binom{3}{0}_d \binom{1}{0}_e \rightarrow \binom{1}{0}_a \binom{1}{2}_b \binom{1}{6}_c \binom{1}{2}_d \binom{1}{0}_e$$

Graphing with the notation:

$$\binom{x}{y}_{a.a.}$$

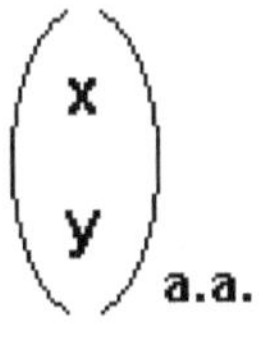

x → ordered waters

y → disordered waters

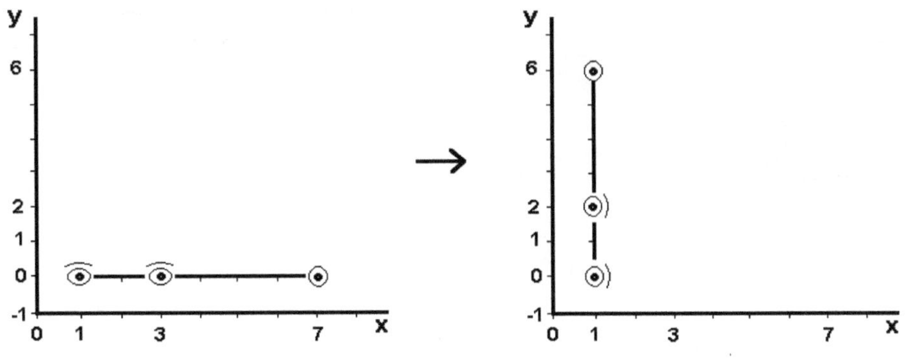

Implausibles : part 2

A third coordinate could be (relative) CM.

y is really the change in water ordering.

If we "nullify the nulls" using our $\binom{x}{y}_{a.a.}$ notation, a propensity for disordering (of waters) is projected (i.e., place initial over final, and condense to remove zeros):

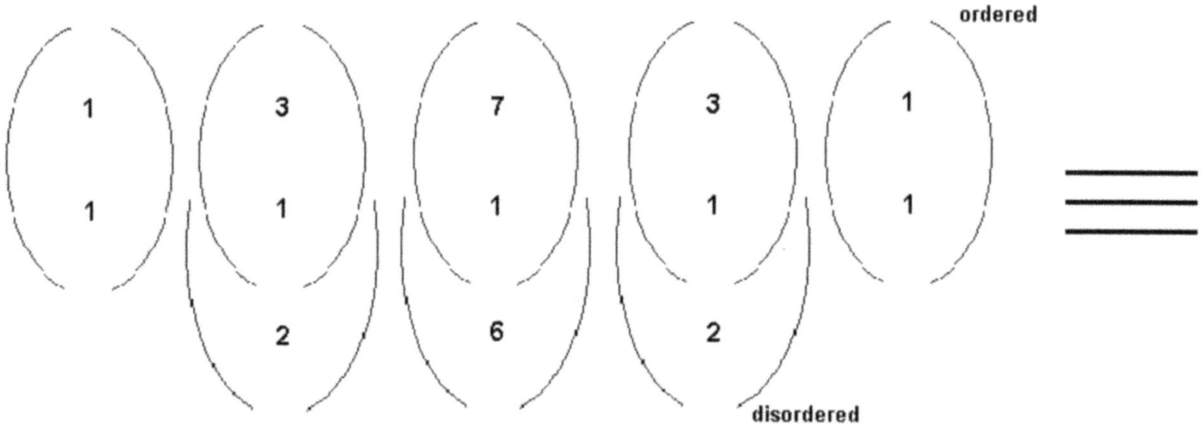
ordered
1
1
3
1
2
7
1
6
3
1
2
1
1
disordered

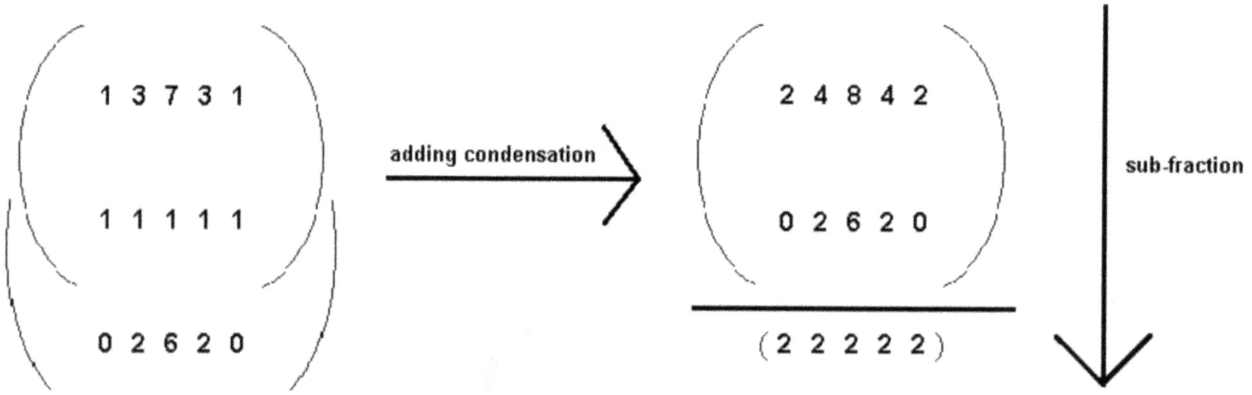
1 3 7 3 1
1 1 1 1 1
0 2 6 2 0
adding condensation
2 4 8 4 2
0 2 6 2 0
(2 2 2 2 2)
sub-fraction

Therefore, the net ordering of waters (for polypeptide) = net disordering of waters (for solvent):

$$(2\,2\,2\,2\,2) \qquad (0\,2\,6\,2\,0)$$

$$2 \cdot 5 = 10 \qquad 2 + 6 + 2 = 10$$

(Eventually, through this process, the polypeptide ends up with 5 ordered waters.)

This is the same as : $(1\,3\,7\,3\,1) + (1\,1\,1\,1\,1) = 20$

$$- [\,(0\,0\,0\,0\,0) + (0\,2\,6\,2\,0)\,] = \underline{-10}$$

$$10$$

But (2 2 2 2 2) is very suggestive.

The molecular interactions determine the (0 2 6 2 0) term; but (2 2 2 2 2) is clearly a discriminant. It is as so, if no charges or polarities let the side-chain retain more waters.

Suppose for a.a. (c), its side-chain has a charge that retains 2 waters (instead of 1):

$$\begin{pmatrix} 1 & 3 & 7 & 3 & 1 \\ 0 & 0 & 0 & 0 & 0 \end{pmatrix} \rightarrow \begin{pmatrix} 1 & 1 & 2 & 1 & 1 \\ 0 & 2 & 5 & 2 & 0 \end{pmatrix}$$

$$\therefore \begin{pmatrix} 1 & 3 & 7 & 3 & 1 \\ 1 & 1 & 2 & 1 & 1 \\ 0 & 2 & 5 & 2 & 0 \end{pmatrix} \rightarrow \begin{pmatrix} 2 & 4 & 9 & 4 & 2 \\ 0 & 2 & 5 & 2 & 0 \end{pmatrix} \downarrow$$

$$\overline{(2\,2\,3\,2\,2)}$$

Now disordered (0 2 5 2 0) ==→ ordered (2 2 3 2 2) = 11 , Δ water = 2

Therefore, a net disordering of (11 – 9 =) 2 waters for the solvent (over ordering for polypeptide) is needed in the solvent "structure" (str.) to allow (or cause) the polypeptide conformational change.

Thus, the

$$\begin{pmatrix} 1\,1\,1\,1\,1 \\ 0\,2\,6\,2\,0 \end{pmatrix}$$

str. is easier to adopt (is more entropic for solvent), with Δ water = 0, than the

$$\begin{pmatrix} 1\,1\,2\,1\,1 \\ 0\,2\,5\,2\,0 \end{pmatrix}$$

str., with Δ water = 2. But adoption is spontaneous for each.

What if the polypeptide remains disordered?

$$\begin{pmatrix} 1\;3\;7\;3\;1 \\ \\ 0\;0\;0\;0\;0 \end{pmatrix} \rightarrow \begin{pmatrix} 1\;3\;7\;3\;1 \\ \\ 0\;0\;0\;0\;0 \end{pmatrix}$$

$$\therefore \begin{pmatrix} 1\;3\;7\;3\;1 \\ 1\;3\;7\;3\;1 \\ 0\;0\;0\;0\;0 \end{pmatrix} \rightarrow \begin{pmatrix} 2\;6\;14\;6\;2 \\ 0\;0\;0\;0\;0 \end{pmatrix} \downarrow$$

$$\overline{(2\;6\;14\;6\;2)}$$

Δ waters = 30
(maximum amount)

Therefore, the solvent (at max. D water) must essentially "mimic" the structure of the (disordered) polypeptide. This would be entropically the least favorable outcome.

$$\begin{pmatrix} 1\ 3\ 7\ 3\ 1 \\ 0\ 0\ 0\ 0\ 0 \end{pmatrix} \longleftarrow \begin{pmatrix} 1\ 3\ 7\ 3\ 1 \\ 0\ 0\ 0\ 0\ 0 \end{pmatrix} \longrightarrow \begin{pmatrix} 1\ 1\ 1\ 1\ 1 \\ 0\ 2\ 6\ 2\ 0 \end{pmatrix} \longrightarrow \begin{pmatrix} 1\ 1\ 2\ 1\ 1 \\ 0\ 2\ 5\ 2\ 0 \end{pmatrix}$$

water mimics structure   **water complements structure**

or
$$\begin{pmatrix} 1\ 3\ 7\ 3\ 1 \\ 0\ 0\ 0\ 0\ 0 \end{pmatrix} \longrightarrow \begin{pmatrix} 1\ 1\ 1\ 1\ 1 \\ 0\ 2\ 6\ 2\ 0 \end{pmatrix}$$

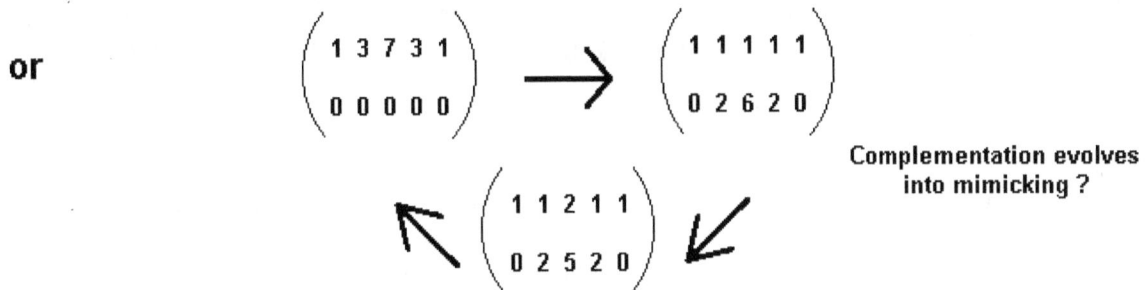

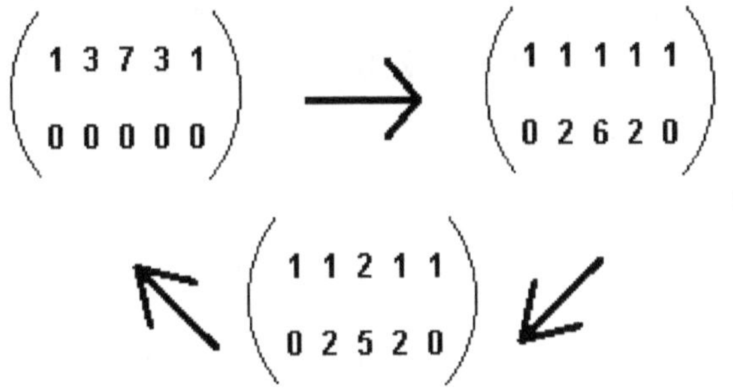

Complementation evolves into mimicking ?

$$\begin{pmatrix} 1\ 1\ 2\ 1\ 1 \\ 0\ 2\ 5\ 2\ 0 \end{pmatrix}$$

$\begin{pmatrix} 1\ 3\ 7\ 3\ 1 \\ 0\ 0\ 0\ 0\ 0 \end{pmatrix}$ **represents many (any) disordered polypeptide states (as far as water bonding is concerned for the sidechains).**

It seems evident that the ordering of the polypeptide (for or towards its first 3D str.) is a process that for the waters seeks to return to the

$$\begin{pmatrix} 1\ 3\ 7\ 3\ 1 \\ 0\ 0\ 0\ 0\ 0 \end{pmatrix}$$

ordered/disordered balance (i.e., increasing ordered and decreasing disordered waters). So somehow the water structure for the disordered and ordered polypeptide strs. might have some sort of equivalence.

But what is the (entropic) force that propels the conversion? Side-chain accommodation (of waters) as much as possible reaches towards the

$$\begin{pmatrix} 1 & 3 & 7 & 3 & 1 \\ 0 & 0 & 0 & 0 & 0 \end{pmatrix}$$

state, but via a route of disordering waters. It is noted that the "arithmetic" of the analysis at least increases the # of ordered waters, more so at the expense of disordered waters provided:

$$\begin{pmatrix} 1 & 3 & 7 & 3 & 1 \\ 0 & 0 & 0 & 0 & 0 \end{pmatrix} \rightarrow \begin{pmatrix} 1 & 1 & 1 & 1 & 1 \\ 0 & 2 & 6 & 2 & 0 \end{pmatrix} \rightarrow \begin{pmatrix} 2 & 4 & 8 & 4 & 2 \\ 0 & 2 & 6 & 2 & 0 \end{pmatrix} \;;$$

$$\begin{pmatrix} 1 & 3 & 7 & 3 & 1 \\ 0 & 0 & 0 & 0 & 0 \end{pmatrix} \rightarrow \begin{pmatrix} 1 & 1 & 2 & 1 & 1 \\ 0 & 2 & 5 & 2 & 0 \end{pmatrix} \rightarrow \begin{pmatrix} 2 & 4 & 9 & 4 & 2 \\ 0 & 2 & 5 & 2 & 0 \end{pmatrix} \;;$$

$$\begin{pmatrix} 1 & 3 & 7 & 3 & 1 \\ 0 & 0 & 0 & 0 & 0 \end{pmatrix} \rightarrow \begin{pmatrix} 1 & 3 & 7 & 3 & 1 \\ 0 & 0 & 0 & 0 & 0 \end{pmatrix} \rightarrow \begin{pmatrix} 2 & 6 & 14 & 6 & 2 \\ 0 & 0 & 0 & 0 & 0 \end{pmatrix}$$

maximal increase (doubling) of ordered waters

Therefore, the order of increasing propensity of polypeptide folding is:

$$\begin{pmatrix} 2 & 6 & 14 & 6 & 2 \\ 0 & 0 & 0 & 0 & 0 \end{pmatrix} , \begin{pmatrix} 2 & 4 & 9 & 4 & 2 \\ 0 & 2 & 5 & 2 & 0 \end{pmatrix} , \begin{pmatrix} 2 & 4 & 8 & 4 & 2 \\ 0 & 2 & 6 & 2 & 0 \end{pmatrix}$$

corresponding to:

$$\begin{pmatrix} 1 & 3 & 7 & 3 & 1 \\ 0 & 0 & 0 & 0 & 0 \end{pmatrix} , \begin{pmatrix} 1 & 1 & 2 & 1 & 1 \\ 0 & 2 & 5 & 2 & 0 \end{pmatrix} , \begin{pmatrix} 1 & 1 & 1 & 1 & 1 \\ 0 & 2 & 6 & 2 & 0 \end{pmatrix}$$

structures.

We can call the propensity charge (or load) as (additionally):

$$\begin{pmatrix} 2+6+14+6+2=30 \\ 0 \end{pmatrix} \equiv \begin{pmatrix} 30 \\ 0 \end{pmatrix}$$

Therefore, the propensity charge runs:

$$\therefore \text{ propensity charge runs: } \begin{pmatrix} 30 \\ 0 \end{pmatrix}, \begin{pmatrix} 21 \\ 9 \end{pmatrix}, \begin{pmatrix} 20 \\ 10 \end{pmatrix}$$

$$\text{and structural load runs: } \begin{pmatrix} 15 \\ 0 \end{pmatrix}, \begin{pmatrix} 6 \\ 9 \end{pmatrix}, \begin{pmatrix} 5 \\ 10 \end{pmatrix}$$

} difference

↓

$$\frac{\begin{pmatrix} 15 \\ 0 \end{pmatrix}, \begin{pmatrix} 15 \\ 9 \end{pmatrix}, \begin{pmatrix} 15 \\ 10 \end{pmatrix}}{\begin{pmatrix} 15 & 6 & 5 \end{pmatrix}} \downarrow$$

"final" ordered waters

Clathrate of Waters Model

Let's assume that the propensity charge provides a clathrate of water molecules mimicking and complementing the side-chains. Obviously, by this model, next to or adjoining side-chains (during the folding process) would work to displace (most of) these clathrating waters. The CM surface solution should have an optimal (or more easily effective) position for this. The polypeptide chain is embedded in the water molecules.

The direction for the position on the CM surface solution to adopt is governed by the water molecules that are to be displaced upon (growing) protein folding. The orientation of the (placed) peptide bond is of course dependent on this position (the bond being rigid and planar).

Therefore, the contribution to term is the coincidence of a (clathrating) water molecule's CM with that on the (a.a.) CM surface solution (this controlling movement for relative positioning of all previous a.a.s in the chain).

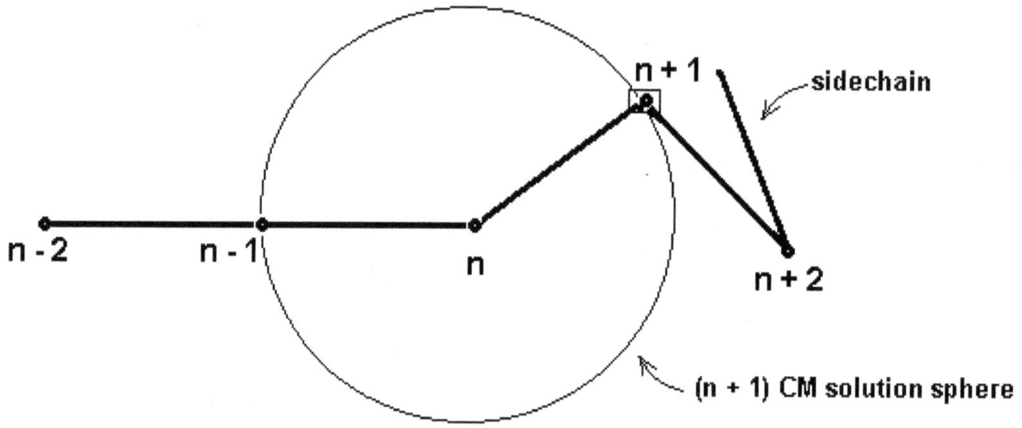

□ = water molecule to be displaced (associated with n + 2 sidechain)

For a spontaneous process there should be only one coincidence per CM surface. (Otherwise, a sampling of 2^{0}, 3^{0} structures would result.)

4) Summation of terms

Since polypeptide folding occurs spontaneously (or starts all at once), the information for the (specific) folding must be found in the (common) disordered conformation. Any disordered conformation contains the same folding information. The information available is the a.a. sequence and the clathrating water structure common to each disordered conformation (thus, dependent on the a.a. side-chains present).

Therefore, we might assume that the locations of the clathrating water molecules guide the angles of latitude and longitude on the CM surface solution (with a radius the length between peptide bond units, n). The summation of these "guides" constructs the (final) 3D str. of the polypeptide. The construction (folding) initiates and occurs at all places along the polypeptide chain simultaneously, and is probably time dependent on the various 2° strs. intermediately formed.

Tethering Model

The (or an) analogy is a puppet of several parts (joints) that when held up by strings (in a particular orientation) always yields to its proper (or recognizable) form. While lying down its proper form might not be apparent (and so can be in various states of disorder), but when held up always adopts its proper (identifiable) conformation (driven by the force of gravity).

If we presume that, from the (disordered polypeptide's perspective, the sea of hydrogen bonded water molecules (of solvent) is infinite, displacement of a.a. related waters involves a pull from this sea (towards incorporation) as could be the tethers promoting protein folding. Then, illustratively, we can construct a sphere of such sea, with the polypeptide at the center. If we initially place the a.a. chain (disordered) in a plane, the tethers occur most strongly to the top and bottom of this plane (as opposed to the two ends of the chain), with the side-chains in alternate array to the sides of the plane (lengthwise).

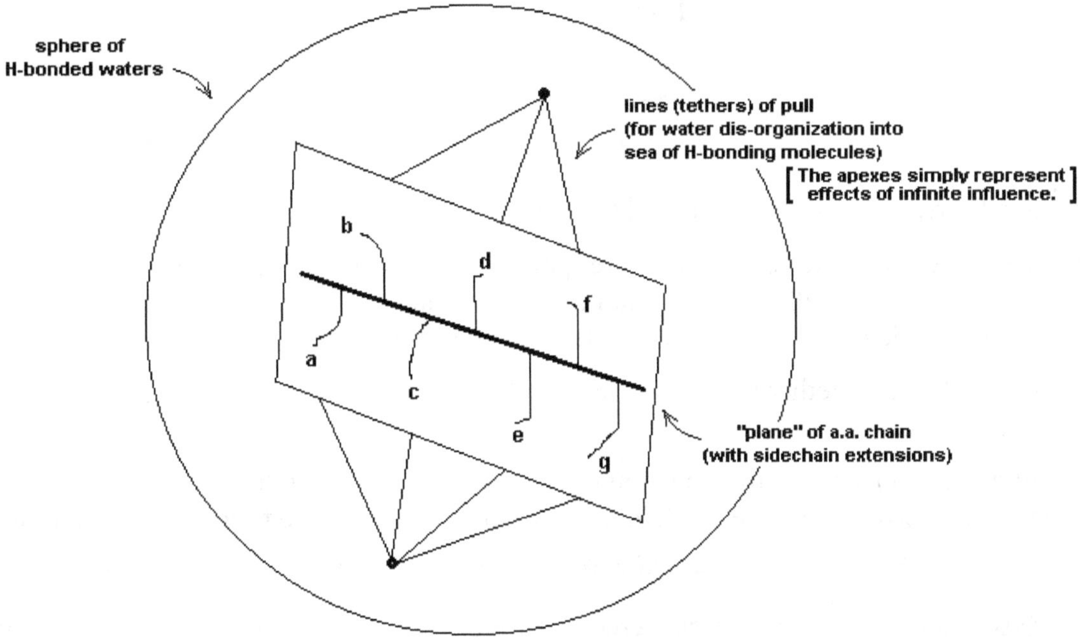

Within the "plane" a propensity (or potential) to dislodge waters (associated with the side-chains) is low. (The total water disordering is stable.)

If (side-chain) c decides to conjoin (hydrophobically) to d's sidechain (say d's sidechain is larger), then the plane is broken with this folding:

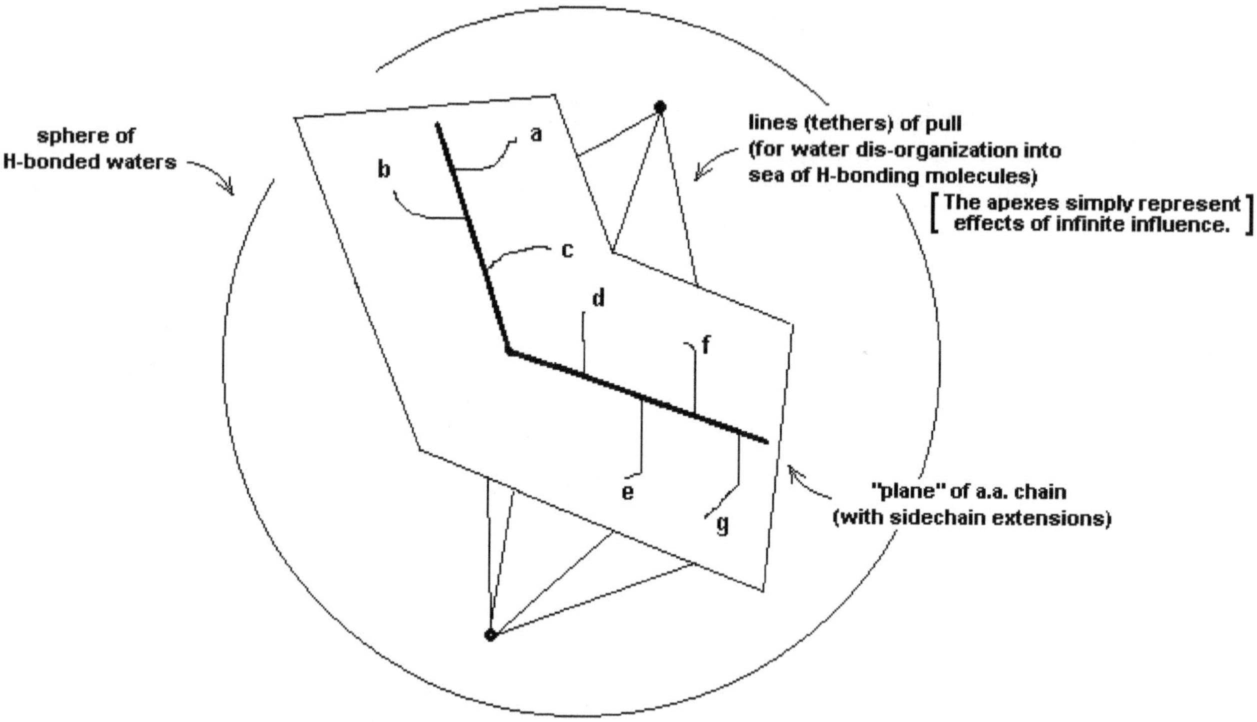

Tethers for a,b,c must be transferred (to other tethers).

But these "hydrophobic conjoinings" are (potentially) occurring for each of the side-chains simultaneously (with tether transfers not only above and below the originating a.a. chain plane, but also to the sides and ends, in perspective to the chain's extent).

Restriction from the tether transfers would be due to the hydrophillic natures of the a.a. side-chains involved.

Therefore, a simultaneous fractionation of the a.a. (chain) plane into many other independent planes characterizes the protein folding. The orientations (relative to the initial chain plane) of the independent planes of course relate to the ϕ, ψ angles adopted by the peptide bonds.

But these adoptions are clearly staggered, in groups of 3 a.a.s (and in both chain directions):

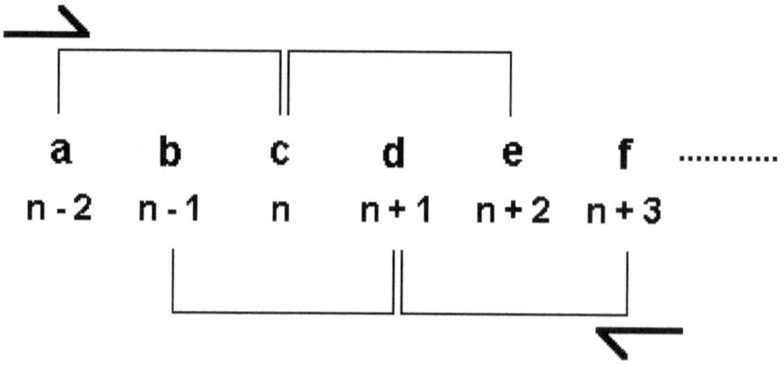

For the need to be simultaneous, we must assume the adoptions do not contradict each other (or prevent each or any other).

While the chain is within the original a.a. plane it is disordered despite any rotational curves of chain with the plane. Thus, from any such structure the same final (folded) 3D str. will result.

For (or in) any given a.a. in the final 3D str. (3° str.), it has two relative planar locations: that which it shares with (n-1) a.a. and that which it shares with (n+1) a.a. These two planes control all previous n (as n-) and all further n (as n+) locations (or positions) of the pre- and post-a.a.s.

Simultaneity of Folding

Simultaneity in Protein Folding: the torsions (or oscillations) of the flexible bonds along the peptide chain occur at (or around) the same time, or within a relatively short time interval compared to the disordered state for the polypeptide. How is this accomplished?

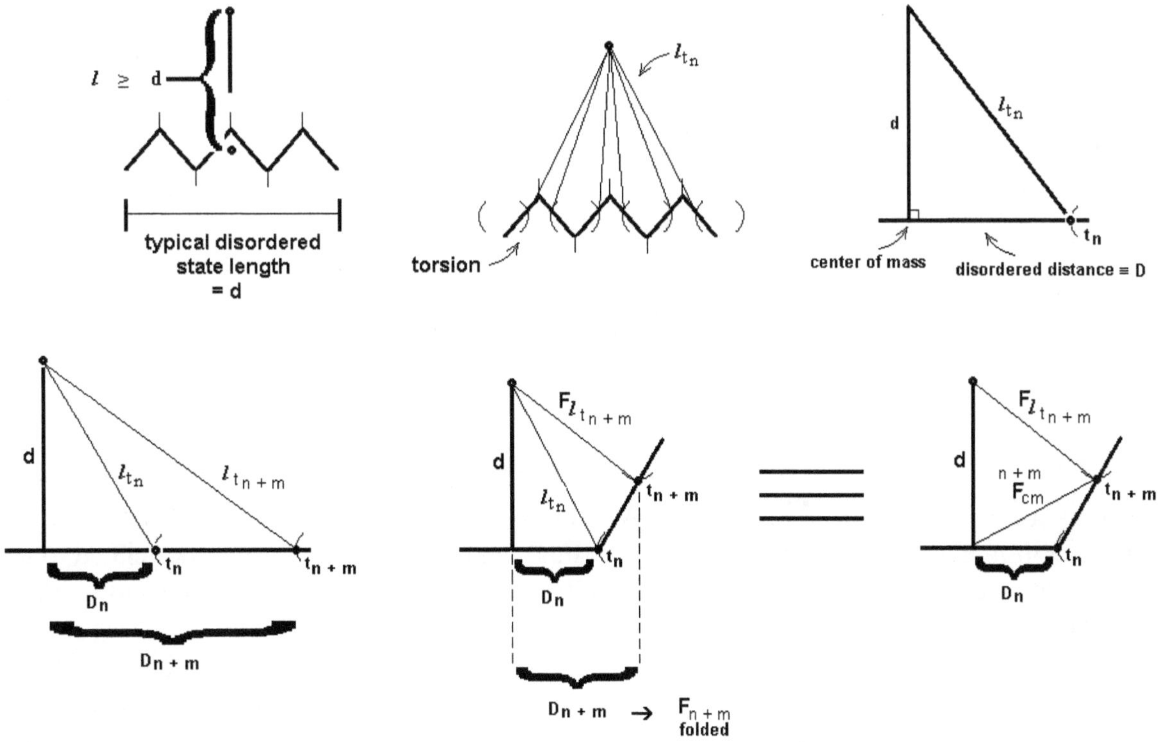

Constraint : ${}^{(n+m)}F_{CM}$ = F_{n+m} (dist. from CM)
compound dist. along disordered CM line

Therefore, F_{CM} is a radius of a sphere (for each "t" unit).

Why this constraint?

Because the polypeptide atoms can only replace water molecule positions. (The a.a.s will only move to replace waters, to increase water entropy.) Assume that if all the "t"s occur at the same time, the CM is maintained due to the expulsion or re-positioning of waters.

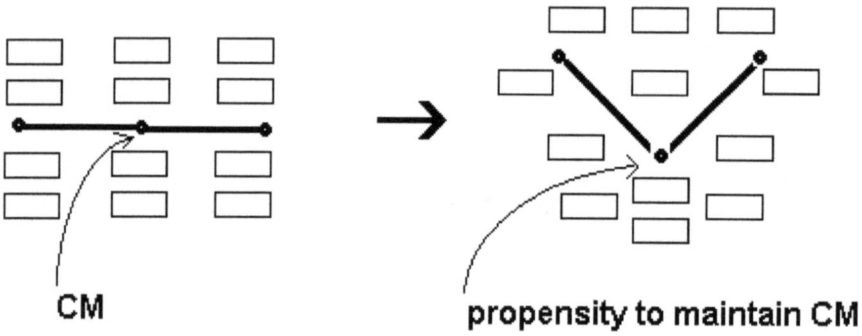

A scenario is that a torsion (from 0 to 180 degrees) reduces the radius (about the CM), and a torsion ($180^0 \rightarrow 360^0, 0^0$) restores toward the (disordered) radius. Thus, there could be a competition here (to retain the disordered radius), and this competition can be related to not only both sides of the torsion point, but also to competitions at all torsion points affecting the particular torsion point considered (or examined).

Implausibles : part 2

Assignment of torsional qualities (per torsion point):

1) The # of waters displaced by a torsion (of side-chain positioning) is analogous to the absence of water for the side-chain; therefore, each side-chain may be assigned a water molecule per charge and/or polar interaction. The absence of water indicates the displeasure of water being there such that an entropic (thermodynamic) emphasis or propensity is favored to remove it.

2) Each torsion has a pre-influence and a post-influence, these influences being conducted by pre (or prior) and post (or further) torsions to be conducted. A competition is then set up between the pre- and post- upon the torsion considered, to enhance or attenuate it by the sum of influences. Thus, if a torsion is given as "t," it may be increased or diminished by the sum of $t_{pre} + t_{post}$, where an enhancement is to favor the t direction and attenuation is to counter (or contradict) this direction. The modification thus changes the disordered radius (from or to the CM) for the position of the unit torsioned with (i.e., side-chain and a.a. residue). This solution can then be accomplished simultaneously for all of the torsions, although the temporal component of the changes can vary from instantaneously (or immediately) to very slow to not adopted (restricted).

3) Since waters counter charges and polar influences, the propensities of side-chains to be neighbors (and allow maximized torsions towards that aim) amount to relative hydrophobicities of the side-chains. Waters, therefore, act as "buffers" to the torsional changes. The waters associated with (side-chain) charges or polar attractions represent extreme cases of this "buffering" from torsion. [The entropic disordering of the water molecules allows for the hydrophobic ordering of the proteiin folding.]

P-gradient Theory

The relationship between side-chain hydrophobicity and protein folding presents an interesting correlation to the torsional event as being due to a "gradient" of hydrophobicity among the immediately involved residues:

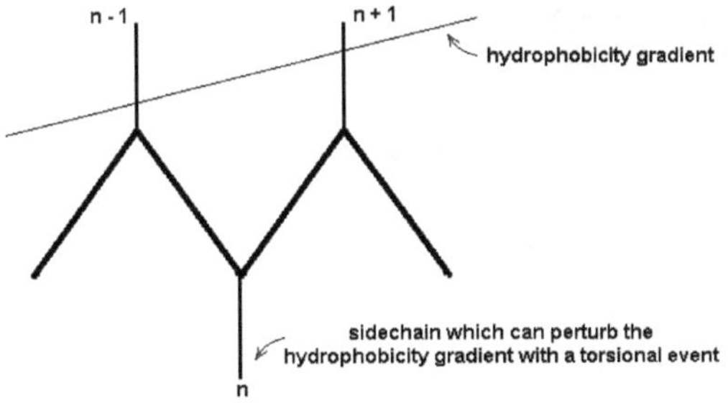

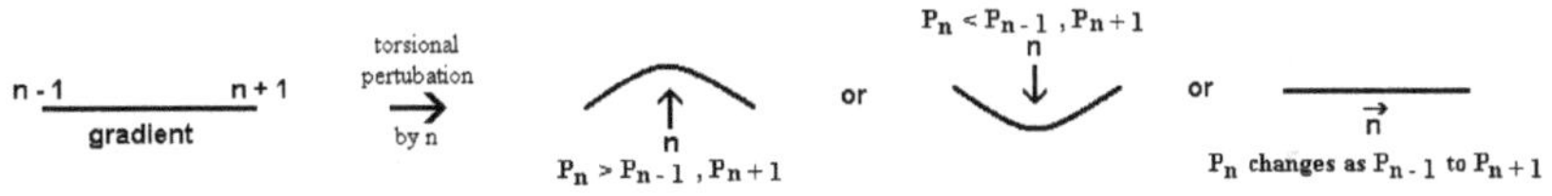

Therefore, this torsional perturbation (P) has aspects of a parabolic maximum or minimum; i.e., (reduced to 2 dimensions) the gradient becomes a parabola (due to the torsional perturbation of residue n).

Thus, given the hydrophobicities of the three residues (under given solvent conditions) as P_{n-1}, P_n, P_{n+1} : P_n occurs at a max or min (or saddle or inflection); the three points define a parabola (or a circle, etc.).

Entropically, the least perturbation would be favored: i.e., P_n changes as P goes from P_{n-1} to P_{n+1} suggests of gradient. This means either $P_{n-1} =< P_n =< P_{n+1}$ or $P_{n-1} => P_n => P_{n+1}$ would be most favored, as $P_n = (P_{n-1} + P_{n+1})/2$ therefore follows as an average point.

Therefore, a divergence of P_n from $(P_{n-1} + P_{n+1})/2$ modifies the (regular) torsion due to n :

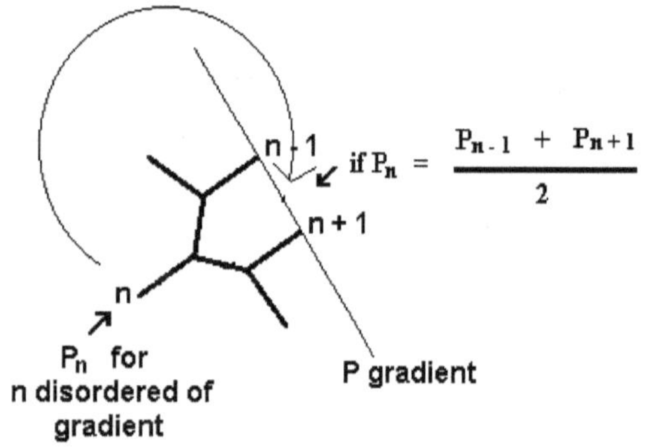

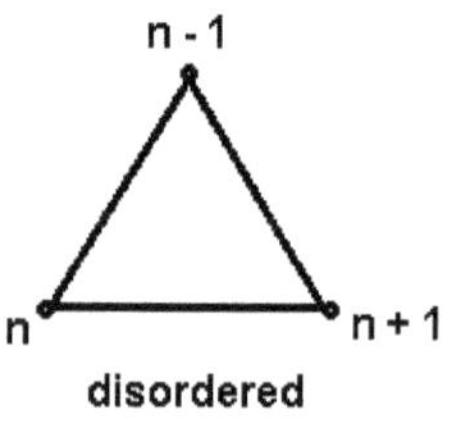

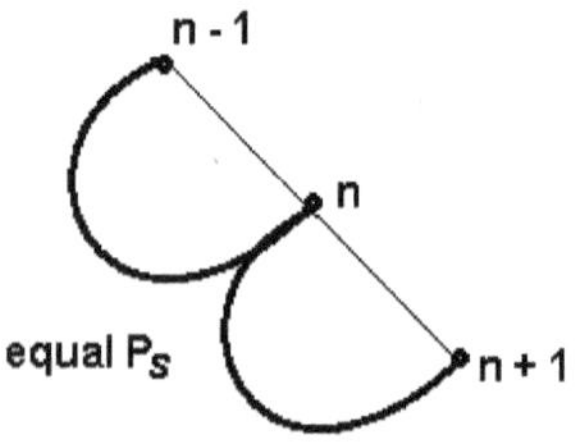

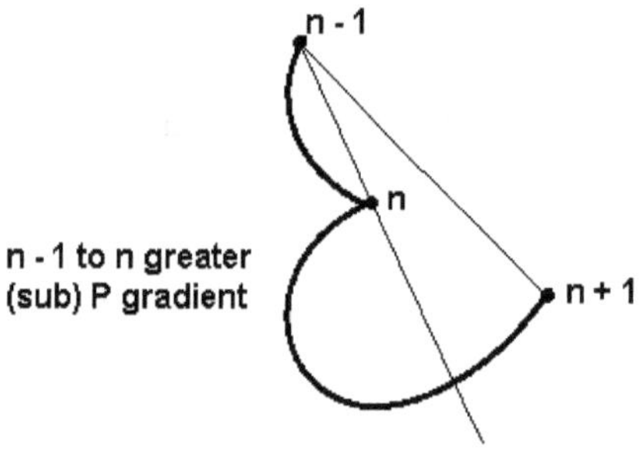

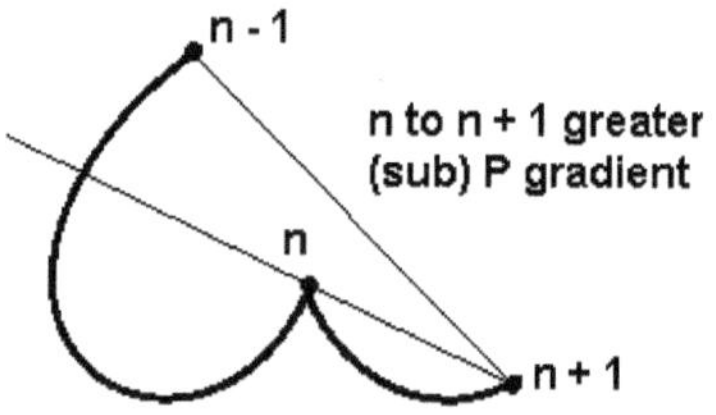

A clear implication is that the (n) torsion point is related to the (peptide) ϕ and ψ angles about it:

e.g.

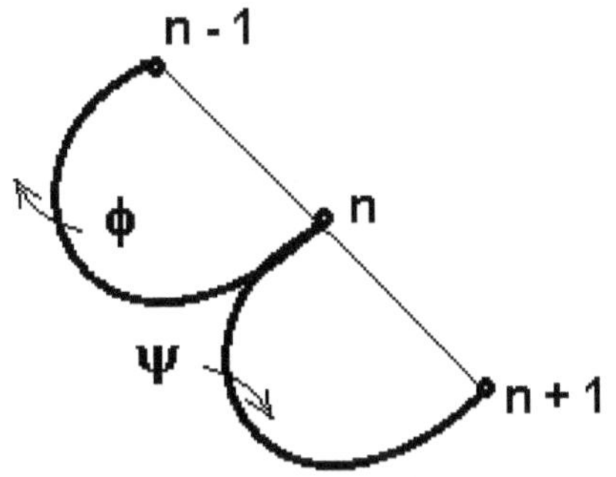

The restriction of "simultenaity" (or being simultaneous) establishes several parameters:

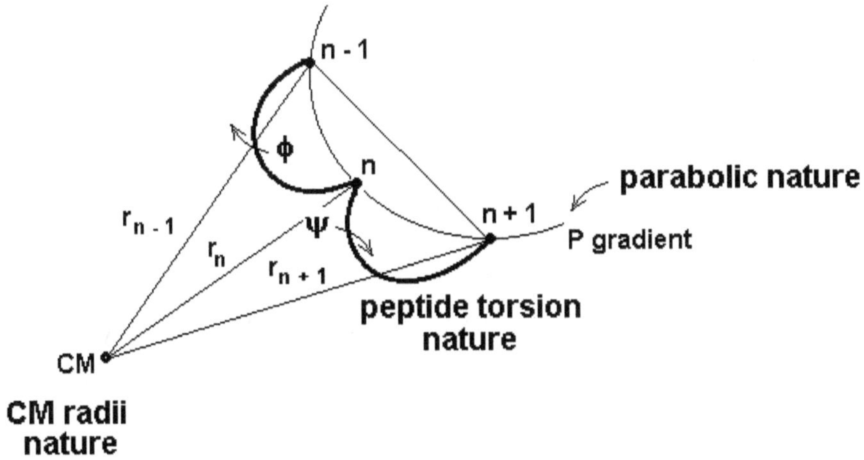

(an idealism)

When $P_n = \dfrac{P_{n-1} + P_{n+1}}{2}$,

n is at the vertex of the parabola.

Map of P gradients:

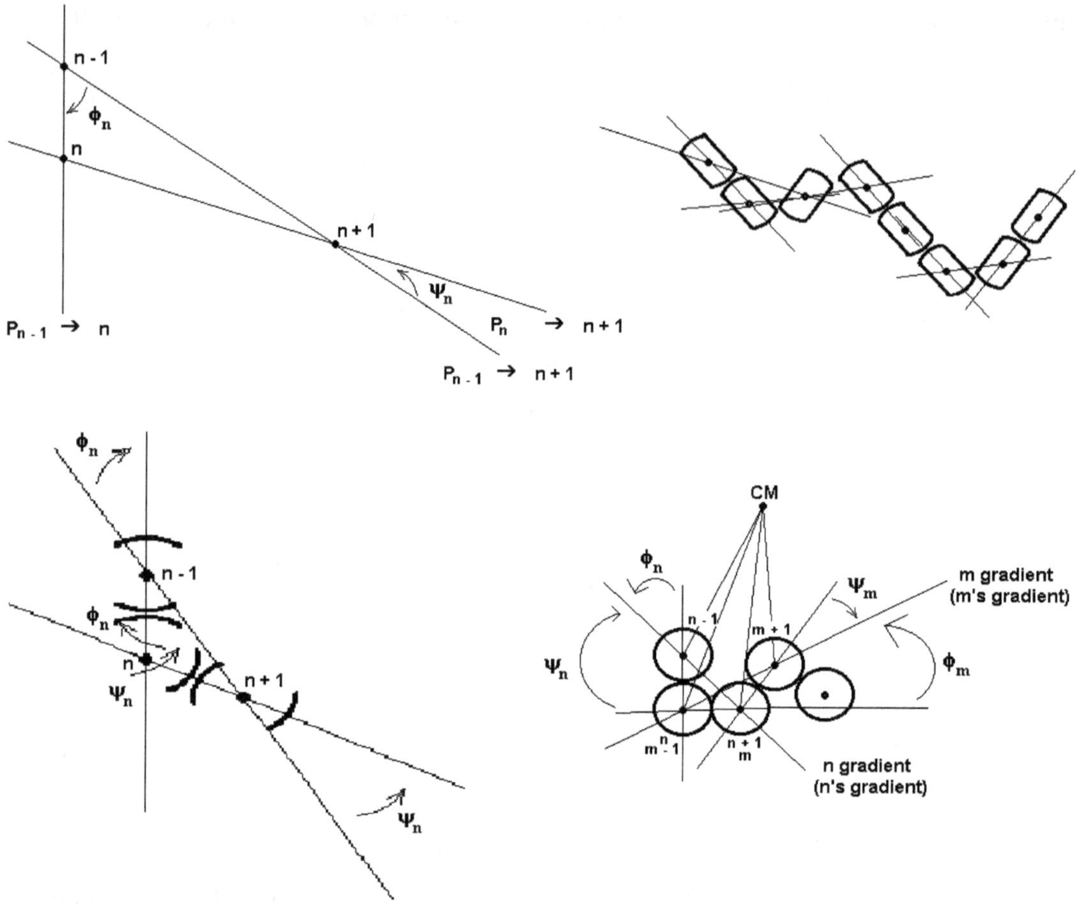

The radii (from CM) must change from disordered state to folded peptide.

We note that, upon "scalping" (scaffolding) a parabola relation to the problem, we have a sufficiency of 3s as variables or quantities for the standard-equation formula:

$$y = ax^2 + bx^1 + cx^0$$

n-1, n, n+1

P_{n-1}, P_n, P_{n+1}

r_{n-1}, r_n, r_{n+1}

x^2, x^1, x^0

ϕ_n, ψ_n, ζ_n

a, b, c

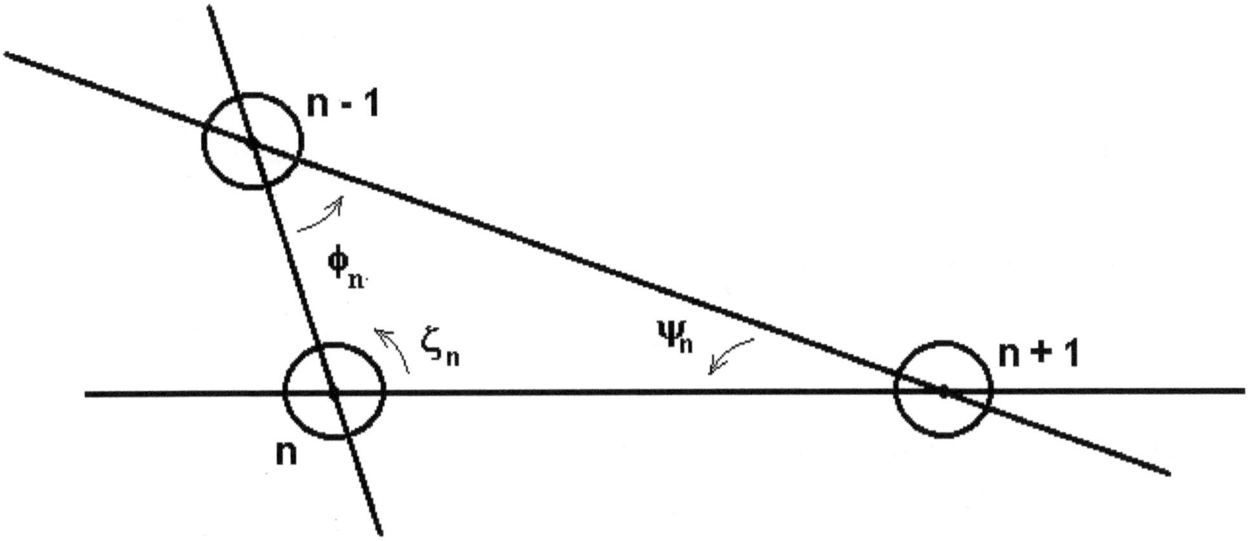

zeta: $\zeta + \phi + \psi = 180^0$

But, physically, what do $(\zeta, \psi, \mathrm{CM})$ mean?

To remove the (idealistic or improper) x^0 term (where $x^0 = 1$ by definition) we can multiply the formula by x, causing a cubic term x^3:

$$x\,(\,y = ax^2 + bx + c\,) \rightarrow yx = ax^3 + bx^2 + cx^1$$

This does not change the parabolic character of the relation. But we see that if y is of c in kind (e.g., units), then we have two terms of like kind: yx, cx (x must remain an independent variable.)

$$\therefore\; yx - cx = ax^3 + bx^2 = x\,(\,y - c\,) = (\,y - c\,)\,x$$

and we have, in effect, 3 coefficients of x:

$$(\,y - c\,)\,x^1$$

$$b\,x^2$$

$$a\,x^3$$

Forcing y to be like c, we can make x == P and have y == r, plotting (r vs x).

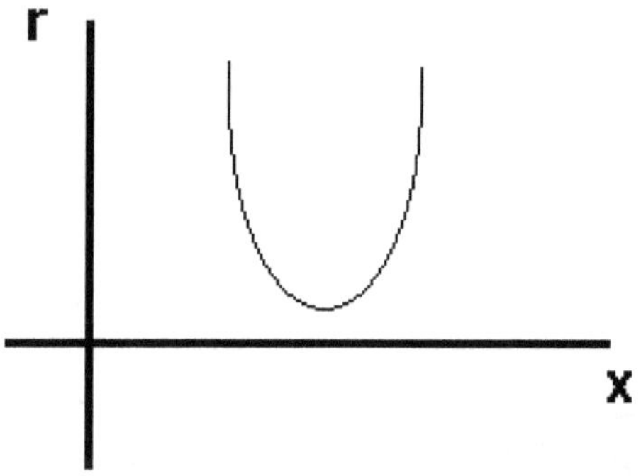

Therefore, c == original r and y == new r.

This yields the equations:

$$y_{n-1} P_{n-1} = aP_{n-1}^{3} + bP_{n-1}^{2} + r_{n-1} P_{n-1} \quad | \; n - 1$$

$$y_{n} P_{n} = aP_{n}^{3} + bP_{n}^{2} + r_{n} P_{n} \quad | \; n$$

$$y_{n+1} P_{n+1} = aP_{n+1}^{3} + bP_{n+1}^{2} + r_{n+1} P_{n+1} \quad | \; n + 1$$

Here we seem to have 5 unknowns (y_{n-1}, y_n, y_{n+1}, a, b) for 3 equations. But idealistically, since n's positioning is dependent on n-1 and n+1, we can force the influence of n's hydrophobicity as to make y_{n-1} and y_{n+1} related to a and b.

Therefore, dividing by P :

$$y_{n-1} = aP_{n-1}^{2} + bP_{n-1} + r_{n-1}$$

$$y_{n+1} = aP_{n+1}^{2} + bP_{n+1} + r_{n+1}$$

[reducing the problem to two equations with two unknowns (a, b)]

Thus, $y_{n-1} = r_{n-1} (P_{n-1} / P_{n})$, $y_{n+1} = r_{n+1} (P_{n+1} / P_{n})$

This must employ nul/los parameterization since P_n can not be zero; then hydrophobicity must begin positively (of scale), perhaps by an exponential or asymptotic (to zero) function.

So, $y_{n-1} = r_{n-1} (P_{n-1} / P_n) = [aP^2 + bP + r]_{n-1}$

$y_{n+1} = r_{n+1} (P_{n+1} / P_n) = [aP^2 + bP + r]_{n+1}$

yields a and b for (as a calculation): $y_n = [aP^2 + bP + r]_n$

Thus, y_n need not equal r_n (based on the a, b influences of P_{n-1} and P_{n+1}). We find, then, that while the function (y, P) is parabolic, the relationship (a, b) is all linear (of variables).

[Or more conventionally: $(P,y) = (P, f(P))$ curved $\rightarrow$ (a,b)]; { (P,y) | a, b }

We have mapped a parabolic function to a linear variation. The conditions to be considered, then, are (for hydrophobicities):

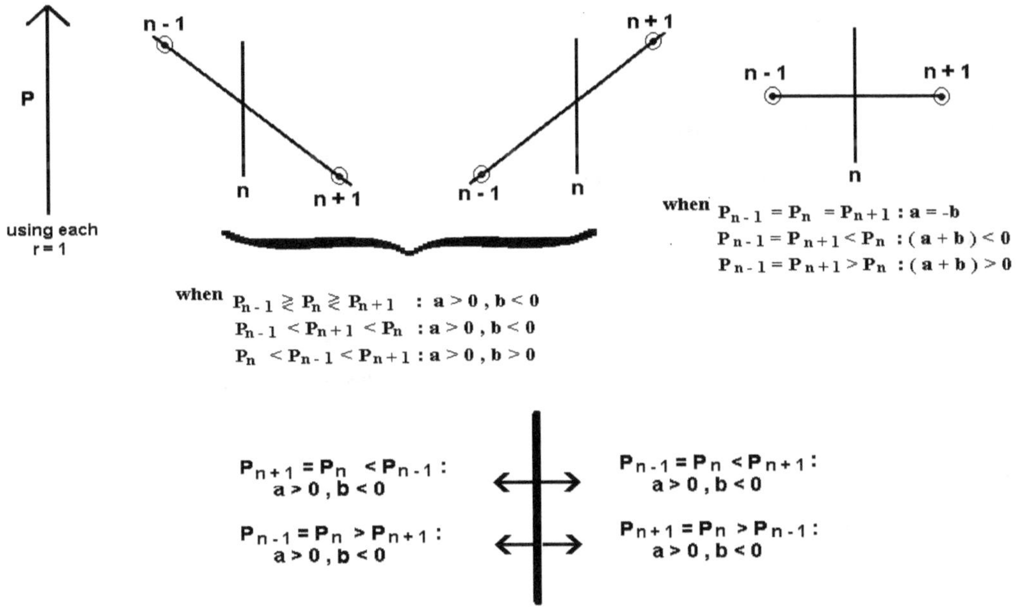

Therefore, when P_n is within the range of P_{n-1} to P_{n+1} : $a > 0$, $b < 0$.

When P_n is outside the range of P_{n-1} to P_{n+1}: $a > 0$, $b > 0$ or $b < 0$.

Thus, a is always positive; and b is always negative unless $P_n < (P_{n-1}, P_{n+1})$.

$$a = (-bP / P^2) + ((y - r) / P^2) , a > 0$$

$$a = -(1 / P)b + ((y - r) / P^2$$

(b,a) does not distinguish between [P_{n-1}, P_{n+1}] or [ϕ, ψ] (from P_n's perspective), as one would expect for a radius.

Note:

1) relationship is idealistic (due to the divisions)

2) P must not be zero

3) slope is negative

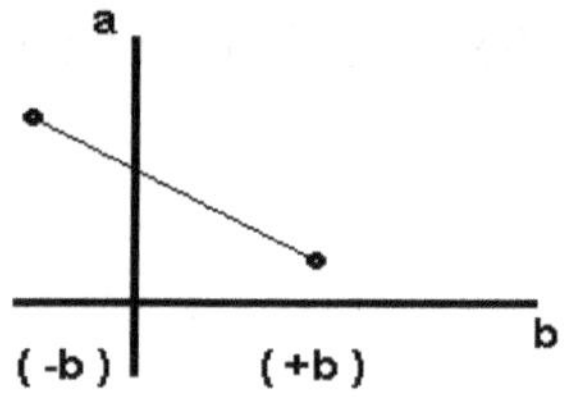

4) a is dependent on manipulation of b (in this mode)

5) slope is minus the reciprocal of the hydrophobicity

6) when y = r (as for y_n), the P^2 influence drops out

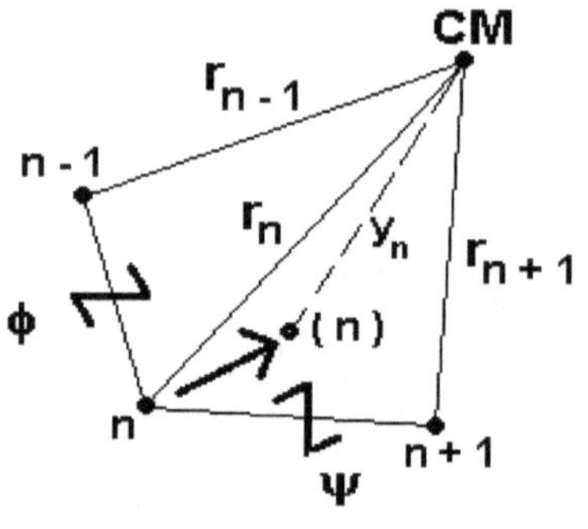

$\Delta\phi$ is related to $[r, y]_n$ and r_{n-1}
$\Delta\psi$ is related to $[r, y]_n$ and r_{n+1}

Let (empirically or intuitively):

$$\phi_{y_n} = \phi_{r_n}\left(\frac{y_n - r_n}{r_{n-1}}\right) + \phi_{r_n}$$

$$\psi_{y_n} = \psi_{r_n}\left(\frac{y_n - r_n}{r_{n+1}}\right) + \psi_{r_n}$$

$$\therefore\ (\phi, \psi)_{y_n} = (\phi, \psi)_{r_n}\left(1 + \frac{y_n - r_n}{(r)_{\substack{n-1,\ n+1 \\ \phi \quad\ \psi}}}\right)$$

This is a same (syn-progressive) unidirectional argument, where the change in $\Delta(\phi, \psi)$ rates establishes the relative relationship between ϕ and ψ .

But (ϕ, ψ) are angles, while radii are distances. Perhaps the radii should be converted to arc lengths by: $s = r\theta$; $\theta = \phi, \psi$

Therefore, as for (s / r), the dimensionless (used as like $r = 1$) fraction $(y_n - r_n)/r$ can be multiplied by 2π (or 360^0).

Therefore,

$$(\phi, \psi)_{y_n} = (\phi, \psi)_{r_n} \left(\frac{1 + 2\pi(y_n - r_n)}{(r)_{\substack{n-1,\ n+1 \\ \phi \quad \psi}}} \right)$$

The hydrophobicity scale, for all residues or a.a.s, must begin at $P > 0$. This is evident for peptides since all a.a.s have (start with) the same peptide bond forming components and the same structural frame for an amino acid (imino as modified for proline). We might, in a simplistic but straightforward way, construct such a scale based on entropic arguments for the disruption of the organized hydrogen bonding clusters of water needed during the protein folding, where the entropy loss from the disordered peptide being folded is (overly) compensated for by the water cluster disruptions. This is as we consider each atom (of the a.a.) for its hydrogen bonding capabilities, each atom being the grouping of the heavy (non-H) atom and its H constituent (when present).

Let us then consider each atom (or grouping) by its atomic number (or # of electrons), and reduce this by each polar interaction or charged interaction, where a polar interaction is worth -1 while a charged interaction is an abstraction and so is worth the entire electron weight of the water molecule.

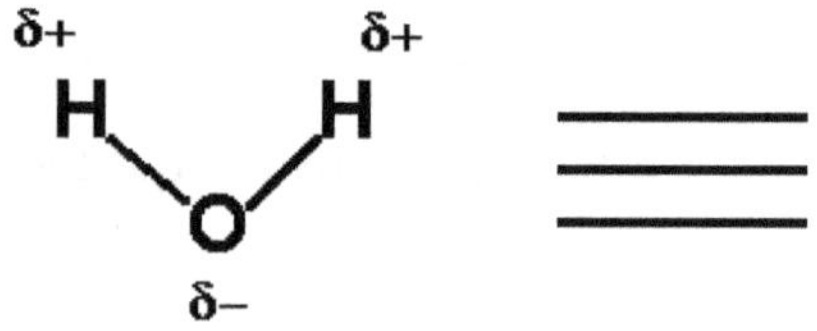

-(8 + 1 + 1) = -10
value for water abstraction

e.g., for

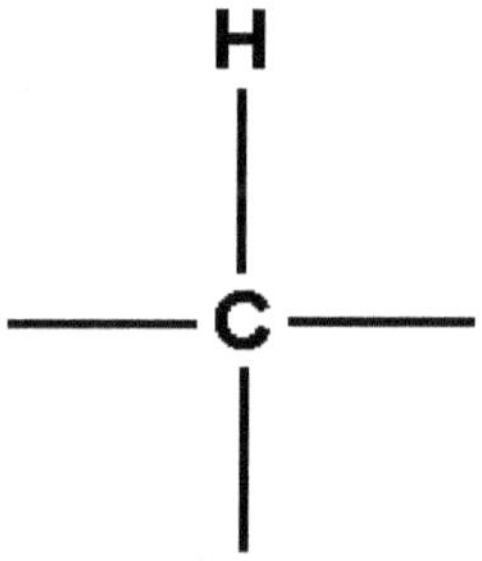

the value of this carbon atom (grouping) is:
6 - 1 polar interaction = +5 for P
+1 hydrogen **+1**
7 **6**
for atom
grouping P

for

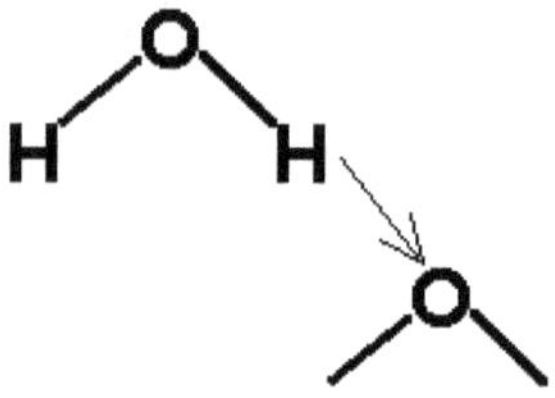

the value of this oxygen atom is:
8 - 10 water abstraction = -2 for P

For glycine (in peptide chain): it has same P value (26) as for a base P value (26; i.e., without R group interactions).

for glycine (in peptide chain)
-- has same P value (26)
as for base P value;
i.e., w/o R group (interactions)

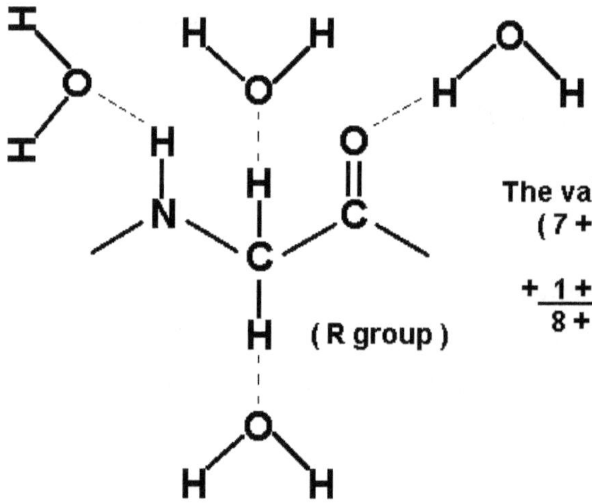

The value for this residue is:
(7 + 6 + 6 + 8) - 4
polar interactions
+ 1 + 2
8 + 8 + 6 + 8 -4 = 26 for P.

The base (starting) value for each a.a. (except proline), then, is

26 – 1 hydrogen + 1 polar interaction (glycine R group; interactions cancel) = 26

For proline (in peptide chain) :

For proline (in peptide chain)

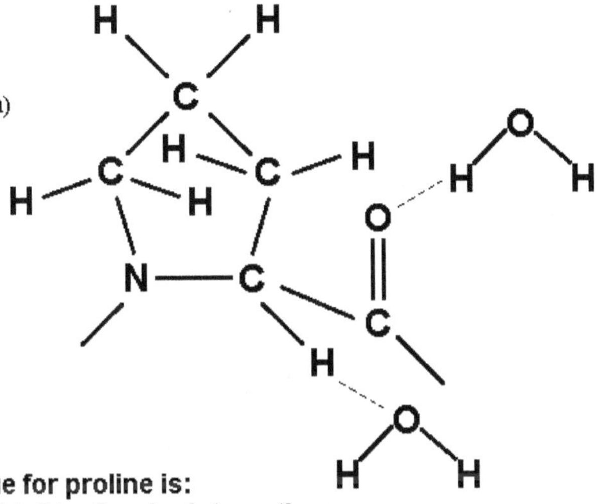

8 polar interactions are possible.

The value for this residue is:

(7 + 6 + 6 + 6 + 6 + 6 + 8) - 8
+ 2 + 2 + 2 + 1 **polar interactions**

7 + 8 + 8 + 8 + 7 + 6 + 8 - 8 = 44 for P .

The base value for proline is:

(7 + 6 + 6 + 8) - 2 polar interactions
+ 1

7 + 7 + 6 + 8 - 2 = 26

When glycine starts the peptide chain, as the amino end, under neutral or acidic pH the amino group will have a positive charge:

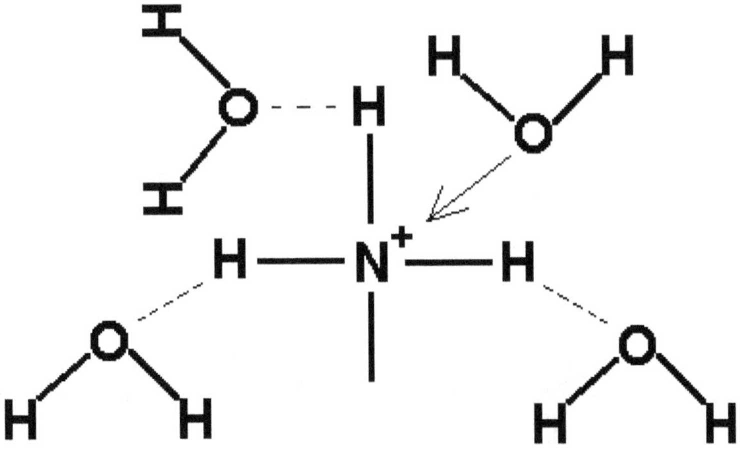

There will be one abstraction (of water molecule) and 3 polar interactions (for the amino group). So the net value for that residue is:

Hydrogen polar interactions
↓

$(7 + 6 + 6 + 8) - 5 - 10 - 1$ ↙ **carbonyl polar interaction**

$+3 \; +2$ ↳ **water abstraction**

$$\overline{10 +8 +6 \; +8 \quad -5 \; - 10 - 1 = 16}$$

Therefore, this is the lowest P value for all the residues of a peptide, and (idealistically) all values can be normalized to this number (16).

Note that all **a.a.** hydrogen to water polar interactions essentially cancel each other out (for P value).

Therefore, (the) value is $(7 + 6 + 6 + 8)$ - 1 (carbonyl polar interaction) $- 10 = 16$.

$(7 + 6 + 6 + 8)$ - 1 is the base value for all internal **a.a.**s $= 26$ (also for proline).

So this can be called a "hydrogen equivalency" scheme.

"Simultenaity" is established when all calculations can be done at the same time (as opposed to ordered influences).

Spontaneity is due to the entropic arguments. This amounts to the disruption of (ordered) water clusters due primarily by water abstraction being greater (in gain of entropy) than the folding of the protein (with loss of entropy). This is, then, due to carbonyl oxygen abstractions (of main-chain backbone or peptide bond carbonyls), amino and carboxyl ends being charged for water abstractions, and charged R groups. Thus, it is due to charged interactions (of abstraction). At neutral pH, when both ends of the peptide are charged, the longer the peptide (with the greater chance of charged side-chains to occur) the more chances of abstraction countering oppositely charged ends of the peptide(s) interacting with each other. This supposes, then, that a peptide with interacting ends and all non-charged internal and end residues remains disordered, although intra-ring closure is a form of ordering. The same can be said of inter-chain linkages.

CIW Effect

But how does the water abstraction increase the entropy more of water (via greater disruption of water clusters) than decrease the entropy of the protein folding? Typically, macromolecules are screened from electrical effects by counter ions (from salts) forming a 'cloud' around them. So we actually have a **c**arboxyl (or amino)-counter **i**on-**w**ater structure to consider: CIW effect.

There are many more counter ions than protein charges, and many more water molecules than counter ions. Therefore, a small change in C (carboxyl or carbonyl, or protein charge) must cause an amplified change in I (salt ions, used in screening), which must cause an even more amplified change in W (water molecules) as far as for the disruption of water clusters. That is to say (or suggest) water clustering is very sensitive to salt presence (and concentration). So, as a postulated example, the loss of one C charge (to water abstraction) frees one counter ion; but the counter ion now increases the ion concentration to effect a greater disruption of water cluster(s) than the C could do alone. This might be related to the physical surrounding (e.g., "coating the clusters") of ions to water clusters, the volumetric (volumeric) geometry of which causes (a relative) few ions to totally disrupt a cluster of many more waters.

Cycles of Folding

How many cycles of "folding" must our method go through to reach a final (folded) peptide chain?

If we consider one cycle per (calculated) CM (i.e., the CM must be recalculated for each cycle), we see that a final or non-changing structure can be derived when $a = b = 0$. But this would insist that all residue hydrophobicities be the same (or become the same), so that the old radii do not change (in extent) from the new. This can not be true for even poly-monotonic peptides since here, while all of the residues are the same amino acid, one or both of the terminal residues must have different hydrophobicities from the (chain) internal ones.

From our empirical searches, we find that a is restricted to positive values, and that $a = -b$ when $P_{n-1} = P_n = P_{n+1}$. This is the same as saying: $aP^2 = -bP$. Again, the avoidance of zero seems key (and keen).

But ($aP^2 = -bP$) is broader of values for (b, a) than just (0, 0), or $-bP = -(y - r)$; and when all $y_n = r_n$, no further change is allowed (calculable).

This occurs when $a = -(1 / P)b$, which does not allow for $a = -b$ unless $P = 1$.

Of course, the CM would not change, but its fixture does not guarantee (lack of) radii changes.

Therefore, for the summation of terms, we have the following scheme:

1) Provide a "disordered" peptide with known (relative) coordinates of atoms and (thus) known (ϕ, ψ) angles, center of mass, and radii from CM to (presumably) each carbon α atom (c_α of each residue a.a.)

2) For each residue, calculate y_n and (therefore) $(\phi, \psi)_{yn}$ (i.e., new radii)

3) Apply $(\phi, \psi)_{yn}$ to each residue

4) Calculate new CM and (thus) new $r_{n-1,n,n+1}$ radii (not necessarily y_n)

5) Cycle back to step 2, continuously until $y_n \sim r_n$ and CM doesn't change

For step 4, why are not the new $r_{n-1,n,n+1}$ the same as y_ns ? Because of the divisions involved (in the calculations) the radii determinations are only idealistic (comparisons or conversions); but to be consistent of the method each new CM determines the (r) radii, while y_n determines (ϕ, ψ).

The charge – polar interaction should be weaker than the charge – charge interaction of ions. Therefore, to break the charge – charge interaction (to allow for folding) requires more energy than maintaining water cluster architectures; i.e., the freed salt ion is more disruptive to clusters than clusters are allowed to maintain the "disordered" peptide structure(s). Thus, some salt might (and may) help to initiate folding, by putting the disordered state in a condition favorable for the transition.

Water Sphere for Protein Molecule

But what is the CM that we are calculating (or following)? It is clear that it responds to the entire system of folding peptide with appropriated and adjusted and possibly surrounding (but stationary) waters. Might we then define (or insist, presume or estimate) the peptide to be within (structurally) an (isolated) involved water cluster? Then we can construct a hypothetical (roughly) sphere (spherical) of volume for its enclosure, the disordered peptide lying in the center plane (geometrically). The peptide, being mostly carbon, should on average be less dense than the equivalent volume of water (which is mostly oxygen).

If we define such a volume, the extreme of which has waters not affected by the protein's folding activities, the (by definition) center of the sphere (holding the disordered peptide) is found roughly in a central plane of lighter density. There, the system's (sphere's) center of mass is above (or below) this plane (favoring the greater mass of the water molecules collectively).

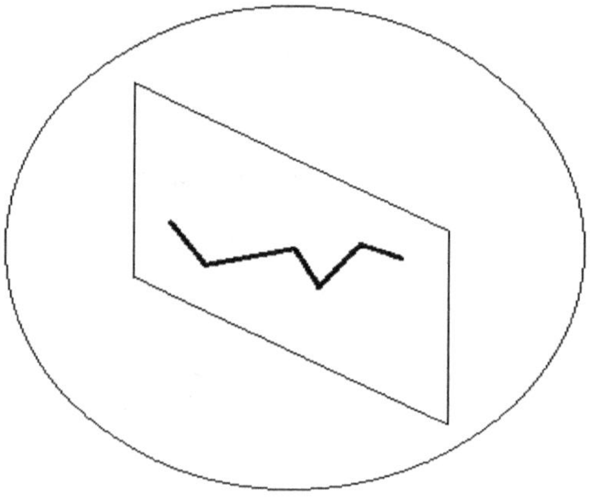

But where should the (spherical) boundary of unaffected waters occur (i.e., what mass of water is affected by the folding)? [Of course, the peptide has a greater molecular weight than any water molecule; but we are considering the waters in the sphere (surrounding the peptide) as a (super-)cluster or molecule (including the peptide) with its own CM.]

The (2X) radius of the sphere could perhaps be estimated by the extended length of the peptide (= #a.a.s x length of 1 a.a.) + 2 waters on each end of the extended peptide

(= 2 x 2 = 4 waters).

Therefore, 2 x radius = (#a.a.s)(a.a. length) + 4(water length)

(the extra water on each end representing the boundary of the unaffected water shell or surface).

This presents the sphere essentially as a collection (constrained) of water molecules with a slightly less dense center (restrained) of the peptide. The constraining yields to the water cluster, and the restraining yields to the peptide folding.

If we consider a water molecule volume not as of the atoms' involved, but rather as of a sphere of radius from the center of the oxygen (nucleus) to the (electronic) extent of a hydrogen (bonded to it), then the volume of the cluster sphere can be made up (roughly) of so many water "spheres" + the volume of the peptide:

> **cluster volume = CV / water "sphere" + peptide volume + "interstitial" (non-mass) volume** , where **CV / water "sphere" == # water molecules** (in volume) in the cluster.

Each water "sphere," then, represents a mass (constituent of the water cluster). Energetically, then, we note there are two manners of cluster volume:

1) without peptide (in the cluster), the interstitial volume is optimal == IVo

2) with peptide (in the cluster); (with "residual" IV symbolized as IVr)

> if IVo >= peptide volume + "residual" IV, then peptide can remain unfolded;
>
> if IVo < peptide volume + "residual" IV, then peptide must (re)fold to draw IVr towards IVo.

In this (IV) model, peptide (volume) is not replacing water volume. Deformation of the water "sphere" (towards that of the actual peptide molecule shape), by ions, temperature, and the nature of hydrogen bonding for the cluster formation, affects IVo and IVr. Each water "sphere" amply represents a probability of orientation for the water molecule.

This is reminiscent of the "phase compulsion" in x-ray crystallography, since each water "sphere" has a determinable phase for orientation of its water molecule, and yet, within the cluster each sphere is associated with a positional (x,y,z) coordinate (i.e., oxygen center) as for the asymmetric unit of a crystal. But here the convolution is for the folding of a protein.

Thus, we can adopt to a probability volume relation:

$$\underset{\text{cluster volume}}{CV} = \sum_{i=1}^{n} P_i(V_{ws}) + \text{peptide volume} + I_{V_r}$$

↑ probability of orientation; ↳ typical water molecule volume (water "sphere"); I_{V_r} interstitial residual volume

where n = # water molecules in the cluster sphere (as defined).

Therefore,

$$CV - \sum_{i=1}^{n} P_i(V_{ws}) = \text{peptide volume} + I_{V_r} > I_{V_o}$$

for protein folding

$$CV - I_{V_r} = I_{V_o}$$

(no peptide present)

$$\therefore I_{V_r} = I_{V_o}$$

$+\Sigma$

Combining (conceptually):

$$CV - \sum_{i=1}^{n} P_i(V_{ws}) = \text{peptide volume} + (CV - I_{V_o})$$

$$-\sum_{i=1}^{n} P_i(V_{ws}) = -I_{V_o} + \text{peptide volume} - \Sigma$$

$$\therefore \sum_{i=1}^{n} P_i(V_{ws}) = I_{V_o} - \text{peptide volume} + \Sigma$$

$$\therefore \text{PepV} = I_{V_o} \quad \text{or} \quad \text{PepV} \gtrless I_{V_o} \quad (> \rightarrow \text{folding}; \; < \rightarrow I_{V_r}\text{, no folding})$$

Assuming that (for a peptide chain of any considerable size or length) it (the condition or state) starts out that PepV + IVr > IVo, then the "energy" (or imperative) for folding comes from the Pi(Vws) terms since this derives from Pi reducing from "1.0" to a minimal value (corresponding to the water molecule "frozen" to a definite orientation and position in the cluster). Therefore, the space (or the room) for folding is provided by the lowering of some (Pi)s. But, entropically, the sum of Pi(Vws) terms must increase to provide for the folding. This increase in the Pi(Vws) sum is equivalent to a disrupting of the order of the water cluster (i.e., "heating" of water molecules).

$$\therefore \text{ as } \quad \Sigma P_i(V_{ws}) \uparrow \quad , \quad (CV - \Sigma P_i(V_{ws})) \downarrow$$

$$\therefore (Pep + Iv_r) \downarrow \quad \text{bringing this closer to } Iv_o$$

Therefore, in a sense, all change of state (here water cluster integrity and peptide folding) is dependent on changes in Pi.

PepV might be considered a constant (for the extended chain), but it is also related (and thus variable) to the 3D geometry of the folding (e.g., reduction in surface area due to compaction). Some IVr might also be incorporated into the PepV (trapped waters within the folded peptide), to reduce IVr (apparent). Thus, IVr might contribute to PepV (without altering PepV), to cause its reduction.

Manner of Water Disruption, for Peptide Folding

Pi is a useful fudging factor, for the model. But what is the reality of the situation?

1) Pi must be definite and determinate, for each water directly associated with any particular stage of the peptide folding.

2) For such determinate waters, their (Pi)s are minimal, to yield waters of definite orientation and position/location (i.e., frozen).

3) As some waters become definite, others must compensate by becoming indefinite, with increasing (Pi)s, thereby increasing the overall sum (ΣPi(Vws)) as an amplification (structurally) of effects (i.e., over-compensation).

4) The energy for the over-compensation is entropic (i.e., as like "$T\Delta S$"): temperature and therefore movement (or adjustment) of water molecules increases, and they "thaw" transiently within the cluster (at each stage of folding, each stage defined by the participation of particular "frozen" waters associated with the peptide).

5) The over-compensation is reflected by:

$$(Pep + lV_r) \downarrow \quad \rightarrow \quad lV_o$$

6) The enthalpy of the cluster is apparently increased (due to "thawing" of water molecules; loss of enthalpy or heat content) bringing the cluster to the lower energy state as result of the peptide folding (i.e., the release of energy due to the adjustment of water molecule positions, and the increase in entropy, still allow for an exothermic transition, until IVo is reached).

But these postulates are merely analogies of thermodynamics. The effects of Pi can be seen directly by the established relationship: as Pi decreases, the available volume for PepV increases during folding. Before the folding initiates, it is assumed that Pi is at a minimum for each water. During folding, for some waters, Pi increases toward "1.0" as the water molecules are "heated" up. This is as the cluster volume is maintained.

The Reaction Pool

A fundamental limitation of traditional mathematics is that it limits the expression of a relation to a 2-dimensional coordination (i.e., equation) such as $a + b = c$, whereas a more accurate (or truer) description could be multidimensional:

$$\begin{array}{c} a + d = c + e \\ \| \\ +b \end{array} \qquad \begin{array}{l} \text{where } a + d \neq a + b \\ \therefore a + d + b \neq a + d = \\ \qquad\qquad\qquad\quad \| \\ \qquad\qquad\qquad\quad +b \end{array}$$

(i.e., $+b$ is independent of the condition $a + d$).

We might try the notation:

$$a + d \longrightarrow c + e$$
$$+b$$

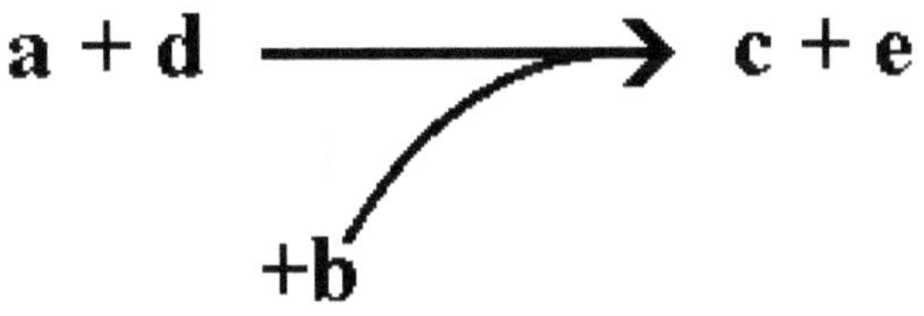

except this implies an addition of b to (a + d), as for a chemical reaction (rxn).

Therefore, one might prefer:

$$a + d = \,| \, c + e$$
$$+ b = \,|$$

with the symbol $\dfrac{=}{=}\,|$ **simplified to** $\supset\!\!\!=$

and then to $\succ$ **or** Σ **or** $>$ **to** $\overset{=}{>}$ **to** $\overset{-}{>}$

(too similar to Σ)

$$\therefore \quad a + d \overset{\displaystyle -}{} \Big\rangle \; c + e$$
$$+ b \; -$$

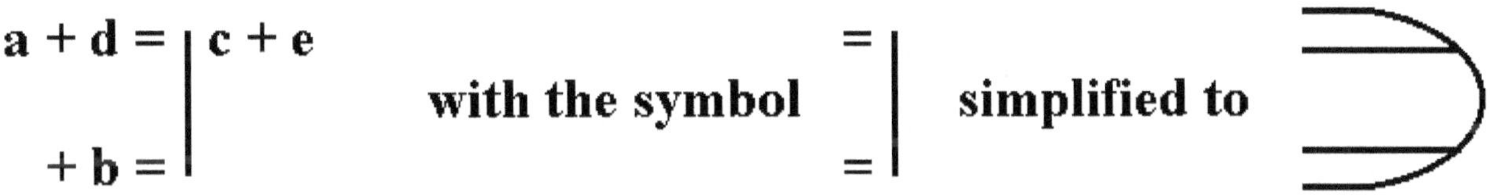

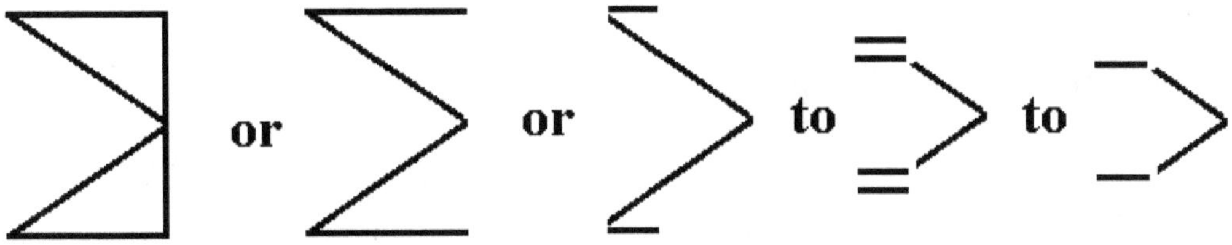

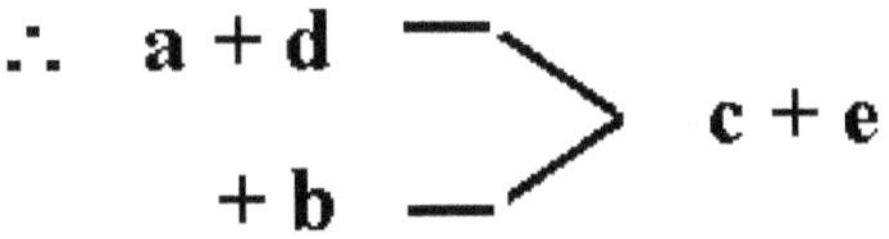

and then to

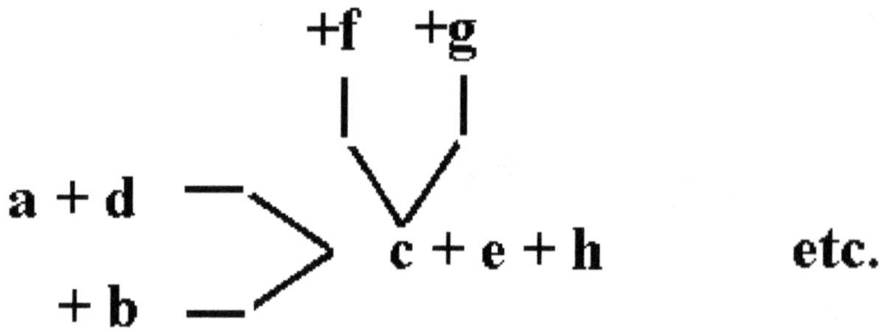

for more dimensions.

This might be simplified (or generalized) to

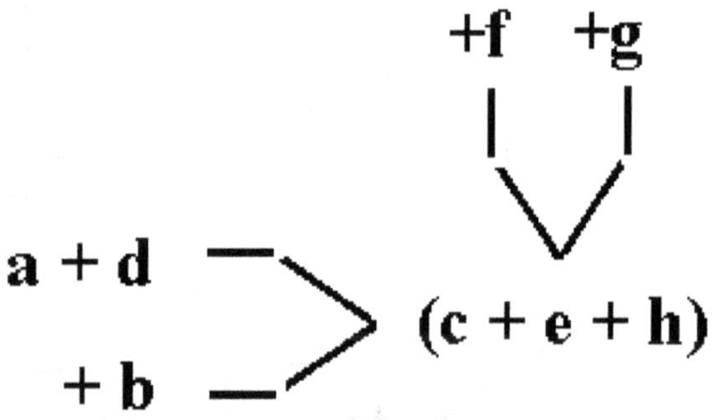

which represents (or can represent):

a + d = c

b = c + e

f = c + e + h

g = c + e + h } **f not necessarily equal to g ; e.g., they approach (c + e + h) from different directions or means/methods.**

f not necessarily equal to g; e.g., they can approach (c + e + h) from different directions or means/methods.

Summing (with c = b – e)

$$
\begin{array}{r}
a + d = c \\
b = c + e \\
f = c + e + h \\
g = c + e + h \\
\hline
a + d + f + g = b + 2c + e + h
\end{array}
$$

The summation is what is observed, for a more complicated component-ing of occurrences.

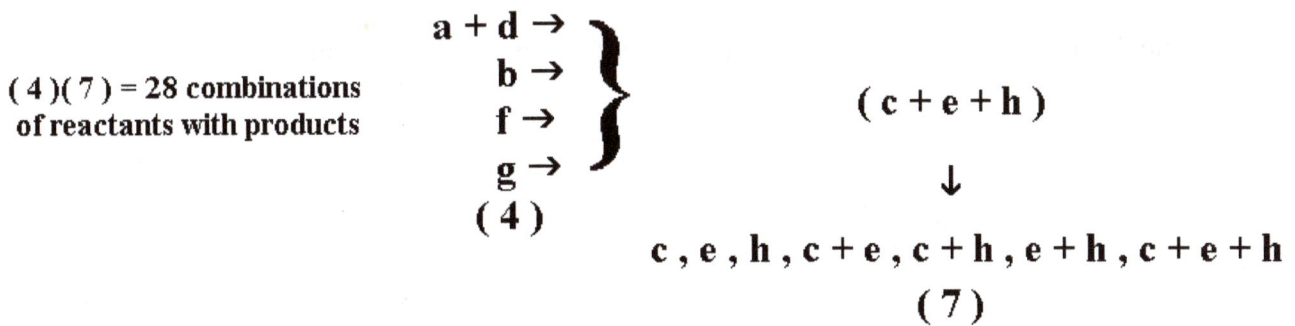

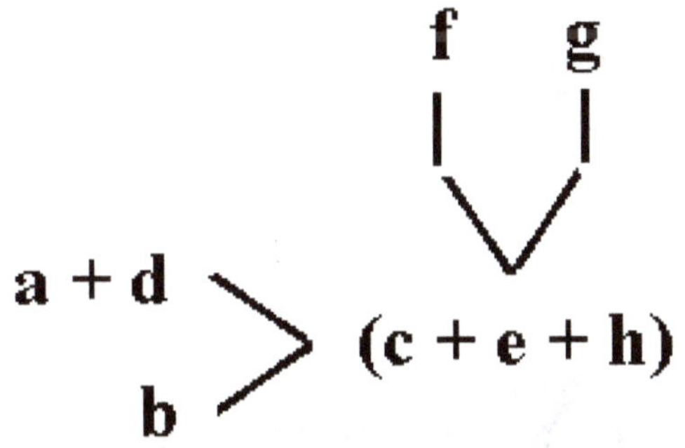

is of course a degenerate symbolization (it can represent different results, or a set of different combinations of results):

(4)(7) = 28 combinations of reactants with products

$$
\left.\begin{array}{r} a + d \rightarrow \\ b \rightarrow \\ f \rightarrow \\ g \rightarrow \end{array}\right\} \quad (c + e + h)
$$

(4)

$$
\downarrow
$$

c , e , h , c + e , c + h , e + h , c + e + h

(7)

We can call (c + e + h), or (....), the reaction pool, for products or results. Empirically, the number of possible results can be found (apparently) by:

n product terms	combinations, c	c =
1	n^o	1
2	$n + n^o$	3
3	$n + n^o + n^1$	7
4	$n + n^o + n^1 + n^2$	25

etc.; c is always odd (when using integers), until $n = 10$ (an inversion point?).

If we replace n^o with "1," the formula is:

$$c = n + 1 + \sum_{x=1}^{n-2} n^x \quad \text{for } n > 2$$

Therefore

n	c
1	1
2	$1 + n$
>2	$1 + n + \sum_{x=1}^{n-2} n^x \equiv 1 + 2n + \left(\sum_{x=1}^{n-2} n^x\right) - n \quad (\text{since } n = n^1)$

$$1 + 2n + \sum_{x=1}^{n-2} n^x \quad \text{for } n > 3$$

Note that an additive combination requires 2 or more products (or results).

Therefore

	n	c
1	1	1
2	2	$1 + n$
3	>2	$1 + n + 1 + n + \sum\limits_{x=1}^{n-2} n^x$
4	>3	$1 + 2n + 1 + 2n + \sum\limits_{x=2}^{n-2} n^x$

valid (since $n = n^1$)

valid
(most general relation, for $n > 2$)

(last relation needed)

The system can be given coordinates as $(r, P) == [r, P(r)]$.

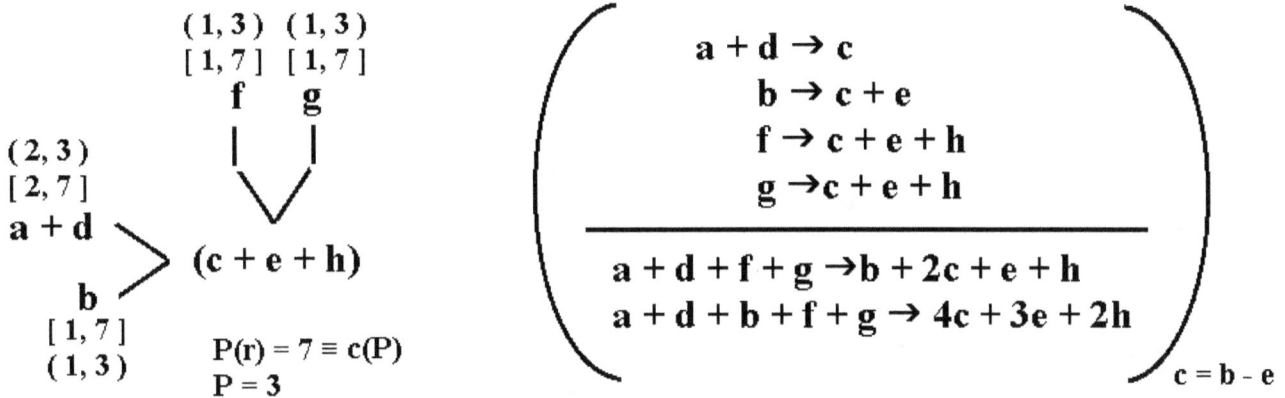

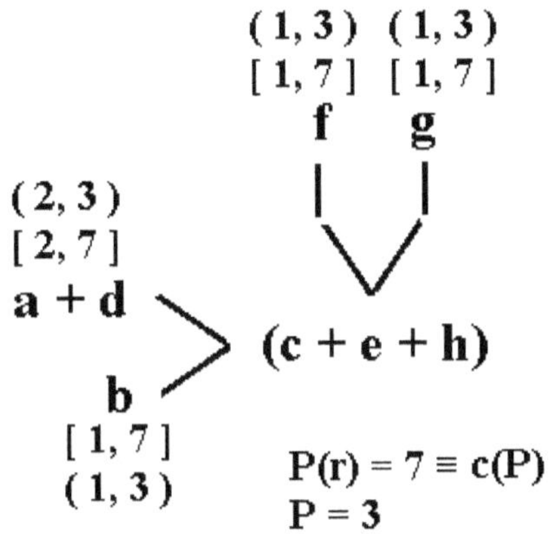

$$(1,3) \quad (1,3)$$
$$[1,7] \quad [1,7]$$
$$f \qquad g$$

$$(c + e + h)$$

$$(2,3)$$
$$[2,7]$$
$$a + d$$
$$b$$
$$[1,7]$$
$$(1,3)$$

$$P(r) = 7 \equiv c(P)$$
$$P = 3$$

$$\left(\begin{array}{l} a + d \to c \\ b \to c + e \\ f \to c + e + h \\ g \to c + e + h \\ \hline a + d + f + g \to b + 2c + e + h \\ a + d + b + f + g \to 4c + 3e + 2h \end{array} \right)$$

$$c = b - e$$

But the $c(r, P)$ combinations are of several types:

	$r \cdot P$	$r \cdot c(P)$	$c(r) \cdot P$	$c(r) \cdot c(P)$
$a + d$	$2 \cdot 3 = 5$	$2 \cdot 7 = 14$	$3 \cdot 3 = 9$	$3 \cdot 7 = 21$
b	$1 \cdot 3 = 3$	$1 \cdot 7 = 7$	$1 \cdot 3 = 3$	$1 \cdot 7 = 7$
f	$1 \cdot 3 = 3$	$1 \cdot 7 = 7$	$1 \cdot 3 = 3$	$1 \cdot 7 = 7$
g	$1 \cdot 3 = 3$	$1 \cdot 7 = 7$	$1 \cdot 3 = 3$	$1 \cdot 7 = 7$
sum	14 +	35 +	18 +	42 $\to$ 109
coordinate	(r, P)	$[r, c(P)]$	$[c(r), P]$	$[c(r), c(P)]$

Therefore, the 4 coordinates (of the reaction pool) are distributed among 109 combinations, assuming (r, P) and [r, C(P)] are equivalent representations (for example).

Accounting for the degeneracy in coordinates, there are 8 rxn. coordinates for 109 combinations.

We can also, of course, include consideration of a "reactant pool" :

$$(a + d, b, f, g) \rightarrow (a + d + b + f + g) \rightarrow r = 5, C_{(r)} = 161$$

r • P	r • c(P)	c(r) • P	c(r) • c(P)	
5 • 3 = 15	5 • 7 = 35	161 • 3 = 483	161 • 7 = 1127	
15 +	35 +	483 +	1127	= 1660

Assuming 1660 includes the 109 previous, with the coordinates increased by 4, a 50% increase in coordinates yields over 15 times as many combinations (here).

So, reactant pools and product pools lead to reaction pools, where the combinations are much greater than the number of reactants and products.

If one magnifies a (pure) water sample to the discrete presence of individual water molecules, one can fairly ask what (substance) occurs within the spaces between the waters (as well as within the "spaces" of individual atoms, as presumed). It is not vacuum (space), since that would imply negative pressure. Perhaps the "substance" relates to the number of combinations allowed for the system.

Let us, then, call this ("substance" of combinations) as a sort of pressure, the energy for which helps to maintain the system. This E_c (energy of combinations) might be able to be related to Pi (for the orientation of a water molecule adopted during the peptide folding process). The probability, P, is of course infinite (for orientation in the water molecule sphere). But the Pi would be distinct (for the peptide folding process), and might be discreet for the allowed orientations to be searched for (by the water). The discreetness applied to Pi is presumably due to the restrictions of ϕ and ψ allowed for the peptide backbone units (a.a.s). Probability is an idealization (like an intensive property). It is mathematical rather than real. Relativity theory eliminated a (need for a) medium for the travel of energy; yet (new) Quantum Mechanics reasserts such a medium, in the form of probability (a mathematical medium). The string upon which a wave travels has mass (i.e., a medium). The "string" of an electromagnetic wave consists of mathematical probability (from the postulates of QM). But perhaps a "real" medium still exists, and is subsumed by measurement (limits, physical, on magnification). This would allow for a (practical rather than idealistic) method for (explaining or demonstrating with insight) the phenomenon of "action at a distance." Protein folding, in a water solvent, might be such an illustrative example (within the water cluster sphere). The "medium mass" would consist of all the reaction pool combinations that may be sampled by the system: i.e., they must occur, each coordinate able to associate with (various) combinations (as being actual rather than virtual, or even virial). The actual (physical) system is, of course, incongruous of the magnitude or vivacity of the effort.

As we have several "reactant containments" in our example (a + d, b, f, g), presumably we can have several "product containments" : (c + e + h, …., ….). So, we may have sums (S) of containments: $S_i r == r_i$, $S_j P == P_j$, where each i has its own characteristic number for r and each j has its own characteristic number for P.

For our example, **i = 4** and **j = 1**	thus $\mathbf{S_{i>1 \rightarrow I+1=n} r_n}$, $\mathbf{S_j = 1 = n}$
$\mathbf{2 == r_1 = a + d , P_1 == 3 = c + e + h}$	We have:
$\mathbf{1 == r_2 = b}$	$\mathbf{r_1, r_2, r_3, r_4, r_5, Pi}$
$\mathbf{1 == r_3 = f}$	$\mathbf{C_{r1}, C_{r2}, C_{r3}, C_{r4}, C_{r5}, C_{Pi}}$
$\mathbf{1 == r_4 = g}$	for
But **i** (being **> 1**) can be supplemented by **1:**	$\mathbf{rP, rC_P, C_rP, C_rC_P}$
$\mathbf{5 == r_5 = a + d + b + f + g}$; thus **i = 5, j = 1**	

Also, conceivably, each reactant might react with itself and other reactants and each product might react with itself and other products.

Therefore, we have:

$\mathbf{r \cdot r}$	$\mathbf{P \cdot P}$	$\mathbf{r \cdot P}$	$\mathbf{r \cdot C_P}$	$\mathbf{C_r \cdot P}$	$\mathbf{C_r \cdot C_P}$
↓	↓	↓	↓	↓	↓
$\mathbf{C_r^2}$	$\mathbf{C_P^2}$	$\mathbf{\sim C}$	**r factored** $\mathbf{\sim C}$	**P factored** $\mathbf{\sim C}$	$\mathbf{\sim C^2}$
		↓			
		$\mathbf{(\sim C^2)^{1/2}}$			

So, the order (of increasing C term) more properly is (with $r = 5, P = 1$) :

$P \bullet P, r \bullet P, r \bullet C_P, C_r \bullet P, C_r \bullet C_P, r \bullet r$;

although categorically, it would (or could) be:

C, r factored C, P factored C, $C_r^{\,2}$, $C_P^{\,2}$, C^2

or

C, P factored C, r factored C, $C^{\,2}$, $C_P^{\,2}$, $C_r^{\,2}$

For our example (system), we have:

$$
\begin{array}{cccccc}
r \cdot P & C_r \cdot P & r \cdot C_P & C_r \cdot C_P & P \cdot P & r \cdot r \\
\downarrow & \downarrow & \downarrow & \downarrow & \downarrow & \downarrow \\
(5 \cdot 3) & (161 \cdot 3) & (5 \cdot 7) & (161 \cdot 7) & (3 \cdot 3) & (5 \cdot 5) \\
\downarrow & \downarrow & \downarrow & \downarrow & \downarrow & \downarrow \\
15 + & 483 + & 35 + & 1127 + & 9 + & 25 = 1694
\end{array}
$$

So, the best order would be:

$$
\left.
\begin{array}{cccccc}
P \cdot P & r \cdot P & r \cdot r & r \cdot C_P & C_r \cdot P & C_r \cdot C_P \\
\downarrow & \downarrow & \downarrow & \downarrow & \downarrow & \downarrow \\
C_P^2 & C & C_r^2 & C & C & C^2
\end{array}
\right\} \; r > P
$$

Finally, this pool itself is subject to itself: $C_{(6)} = 1561$ terms.

Just the multiplication of the 6 terms (above) ⟶ a number $> 6.43 \times 10^{10}$ (for one term).

Therefore, the limiting case appears to be at r = P = 1 , where the sum (Σ) of terms = 1561, and the product (Π) (from the) sum of terms = 1 .

For P = 0 (i.e., no reaction, thus no product or result), r = r; but the product of the sums = 0 when r = 1; the P = 0 (single reactant, no reaction) sum of terms = 1 .

Continuing with the restriction, or limitation, of linear change,

e.g., $r \rightarrow P$

and allowing for reactants to become products, and visa versa, geometric representations can be condensed as follows (for examples):

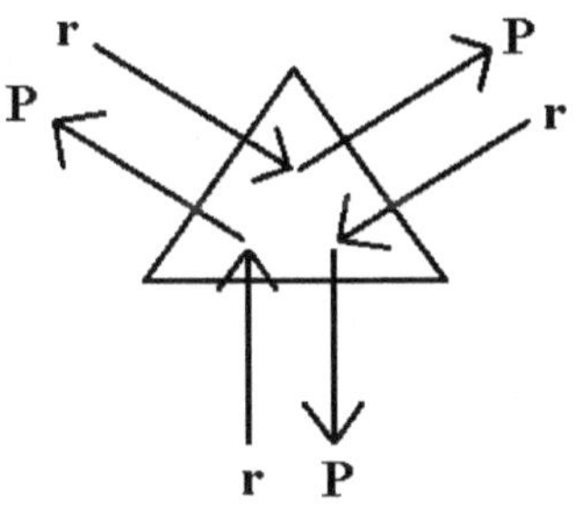

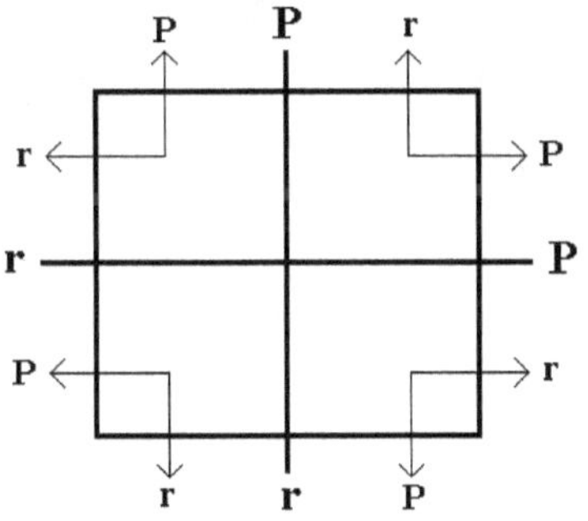

Presumably

	5 sides	6 sides
	5!	**6!**

 # rxns:

$3(r \overset{2}{\longleftrightarrow} P) = 6 = 3! \} C_6 = 1561$

$4(3(r \overset{2}{\longleftrightarrow} P)) = 12 . 2 = 24 = 4!$, $C_{12} = 67546214400$

$C_{[5(4)]} = C_{20}$ $\quad$ $C_{[6(5)]} = C_{30}$

3D

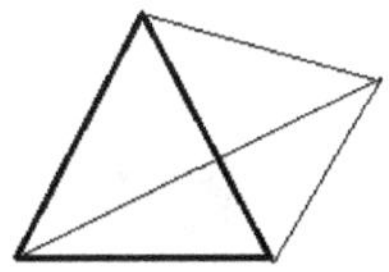

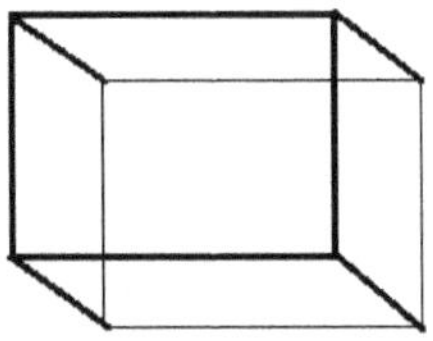

5(4) = 20 sides ∴ 20! $\quad$ 6(5) = 30 sides ∴ 30!

2 modes:

3(2) lines , 6 sides ∴ 6! , $C_6 \cdot 4$

4(3) lines , 12 sides ∴ 12! , $C_{12} \cdot 6$

$C_{20} \cdot 8$ $\quad$ $C_{30} \cdot 10$

or

(planes or lines)

(3 + 1 =) 4 planes ∴ 4! ($C_4 = 25$ $C[4(3)]$

(4 + 2 =) 6 planes ∴ 6! $C_6 = 1561$ $C[6(5)]$

(5 + 3 =) 8 planes ∴ 8! $C_8 = 299601$ $C[8(7)]$

(6 + 4 =) 10 planes ∴ 10! C_{10}) $C[10(11)]$

The pool initiates with:

1 point ● ═══ **a reactant, no reaction, ∴ r**

2 points ● ● ⟶ **makes a line : r ⟵⟶ P or r •⟶• P**

3 points ⟶ △ **(1st 2D figuration)**

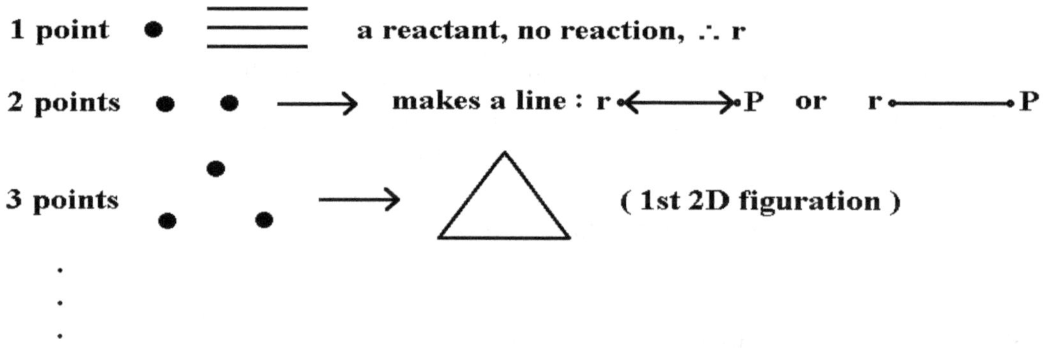

Values (calculations) can be modified by including factors considering the relative size (lengths or areas) of sides (lines) or planes (faces), for (especially) irregularly shaped figures or objects (replacing " → " or " = ").

The reaction pool can be conceivably extended by combining figures via joining lines and/or faces, or higher order combinations such as prismatic or rectangular slots (grooves) and joints, etc.

Our example:

$$a + d \searrow$$
$$\qquad \Big> (c + e + h)$$
$$b \nearrow$$

$$f \searrow \quad g \swarrow$$
$$\vee$$

$$a + d \rightarrow c$$
$$b \rightarrow c + e$$
$$f \rightarrow c + e + h$$
$$g \rightarrow c + e + h$$

might be represented as:

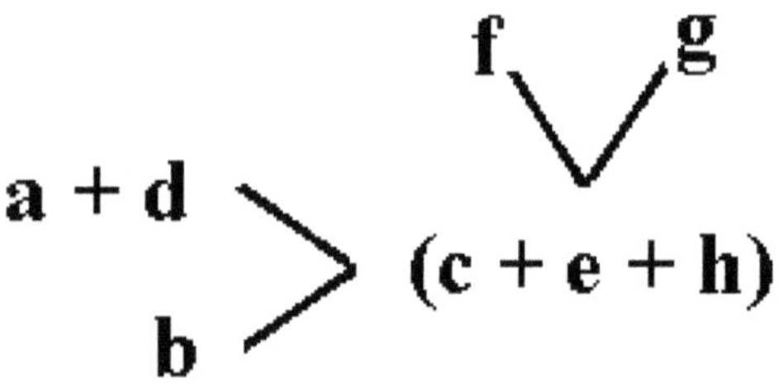

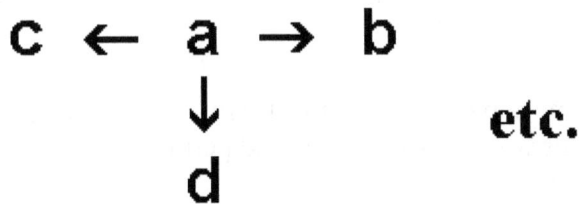

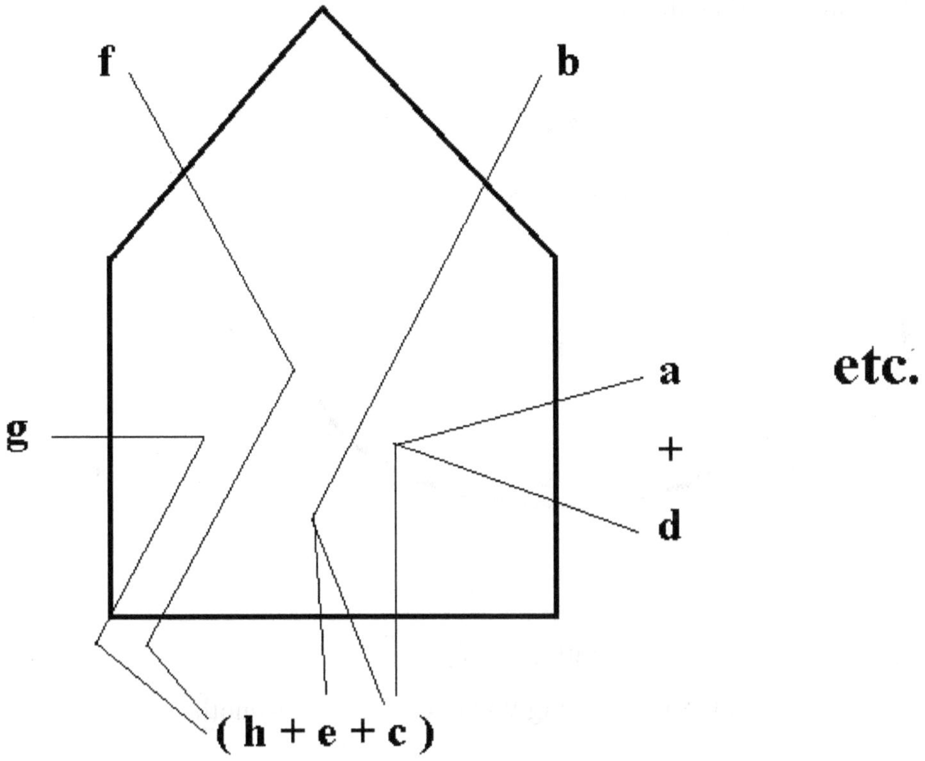

A linear reaction could be: a + d → c

Bi-linear: c ← a → b

Multi-linear:

c ← a → b
↓
d

etc.

A non-linear rxn. could be depicted:

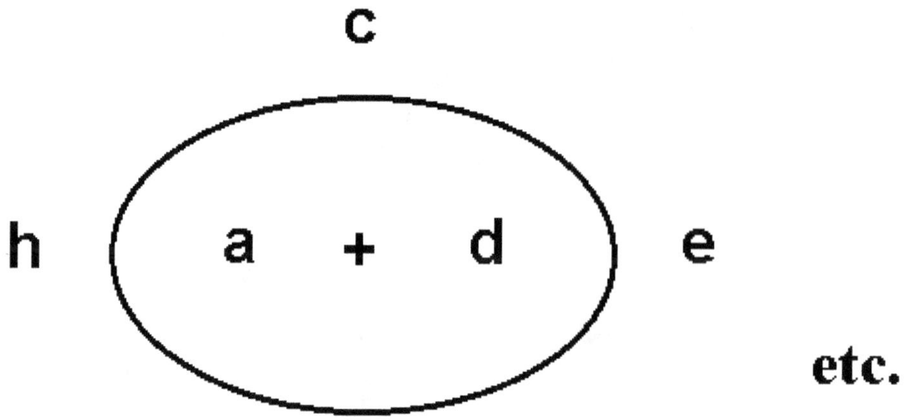

etc.

But what does "non-linear" mean (effectively)?

We can simplify it (as a term) by saying that one rxn. causes another:

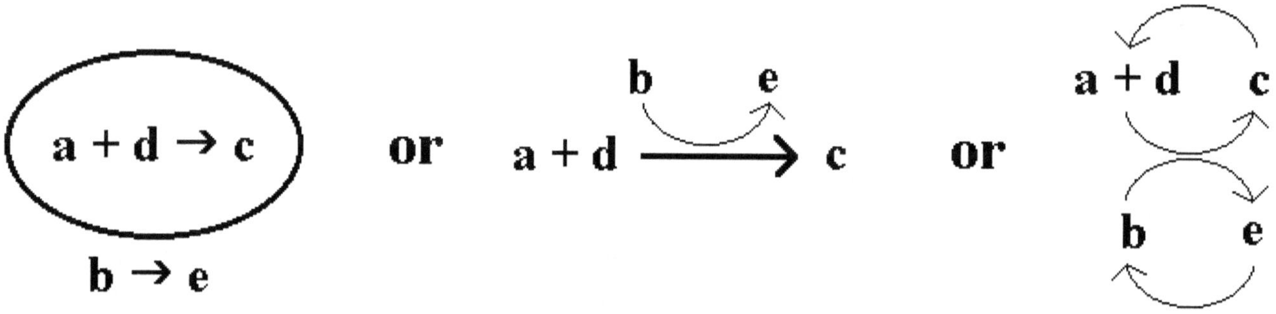

where either rxn. itself may be linear.

If either or both rxns. are non-linear, then one can cause higher order reaction results (or influences) for the other, such as periodic (or wave-like) changes and perturbations:

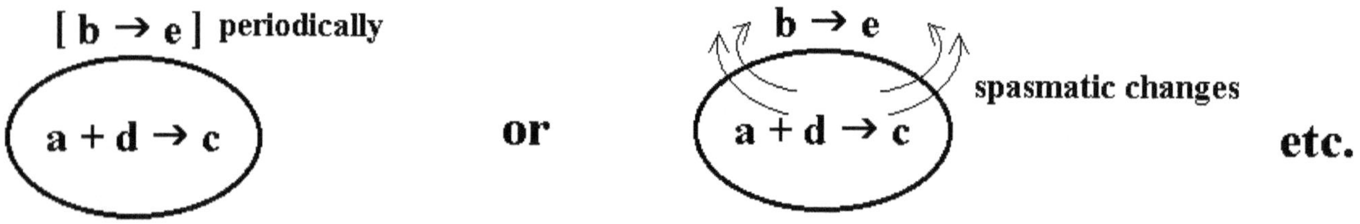

The protein folding (within a water cluster) might be an example of the coupling of reactions non-linearly (or higher order), with the folding of the peptide affecting the orientations of surrounding water molecules (and those in the neighborhood), and visa versa:

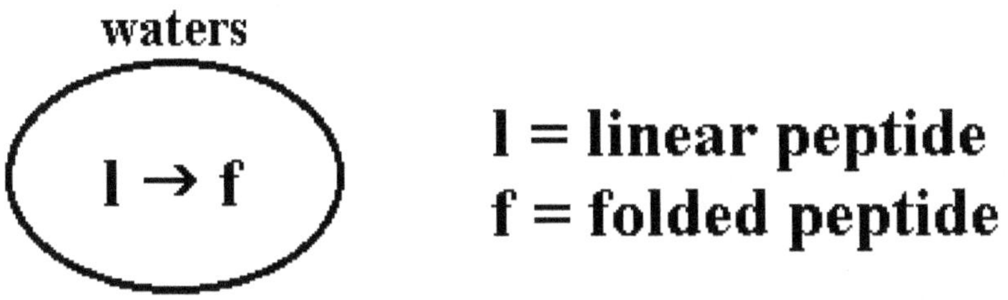

Modeling the Folding

Now, we may model this differently, assigning the mechanics of manipulating the peptide bond (ϕ, ψ), and other bonds, to the time or duration of the folding needed, so that one need only determine (ϕ, ψ) values for the unfolded peptide and each folded intermediate to be considered (including the final folded conformation). This obviates a need to consider (ϕ, ψ) during the folding process. The peptide chain is now simply treated as flexible strings (that are connected).

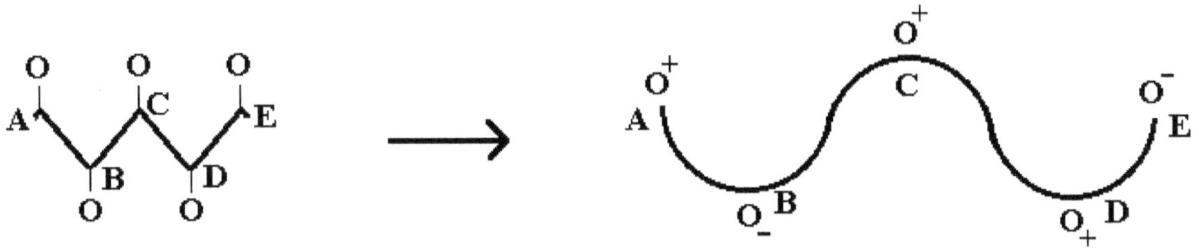

If the partial charges of the water to peptide are satisfied, the water is given the charge of that which is left unsatisfied;

e.g.,

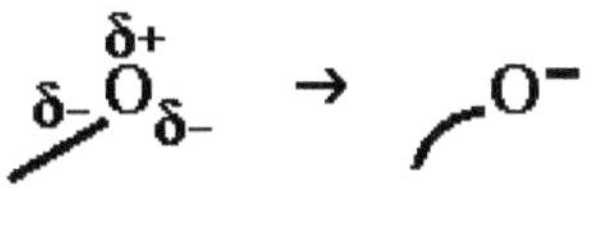

Here, we appear to have a competition, concerning whether (a.a. moiety) A conjuncts with B or E (to satisfy charges), based on the hydrophobicities of A and B and E, and the hypothetical chains of waters allowed between A ← → B and A ← → E, where the # of waters for each chain is weighted (for the folding process) by $C_{(\#\ of\ waters)}$ while the hydrophobicities are weighted by the distances between (A, B, E), and the hydrophobicity values (higher H values favor conjunctions).

We can picture the water chain between two residues as forming the circumference arc of a circle between, for example, A and E.

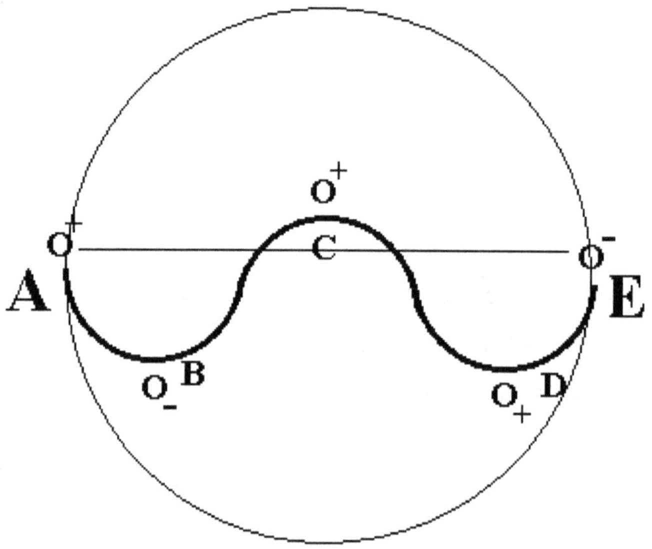

If the distance (used) between A and E is taken as the diameter, then the water chain consists of as many waters (that) can (circumferentially) fit in ½ of the circle's circumference.

The circle's circumference is $2\pi r$; therefore, the # of waters occur within $(\frac{1}{2})2\pi r = \pi r$.

Elimination of this chain of (orientated) waters, with weight $C_{(\#waters)}$, favors the entropy.

Hydrophobic compatibility (between A and E) favors this loss. But the ease of elimination (shortening the chain) increases with a decreasing distance between residues (less waters to eliminate). So, the closer the residues, the more effective the hydrophobic compatibility, while the farther the greater the potential entropic gain.

A compromise might be constructed by disturbing the circle towards "ellispticity" (ellipsoidality), to the extent that the chain length becomes the (original circle's) diameter. This means the length goes from $\pi r \rightarrow 2r$.

An integer # of waters to the chain's length will generally leave an extra length (less than a water's sphere diameter), δ (delta) ("loosening" the water chain). This might initiate ellipsoidality.

So which will occur (or occur first): $A \leftarrow \rightarrow B$ or $A \leftarrow \rightarrow E$?

Since time is the constraint for folding, which ever occurs sooner will occur (or occur first), subject to the (water) entropic and hydrophobicity (compatibility) considerations.

Here, if $A \leftarrow \rightarrow B$ occurs (before allowing $A \leftarrow \rightarrow E$), presumably assuaging the (partial) charges (on waters) for $A \leftarrow \rightarrow B$ will eliminate (or prevent or usurp) $A \leftarrow \rightarrow E$.

The folding process, then, can be approximated as follows (for the moment discounting excluded volumes):

O $\equiv$ **- and +** **or** **2+**
water with (1+ satisfied) (1- satisfied)
δ charges

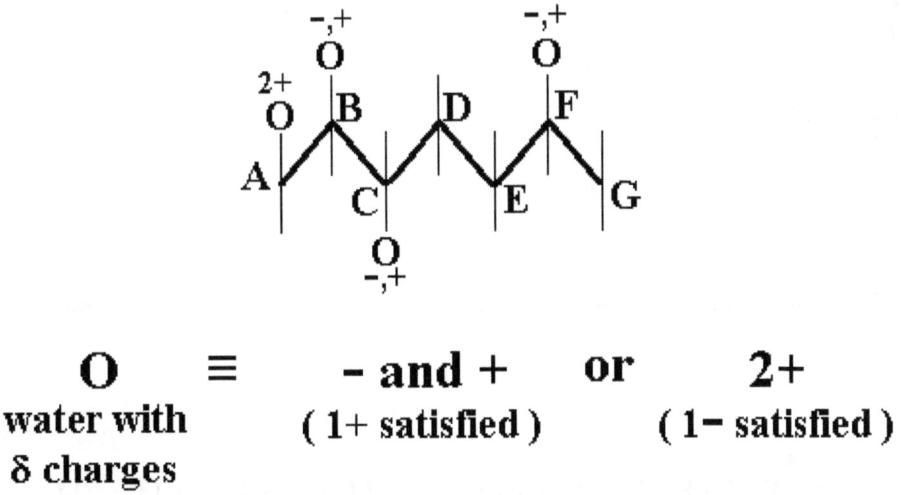

Implausibles : part 2

We first consider only a common residue atom (type), say C-alpha (c_α) :

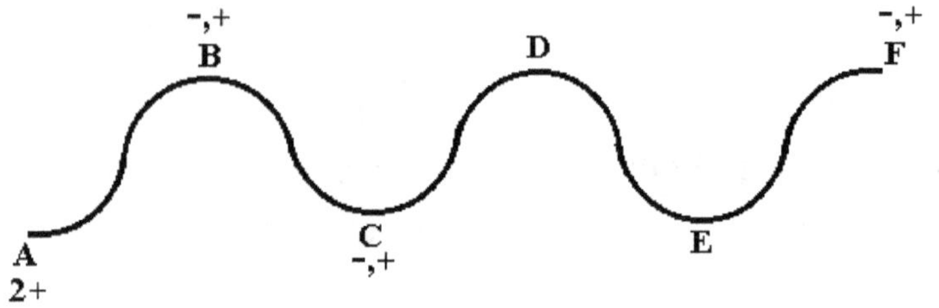

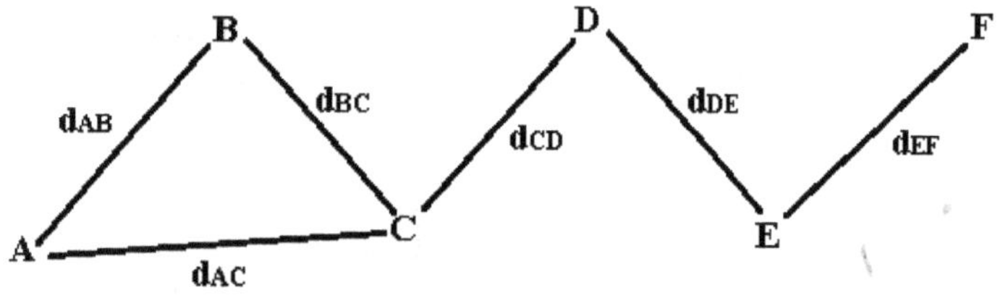

d ≡ diameters
(of clusters for half-circumferences, and also distances betw. residues)

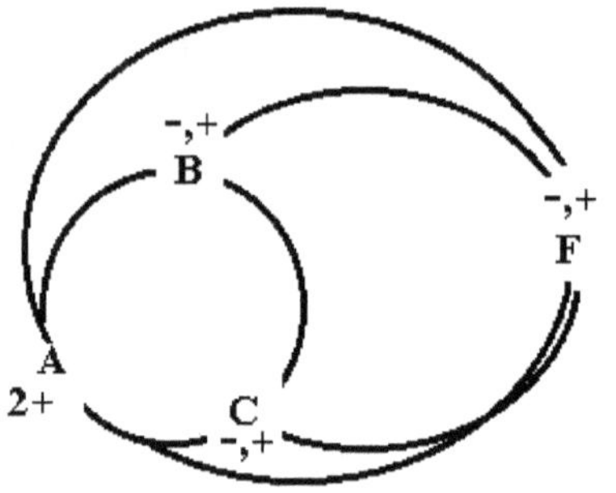

water half-circumferences (correlated to d via πr)
→ $C_{\#waters}$ (hypothetical on each half-circumference): $C_{AB(\#)}$, ...

Water half-circumferences (correlated to d via πr) → $C_{(\#\ waters)}$ (hypothetical on each half-circumference): $C_{AB(\#)}$, ...

H(A, B, C, D, E, F) hydrophobicities → H couplings: H_{AB}, H_{AC}, ...

Implausibles : part 2

We now have a conceivable table to help us determine (or estimate) which movement (of relative c_α) occurs first (priority of order) to satisfy $C_\#$ → the more waters displaced between residues (including those of d):

	A	B	C	D	E	F
A		$C_{AB(\#)}$ H_{AB} d_{AB}	$C_{AC(\#)}$ H_{AC} d_{AC}	$\begin{pmatrix}H_{AD}\\ d_{AD}\end{pmatrix}$	$\begin{pmatrix}H_{AE}\\ d_{AE}\end{pmatrix}$	$C_{AF(\#)}$ H_{AF} d_{AF}
B			$C_{BC(\#)}$ H_{BC} d_{BC}	$\begin{pmatrix}H_{BD}\\ d_{BD}\end{pmatrix}$	$\begin{pmatrix}H_{BE}\\ d_{BE}\end{pmatrix}$	$C_{BF(\#)}$ H_{BF} d_{BF}
C				$\begin{pmatrix}H_{CD}\\ d_{CD}\end{pmatrix}$	$\begin{pmatrix}H_{CE}\\ d_{CE}\end{pmatrix}$	$C_{CF(\#)}$ H_{CF} d_{CF}
D					$\begin{pmatrix}H_{DE}\\ d_{DE}\end{pmatrix}$	$\begin{pmatrix}H_{DF}\\ d_{DF}\end{pmatrix}$
E						$\begin{pmatrix}H_{EF}\\ d_{EF}\end{pmatrix}$
F						

Obviously, as $C_\#$ increases the entropy quickly increases with water number, but the half-circumference chain is more stringent of order (to maintain). The stringency must be related to hydrophobicity. (Distance is assumed, for the cluster as ordered waters).

For example, if the AC coupling is most favorable (of that given), then a rigid body rotation from A to (C....F) will occur, to (relatively) bring C to the B position. If AB coupling is most favorable, no rotation is needed (concerning c_α). If AF coupling is most favorable, a rigid body rotation might not be physically possible (geometrically).

Each ordered (or prioritized) rotation (or torsion) to adjust c_α positions represents a state of folding for the peptide.

But there is an interesting dichotomy between the hydrophobic influence (towards folding) and the entropic gain (of disordered waters). The rapid increase in $C_{(\#\ waters)}$, with each water added to the half-circumference chain make-up, relates to a greater distance between affected residues, while hydrophobicity effects (or coupling) is stronger the closer the affected residue (to physical limit). Essentially, distance increases much more slowly than $C_{(\#\ waters)}$. Yet, hydrophobic coupling (of side-chains) decreases with distance (between them).

An obvious way to slow down the $C_{(\#\ waters)}$ factor is to take the (natural) logarithm, whereby the exponent to (base) e yields $C_{(\#\ waters)}$; thus, $\ln C_{(\#\ waters)}$.

This would be analogous to the statistical definition of entropy, from the term $\ln\Omega$, where Ω represents the # of ordered states permitted of a system.

(The) $\ln C_{\#}$ is allowed to increase on a scale compatible (or comparable in order) with distance, d. And $d\uparrow$ decreases the hydrophobic effect: thus, $(\ln C_{\#})\uparrow \bullet H_{(d\uparrow)}\downarrow$ may show a balancing of effects.

An alternative might be to divide by $C_{\#}$, to produce a very small factor (with increasing d) reducing H (with distance). H coupling would soon become negligible, with increasing distance, although an idealization is introduced (by division, to create an intensive quantity from extensive ones).

$C_{\#}$ and $\ln C_{\#}$ are already correlated (through πr) with d.

The hydrophobic coupling can simply be the sum of hydrophobic terms:

e.g., $H_{AB} = H_A + H_B$

Therefore, we have:

$$\ln C_{\#(AB)} \qquad\qquad H_{AB} = H_A + H_B$$

$$\downarrow \qquad\qquad\qquad \uparrow$$

$$\# = \frac{\pi r}{d_{water} / \text{water molecules}} \longrightarrow d_{AB} = \frac{\pi r}{(\pi/2)} = \frac{(\#)d_{water} + \delta_{delta}}{\pi/2} = \frac{2((\#)d_{water}) + \delta}{\pi}$$

Therefore, a comparison term, for prioritization of folding torsions could be:

$$\frac{(\ \ln C_{H_{AB}}\ )(\ H_A + H_B\ )}{(\,2(\,(\,\#\,)d_{water} + \delta\,)\,)\,/\,\pi} \equiv \frac{(\ \ln C_{\#}\ )H\pi}{2(\,(\,\#\,)d_{water} + \delta\,)}$$

where H ≡ sum of hydrophobicities,

$$\delta \equiv \pi(\,d\,/\,2\,) - (\,\#\,)d_{water}$$

But if there are no waters to be considered, $\ln C_0 = \ln 0$ as undefined (# = 0), and we have:

$$\frac{H\pi}{2\delta} = \frac{H\pi}{2\pi r} = \frac{H}{2r} = \frac{H}{2d\,/\,2} = \frac{2H}{2d} = \frac{H}{d}$$

If $C_{\#} = 1$, the entropic term is zero: $\ln 1 = 0$; therefore, the comparison term is zero. The side-chains are separated by 1 water molecule; thus, no torsion is necessary (the hydrophobic coupling is already optimal, for them).

The comparison term is idealistic and intensive, and basically comes down to hydrophobicity partitioned by distance (between residues), since $\ln C_{\#}$ and π and 2 and # are only numerical terms. It may be preferable to a typical inverse square law for distance.

Going back to our example, if residue C is rigid body torsioned into the B position:

$(\ 2+_{(A)}\ ) + (\ -, +_{C}\ ) \rightarrow (\ 1+_{A}\ ) + (\ 1+_{C}\ )$ for AC coupling and satisfying (partial) charges (two waters re-ordered).

Appendix 5 gives an example of results employing this manner of (calculated) peptide folding.

Relation of $C_{\#}$ to the physical "Fine Structure" Constant

[On a side-note] It is interesting that, for the denominator of the fraction used to approximate the value of the "fine-structure" constant (Sommerfield constant), which in physics is a fundamental physical constant characterizing the strength of the electromagnetic interaction between elementary charged particles (α):

$$\alpha = \frac{1}{4\pi\varepsilon_0}\frac{e^2}{\hbar c} = \frac{\mu_0}{4\pi}\frac{e^2 c}{\hbar} = \frac{K_e e^2}{\hbar c} = \frac{c\mu_0}{2R_K} \approx \frac{1}{137}$$

$$\alpha^{-1} = \frac{1}{\alpha} = \begin{matrix} 173(35) \\ 137.035999139(31) \end{matrix}$$

the (ordered) digits are the same as the first 3 terms of the $C_{\#}$ summation: 1, 3, 7

$\alpha = 7.2973525664(17) \times 10^{-3}$ (from SI units, dimensionless);

$1/137 = 7.299270073 \times 10^{-3}$

Therefore,

$$\alpha \approx \frac{n_0}{\begin{matrix} 100(n_0) & n = 1 \\ +\,10(n_0 + n) & n = 2 \\ +\;\;1(n_0 + 2n) & n = 3 \end{matrix}}$$

Combination of Terms and a new (numerical) Base

Natures of $C_\#$

$C_\#$ is a summation of the possible combinations of terms, given the number of separate terms to be considered. Its growth is exponential (of character) with this number. Each combination is unique of ordering (although some terms may be equivalent of significance or magnitude of variable).

If one wishes to conduct a search of specific combinations, it is clear that, compared to finding (the existence of) a combination of single variable, that of multiple variables could take a factor of time exponentially raised by the number of variables for the combination:

If C takes t time to find, then CD would take t^2 as long to find (or couple), etc. Thus, a combination of n variables would take t^n time (to occur, constrain or distinguish).

Equating a search with a *distance*, d, and a time variable with a *magnitude* (or factor) of $t^n = m$ (for t), then a "velocity" may be constructed as $d / (mt)$, assuming each combination is d distant (of search) for a given $C_\#$ set.

Therefore, the larger the combination (of number of variables), the larger is mt and the smaller is the velocity d / mt (i.e., it is slowest to find the combination composed of the most number of variables, the last combination).

For combinations of many variables, the (search) velocity becomes exceedingly slow (long) very quickly (an $n > 32$ can exceed the calculation of most computers, to create a denominator too large to be rendered and thus a velocity or fraction too small to be conceived).

Then accordingly, $C_\#$ divided by its last combination velocity can produce a very large number:

$$C_\# / (d / mt) = mtC_\# / d$$

But, divisions are idealistic (and possibly converting), and the quantity is actually: t^nC_n , where we allow $d = 1$ == as a search parameter equivalent for each combination.

Therefore, essentially, we have a new base (t), where:

$$\log_t t^n = n \text{ , } \log_t t^nC_n = n + \log_t C_n = n + (\log_{10} C_n) / \log_{10} t$$

Obviously, t must have a minimal (positive) value, since it can not be $t = 0$.

Three unique points (x,y) can make a parabola (quadratic function). The first three points for $C_{n=1 \to 3}$: (1, 1) , (2, 3) , (3, 7) for (n, C_n) yield the relation:

$y = x^2 - x^1 + 1(x^0) == x^2 - x^1 + x^0$.

From the first 4 points (i.e., including (4, 25)), the relation is:

$y = 2x^3 - 11x^2 + 21x - 11(x^0)$.

[Again, for both relations all of the coefficients of x are whole numbers.]

Adding the fifth point (5, 161) yields an x^4 function with non-whole number coefficients.

Points: (3, 7) , (4, 25) , (5, 161) → $y = 59x^2 - 395x + 661(x^0)$

Points: (2, 3) , (3, 7) , (4, 25) → $y = 7x^2 - 31x + 37(x^0)$

So, from the first (n =) 5 points, the parabolic interpolated relations are:

(1, 1) , (2, 3) , (3, 7) $y = x^2 - x + 1$

(2, 3) , (3, 7) , (4, 25) $y = 7x^2 - 31x + 37$

(3, 7) , (4, 25) , (5, 161) $y = 59x^2 - 395x + 661$

[Whole number coefficients from 3 (ordered) points result until trying:

$(6, C_8)$, $(7, C_7)$, $(8, C_8)$]

For (integers) n = 1 → 9, C_n is always odd; then, from n = 10 → 22 (limit of numerical testing for program "rxn_pool", using FORTRAN floating-point precision), C_n is always even (The limit for computer calculation, before yielding "Infinite," is n = 28). Therefore, n = 10 yields an inversion point (from odd to even C_n), as opposed to an inflection point.

A transcendental nature is suggested.

If it is similar to e [(= Σ sum from n = 0 → ∞ of 1 / n!)

$$\mathbf{e = \left[\sum_{n=1}^{\infty} \frac{1}{n!} \right] + 1}$$

then we can define:

$$(d \leftarrow d' \leftarrow)\ \Psi = \sum_{n=1}^{\infty} \frac{n}{C_n} \quad (\rightarrow ȸ)$$

[For convenience of typing, we can use the symbol " Ψ ", because it looks similar. But it is actually a disfigured "d," the base occurring before "e."]

Due to the divisions, this number (or base) is idealistic; i.e., it is transformative of quality.

The number can be approximated by the first few n terms (n = 1 → 10) :

$$\Psi \cong \frac{1}{1}+\frac{2}{3}+\frac{3}{7}+\frac{4}{25}+\frac{5}{161}+\frac{6}{1561} + \frac{7}{19615}+\frac{8}{299601}+\frac{9}{5380849}+\frac{10}{111111120}+ \ldots..$$

= 2.290523020324.... (from calculator)

[By comparison, e = ~ 2.718281828459(0)....]

In the same manor, the minor factor (or "reduced" base) can be defined as:

$$(ȸ_m \leftarrow)\ \Psi_m = \sum_{n=1}^{\infty} \frac{1}{C_n}$$

$$\cong \frac{1}{1}+\frac{1}{3}+\frac{1}{7}+\frac{1}{25}+\frac{1}{161}+\frac{1}{1561} + \frac{1}{19615}+\frac{1}{299601}+\frac{1}{5380849}+\frac{1}{111111120}+ \ldots..$$

A table comparison illustrates the natures of these numbers (bases):

n	C_n	$\sqcup^n_m$	$\sqcap^n$	e^n
1	1	1.5231	2.2905	2.7183
2	3	2.3198	5.2464	7.3892
3	7	3.5333	12.0169	20.0859
4	25	5.3816	27.5246	54.5996
5	161	8.1968	63.0451	148.4181

Therefore, the "rise in function" for each is:

$$\sqcup^n_m \,/\, \sqcup^{n+1}_m \;\cong\; 0.65655 \;\equiv\; 1\,/\,\sqcup^{n=1}_m \;=\; \sqcup^{-1}_m$$

$$\sqcap^n \,/\, \sqcap^{n+1} \;\cong\; 0.43659 \;\equiv\; 1\,/\,\sqcap^{n=1} \;=\; \sqcap^{-1}$$

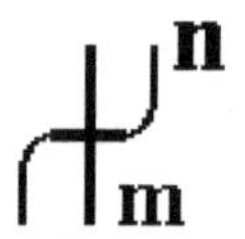

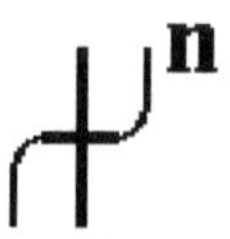

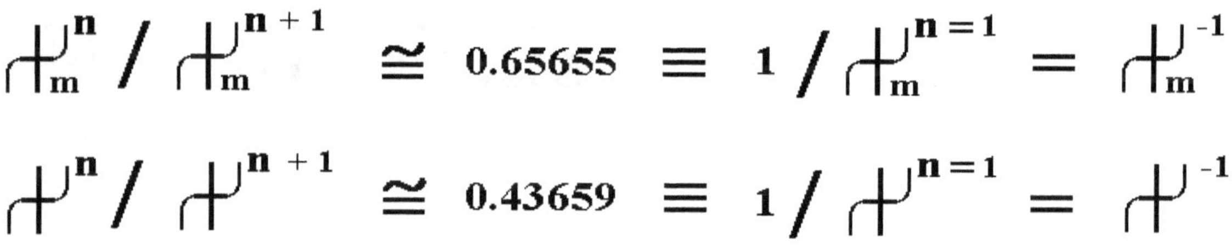

$$e^n \,/\, e^{n+1} \;\cong\; 0.36788 \;\equiv\; 1\,/\,e^{n=1} \;=\; e^{-1}$$

The "n-weight" might be found from:

$$n_w = \frac{\sqcap^n}{\sqcup^n_m}$$

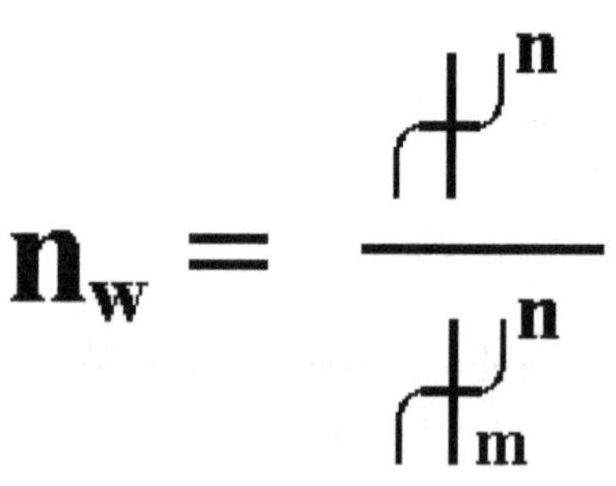

e.g., for

$n = 1$, $n_w = 1.5038$

$n = 2$, $n_w = 2.2616$

3 , 3.4010

4 , 5.1146

5 , 7.6914

When using a polynomial interpolation program (basis method used: monomial, Lagrange, or Newton) for points (Ψ^n, C_n) $n = 1 \rightarrow 5$, values to the tenthousandths (result):

Ψ^n	C_n
2.2905	**1**
5.2464	**3**
12.0169	**7**
27.5246	**25**
63.0451	**161**

The interpolated curve (to $x^{n-1} = x^4$) intersects Ψ^n axis at two points: the first exactly (or very nearly) at $\Psi^n = 1$ (1st intercept $\Psi^n \sim 1.03125$; exact for $n = 4$); $\therefore$ $n = 0$ and $C_n = 0$; the second about mid-way between $\Psi^n = 96$ and $\Psi^n = 128$.

The curve is negative (of Ψ^n) before $\Psi^n = 1$ and after the second intercept.

The curve rises of positive C_n and Ψ^n up to around $C_n \sim 250$ at a maximum, then drops precipitously to the second intercept point (for $C_n = 0$). This tends to limit C_n to values from zero to the 2nd intercept (maybe $\Psi^n \sim 110$?).

The next C_n, of course, is 1561 ($\Psi^6 = 144.4048$). Including this data point yields a curve still of x^4 (rather than x^5), and with the (27.5246, 25) point poorly fitted. Thus, the polynomial is less exact of the data points (e.g., (63.05, 161) is away from curve).

Using $(\Psi_m{}^n, C_n)$ points, problems with exactness don't seem to occur (as readily), although the first $C_n = 0$ intercept is not at $\Psi_m{}^n = 1$.

The (Ψ^n, C_n) polynomial curve looks like:

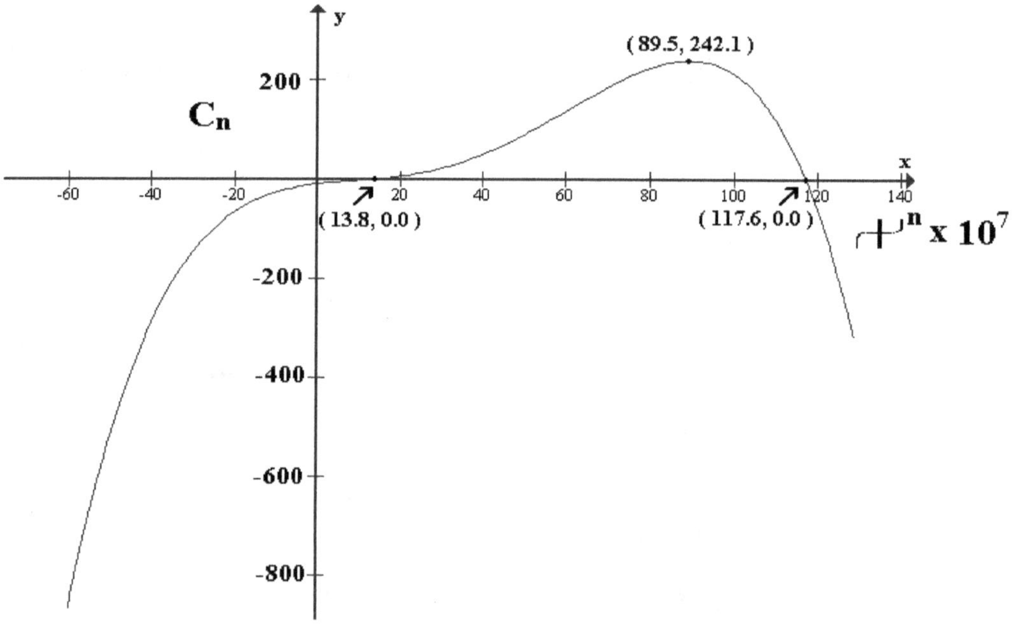

The results might suggest (mathematically) a maximum obtainable value for C_n (when Ψ^n is positive). This limits the positive base range (Ψ) that is usable.

The polynomial coefficients are: $P(x) = a_4x^4 + a_3x^3 + a_2x^2 + a_1x^1 + a_0x^0 == C_n$

$x == \Psi^n$	$a_0 \rightarrow 4$
x^0	-0.9116613758
x^1	0.9270150259
x^2	-0.0449901267
x^3	0.0020619031
x^4	-0.0000148367

Graphically, using the coefficients given: max = (89.25, 247.9);

Ψ^n intercepts are 1.03125, 117.525 .

Aside from the ill-fit of the (27.52, 25.0) point, inclusion of more data points (Ψ^n, C_n) has the effect of lowering (with oscillation) the C_n maximum and lowering (likewise) the Ψ^n intercept (2nd intercept); the 1st intercept remains at $\Psi^n = 1$, and the highest power of x remains x^4 (or near; therefore the shape of the curve remains the same).

Limiting n to 5 (or less than 6) seems to be a peculiar (physical) restriction. Perhaps it is empirical, for this range of n and its "c-like" objects.

For (Ψ^n, C_n) n = 5 : a graphing program that shows coordinates (using a line-ruler) of max/min and intercepts yields (from the given a coefficients, with graphical rounding of numbers)

maximum: (89.20319, 247.92345)	(89.20, 247.92)
Ψ^n intercept: (1.03, 0) == (1.03422, 0.00122)	(1.03, 0.00)
Ψ^n intercept: (117.67924, 0.00130) == (117.68, 0)	(117.68, 0.00)
C_n intercept: (0, -0.91) == (-1.43201×10^{-4}, -0.911794)	(0.00, -0.91)

For ($\Psi_m{}^n$, C_n) :

n = 1 → 3 is parabola-like (up) (highest x term is x^2; thus, parabola up)

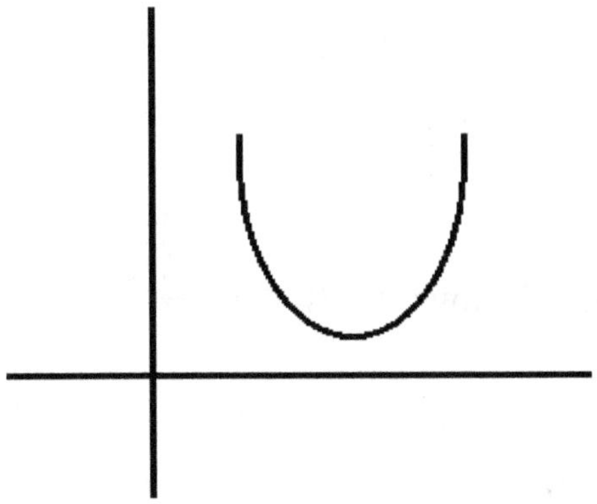

$n = 1 \rightarrow 4$ is like

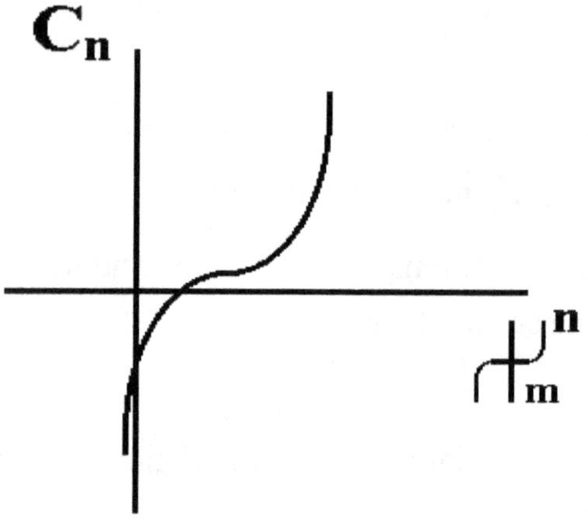

(highest x term is x^3)

$n = 1 \rightarrow 5$ is like (up) paraboloid (highest x term is x^4)

$n = 1 \rightarrow 6$ favors $n = 1 \rightarrow 4$ shape (highest x term is x^5)

$n = 1 \rightarrow 7$ is (up, distorted) paraboloid-like (highest x term is x^6)

$n = 1 \rightarrow 8$ favors $n = 1 \rightarrow 4$ shape (highest x term is x^7)

$n = 1 \rightarrow 9$ favors $n = 1 \rightarrow 4$ shape (highest x term is x^7)

$n = 1 \rightarrow 10$ favors $n = 1 \rightarrow 4$ shape (highest x term is x^7)

For (e^n, C_n) :

$n = 1 \rightarrow 3$ is parabola down

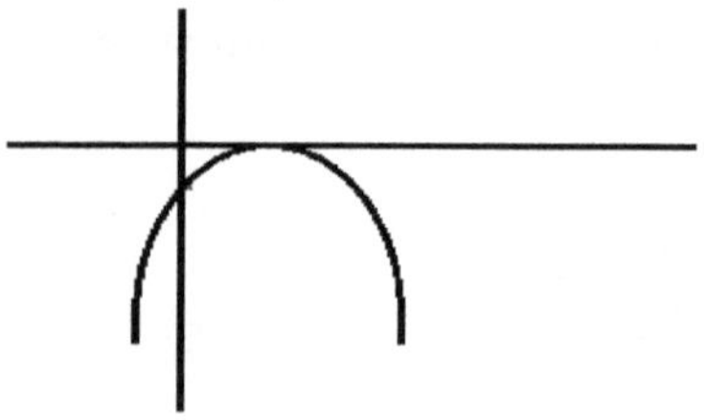

(highest x term is x^2) ;((Ψ^n, C_n) is also parabola down for $n = 1 \rightarrow 3$)

$n = 1 \rightarrow 4$ is like

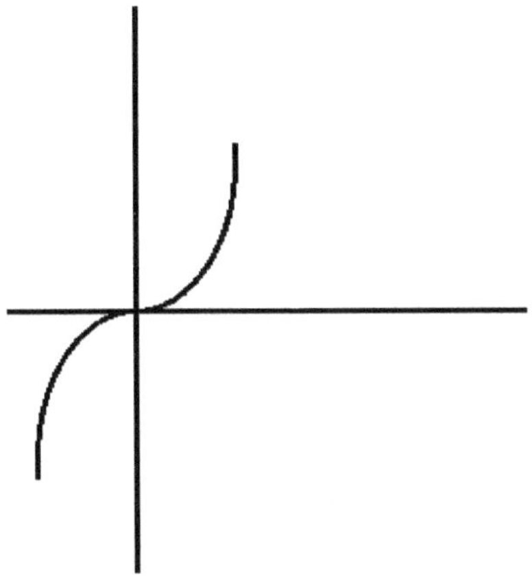

Highest x term is x^3, similar to (Ψ^n, C_n) for $n = 1 \rightarrow 4$

$n = 1 \rightarrow 5$ is like (Ψ_n, C_n) $n = 1 \rightarrow 5$ in shape, with $(148.42, 161)$ near maximum. Highest x term is x^4 .

$n = 1 \rightarrow 6$ is like $n = 1 \rightarrow 5$ in shape, though diminished and $(56.6, 25)$ poorly fitted and $(148.42, 161)$ totally off (far) from curve. Thus, it also favors (Ψ^n, C_n) ; (highest x term is x^4)

$n = 1 \rightarrow 7$ is like $n = 1 \rightarrow 6$ in shape and manner of points: $(148.42, 161)$ and higher far away from curve. (highest x term is x^4)

Therefore, Ψ^n more favors (in character) e^n (than $\Psi_m{}^n$) :

$$\Psi = \sum_{n=1}^{\infty} \frac{n}{C_n}$$

$$e = \left[\sum_{n=1}^{\infty} \frac{1}{n!} \right] + 1$$

$$\Psi_m = \sum_{n=1}^{\infty} \frac{1}{C_n}$$

even though $\Psi_m^{(n)}$ is calculated more as like (in form) $e^{(n)}$.

The serviceable domain of the Ψ^n base, where $\Psi^n >= 0$ and $C_n >= 0$, yields to a range of n from 0 :

$1 = \Psi^0 (n = 0) , C_n = 0$

to an n formally found:

(roughly) $118 = \Psi^n = (2.2905)^n$, $C_n = 0$;

therefore,

log(118) = n•log(2.2905)/log10 == log(118)

[i.e., $\log(118 = \Psi^n) = \log 118 = \log\Psi^n = n\log\Psi$] .

∴ (log118)(log10)/log2.2905 = n = log(118) • 1 / log2.2905 = 5.756 (i.e., < 6) .

Where C_n is negative, Ψ^n may run as:

-∞ to +∞ , sans **[1; (C_n = 0)] < Ψ^n < [118; (C_n = 0)]** ;

but this run (of negative C_n) is much faster than for Ψ^n.

What would a negative (i.e., not serviceable) C_n signify? Such combinations are not physically imaginable. But they might be adoptable to comparisons with the (physically) real (or extant).

So, for positive Ψ^n and positive C_n, we restrict (by polynomial interpolation)

$n : 0 \rightarrow 5756$	||	{ 0, 1, 2, 3, 4, 5 }
$C_n : 0 \rightarrow 248$	||	{ 0, 1, 3, 7, 25, 161 }
$\Psi^n : 1 \rightarrow 118$	||	{ 1.000, 2.291, 5.246, 12.017, 27.525, 63.045 }
mathematical		of physical conception (operation)

How, then, do we get to n = 6 ?

From our graphical restrictions, with $\Psi^6 = 144.40$, C_n would have to be a very negative number. (Graphically, it occurs at around -1049 .)

We can now suppose "additive coordinate" plots, where points on the (interpolated) curve are connected by a straight line, and the sum of the coordinates (of these points) can be compared to the domain and range of the curve. If (they are) within, the demonstration is feasible (for the corresponding n). But all points crossing the straight line must be considered (for the final coordinate sums) :

e.g.,

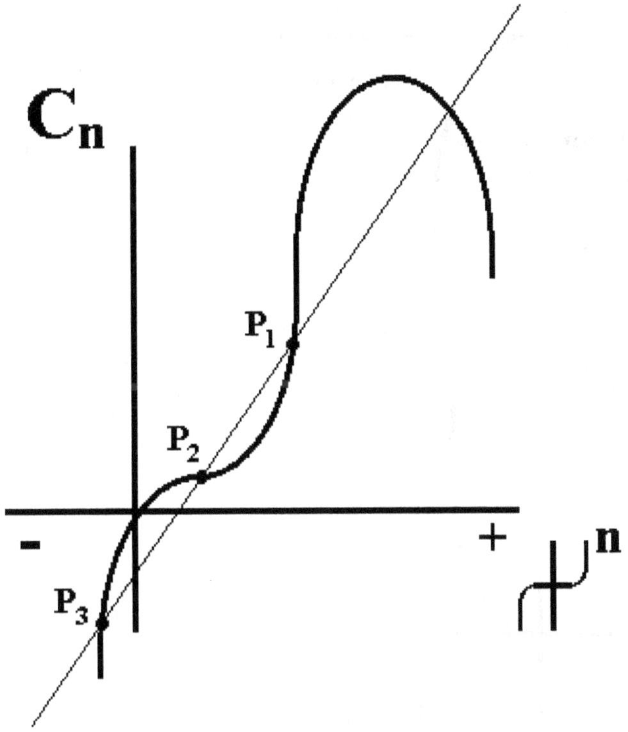

$P_1 == (\Psi_1^{\ n1}, C_{n22})$

$P_2 == (\Psi_2^{\ n2}, C_{n22})$

$P_3 == (\Psi_3^{\ n3}, C_{n33})$

$P_{sum} == (\Psi_1^{\ n1} + \Psi_2^{\ n2} + \Psi_3^{\ n3}, C_{n11} + C_{n22} + C_{n33})$

$n_{sum} ==$ (n1 + n2 + n3)'s influence on Ψ

But here, since Ψ^n can go to $-\infty$ there must be (at least) a 4th point.

$\therefore P_{sum} = (\Psi_{sum}^{\ nsum}, C_{nsum}) = (\Psi_1^{\ n1} + \Psi_2^{\ n2} + \Psi_3^{\ n3} + \Psi_4^{\ n4}, C_{n11} + C_{n22} + C_{n33} + C_{n44})$

If P_{sum} occurs within the range (C_{nsum} <= C_{nmax}) and domain (-∞ to +∞) of the curve such that it is in a continuous area (here: on or "under" the curve), the P_{sum} is feasible and its corresponding n is obtainable. Here, due to C_{nmax}, discontinuous areas are formed (by C_{nmax}):

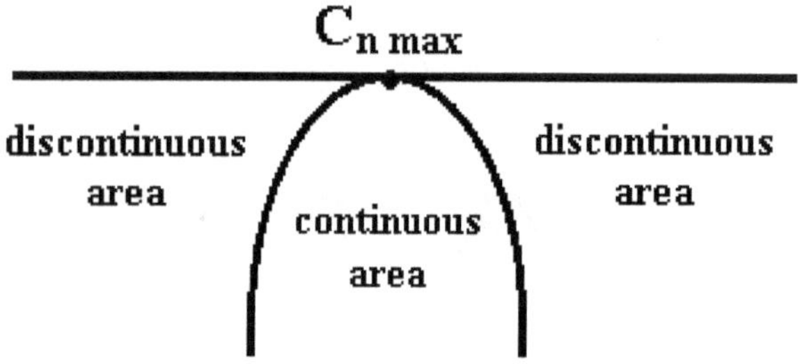

For a plot like:

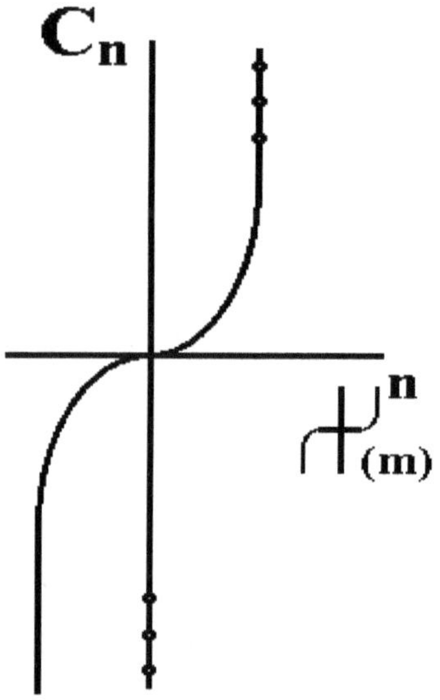

without a C_{nmax}, the areas are ambiguous of continuity (yet separate) and must be distinguished by operative definitions. But essentially, all P_{sums} (and all n_s) are feasible.

The continuous area postulate can be extended to the extreme. If the curve is, or contains, a straight line, then the # of points is (or can be) infinite and the P_{sum} coordinates are infinite (one or both).

Areas may remain continuous but separate:

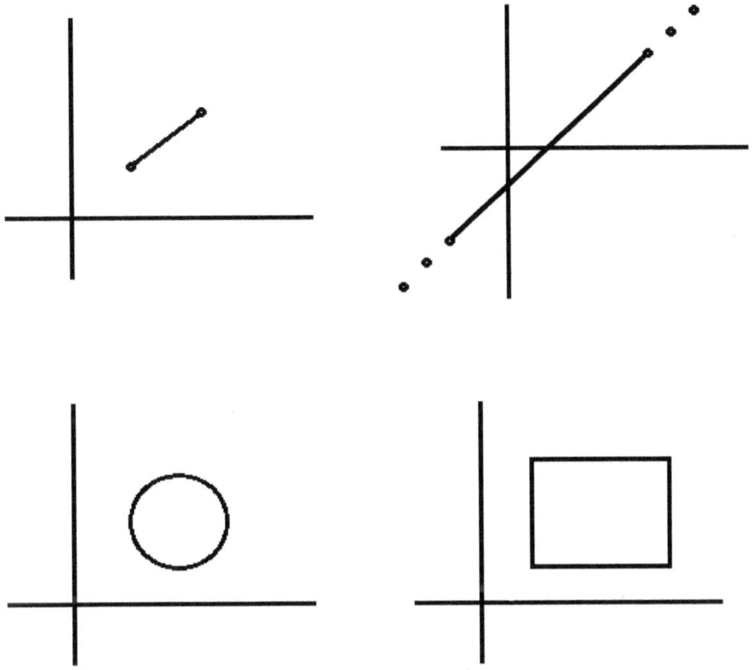

etc.

whereby an "infinite" P_{sum} point may be allowed (or "found"). A distinction occurs when a (sharp) max or min vertex is made:

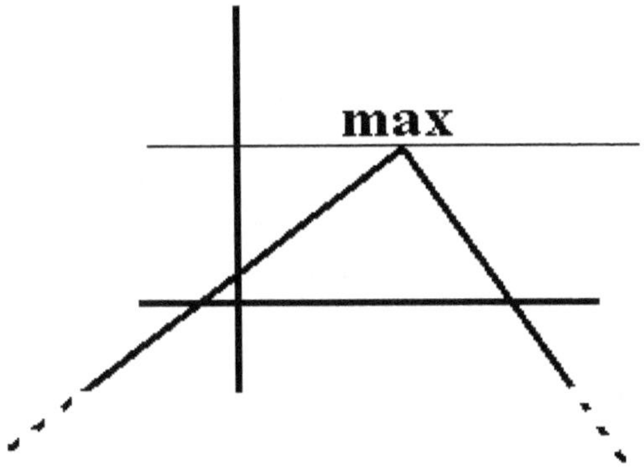

where infinite areas can be discontinuous and thus divided (partitioned), restricting P_{sum} to one of them.

In all cases, however, a resulting n (while infinite) is still "obtainable."

How do the (interpolated) Ψ^6 and the operational (i.e., exact by definition) Ψ^6 compare?

For ${}_i\Psi^6$: ${}_i\Psi^6 = 144.40$, $C_n = {}_iC_6 \sim = -1049$.

For ${}_o\Psi^6$: ${}_o\Psi^6 = 144.40$, $C_n = {}_oC_6 = 1561$.

If we draw a line between the two points (Ψ^n, C_n) ,

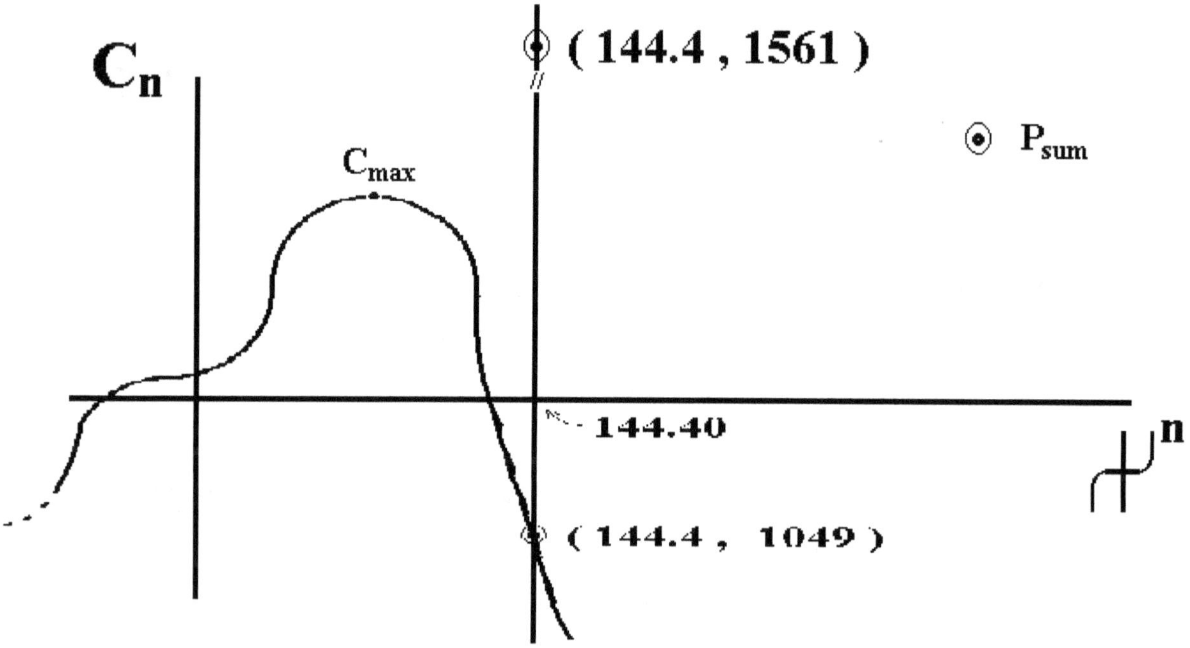

the line crosses the (interpolated) curve only (presumably) once.

The (implied) P_{sum} = (144.4 + 144.4, 1561 – 1049) = (288.8, 512) .

Interestingly: $512 = 2^9 = \Delta C_6$ or ${}_{(i+o)}C_n$. The P_{sum} occurs in a dis-allowed region (due to the $C_{max} = 248$).

This P_{sum} is only implied; because it is implied it belongs to a different physical curve. (An actual P_{sum} would have formed from only points on the interpolated curve.)

It is interesting how a division can be related to an addition (or subtraction):

-1049 / 1561 = -0.672 , -0.672 + 1 = +0.328 ,

(1561)(0.328) = 512 = 1561 – 1049 (a subtraction).

How can ${}_{i}\Psi^{6}$ be "corrected" to ${}_{o}\Psi^{6}$?

If we make a triangle of points C_{max} : ${}_{o}\Psi^{6}$, ${}_{i}\Psi^{6}$, we find that it is almost isosceles:

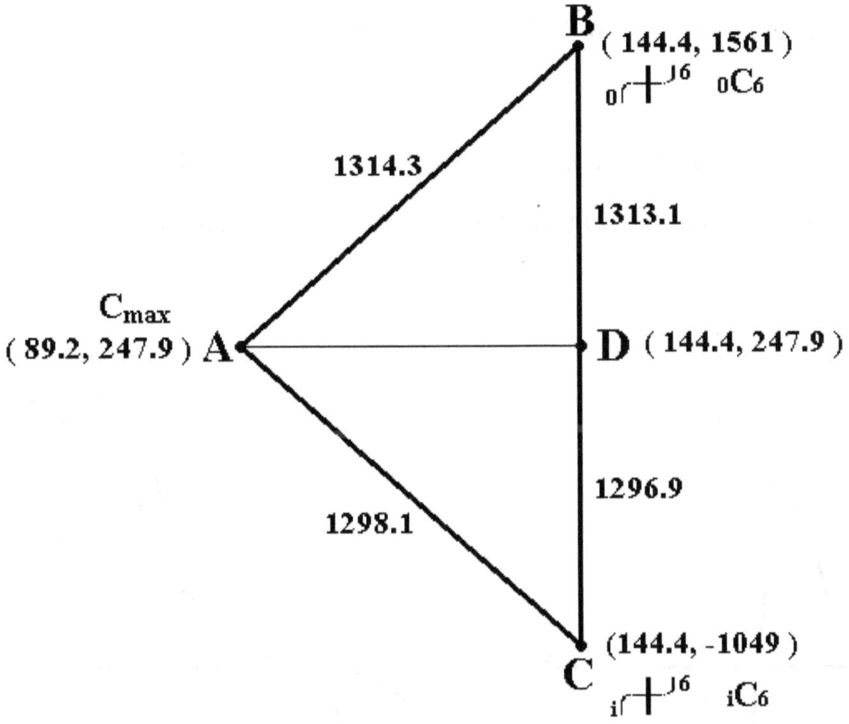

area of △ ABD = (1/2)bh = (1/2)(1313.1)(55.2) = 36,241.56

area of △ ADC = (1/2)bh = (1/2)(1296.9)(55.2) = 35,794.44

ratio of areas = 0.988

If we extend DC (downward) by (line)BD – (line)DC = 1313.1 – 1296.9 = 16.2 (forcing AC to go from 1298.1 → 1314.3 for the right triangle): the two Δ areas will be equal, Ψ^{6} remains at 144.4, and ${}_{i}C_{6}$ is off-curve to -1049 + (-16.2) = -1065.2 (on-curve, Ψ^{n} would be about 144.68, or +0.28 from ${}_{i}\Psi^{6}$). So, the off-curve new point for C : (144.68, -1065.2) , would correspond to a change in n as follows:

log144.68 = nlog2.2905 / log10 = nlog2.2905

$\therefore$ n = log144.68 / log2.2905 = 6.0023 ,

or a change in n of 6.0023 – 6.0000 = 0.0023 = 2.3 x 10^{-3} .

This is a very small change in n, to yield equal areas.

The curve can, of course, now be reflected about AD , to bring C (and its infinity, +∞) up to B. Therefore, the method seems to be reflection from C_{max}.

The region (or rxn. coordinate) between C_{max} and Ψ^6 is, of course, also changed, so that the modification is from C_{max} to all rising (increasing) Ψ^n. This removes C_{max} from being a maximum.

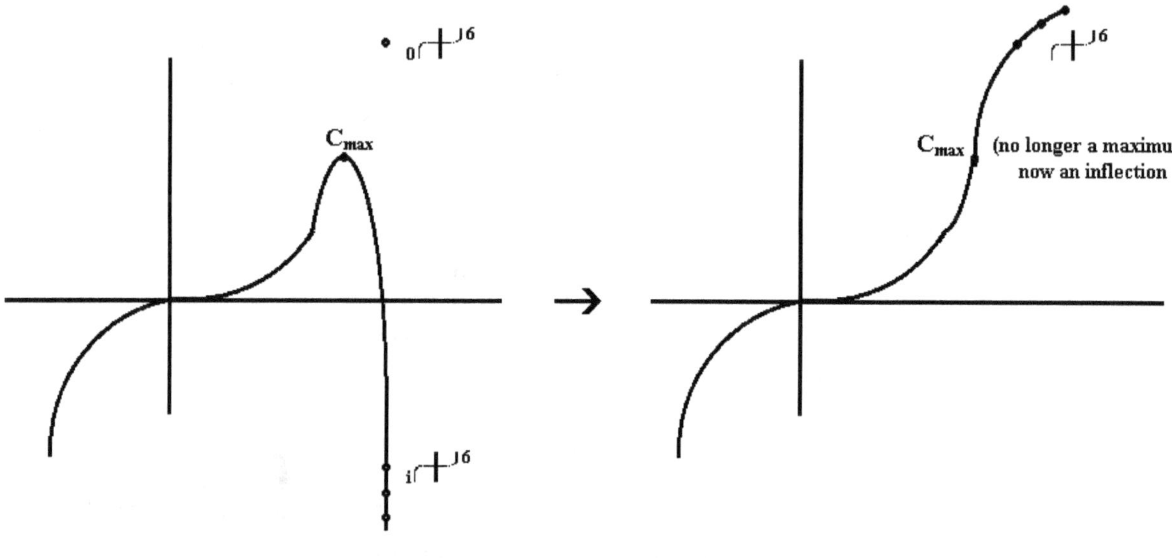

The conversion of C_{max} to an inflection is the functional equivalent of putting a "kink" in a line, with its opposite face (of line) now forward. It is a torsion. Or, it is a bending (keeping the line faces unperturbed or unaltered).

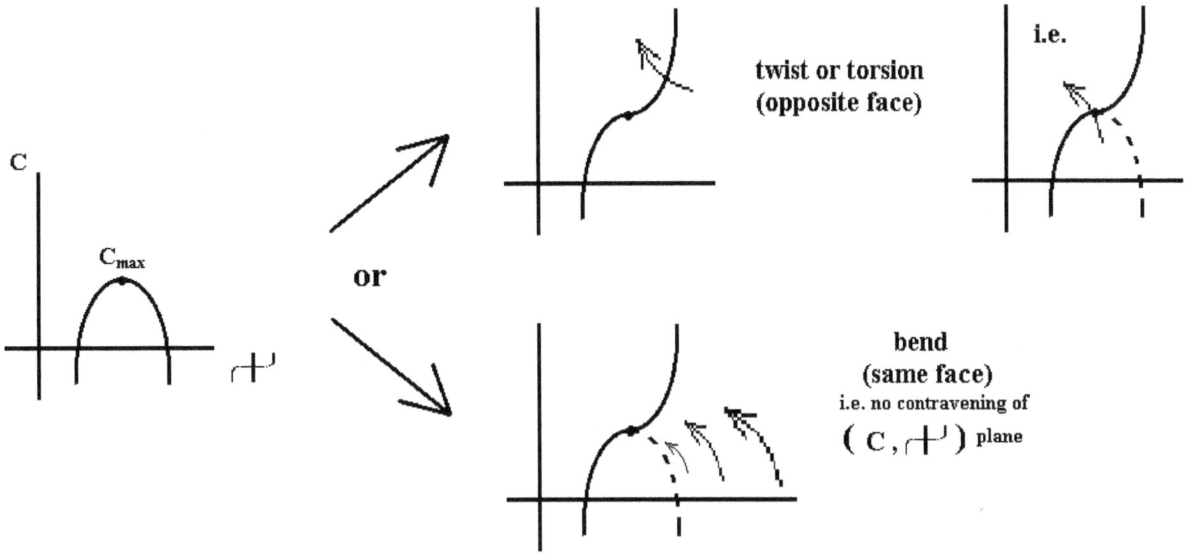

The distinction imparts, effectively, another dimension to the (Ψ^n, C_n) points: i.e., (Ψ^n, C_n) face forward ("front") and (Ψ^n, C_n) face backward (back"), with the formation of a kink: $C_{max} \rightarrow C_{kink}$ (or ${}_kC_n$).

The transformation $(\Psi^n, {}_{max}C_n)_f \rightarrow (\Psi^n, {}_{kink}C_n)_b$ might be considered purely mathematical (as for a reflection to not distinguish f from b).

But both transformations

$$(\Psi^n, {}_{max}C_n)_f \rightarrow (\Psi^n, {}_kC_n)_b$$

$$(\Psi^n, {}_{max}C_n)_f \rightarrow (\Psi^n, {}_kC_n)_f$$

may adopt (or represent) mathematical and/or physical (real) attributes.

The distinction is relevant (important) if the method of "Ψ^n correction" pertains to a physical process. The $(max)_f \rightarrow (kink)_f$ amounts to (or is analogous to) a melting and re-annealing, while $(max)_f \rightarrow (kink)_b$ to a twisting. Which is more energetically feasible depends on the relative difficulties of (Ψ, C) plane contravention.

But, the polynomial interpolation is hypothetical for mathematically predictive purposes, rather than mathematically operative executions.

At (graphically seen) :

C_{max}, ${}_{(i)}\Psi^{ni} = 89.2$; $\therefore$ $n_{(i)} = \log(89.2) \bullet 1 / \log(2.2905) = 5.42$ (or 5.41873) .

${}_{(i)}C_n = 247.9$ ["rxn_pool" yields ${}_{(o)}C_{5.419} = 200.3$, since the program allows for non-integer ns > 3 .]

Therefore, the "kink" can be functionally related to:

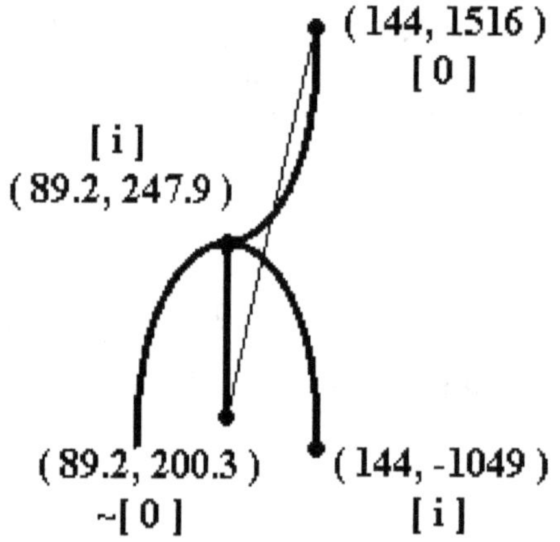

Point (89.2, 200.3) is only quasi-operational (~o), since operationally only integer values of n are used (to compute C_n).

A "Fine-structure Constant" Derivation

Now, we may come to the proposal that C_n reflects (of properties) in some way the value of the "fine-structure constant." We have noted that the reciprocal of this constant (i.e., α^{-1}) has digits which favor (in kind) the values of C_n with increasing n. With the risk of promoting a convoluted argument, we note that C_n from $n = 1 \rightarrow 10$ has the values:

1, 3, 7, 25, 161, 1561, 19615, 299601, 5380849, 111111120 .

α^{-1} has the value: 137.0359991 (or last digit is 2, from various refined measurements). The digits 1, 3, 7 are obviously given.

But what of those beyond (to the right of) the decimal point? How can we represent the multi-digited values of C_n to yield to the single digits of α^{-1} ?

First, we may simplify C_n to single digit values by simply adding digits until a single digit results:

e.g., $25 == 2 + 5 = 7$, etc.

This yields (from C_n) the digits (as ordered with increasing n) :

1 3 7 : 7 8 4 4 9 1 9

There is the apparent transition between the two sevens, where digits to the right should yield to fractional values.

The zero (of α^{-1}) would appear to come from: $7 - 7 = 0$; so we will follow this "directrix" (or manner of leading) :

0 | **0** **(scoring)**

↑

1 3 7 ⁝ 7 8 4 4 9 1 9

7 - 7 = **0**

The next (α^{-1}) digit would then appear to be from:

7 + 7 - 8

14 - 8

V

1 + 4 - 8

∴ 5 - 8 = -3

We assume -3 adopts its absolute value.

Thus,

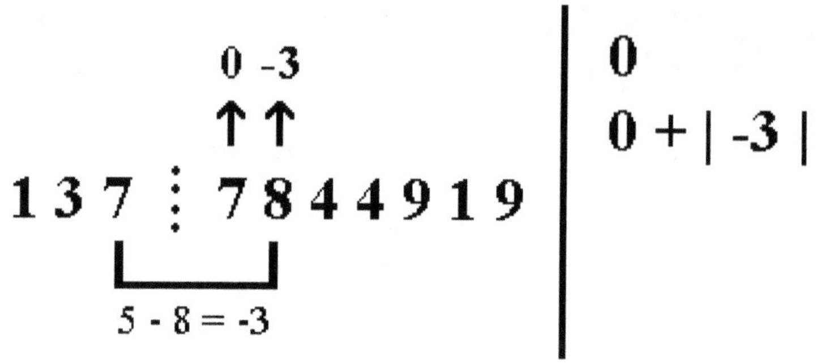
0 -3
137 7844919
5 - 8 = -3
0
0 + | -3 |

Therefore, we assume -3 adopts its absolute value.

Maintaining a symmetry (about digits), we extend one digit to the left and one to the right, for the 'directrix' calculations:

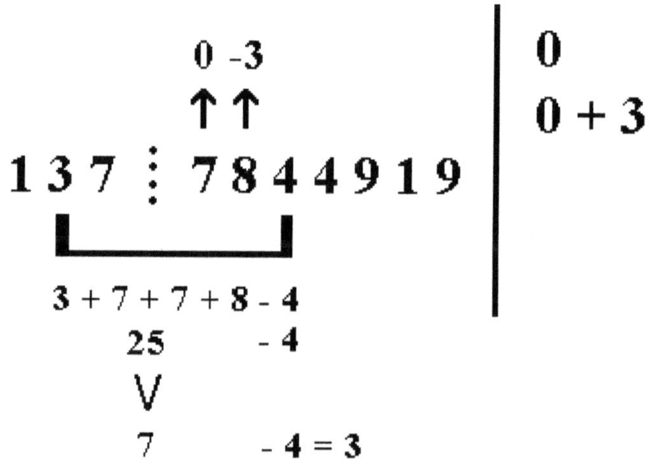
0 -3
137 7844919
3 + 7 + 7 + 8 - 4
25 - 4
V
7 - 4 = 3
0
0 + 3

To accommodate α^{-1}, we must apparently include the next left digit, to the 'directrix' operation:

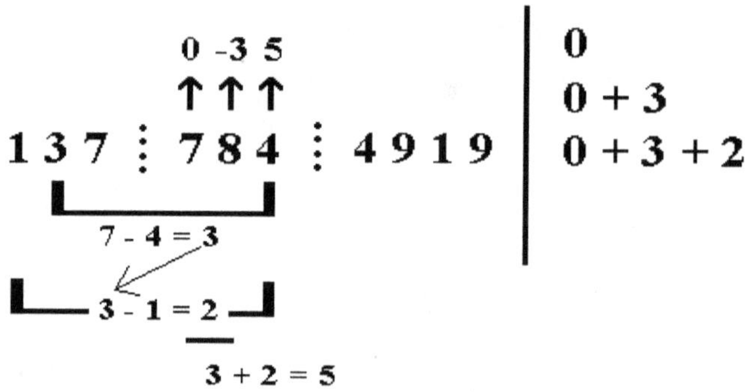
0 -3 5
137 784 4919
7 - 4 = 3
3 - 1 = 2
3 + 2 = 5
0
0 + 3
0 + 3 + 2

Now, we come to another transition, since there are no more left digits to employ. So, simple additions are used:

$$
\begin{array}{ccc}
0 & -3\ 5 & 9\ 9\ 1\ 1 \\
\uparrow\uparrow\uparrow & \uparrow\uparrow\uparrow\uparrow \\
1\ 3\ 7 : 7\ 8\ 4 : 4\ 9\ 1\ 9
\end{array}
$$

$$5$$

$$5 + 4 = 9$$

$$9 + 9 = 18$$
$$\lor$$
$$1 + 8 = 9$$

$$9 + 1 = 10$$
$$\lor$$
$$1 + 0 = 1$$

$$1 + 9 = 10$$
$$\lor$$
$$1 + 0 = 1$$

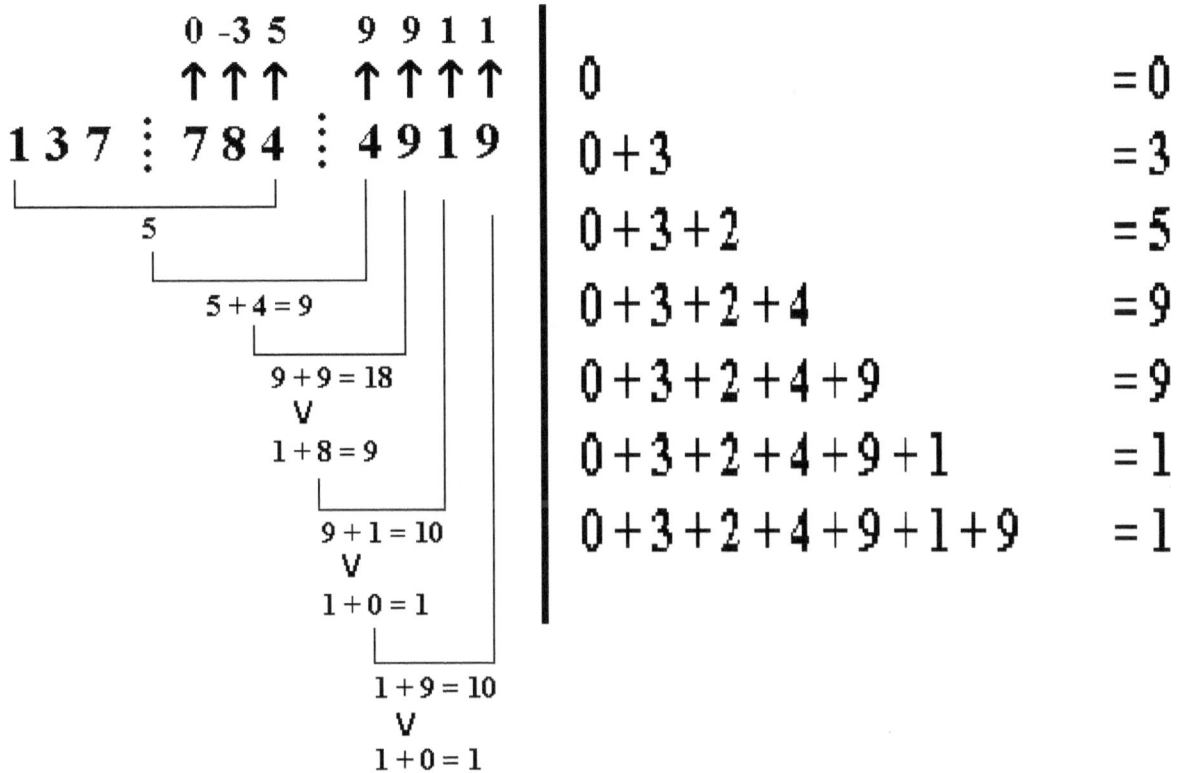

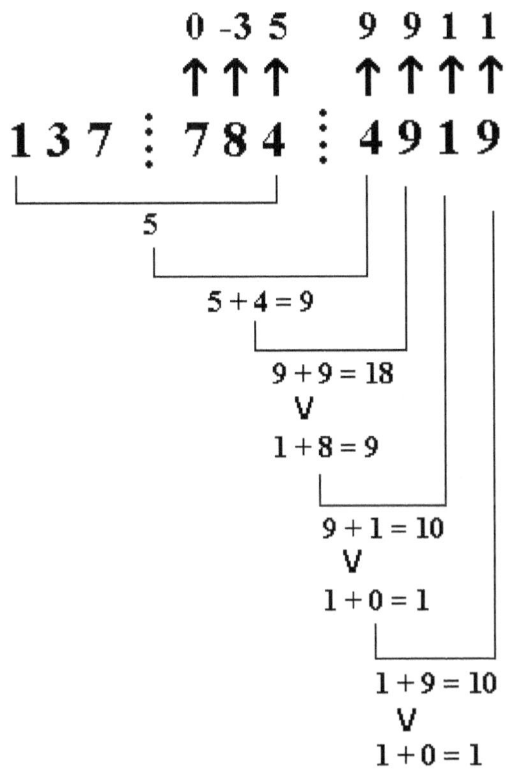

0	$=0$
$0+3$	$=3$
$0+3+2$	$=5$
$0+3+2+4$	$=9$
$0+3+2+4+9$	$=9$
$0+3+2+4+9+1$	$=1$
$0+3+2+4+9+1+9$	$=1$

The result (from $n = 1 \rightarrow 10$) is : 137.0359911 , compared to $\alpha^{-1} = 137.0359992$.

The manipulations, adjustments, and accommodations used to derive from C_n values an interesting approximation (or derivation) of the experimentally determined "fine-structure constant" seem in many respects to suggest an archaeological process. One can submit as to a postulate that when these modifications or discoveries are made only by numerical additions and subtractions, then (in some real nature) they may represent actual physical properties, assignments, activities or phenomena that can yield to the desired results. Numerical divisions, however, are more suspicious (due to changes in quality or kind, which might not be followed through towards actualities and observations). So, one must be careful to obtain reductions (divisions of parts) which nevertheless maintain a constant (i.e., definable) system (e.g., a pie still remains, though it can be divided into wedges of pie). Therefore, revelations from additions and subtractions are (conceptually) preferable than the need to resort to divisions.

Physics of Combinations

But, how can we use Ψ^n (for a physics) ?

Let us propose a conceptual problem:

Say we have 3 points arranged as such,

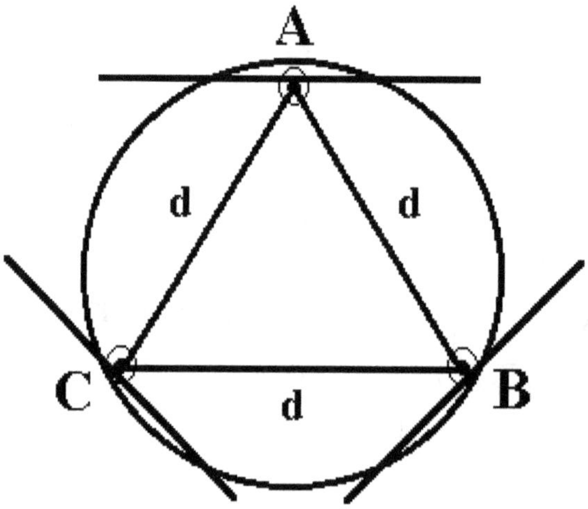

The physical characteristic is that a route starts from A and must end at A. Let us say that for the first analysis one is only at A at the beginning and at the end of the route. The distances between each of the points is always d. So, a route can be $A \rightarrow B \rightarrow C \rightarrow A$, etc. The physical restraint is that the time for each and any route must be at least t. The question asked is what is the maximum speed for a route. If we let the points (in a route) between "A"s (starting and ending) to number n, then the total distance for a route must be:

$$\Sigma d = (n + 1)d$$

Therefore, the maximum speed of a route is $Sp = (n + 1)d / t$

e.g.,

route	speed	route (repetitive)	speed
A → B → A	**Sp = 2d / t**	**A → B → C → B → A**	**4d / t**
A → C → A	**2d / t**	**A → C → B → C → A**	**4d / t**
A → B → C → A	**3d / t**	**A → B → C → B → C → A**	**5d / t**
A → C → B → A	**3d / t**	**A → C → B → C → B → A**	**5d / t**

etc.

Since $Sp = (n+1)d/t$, we can employ Ψ^n as follows:

$$Sp = (n+1)d/t = (dn+d)/t$$

$\therefore$ Let

$$\Psi^{Sp} = \Psi^{(dn+d)/t} = \sqrt[t]{\Psi^{dn+d}} = (\Psi^{dn+d})^{1/t} = (\Psi^{dn} \bullet \Psi^{d}) - \Psi^{t}$$

$$(\Psi^{Sp})^{t} = \Psi^{dn+d} = \Psi^{Sp \bullet t}$$

$$\Psi^{Sp \bullet t} = \Psi^{dn+d} = \Psi^{dn} \bullet \Psi^{d}$$

$$\frac{\Psi^{Sp \bullet t}}{\Psi^{d}} = \Psi^{dn} = (\Psi^{d})^{n} = (\Psi^{n})^{d}$$

$$\Psi^{Sp} = (\Psi^{dn} \bullet \Psi^{d}) - \Psi^{t}$$

$$\Psi^{Sp} + \Psi^{t} = \Psi^{dn} \bullet \Psi^{d}$$

$$\frac{\Psi^{Sp} + \Psi^{t}}{\Psi^{d}} = \Psi^{dn} = \frac{\Psi^{Sp \bullet t}}{\Psi^{d}}$$

Therefore, we have the relation:

$$\frac{\Psi^{Sp} + \Psi^{t}}{\Psi^{d}} = (\Psi^{n})^{d}$$

$$\therefore \quad \left(\frac{\Psi^{Sp} + \Psi^{t}}{\Psi^{d}}\right)^{1/d} = \Psi^{n}$$

or

$$\Psi^{Sp} = (\Psi^{n})^{d} \bullet \Psi^{d} - \Psi^{t}$$

(as determined above)

Note the evolution of n, for the routes:

n (routes)	n =
A → B → A	1
A → C → A	
A → B → C → A	2
A → C → B → A	
A → B → C → B → A	3
A → C → B → C → A	
A → B → C → B → C → A	4
A → C → B → C → B → A	
⋮	

Clearly (physically) n can not go to infinity, since a route is constrained by t. So, there must be a maximum n for a given t (and d), and therefore a maximum Sp.

We employ Ψ, instead of another base, since Ψ^n yields C_n.

Interestingly, if $Sp = d / t$ and we let $t = 1$ (for our problem, Sp can never be d / t but must be at least $2d / t$) we find that (regardless of base >= 1), as d increases n goes to zero:

$$\left(\frac{X^{Sp}+X^{t}}{X^{d}}\right)^{1/d} = \frac{(X^{Sp}+X^{t})^{1/d}}{X} \cong \frac{(X^{d}+X)^{1/d}}{X} \qquad (\text{for } t = 1)$$

$$\frac{(X^{d}+X)^{1/d}}{X} = X^{n} \xrightarrow[X \geq 1]{d\uparrow} 1$$

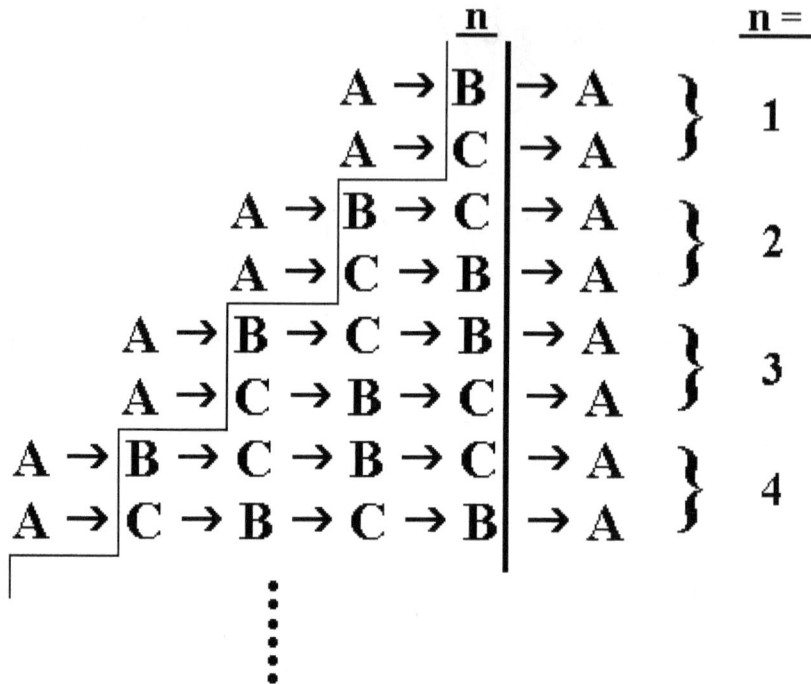

Therefore, $\log_x x^n = n = \log_x 1 = 0$ (n = 0)

For Sp = 2d / t (t = 1, regardless of d), as d↑, $x^n \to x^1$ (n = 1)

For Sp = 3d / t, $x^n \to x^2$ (n = 2)

= 4d / t, $x^n \to x^3$ (n = 3)

= 5d / t, $x^n \to x^4$ (n = 4)

Therefore, for Sp = ad / t, $x^n \to x^{a-1}$; thus, $n \to a - 1$ (n = a – 1)

But a is simply n + 1 ; ∴ $x^n \to x^{a-1} = x^{(n+1)-1} = x^n$

Therefore, n = a – 1 .

This is reasonable, since (for our model) the shortest route (2d) has n = 1 (as can be seen by the route-evolution diagram). The "speed" is of course for the process of going from and returning to A.

Since d can take any (positive) value (> 0), and t can be constrained to equal 1 (of some time unit used consistently), then Sp is entirely reflected of puissance (or strength and magnitude) by n.

In fact, if both d and t equal 1, then

$$\left(\frac{X^{Sp} + X^{t}}{X^{d}}\right)^{1/d} \equiv \frac{X^{Sp} + X}{X} = \frac{X^{Sp}}{X} + 1 = X^{n}$$

∴ $x^{Sp-1} + 1 = x^n$ or $x^{Sp-1} = x^n - 1$ (where Sp = ad / t = a • 1 / 1 = a).

∴ $x^{a-1} = x^n - 1$; but here n does not equal a – 1 , or Sp does not equal a :

$\log_x x^{a-1} = \log_x(x^n - 1)$ does not equal $\log_x x^n - \log_x 1$; $a - 1 = \log_x(x^n - 1) =?\ n$.

So, there appears to be an incongruity with Sp = (n + 1)d / t :

$$X^{n} = \left(\frac{X^{Sp} + X^{t}}{X^{d}}\right)^{1/d} \cong \frac{\left(X^{Sp} + X\right)^{1/d}}{X} \cong \frac{\left(X^{ad} + X\right)^{1/d}}{X}$$

when t = 1 → n = a – 1 , for Sp = ad / t ~= ad .

But

$$\mathbf{X^n = \left(\frac{X^{S_P} + X^t}{X^d}\right)^{1/d} \cong \frac{X^{S_P} + X}{X} = X^{S_P - 1} + 1}$$

when $d = 1$, $t = 1 \rightarrow n =/= a - 1$, for $Sp \sim = a(1/1) = a$.

$\therefore x^{Sp-1} = x^n - 1 \rightarrow x^{a-1} = x^n - 1$ or $a - 1 = \log_x(x^n - 1)$.

We can also try (an analysis) when $Sp = 1$; i.e., $d = t$ for $a = 1$, or $ad = t$:

$$\mathbf{X^n = \left(\frac{X^{S_P} + X^t}{X^d}\right)^{1/d} \cong \frac{(X + X^t)^{1/d}}{X} \cong \frac{(X + X^{ad})^{1/d}}{X}}$$

$$\therefore\ \mathbf{\log_x X^n = n = \log_x\left[\frac{(X + X^{ad})^{1/d}}{X}\right] = (1/d)\log_x(X + X^{ad}) - \log_x X}$$

$$\therefore\ \mathbf{n = \frac{\log_x(X + X^{ad})}{d} - \log_x X}$$

$$\mathbf{= \frac{a\log_x(X + X^{ad})}{t} - \log_x X^1}$$

$$\mathbf{= \frac{a\log_x(X + X^{ad})}{t} - 1}$$

Here, if (ad =) t = 1, then:

$$n = \frac{a \log_x \left(X + X^{ad} \right) - 1}{1} = a \log_x \left(X + X^1 \right) - 1$$

$$= a \log_x \left(X + X \right) - 1 = a \log_x \left(2X \right) - 1$$

$$= a(\log_x X + \log_x 2) - 1 = a(1 + \log_x 2) - 1$$

$$= a\left(1 + \frac{\log_{10} 2}{\log_{10} X}\right) - 1$$

Now,

$$a = \frac{(n+1)}{\left(1 + \frac{\log_{10} 2}{\log_{10} X}\right)} \qquad \therefore \; d = \frac{\left(1 + \frac{\log_{10} 2}{\log_{10} X}\right)}{n+1}$$

since ad = 1 = t = Sp .

So, we have three (magnitude) characteristics of a , from $Sp = a(d / t)$:

$a = n + 1$ when $t = 1$ and $Sp = ad$

$a = 1 + \log_x(x^n - 1)$ when $d = t = 1$ and $Sp = a$

$a = (n + 1) / (1 + \log_{10}2 / \log_{10}x)$ when $Sp = ad = t = 1$

These are magnitude characteristics (as opposed to unit types; i.e., we disregard units here).

In all three cases, effectively $Sp = ad$, or $|Sp| = |ad|$.

For employment of base Ψ, with Sp constrained to 1 (by a definition), then:

$a = (n + 1) / (1 + \log_2 / \log_{10}\Psi) \sim= (n + 1) / 1.83635$

For example, if d is given the Plank Length $l_P = 1.61619997 \times 10^{-35}$ meters , and t is kept to $(t =)$ 1 second, then

$a == 1 / d = 6.187353165 \times 10^{+34}\ m^{-1}(1\ m)$, and

$n = (1.83635)(6.187353165 \times 10^{34}) - 1 \sim= 1.13621 \times 10^{35}$

($Sp = c = 1$ by constraint, as much as l_P is dependent on the speed of light c.)

C_n would of course be much larger than n (here). How can one hope to calculate, or determine any significance to it?

One can annotate with Ψ^n:

$$\Psi^{1.13621 \times 10^{35}} = \Psi^{1.13621} + \Psi^{10^{35}} = \Psi^{1.13621} + 35 \cdot \Psi^{10}$$

$= 2.5643 + 35 \bullet (3975.086733) = 139130.5999 \sim= 1.39131 \times 10^5 \sim 1.4 \times 10^5$.

It might be a coincidence, due to the similarities of the coefficients, but:

Ψ^$1.13621 \times 10^{35} / k_B = 1.39131 \times 10^5 / 1.3806485279 \times 10^{-23}$ J/K
$= 1.0077 \times 10^{+28}$ K/J .

k_B, the Boltzmann constant, is not directly related to the Plank length, nor c, but is completely related to (the) statistical entropy (and probability).

$S = k_B \ln W$, where W = # distinct microscopic states of a system (with macroscopic constraints).

A similarity, then, might be expected (between Ψ^n and k_B),

since in a way C_n is similar to W.

The arguments are fanciful, but :

Ψ^(1.1362 x 10^{35}) =/= $\Psi^{1.13621}$ + Ψ^10^{35} =/= $\Psi^{1.13621}$ + 35 • Ψ^{10}

$\Psi^{1.13621}$ + Ψ^10^{35} =/= Ψ^(1.13521 + 10^{35})

$\Psi^{1.13621}$ • Ψ^10^{35} = Ψ^(1.13621 + 10^{35})

Ψ^(1.13621 x 10^{35}) = Ψ^(1.13621 • 1035) = ($\Psi^{1.13621}$)^10^{35} ~ = (2.5642)^10^{35}

(2.5642)^10^{35} == $(2.5642^{10})^{35}$ ~ = $(1.2289 \times 10^4)^{35}$

$(1.2289 \times 10^4)^{35}$ == $(1.2289^{35})(10^4)^{35}$ ~ $(1.35856 \times 10^3)(10^{140})$

$(1.35856 \times 10^3)(10^{140})$ = 1.35856×10^{143} = Ψ^n

C_n would be still larger. (Determined graphically?)

But, if $S = k_B \ln W$:

What does $k_B \log_Y C_n$ portend ?

What does $k_Y \log \Psi C_n$ portend ? ($S_Y / \log_Y C_n = k_Y$)

What does $k_B \ln C_n$ portend ? (How is C_n related to W ?)

What does $k_B \log_{10} C_n$ portend ?

The reduction: $a \bullet b^c \rightarrow a + c \bullet b$

is interesting, and perhaps can be used for a mathematics. It essentially means (or commands) the replacement of multiplication with addition throughout a function.

We are attempting to encompass physical (i.e., real) processes, so we will allow reflection for the operations:

$$\mathbf{a \cdot (b_1 + b_2 + \ldots.\ b_c) \rightarrow a + b^c \rightarrow a \cdot c \cdot b \leftrightarrows a + b + c}$$

$$\uparrow$$

$$\mathbf{a \cdot b^c \rightarrow a + c \cdot b \leftrightarrows a \cdot (c + b) \rightarrow (a + c)(a + b)}$$

$$\nwarrow$$

$$\mathbf{\rightarrow a \cdot c + a \cdot b \rightarrow (a + c)(a + b) \rightarrow a \cdot c + a \cdot b}$$

Then, for each of these reflective progressions we can assign an operator:

$$\mathbf{X^{a \cdot bc} \rightarrow X^a + X^{c \cdot b} \rightarrow X^{a \cdot (c + b)} \leftrightarrows X^a + X^{c \cdot b} \quad or \quad X^{a + c} \cdot X^{a + b}}$$

$$\uparrow\downarrow$$

$$\mathbf{X^{ac} + X^{ab}}$$

e.g.,

The arrows represent the flexible "reductions," rather than equalities.

$$\mathbf{X^a + X^{c \cdot b} \rightarrow X^a \cdot (X_{b_1} + X_{b_2} + \ldots.\ X_{b_c}) \rightarrow X^a + X^{b^c} \rightarrow X^a \cdot X^{c \cdot b} \leftrightarrows X^a + X^{b + c}}$$

$$\mathbf{\underbrace{b^c = b_1 \cdot b_2 \cdot b_3 \cdot \ldots.\ b_c}_{\downarrow} \leftrightarrows \underbrace{b_1 + b_2 + b_3 + \ldots.\ b_c = c \cdot b}_{\downarrow}}$$

$$\mathbf{c \cdot b \leftrightarrows c + b \leftrightarrows c \cdot b \leftrightarrows c + b}$$

Each "reduction" can go off through an equality (to end the progression).

e.g.

$$X^{a \cdot b^c} = (X^a)^{b^c}$$

$$\downarrow \not\uparrow$$

$$X^a + X^{c \cdot b} = X^a + (X^b)^c$$

e.g.

$$X^a \cdot X^{c \cdot b} = X^{a + c \cdot b}$$

$$\downarrow \uparrow$$

$$X^a + X^{b + c} = X^a + X^b \cdot X^c$$

Therefore, interesting (or applicable or useful) transitions can be found (or formed):

$$X^{a \cdot b^c} \rightarrow X^{a + c \cdot b} = X^a \cdot X^{c \cdot b} \rightarrow X^a + X^{b + c} \leftrightarrows X^a \cdot X^{bc} = X^{a + bc} \rightarrow$$

$$X^{a \cdot (b + c)} = X^{a \cdot b + a \cdot c} \leftrightarrows X^{(a + b)(a + c)} = X^{a^2 + ab + ac + cb} = X^{a^2} \cdot X^{ab} \cdot X^{ac} \cdot X^{cb}$$

$$X^{a^2 + ab + ac + cb} = X^{a(a + b + c) + cb} = X^{a(a + b + c)} \cdot X^{cb} \rightarrow$$

$$X^{a + (abc)} + X^{c + b} = X^a \cdot X^{abc} \cdot X^b \cdot X^c$$

$$X^a \cdot X^{abc} + X^b \cdot X^c \stackrel{\sim}{\rightarrow} (X^a + X^{abc})(X^b + X^c) \rightarrow X^a \cdot X^{a + b + c} + X^b \cdot X^c$$

$$\leftrightarrows (X^a + X^{a + b + c})(X^b + X^c) \rightarrow X^a \cdot X^{abc} + X^b \cdot X^c$$

etc.

This method (or "maths") can be used to increment exponents:

e.g.,

$$X^a = X^{a \cdot 1} \rightarrow X^{a+1} = X^a \cdot X^1 \rightarrow X^{a+1} + X^1 = X^a \cdot X^1 + X^1 \quad \textbf{etc.}$$

$$X^1 = X^{1 \cdot 1} \rightarrow X^{1+1} = X^2 = X^{2 \cdot 1} \rightarrow X^{2+1} = X^3 \quad \textbf{etc.}$$

$$\downarrow$$

$$(X^1)(X^1) \rightarrow X^{1+1} + X^{1+1} = 2X^2 \rightarrow 2 + X^{2+1} \quad \textbf{etc.}$$

$$2 + X^{2+1} = 2 \cdot 1 + X^3 \rightarrow (2+1) \cdot X^{3+1} = 3X^4 \quad \textbf{etc.}$$

$$X^a \cdot X^1 \rightarrow X^{a+1} + X^{1+1} = X^{a+1} + X^2 = X^a \cdot X^1 + X^2 \rightarrow (X^{a+1} + X^{1+1})(X^{2+1})$$

$$= (X^{a+1} + X^2)X^3 \rightarrow (X^{a+1} \cdot X^{2+1}) + X^{3+1} = X^{(a+1)+3} + X^4 \quad \textbf{etc.}$$

$$\rightarrow (X^{(a+1)(1+1)} \cdot X^{2+1}) + X^4 = X^{2(a+1)+3} + X^4 \quad \textbf{etc.}$$

We can "progress" the arrow symbol to :

$$\ni$$

which can be read "allows," through progressive reflective reduction.

So, while

$$\ulcorner\!\!+^{1.13621 \times 10^{35}} = \mathbf{1.35856 \times 10^{143}}$$

$$\ulcorner\!\!+^{1.13621 \times 10^{35}} \ni \mathbf{1.39131 \times 10^{5}}$$

$$\frac{\ni}{=} \approx \mathbf{1.024(106407) \times 10^{-138} = 1{,}024(.106407) \times 10^{-141}}$$

i.e., some physical means allows for this (as a result).

Therefore, this is (roughly) $2^{10} \times 10^{-141}$.

Or,

$2^x = 1024.106407$

$x = \log_2 2^x = \log_2 1024.106407 = \log_{10} 1024.106407 / \log_{10} 2 = 10.00014991$

$\Psi^y = 1024.106407$

$\log_\Psi \Psi^y = y = \log_\Psi(1024.106407) = \log_{10} 1024.106407 / \log_{10} \Psi = 8.363587646$

$\therefore \sim \Psi^{8.36359} \times 10^{-141}$ ($C_y \sim 388747.8$, from "rxn_pool").

[For $e^z = 1024.106407$, $z = 6.931575713$]

But this (numerical) analysis relies on the acceptance of l_p (Plank length) for our model problem (which is conceptual).

A mathematical solution (or lack of) is not proof of a physical resolution (or observance). They such solutions are more like arguments of (or for a) philosophy. But proofs are physical (results).

The famous paradox of dividing a length in half, and each half further, to infinity, is solved by saying that the (total) divisions would take too long. So, physically, there is a limit not only to the divisions but to the time for accomplishing the divisions.

For our model (problem), we are dividing the time allotted for each (any) route, for we have (essentially):

$Sp == (n + 1)d / (t / C_n) = C_n(n + 1)d / t$

As $n\uparrow$, $C_n\uparrow$ even more, so that $Sp\uparrow$ even more than that. But t is restricted to the least time for the (any) route, and d is taken (here) as a constant distance (or unit distance). For physicality (i.e., to avoid infinity), then, much as for the interpolated polynomial plots for (Ψ^n, C_n), C_n must reach a (real or actual) C_{max}. This will provide for a maximal speed. The n at C_{max} can be called n_{max}.

Therefore, fastest speed is $Sp \sim= C_{max}(n_{max} + 1)d / t$, or around here (depending on how the $(n + 1)$ factors affect the neighboring C_n values).

C_{max} may be found from its (physical) relationship to Ψ^{max}, i.e. the dependence of C_n on Ψ^n (physically; mathematically, or arithmetically, they may both go to $+\infty$).

Apparently, this relationship must be constrained to the particular model used (in its employment). Clearly, for our model, the "weight" of a route decreases with the number of steps taken (before return to A):

e.g., route $A \rightarrow B \rightarrow A$ "weighs" more than

$A \rightarrow B \rightarrow C \rightarrow A$ "

$A \rightarrow B \rightarrow C \rightarrow B \rightarrow A$ "

$A \rightarrow B \rightarrow C \rightarrow B \rightarrow C \rightarrow A$ " etc. ,

where "weight" == contribution of route to the process' success or fulfillment. The greater the number of steps (for a route), the more division of t, so the faster the speed, but also the less likely (or necessary) the adoption of the route for the process. But one can also argue that the faster the speed, the greater the energy employed (for the route) during the process.

The model, then, is capable of n (discreet) speeds. And each such speed can be arbitrarily weighted by Ψ^n, such that a relationship (Ψ^n, C_n) issues. This mapping (of Ψ^n) to the model allows for the physical (e.g., interpolated of polynomial) possibility of a C_{max} to develop:

(Ψ weighted Sp is) ${}^{\Psi}Sp = C_n(n + 1)d / \Psi^n t$

In fact, an "average speed" can be calculated (or estimated):

$$Sp_{avg} = \frac{\Sigma\, {}^{\Psi}Sp}{n} = \frac{\sum_{i=1}^{n} \frac{C_i(i + 1)d}{\Psi^i t}}{n}$$

Mathematically (arithmetically), n can still go to $+\infty$ yielding an average speed of essentially zero (via division by $+\infty$).

Notably, when $n = 1$, the weighted speed becomes:

${}^{\Psi}Sp = C_1(1 + 1)d / 2.2905t = 1(2)d / 2.2905t = 2d / 2.2905t = 0.873d / t$
$(\sim 1.3131d / t$, from $\Psi_m^{\ 1}$) ; $\Psi^1 \sim= 2.2905$

The weighted Sp, of course, continues to increase dramatically with rising n.

Therefore, a useful relationship to draw on (or consider, contemplate) is:

$$^{\Psi}Sp \bullet \Psi_{nmax} = C_{nmax}(\, n_{max} + 1\,)d \,/\, t = Sp_{(max)} \quad (== \text{speed of } C_{max}\,)\,.$$

Combinational Energy of a System

But what is the physical attribute (of our model) that can lead to (or contribute to) a C_{max}?

Obviously it is due to polynomials of higher powers (of Ψ^n), their derivations assumed to be the same as for a (or the) correct function, for (Ψ, C).

Since each route is seen to be the representation of a (characteristic) energy (and therefore, speed), the total energy of the model or system (that may be observed) is multiplicative (say) to powers of Ψ^n (where Ψ^n represents an independent, though knowable or calculable, variable).

Therefore, we can arrange that the energy of the system is to be related as follows, with $x = \Psi^n$ $(y = C_n)$:

$$\mathbf{E} \cong \sum_{i=1}^{n} \frac{C_i(\,i+1\,)d}{\Psi^{i}t} \equiv \sum_{n=1}^{\infty} \frac{C_i(\,n+1)d}{(\Psi^{n})^{n}t} = \sum_{n=1}^{\infty} \frac{C_i(\,n+1)d}{\Psi^{n^2}t}$$

The $x^i == (\Psi^n)^i = (\Psi^n)^n = \Psi^{n \bullet n} = \Psi^{n\wedge 2}$ is essentially quadratic (x^2), since $i = n$.

But the greater the energy and thus speed of a route, the lower (or more difficult) its contribution to the total system energy (and the lower its "weight").

A quadratic relationship implies a parabolic function (with maximum or minimum vertex).

So, we have (defined, for summary) "sculptural" mathematics:

the speed of a route is $Sp = C_n(\, n + 1\,)d \,/\, t$

the weight of a route is $1 \,/\, \Psi^n = \Psi^{-n}$

the weighted speed of a route is $\mathbf{{}^{\Psi}Sp = C_n(\, n + 1\,)d \,/\, \Psi^n t}$ for $\mathbf{{}^{\Psi}Sp}$;

$\therefore$ $\mathbf{C_n \,\alpha\, \Psi^n}$ (by definition they are inversely proportional) ;

the energy contribution of a route is $E_n \sim= C_n(n + 1)d / \Psi^{n\wedge 2}t$

$$\sim == \Psi^{-n} \bullet C_n(n + 1)d / \Psi^{n}t \sim == \Psi^{-n} \bullet {}^{\Psi}Sp_{(n)}$$

If we assign $(\Psi, C) == (x, y)$, then:

$y = [E_n t/(n + 1)d]x^2$ with x always greater than zero.

Note that the size (numerical quantity or magnitude) distribution is:

$n < \Psi^n < C_n < \Psi^{n\wedge 2}$, or $n < x < y < x^2$

Mathematically, it is C_n that is used to define Ψ^n. For our arguments the dependencies are reversed (in order to find a C_{max}): $n < C_n > \Psi^n < \Psi^{n\wedge 2}$.

The (parabolic) vertex for each E_n contribution of (x, y) occurs at $x = 0$, which is not within the domain of Ψ^n. That would argue C_{max} is never obtained.

Let us study the discernment of the (relational) definitions:

$${}^{\Psi}Sp = C_n(n + 1)d / \Psi^{n}t \ , \Psi^n \sim\sim n / C_n$$

Therefore,

$${}^{\Psi}Sp \sim\sim C_n(n + 1)d / \Psi^{n}t = C_n(n + 1)d / (n / C_n)t = C_n^{\,2}(n + 1)d / nt =$$
$$C_n^{\,2} \bullet (n+1)/n \bullet d/t$$

$$E_n \sim= C_n(n + 1)d / \Psi^{n\wedge 2}t \ , \Psi^{n\wedge 2} \sim == \Psi^n \bullet \Psi^n \sim\sim \Psi^n \bullet n/C_n$$

[let $\Psi^{n\wedge 2} \rightarrow (\Psi^n)^2$ instead of $\Psi^{n\wedge 2} = \Psi^{n \bullet n} = (\Psi^n)^n$]; therefore,

$$E_n \sim= C_n(n + 1)d / \Psi^{n\wedge 2}t \sim\sim C_n(n + 1)d / \Psi^n(n/C_n)t =$$
$$C_n/\Psi^n \bullet (n+1)/n \bullet C_n \bullet d/t$$

$$\sim\sim \text{(for large } n\text{)} \ C_n/\Psi^n \bullet 1 \bullet C_n \bullet d/t = C_n^{\,2}/\Psi^n \bullet d/t = C_n \bullet C_n/\Psi^n \bullet d/t$$

${}^{\Psi}Sp / (n + 1) = C_n d / \Psi^{n}t = C_n/\Psi^n \bullet d/t$; therefore, $E_n \sim\sim {}^{\Psi}Sp \bullet C_n / (n + 1)$

["$E_n \sim\sim$" indicates the relation does not yet result in energy units.]

So, for large n:

$^{\Psi}Sp \sim\sim C_n \bullet C_n \bullet d/t$ and $E_n \sim\sim {}^{\Psi}Sp \bullet C_n / (n+1)$, or

$\mathbf{E_n \sim\sim d \bullet C_n \bullet C_n \bullet C_n / [t \bullet (n+1)] = C_n^{\,3} / (n+1) \bullet d/t, {}^{\Psi}Sp \sim\sim C_n^{\,2} \bullet d/t}$

Therefore, for a given velocity or speed, (d / t),

$^{\Psi}Sp$ rises by $C_n^{\,2} / 1$,

E_n rises by $C_n^{\,3} / (n+1) \sim C_n^{\,3} / n = C_n / 1 \bullet C_n^{\,2} / 1 == {}^{\Psi}Sp \bullet C_n / n$

But more actually, by these definitions:

$E_n \sim= C_n(n+1)d / \Psi^{n\wedge 2}t = C_n(n+1)d / \Psi^{n \bullet n}t$

$== \Psi^{-n} \bullet C_n(n+1)d / \Psi^{(n-1)n}t$

$\sim\sim C_n / n \bullet C_n(n+1)d / \Psi^{(n-1)n}t = C_n^{\,2} \bullet (n+1) / n \bullet 1 / \Psi^{(n-1)n} \bullet d/t$

[where $\Psi^n \sim\sim n / C_n$]

Note: $1 / \Psi^{n\wedge 2} = 1 / \Psi^{n \bullet n} = 1 / (\Psi^n)^n == 1 / [(\Psi^n)_1(\Psi^n)_2 \ldots (\Psi^n)_n = (\Psi^{-n})_1 \bullet 1 / (\Psi^n)_2(\Psi^n)_3 \ldots (\Psi^n)_n] == \Psi^{-n} \bullet 1 / (\Psi^n)^{n-1} == \Psi^{-n} \bullet 1 / (\Psi^n)_{2\to 1}(\Psi^n)_{3\to 2} \ldots (\Psi^n)_{n\to n-1}$

or

$1 / \Psi^n \bullet 1 / (\Psi^n)^{n-1} == 1 / \Psi^{n+n(n-1)} = 1 / \Psi^{n+n\wedge 2-n} = 1 / \Psi^{n\wedge 2} == 1 / \Psi^{n \bullet n}$

Therefore,

$E_n \sim= C_n(n+1)d / \Psi^{n\wedge 2}t \sim\sim C_n^{\,2} \bullet (n+1) / n \bullet 1 / \Psi^n(n-1) \bullet d/t$

$== C_n \bullet (n+1) / n \bullet C_n / \Psi^n \bullet 1 / \Psi^{n(n-2)} \bullet d/t =$

$(C_n / \Psi^n \bullet d/t) \bullet C_n \bullet (n+1)/n \bullet 1 / \Psi^{n(n-2)}$.

$^{\Psi}Sp / (n+1) = C_n / \Psi^n \bullet d/t$

Therefore,

$E_n \sim\sim {}^{\Psi}Sp / (n+1) \bullet C_n \bullet (n+1)/n \bullet 1 / \Psi^{n(n-2)}$

$= {}^{\Psi}Sp \bullet C_n / n \bullet 1 / \Psi^{n(n-2)}$;

for large n, $^{\Psi}Sp \sim\sim C_n^{\,2} \bullet (n+1) / n \bullet d/t \sim C_n^{\,2} \bullet d/t$

Therefore,

$$E_n \sim\sim d / t \bullet C_n^{\,2} \bullet C_n / n \bullet 1 / \Psi^{n(n-2)} = C_n^{\,3} / n \bullet 1 / \Psi^{n(n-2)} \bullet d / t \ .$$

Thus, ${}^{\Psi}Sp$ rises by $C_n^{\,2} / 1$ (for a given d / t)

and E_n rises by $C_n^{\,3} / n \bullet 1 / \Psi^{n(n-2)} = C_n / n \bullet C_n^{\,2} / 1 \bullet 1 / \Psi^{n(n-2)}$

$== {}^{\Psi}SpC_n /n \bullet 1 / \Psi^{n(n-2)}$.

More generally:

$$E_n \sim\sim [\, C_n^{\,2} \bullet (\,(n{+}1)/n\,) \bullet d / t \,] / (\, n + 1 \,) \bullet C_n \bullet (\, n + 1 \,) / n \bullet 1 / \Psi^{n(n-2)}$$

$$= C_n^{\,3} \bullet (\, n + 1 \,) / n^2 \bullet 1 / \Psi^{n(n-2)} \bullet d / t \,.$$

Methods of Discernment

But, what is an "energy" contribution to the process (for our model)? There is no mass involved, and the mathematical manipulations are obviated by the fact of the process' being or extant occurring (i.e., only a result is observed or is of comment). The mental thought, thereby, for the process to contemplate is an "effort" to characterize (attributes) for the process its operative existence. Therefore, "E" is for the efforts of the speeds (and velocities) constrained within a time (t) due a route, and this effort remains in the units of speed (as d / t).

[More canonically, one can (arbitrarily) "unitize" components of the (basic) relation to yield an energy using the following associations:

Energy == work = force x distance = mass x acceleration x distance

= (m)(distance/time^2)(distance) = (m)(d/t^2)(distance)

== (a mass)(a distance / a time)(d / t) ; so, we can say

C_n represents a mass, (n + 1) represents a distance, and $\Psi^{n\char`^2}$ represents a time.

Thus, $C_n \bullet (\, n + 1 \,) / \Psi^{n\char`^2} \bullet (\, d / t \,) \rightarrow$ an "energy" E_n , with these analogies.

A more restrictive unitization might be:

$E_n = C_n \bullet (n+1) / \Psi^{n\wedge 2} \bullet (d/t) ==$

$C_n \bullet (n+1) / \Psi^{n\wedge 2} \bullet a(n+1) / b\Psi^{n\wedge 2} =$

$(a/b) \bullet C_n \bullet (n+1)^2 / (\Psi^{n\wedge 2})^2 = (a/b) \bullet C_n \bullet (n^2 + 2n + 1) / \Psi^{2n\wedge 2}$,

with the units restricted to the appropriate factors a and b.]

E_n, then, is simply Sp "double" weighted (or $^{\Psi}$Sp further weighted), by these (arbitrary) definitions. The algebraic minutiae can only be treated as descriptive, to make the model particular (of its parts). The various discernments of E_n are generative, of a recursion of calculations, with E_n gradually increasing:

e.g., let n = 3 ; thus $C_n = 7$, $^{\Psi}Sp = 2.330(d/t)$

$E_n \sim= C_n(n+1)d / \Psi^{n\wedge 2}t \rightarrow 0.016(d/t)$

$E_n \sim\sim C_n^{\ 2} \bullet (n+1)/n \bullet 1/\Psi^{n(n-1)} \bullet d/t \rightarrow 0.452(d/t)$

$E_n \sim\sim C_n^{\ 3} / n \bullet 1/\Psi^{n(n-2)} \bullet d/t \rightarrow 9.514(d/t)$

$E_n \sim\sim C_n^{\ 3} \bullet (n+1)/n^2 \bullet 1/\Psi^{n(n-2)} \bullet d/t \rightarrow 12.686(d/t)$

The "evolution" is apparently:

$C_n^{\ 1} \bullet (n+1)/1 \bullet 1/\Psi^{n \bullet n} \rightarrow C_n^{\ 2} \bullet (n+1)/n \bullet 1/\Psi^{n(n-1)} \rightarrow$

$C_n^{\ 3} \bullet 1/n \bullet 1/\Psi^{n(n-2)} \rightarrow C_n^{\ 3} \bullet (n+1)/n \bullet (1/n) \bullet 1/\Psi^{n(n-2)}$

This "evolution of effort" has predictive elements (or characters):

$(n+1)/1 \rightarrow ((n+1)/1)(1/n) \rightarrow ((n+1)/1)(1/n)(1/(n+1)) \rightarrow$

$((n+1)/1)(1/n)(1/(n+1))(1/n)((n+1)/1)$.

So, one might predict (for example, at n = 3):

$C_n^{\ 3+1} \bullet ((n+1)/1)(1/n)(1/(n+1))(1/n)((n+1)/n)(1/n)(1/(n+1)) \bullet 1/\Psi^{n(n-2-1)}$

$\rightarrow C_n^{\ 4} \bullet (1/n^3) \bullet 1/\Psi^{n(n-3)} \rightarrow 88.925$

and

$$C_n^{4+1} \bullet ((n+1)/1)(1/n)(1/(n+1))(1/n)((n+1)/1)(1/n)(1/(n+1))(1/n)$$

$$\bullet ((n+1)/1) \bullet 1/\Psi^{n(n-3-1)}$$

$$\rightarrow C_n^{5} \bullet (n+1)/n^4 \bullet 1/\Psi^{n(n-4)} \rightarrow C_n^{5} \bullet 4/81 \bullet 1/\Psi^{-3} = 9973.694$$

etc.

But the jump from the C_n^4 terms to the C_n^5 terms is tremendous (due to the negative exponent for Ψ). It is like crossing the inflection point of a titration curve.

[For the C_n^6 terms, the yielded value is only 69,913.934 (i.e., less than 10 times greater than for C_n^5).]

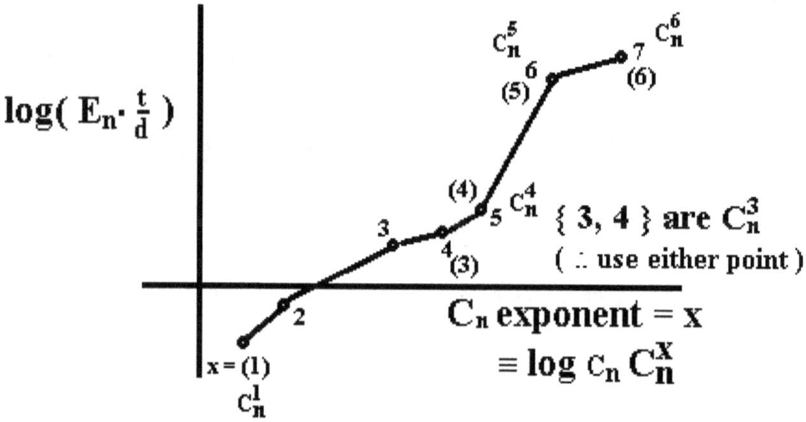

log / log plot of E_n discernments

∴ Discernment methods show and emphasize theoretical (mathematical) behaviors of a quantity (algebraically manipulated). But they also demonstrate a possible mathematics about the quantity. (For example, they can be used to effect predictions, without assessment of correctness.) Repetitive or recursive patterns may be looked for or discerned. (Mathematical methods may be "brokered" to achieve this.)

We see here that E_n continues to increase (without bound) for a particular n, via Ψ^n and C_n.

The "brokering" of mathematical methods refers to the distortion or "error-ing" of a technique, which can be done consistently throughout an extended operation or inconsistently if well annotated. Here, the use of (n / C_n) as a simplification of

$\Sigma(n / C_n)$ is useful, but might not be as interesting as employing (once demonstrated or recognized) a common mistake of converting $(x^n)(x^n) \rightarrow x^{n^2}$ instead of x^{n+n} ;

for $x^{n^2} == (x^n)^n$, or (x^n) multiplied of itself n times.

A comparison of the results for E_n are illustrative. With the error we have $E_n \sim\sim {}^{\Psi}Sp \bullet C_n / (n + 1)$ instead of the correct: $E_n \sim\sim {}^{\Psi}Sp \bullet C_n / n \bullet 1 / \Psi^{n(n-2)}$.

When n is large they (both results) are clearly related by the ${}^{\Psi}Sp \bullet C_n / n$ term, and distinguished by the (significant) $1 / \Psi^{n(n-2)} == \Psi^{2n - n^2}$ term.

Therefore, the common factor (amongst the two results) of ${}^{\Psi}Sp \bullet C_n$ is most notable, and the factor ${}^{\Psi}Sp \bullet C_n / n$ is employable (for large n).

The $\Psi^{2n - n^2}$ term is apparently responsible for the inflection (or transition) crossing seen in the $(\log_{C_n} C_n^{\,x}, \log[E_n \bullet t / d])$ plot. The negative exponent for Ψ (i.e., $2n$-n^2), as n increases, tremendously "explodes" the E_n value, since n^2 rises much faster than 2n.

The (this) term also becomes necessary of the predictive value of the approach (to a discernment for E_n), since $C_n^{\,x+a}$ is (here) as important as $\Psi^{an - n^2}$.

The conversion of $(1 / n) \rightarrow (1 / n) \bullet (n+1 / 1)(1 / n)$ and the alternating use of factor pairs: $(n+1 / 1)(1 / n)$ and $(1 / n+1)(1 / n)$, are rapidly (and easily) employed for predictive usage.

This does not mean (nor demonstrate) that the predictions are correct, but they (or this method) display patterns which may be recognizable during actual (physical) situations, much like a "fingerprinting" for recurrences. The prediction is of course an (educated or inferred) assumption for the evolution of terms (rather than a deduction from those terms previous).

The use of alternating factor pairs (as above) is of intra-$C_n^{\,x}$ prediction, while the a incrementing of C_n and $\Psi^{n(n-a)}$ is of inter-n prediction. Therefore, the intra-$C_n^{\,x}$ prediction converts points to the more general (of discernments), while the inter-n prediction creates points for the discernment. The points are, essentially, of (Ψ^n, C_n) character.

'Discernments' methods (as theory or illustration) are allowed because:

1) all physics are un-erring (no physical occurrence is in error)

2) the idealism of math provides for no "error-able" paths (any mathematical pathway and construction of operations can be provided for).

In fact, errors may uncover practices (or reveal usable assumptions). Presumptions describe practices (physics), as the physics remain unconcerned (nor consulting) of them.

The expiation of a $C_n^{\ x}$ term (intra-$C_n^{\ x}$ prediction) for predictive extension might seem somewhat artificial (for the inter-n predictions). But this is overpowered by the usefulness of its employment. (It follows an apparent trend.)

So, from these discernments we may study (or at least try to project) the natures of particular C_n (and thus n) values. For example, they suggest a transition occurs between n = a and a > n (due to the Ψ^{an-n^2} factor; but see below, there is no such effect on the slopes):

a = n $\qquad \Psi^{an - n^2} = \Psi^{n^2 - n} = \Psi^0 = 1$

a = n + 1 $\qquad \Psi^{an - n^2} = \Psi^{(n+1)n - n^2} = \Psi^{n^2 + n - n^2} = \Psi^n$

[For a = n + b , $\Psi^{an - n^2} = \Psi^{bn}$]

[For a = n – 1 , $\Psi^{an - n^2} = \Psi^{-n}$; $0 < \Psi^{-n} < 1$]

Therefore, a transition occurs between (terms for) $C_n^{\ n+1}$ and $C_n^{\ n+1+1}$ (n = 3; therefore between terms $C_3^{\ 4}$ and $C_3^{\ 5}$) ;

i.e., the steepness of the slope $\Delta\log(E_n \bullet t/d)_{Cnx}/\Delta\log_{Cn}C_n^{\ x}$ is (logarithmically) greatest (up to this point) between $C_n^{\ n+1}$ and $C_n^{\ n+1+1}$, and the greater is n the greater the slope at this transition (since it is due to the change in magnitude of $\Psi^{an - n^2}$ going from $1 \rightarrow \Psi^n$).

This "transition junction" occurs differently (with respect to n) for different n, and is dependent on the $\Psi^{an - n^2}$ factor (for the $C_n^{\ x}$ boundaries);

e.g..	for	$n = 4, C_n = 25$:	the steep transition occurs between	$C_n^{\ 4}$ and $C_n^{\ 5}$
		$n = 5, C_n = 161$:	"	$C_5^{\ 4}$ and $C_5^{\ 5}$
		$n = 2, C_n = 3$:	"	$C_2^{\ 4}$ and $C_2^{\ 5}$
		$n = 1, C_n = 1$:	"	$C_1^{\ 4}$ and $C_1^{\ 5}$
		$n = 6, C_n = 1561$:	"	$C_6^{\ 4}$ and $C_6^{\ 5}$

At least for n = 1 → 6 (or 7), the 1st most steep slope occurs between $C_n^{\ 4}$ and $C_n^{\ 5}$:

(i.e., $C_n^{\ 6}/C_n^{\ 5}$ is less than $C_n^{\ 4}/C_n^{\ 5}$) .

n	1	2	3	4	5	6	7
C_n^5/C_n^4	4.6	23.6	112.2	860.1	1.2 x 104	2.6 x 105	7.4 x 106
C_n^6/C_n^5	1.15	2.62	7.01	34.40	338.3	5.3 x 103	1.16 x 105

The C_n^x terms here are the coefficients for E_n : $E_n \sim\sim (C_n^x \text{ term})(\, d / t \,)$

These coefficients simply follow from the/this predictive model, and this is probably most due to the $(\, n+1 / 1 \,)(\, 1 / n \,)(\, 1 / n + 1 \,)(\, 1 / n \,)$.... pattern (for the relative slopes).

The $C_n^4 \rightarrow C_n^5$ transition is purely predictive (since the C_n^4 terms on are due to predictions from C_n^1, C_n^2, C_n^3 E_n discernments).

The method is probably as intellectually valid as polynomial interpolation, for yielding higher powers of a variable or coefficient, but of more sophistication.

Now, we can continue the discernment, from C_n^3, to find the C_n^4 power without prediction, and then compare:

$E_n \sim\sim C_n^3 \bullet (n+1)/n^2 \bullet 1/\Psi^{n(n-2)} \bullet d/t = (C_n/\Psi^n \bullet d/t) \bullet (n+1)/n^2 \bullet C_n^2 \bullet 1/\Psi^{n(n-3)}$.

${}^{\Psi}Sp/(n+1) = C_n/\Psi^n \bullet d/t$; therefore, $E_n \sim\sim {}^{\Psi}Sp/(n+1) \bullet (n+1)/n^2 \bullet C_n^2 \bullet 1/\Psi^{n(n-3)}$.

${}^{\Psi}Sp \sim\sim C_n^2 \bullet (n+1)/n \bullet d/t$, $(\Psi^n \sim\sim n/C_n)$; therefore, $E_n \sim\sim$ $[C_n^2 \bullet ((n+1)/n) \bullet (d/t)/(n+1)] \bullet C_n^2 \bullet 1/\Psi^{n(n-3)} \bullet (n+1)/n^2 =$ $(n+1)/n^2 \bullet C_n^2/n \bullet d/t \bullet C_n^2 \bullet 1/\Psi^{n(n-3)} = C_n^4 \bullet 1/n \bullet 1/\Psi^{n(n-3)}$ $\bullet\, d/t \bullet (n+1)/n^2 = C_n^4 \bullet (n+1)/n^3 \bullet 1/\Psi^{n(n-3)} \bullet d/t$.

The prediction uses a $(\, 1 / n^3 \,)$ factor, instead of the recursion's $(\, n+1 / n^3 \,) \sim (\, 1 / n^2 \,)$.

Therefore, a better (more correct, but still assuming the $\Psi^n \sim\sim n/C_n$ approximation) prediction appears to be:

$E_n \sim\sim C_n^x \bullet (n+1)/n^{x-1} \bullet 1/\Psi^{n(n-x-1)} \bullet d/t$.

This allows for $a = x - 1$: thus $E_n \sim\sim C_n^x \bullet (n+1)/n^a \bullet \Psi^{an - n\text{^}2} \bullet d/t$.

[Compared to the previous prediction method, because of the consistency of the n term factor $(n+1)/n^a$: the C_n^x term ratios yield a constant (ratio of coefficients);

e.g., for $n = 3$, $k \sim= 28$ (or 28 times); for $n = 4$, $k \sim= 172$; etc.]

All prediction (methods) are valid as they prescribe (and describe) a possible physics (real physical manner). (All theories are allowable, w/o proofs.) A physics does not do math, but can be (crudely) described by maths.

This better prediction (method) appears to provide, for all x, some effort (E_n) with speed (d / t), as the coefficient to speed remains greater than zero (n =/= 0).

[When n = 0, $E_n \sim\sim C_n^{\ x} \bullet (0+1)/0^a \bullet \Psi^0 \bullet d/t$, $(0+1)/0^a = 1/0 \rightarrow +\infty$ (or is undefined). Does $C_0^{\ x} == 0$, or is it undefined?]

The arithmetic is ambiguous. If $C_0^{\ x} = 0$, then zero times anything is zero; thus, $E_n = 0$. But zero can be factored out (if the maths allow):

$C_0^{\ x} \bullet 1/0 \bullet 1 == 0 \bullet 1/0 \bullet 1 == 1 \bullet 1 = 1$; then $E_n = (1) \bullet d/t = d/t$.

Therefore, a problem with (interpretation of) zero arrives (for its manipulation).

The contention is that a (truly) physical process can adopt either prediction method (or others). The 2nd method totally obscures the change in slope(s) predicted by the 1St. Therefore, the 2nd method makes the relationship (between E_n and C_n, Ψ^n) "regular" (unvarying slope). There is no transition (change of slope) between $C_n^{\ 4}$ and $C_n^{\ 5}$ for the 2nd method, as there is for the 1st. The 2nd method seems to remove the transition effect of a = n (for the slope). [In fact, though postulated, the a = n effect seems obscured even for method 1 (with changing slopes).] For method 1, the $C_n^{\ 4} \rightarrow C_n^{\ 5}$ slope is bounded above and below by equal slopes (i.e., $C_n^{\ 3} \rightarrow C_n^{\ 4}$ slope = $C_n^{\ 5} \rightarrow C_n^{\ 6}$ slope). So, changes in slope are due to variances of n term factor. For method 2 the form of this factor doesn't change, so neither does the slope.

Another discernment apportions (allows) E_n to be essentially speed modified by powers of C_n, since:

$\Psi^n \sim\sim n / C_n$; therefore, $1 / \Psi^n \sim\sim C_n / n$

Thus, $E_n \sim\sim (C_n^{\ x})(\text{ n term / n })(C_n^{\ y}) \bullet d / t$, or $C_n^{\ x+y} \bullet (\text{an n term}) \bullet d / t$

This is a simpler definition of effort (for the model), each route confined to a span of time t.

Therefore, the function (here) of Ψ^n is to provide powers for C_n. Thus, "weighted" Sp enhances C_n through exponential powers (compared to unweighted Sp). This follows for the definition of Ψ^n (in relation to C_n).

Speed of Light Comparisons

Again, it occurs like a curiosity from numerology, but if we calculate ${}^{\Psi m}Sp$ using $n = 11$ ($C_{11} = 2{,}593{,}742{,}336$), and allow $d = 1$ meter, $t = 1$ second, then:

${}^{\Psi m}Sp = C_n(n + 1)d / \Psi_m{}^n t == C_{11}(12) / \Psi_m{}^{11} \bullet 1\ m_{eter} / 1\ s_{econd} =$ 3.04162423×10^8 m/s ,

which is somewhat close to the accepted vacuum speed of light:

$c = 2.99792458 \times 10^8$ m/s .

We make a "vacuum" in a container with walls. Where the space meets the vacuum wall there is no vacuum. So it (the vacuum) is not "pure." [One might argue that it is not physically possible to achieve a true vacuum, so any measurement of c will be less than numerically possible. It also does not escape notice that at $\Psi_m \sim= 1.5231$ and $e \sim= 2.7183$ then $\Psi_m{}^e = 3.1384$, which is close to $\pi \sim= 3.1416$; the proper base whose e exponent yields or brings it to π is: $(\pi)^{1/e} \sim= 1.5237$.]

Aspects of Combinatorial Speed

We posited earlier that, for E_n, unitage of C_n (coefficient) was (of type) mass,

that of $(n + 1)$ was distance, and that of $\Psi^{n\hat{}2}$ was time. Likewise (in a like manner), for ${}^{\Psi m}Sp$, unitage of $\Psi_m{}^n$ (coefficient) must be mass times distance (of a motion, $m\bullet d$), in order to yield a (proper) speed. If the mass motion is of a constant velocity, then we have a momentum. If the mass is accelerating, then a force develops. So we have (for our model) unit defining characteristics for both $\Psi^{n\hat{}2}$ and $\Psi_m{}^n$, with perhaps Ψ^n being similarly related.

$\Psi_m{}^n$ is most directly related to C_n, so $C_n / \Psi_m{}^n$ contributes directly to powers of C_n.

Some aspects of ${}^{\Psi m}Sp$:

Since it is a speed (or intellectual movement or motion), t is catered to never be zero, and Ψ^n by its (exponential for base) nature is always positive and greater than zero.

Therefore, ${}^{\Psi m}Sp = C_n(n + 1)d / \Psi_m{}^n t$ is presumably always greater than zero (and if directional, as for a velocity, then directed by the “sign” of t or the direction(s) of d).

Still, a division represents a transformation of quantity qualities; and with

Ψ_m^n **always** > 0 , it is more appropriate to conform the relation as:

${}^{\Psi m}Sp \bullet t = C_n(n + 1)d / \Psi_m^n$, representing distance(s).

But why should 1 meter devolving to c, for t = 1 second, be more characteristic than any other distance measure defined for unity (as well as for time measurements or extents)? It would appear, then, that n = 11 (or n + 1 = 12) is specific for meter/second units (unless this development is incidental). We might contort the representation, with the implied insistence that d = 1 (meter) and t = 1 (second), that

$\{d\}_{meter} = \{t\}_{second} = 1$, and therefore $\{\Psi_m^n Sp\} = \{C_n(n+1)/\Psi_m^n\} \rightarrow n = 11$

or $\{{}^{\Psi m\,n}Sp\}/\{t\}second = \{C_n(n+1)/\Psi_m^n\} / \{d\}meter \rightarrow n = 11$

(This latter representation is more transformational, due to the divisions.)

Therefore, $\{{}^{\Psi m\,n}Sp \bullet \Psi_m^{11}\} = \{C_{11}(12)\} == c \bullet \Psi_m^{11}$ (c as meters/second).

Hence, it does not escape notice that:

$\{C_{11}/\Psi_m^{11}\} = \{{}^{\Psi m\,n}Sp/12\} == c / 12$.

The formalism, therefore, of C_n / Ψ^n and C_n / Ψ_m^n may be exemplary of the utility of C_n modes and modalities, and is (by its divisional nature) both transformative and symbolic (or “symbiotic') of powers of C_n to employ.

Since C_n and $\Psi_{(m)}^n$ are both (pure or only) numbers, the division “melds” these numbers to a “legitimacy,” as compared to a disparate division such as d / t , which can only be a representation that nevertheless requires a distinction between (units of) distance and time. The lack of recognizing “legitimacies” of terms, or of ignoring illegitimacies, yields to unacceptable notions (physically), such as the increase of a mass to infinity (by reaching the speed of light), and a mass-less photon nevertheless exhibiting a momentum. It must be remembered (and maintained) that physically there are no nulls nor infinities (in reality), and that such quantities only occur mathematically through measurements. But all measurements require (the employment) of definite and real masses, and as such can only be representative of real masses regardless of the mathematical idealisms developed. Therefore, treating something like time as a same property as (x, y, z) space (to yield space-time coordinates) must still be considered illegitimate (of coupling), mathematically remedied through various “fudges” of acceptance for convenient descriptions of (mathematical and through this physical) notions (i.e., time and space do not “meld” as do pure numbers).

An obvious mathematical function (or operation), to modify a relationship, then occurs as:

$(C_n / \Psi^n)(n + 1)$ and $(C_n / \Psi_m^{\ n})(n + 1)$,

which might be symbolized as $\boldsymbol{N}$ and $\boldsymbol{N}_m$.

One can also try:

$(n! / e^n)(n + 1) == \boldsymbol{N}\bullet e$, for any significance.

A more proper notation, for these operations, may then be: $\boldsymbol{N}_{\Psi}$, $\boldsymbol{N}_{\Psi m}$, $\boldsymbol{N}_e$.

Complexities and Concentric Circles Plotting

A novel type of coordinate system, distinct from say Cartesian and polar, can be constructed to highlight "legitimacy" and illegitimacy (or "dis-legitimacy") relationships. This one is of simple concentric circles:

e.g., for (x, y) :

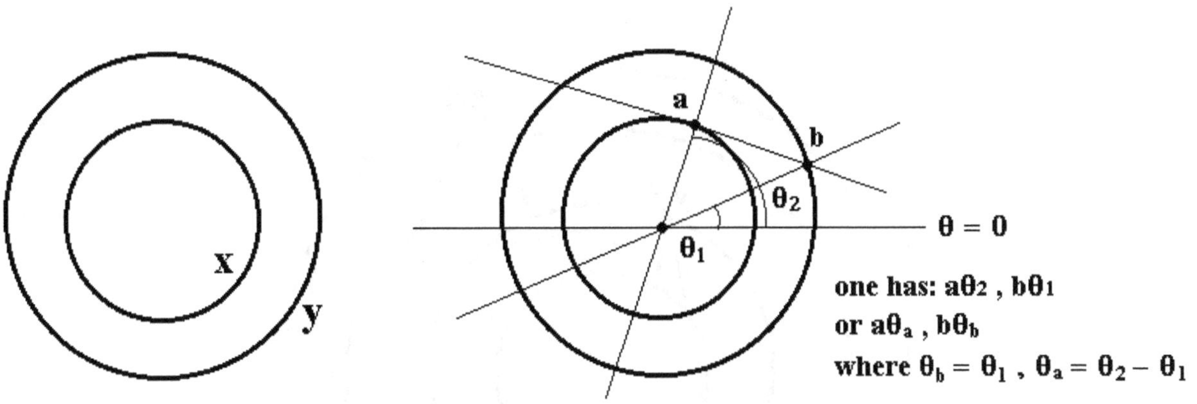

With a units == b units (in kind or proportionality), a straight line can always be drawn between a and b. But this is true even if the units are different, since there are only two points (and that can always define a line). This is why a disparate relationship like (d / t) can "seem" legitimate. For dis-legitimate relations, then, one assumes that the line between points is a curve (thus, not necessarily straight). The base line, to define measure of θs, need not be at zero (and for nul/los, would not be), but rather at some convenient number (or expression) that relates to the points plotted. For example, for an (a, b, c) point the base line can be formed from (a to b), to relate to c, with some starting (or initial) θ value associated to the base (line).

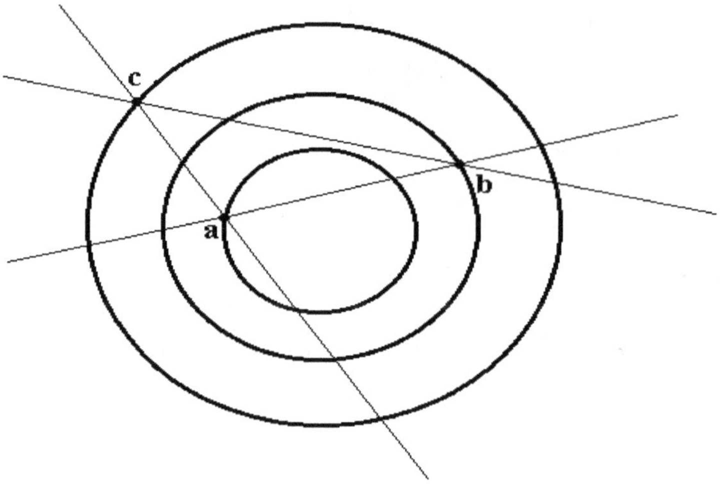

Again, if common units are used for a, b, c, then the lines connecting them are straight.

So now, for something like space-time coordinates: (x, y, z, t) , the lines connecting x, y, z are straight; while (for) lines from t to (plane x, y, z), x, y, z are assumed to be (undetermined) curves:

e.g.,

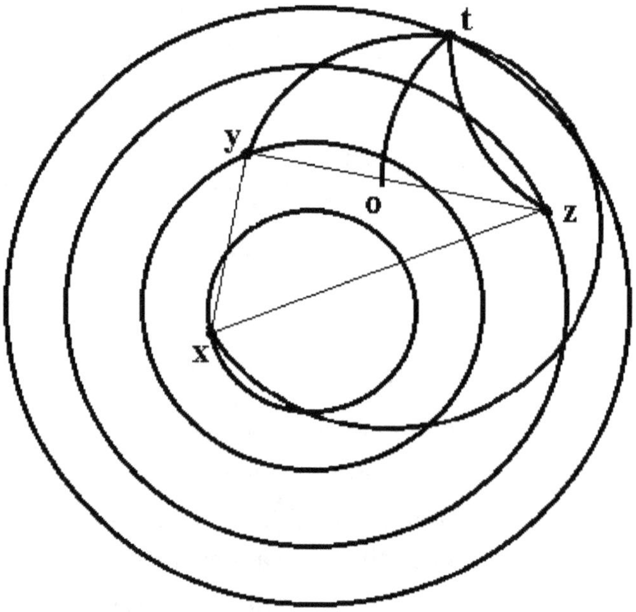

To convert curved to straight lines (to make the relations "seem" legitimate) the concentricity of the t circle(== axis) must be altered (i.e., distorted).

Therefore,

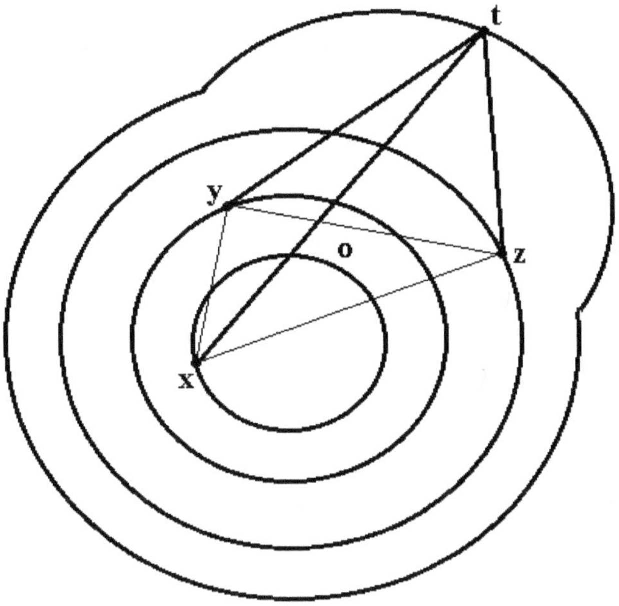

The curved lines, or the t-distortion(s) can be related to (perhaps) *N* operations:

(x, y, z)*N*t , x*N*t , y*N*t , z*N*t .

Here, (x, y, z), x, y, z serve a similar function as d for our model (i.e., distances):

thus, d*N*t == *N* • d / t , x*N*t == *N* • x / t , y*N*t == *N* • y / t , z*N*t == *N* • z / t , (x, y, z)*N*t == *N* • (x, y, z) / t

This assumes that the curves are due to speeds (and velocities). Therefore, some value of n "solves" for the correct curves relating (x, y, z) to (t). n leads to a set of combinations (between x, y, z and t) proper for coordinates (x, y, z, t) to be proper (or valid), to a considered system.

It is clear from the Concentric Circles Plotting (**CCP**) that:

1) each coordinate is only relative to the others (directionally)

2) each coordinate is always positive

3) each coordinate is never zero, because each always occurs (is extant) as functional (thus nul/los).

It is also interesting how an area (e.g., x, y → x • y) is condensed to a line, and a volume (e.g., x, y, z → x • y • z) is condensed to a plane. An innumerable number of coordinate types (circles == dimensions) may be plotted.

(some) ***N*** coefficients :

	n = 1	2	3	4	5
C_n	1	3	7	25	161
n!	1	2	6	24	120
Ψ^n	2.2905	5.2464	12.0169	27.5246	63.0451
$\Psi_m{}^n$	1.5231	2.3198	3.5333	5.3816	8.1968
e^n	2.7183	7.3892	20.0859	54.5996	148.4182
N_Ψ	0.4366 x 2	1.7155	2.3301	4.5414	15.3224
$N_{\Psi m}$	0.6566 x 2	3.8796	7.9246	23.2273	117.8509
N_e	0.3679 x 2	0.8120	1.1949	2.1978	4.8512

Of course, (x, y, z) represents a point in 3D space.

Using Concentric Circles Plotting, this is exploded to (at most) a triangular plane fragment. For a fourth coordinate, e.g. t, which is dis-legitimate, its relationship to (x, y, z) is revealed as more complicated (or sophisticated) than between two quantities. For a legitimate 4th coordinate, a volume seems indicated to evolve. We must distinguish, however, between coordinates and points. (x, y, z) and (x, y, z, t) are both points, but they have (respectively) increasing coordinate complexities. ***N*** operations are (forms of) measurements of those complexities (in terms of combinations).

Complexity Relationships

We see (again) that Ψ^n more favors e^n (and that N_Ψ coefficients more favor N_e coefficients) than $\Psi_m{}^n$ (and $N_{\psi m}$). Some interesting comparisons ensue:

If we consider complexity due to n!, then, say for n = 2 ,e.g. x / t, compared to n = 4 , e.g. (x, y, z) / t , n = 4 is (24 / 2 =) 12 times more complex than n = 2 .

If we consider complexity due to C_n, then n = 4 is (25 / 3 =) 8.3 times more complex than n = 2 (i.e., less complex than for n!).

But when considering ***N*** coefficients (here $N_{\psi m}$),

n = 4 is (23.2273 / 3.8796 =) 5.9870 times as complex than n = 2 ;

while for N_{e},

n = 4 is (2.1978 / 0.81205 =) 2.7067 times as complex than n = 2 .

[N_{Ψ}: n = 4 is (4.5414 / 1.7155 =) 2.6473 times as complex than n = 2 .]

We also note that (for example):

(x), (x, y), (x, y, z) can each be conduced to a distance

$$(x^2)^{1/2}, (x^2 + y^2)^{1/2}, (x^2 + y^2 + z^2)^{1/2}$$

to maintain a "speed" relationship with t

$$d_{(x)} / t, d_{(x, y)} / t, d_{(x, y, z)} / t.$$

But this is only formulaic manipulation, to maintain the d / t relationship. Still, complexities are reduced (by such manipulations). Yet, the more complex N coefficients can be applied to the derived (but less complex) d.

[This is akin to the question of a particle behaving as a wave, and visa versa. The particle is (of course) not a wave, but in large quantity (when it is its own medium, like the molecules of a body of water) the collection may take on wave characteristics. The complexity of the wave behavior can still (via coefficients) be "transmuted" or deemed (defined) for each individual particle. The (N) coefficients, therefore, characterize the (wave) system to which each particle belongs (or behaves of).]

Let's apply comparisons of N complexities of d / t relationships to two actual (x, y, z, t) points.

Choosing $N_{\Psi m}$ with n = 2, 4 ; therefore (using) $\Psi_{m}^{n=2,4}$.

Say (for) points (1, 2, 3, 4) and (5, 6, 7, 8) :

$\mathbf{d_{(1,2,3)} = (1^2 + 2^2 + 3^2)^{1/2} = 3.74}$; therefore, $\mathbf{d_{(1,2,3)} / t = 3.74 / 4 = 0.94 == d_{(1,2,3)} / 4}$

$\mathbf{d_{(5,6,7)} = (5^2 + 6^2 + 7^2)^{1/2} = 10.49}$; therefore, $\mathbf{d_{(5,6,7)} / t = 10.49 / 8 = 1.31 == d_{(5,6,7)} / 8}$

Assuming complexities (for a system) can be additive, (then)

(1, 2, 3), n = 2 : t = 4, $N_{\Psi m}$ = 3.88 ; therefore, (3.88 / 4)(1 + 2 + 3) = 5.82

(5, 6, 7), n = 2 : t = 8, $N_{\Psi m}$ = 3.88 ; therefore, (3.88 / 8)(5 + 6 + 7) = 8.73

(1, 2, 3, 4), n = 4 : t = 4, $N_{\Psi m}$ = 23.23 ; therefore, **(23.23 / 4)(0.94 x 4) = 5.46 x 4 = 21.84**

(5, 6, 7, 8), n = 4 : t = 8, $N_{\Psi m}$ = 23.23 ; therefore, **(23.23 / 8)(1.31 x 8) = 3.80 x 8 = 30.40** (or **30.43**)

Therefore, the complexities rise as :

$(1 + 2 + 3) \rightarrow (5 + 6 + 7) \rightarrow (1, 2, 3, 4) \rightarrow (5, 6, 7, 8)$.

For the $N_{\Psi m}$ complexity to move from $(1, 2, 3, 4)$ to $(5, 6, 7, 8)$:

$d_\Delta = [\,(5-1)^2 + (6-2)^2 + (7-3)^2\,]^{1/2} = [\,3(4^2)\,]^{1/2} = 6.93$; $\Delta t = 8 - 4 = 4$, $n = 4$

Therefore, complexity == (23.23 / 4)(6.93) = 40.24 (or 40.25)

Thus, the complexities rise as: $(1 + 2 + 3) \rightarrow (5 + 6 + 7) \rightarrow (1, 2, 3, 4) \rightarrow$

$(5, 6, 7, 8) \rightarrow (5, 6, 7, 8) - (1, 2, 3, 4)$.

The complexities are in units of the relationship: d / t .

A plot of the rise in complexity might look like:

$(5, 6, 7, 8) - (1, 2, 3, 4) == (1, 2, 3, 4) \rightarrow (5, 6, 7, 8)$

$(1 + 2 + 3) == (1 + 2 + 3 + 4)$ or $(1, 4) + (2, 4) + (3, 4)$ [as (d, t)]

$(5 + 6 + 7) == (5 + 6 + 7 + 8)$ or $(5, 8) + (6, 8) + (7, 8)$

"point ordered" speeds (d/t)	$\left(\frac{1}{4} + \frac{2}{4} + \frac{3}{4}\right)$	$\left(\frac{5}{8} + \frac{6}{8} + \frac{7}{8}\right)$	(1, 2, 3, 4)	(5, 6, 7, 8)	(1, 2, 3, 4) → (5, 6, 7, 8)
summed (d/t)	$\frac{(1+2+3)}{4} = 1.50$	$\frac{(5+6+7)}{8} = 2.25$	0.94	1.31	6.93 = 1.73
n	2	2	4	4	4
$N_{\Psi m}$	3.88	3.88	23.23	23.23	23.23
complexity (N • d/t)	5.82	8.73	21.84	30.43	40.25

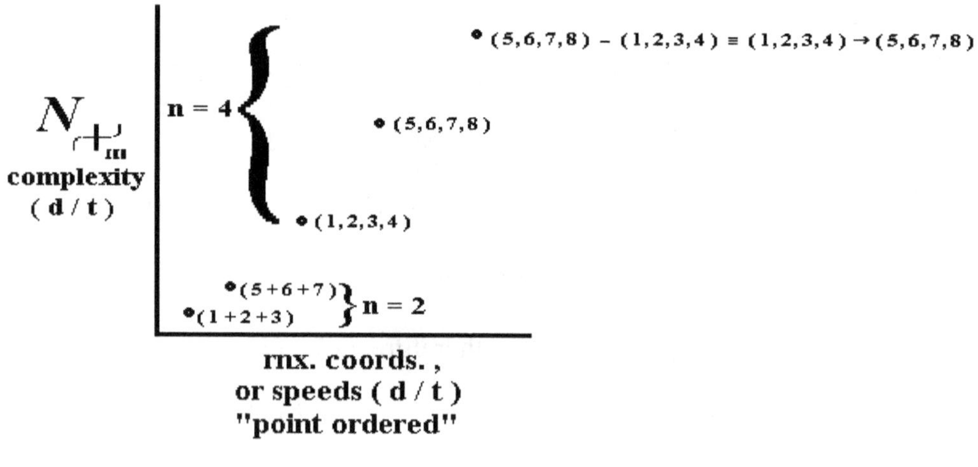

Therefore, the use of N tends to regularize (the increase of) the complexities of a given relationship (here, d / t speeds from coordinates and points), acting much like a linear slope (factor).

Correspondingly, using additive speeds:

$(1 + 2 + 3)$, $t = 4 \rightarrow (5 + 6 + 7)$, $t = 8$ at $n = 2$, $\Delta t = 4\ (= 8 - 4)$

yields $[\ (5 + 6 + 7) - (1 + 2 + 3)\ ] / \Delta t = 12 / 4 = 3$;

Therefore, $N_{\Psi m}$ complexity == (3.88)(3) = 11.64

(i.e., it runs from 5.82 → 8.73 → 11.64 for additive speeds).

We note that, upon plotting the points (6) with complexity = y and x = each point (ordered) as a multiple of 10 (so that x for point 6 = 60, etc.), this point 3 (for (1, 2, 3) → (5, 6, 7)) lies close to the averaged straight line formed from points 4 → 6 (i.e., n = 4) and , of course, lies part of the line for points 1 → 3 (i.e., n = 2) :

slope from points 4 → 6 : (40.25 – 21.84) / (60 – 40) = 0.9205 ;

thus, for point 3 :

0.9205 = (40.25 – 11.64) / (60 – x) , x = 60 - [(40.25 – 11.64)] / 0.9205 = 28.92 (i.e., close to x = 30).

Therefore, point 3 is a transition from n = 2 to n = 4, and shares slopes (for n = 2 and n = 4)

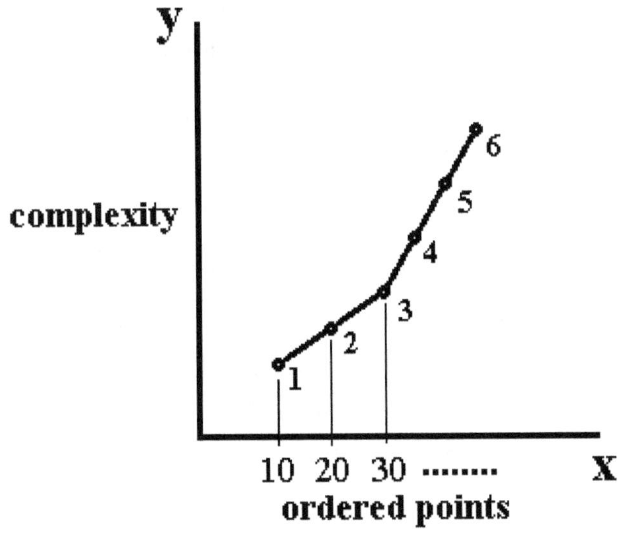

It is like a "hinge-point" or link of slopes ('Gelinkpunkt' or 'Pistengelink').

Point 3 is at 100%(28.92 / 30) = 96.4% (of 30) of the idealized position based on the slope for points 4 → 6 . If we now remove point 3, so that there are 5 points total, and allow that they are all equally spaced along x coordinate (such that again each point represents a multiple of 10), we find point 2's relationship relative to points 3 → 5 :

slope from points 3 → 5 : (40.25 – 21.84) / (50 – 30) = 0.9205

Therefore, for point 2

0.9205 = (40.25 – 8.73) / (50 – x) , x = 50 - [(40.25 – 8.73)] / 0.9205 ; thus, x = 15.76 == 78.8% of 20 .

Likewise, if we now remove point 2, so that there are 4 points total, (in the same manner) the relationship of point 1 to points 2 → 4 is found:

slope from points 2 → 4 : (40.25 – 21.84) / (40 – 20) = 0.9205

Therefore, for point 1

0.9205 = (40.25 – 5.82) / (40 – x) , x = 40 - [(40.25 – 5.82)] / 0.9205 ; thus, x = 2.60 == 26% of 10 .

So we see that, for points 1 → 6, points 1 → 3 approach the slope (per point) of points 4 → 6 much more rapidly (%-wise) than their own (1 → 3) slope (== 0.291) rises. Therefore, the (n = 2, 4) convergence (of complexities) is "forceful" or purposeful.

(For the 6 points,) That point 3 does not exactly occur within the sloped line produced by points 4 → 6 suggests that point 3 is at a "hinge-region" for the transition for n = 2 to n = 4 .

If we use Concentric Circles Plotting as a representation of a point, say (x_1, y_1, z_1, t_1), with t_1 dis-legitimate of (x, y, z), then we can have

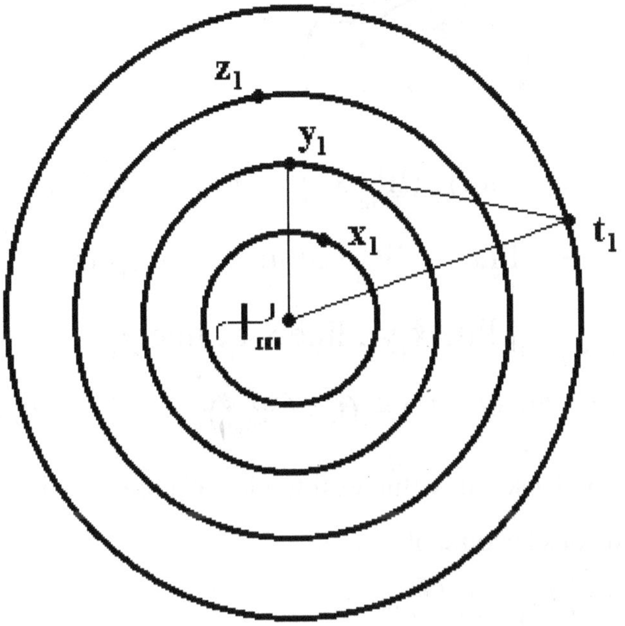

to consider the speed (relation):

$v = y_1 / t_1$ (as an example) .

The center $== \Psi_m$.

The triangle $\Delta(\Psi_m\ y_1\ t_1)$ represents $(n + 1)$, center $== \Psi_m = 1$

$n == y_1 + t_1 ==$ # lines to Ψ_m (center) → exponent of $\Psi_m = n$ (thus, $\Psi_m{}^n$)

Therefore, $n = 2$; $C_n = C_2 == y_1 + t_1 +$ line $y_1 t_1 = 3$.

Therefore,

$$N\Psi_m == (C_n / \Psi_m{}^n)(n + 1) = (C_2 / \Psi_m{}^2)(2 + 1) = (3 / \Psi_m{}^2)(3) = 9 / \Psi_m{}^2 = 3.88$$

To consider the speeds: x_1 / t_1 , y_1 / t_1 (as $n = 3$), the diagram is:

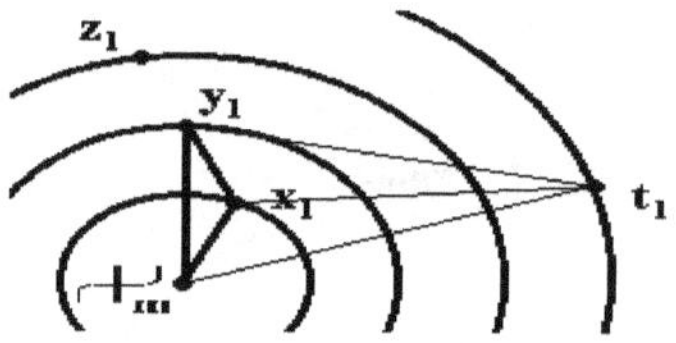

The quadrilateral (or volume) □(Ψ_m x_1 y_1 t_1) represents (n + 1) ,

$n == x_1, y_1, t_1$ == # lines to Ψ_m (center) → exponent of $\Psi_m = 3$ (thus, $\Psi_m^{\ 3}$)

$C_n = C_3 == x_1, y_1, t_1$, line x_1y_1, line x_1t_1, line y_1t_1, $\Delta x_1y_1t_1 = 7$

$\mathcal{N}\Psi_m == (C_n / \Psi_m^{\ n})(n + 1) = (C_3 / \Psi_m^{\ 3})(3 + 1) = (7)(4) / \Psi_m^{\ 3} = 7.92$

Note that CCP condenses all volumes into planar representations.

In fact, all dimensions become planar.

While the speeds x_1 / t_1 , y_1 / t_1 are represented by lines:

x_1t_1 , y_1t_1 , the speed d_{xy} / t_1 [$d_{xy} == (x_1^{\ 2} + y_1^{\ 2})_{x,y}^{\ 1/2}$]

is represented by an area: $\Delta x_1y_1t_1$ (area bounded by lines x_1t_1, y_1t_1, x_1y_1) .

$d_{xy} == (x^2 + y^2)_{x,y}^{\ 1/2}$ represents the projection for line x_1y_1

Therefore, CCP simplifies such projections; i.e., projection coordinates are condensed, as for an (a, b) plane [an (a, b) plane common to the CCP] :

line $x_1y_1 == [(a_{y1} - a_{x1})^2 + (b_{y1} - b_{x1})^2]^{1/2} = d_{xy}$

[(a, b) is the common (== only) plane for the CCP]

Therefore, CCP is a form of projection plotting (planar projections).

CCP Modulations

What, then, might $\Delta(\Psi_m\ x_1\ y_1)$ represent? a form of Ψ_m-speed? or Ψ_m modulation?

All of the circles, of a CCP, are metaphorically (functionally or conceptually) of the same size (or extent). Therefore, one can replace t_1 with Ψ_m to have:

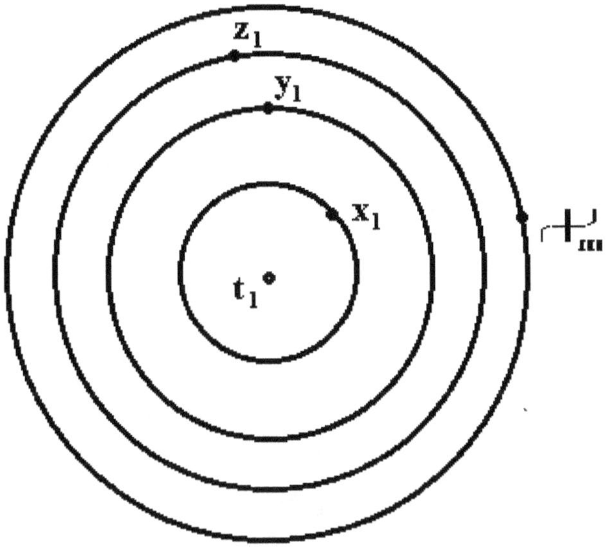

etc. (i.e., other replacements, for the center;e.g., y_1 as center).

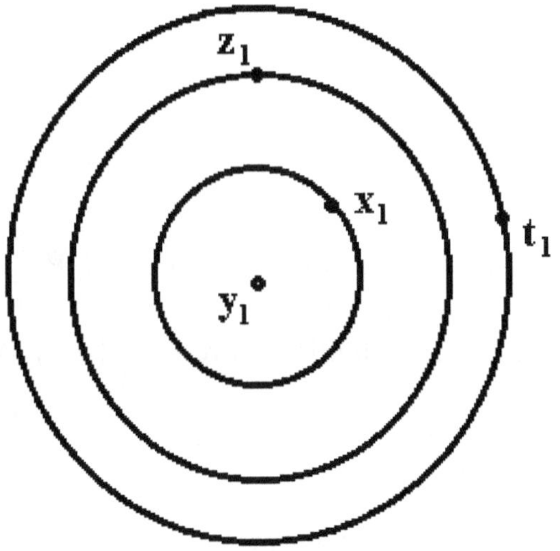

[A legitimate (term or coordinate) removes a dis-legitimate (one), rather than exchanges with.]

Therefore, we have (conceptualized) a functional reduction (of operations) using CCP, in this example:

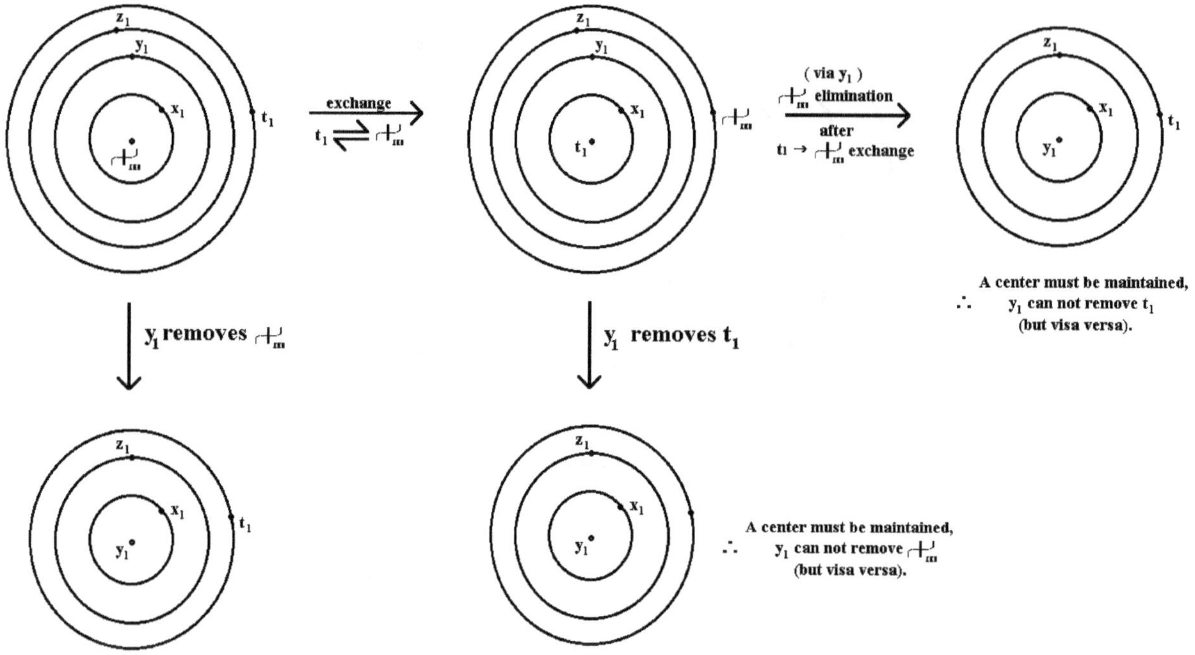
z_1
y_1
x_1
t_1
exchange
(via y_1)
elimination
after
exchange
y_1 removes
y_1 removes t_1
A center must be maintained,
∴ y_1 can not remove t_1
(but visa versa).
A center must be maintained,
∴ y_1 can not remove
(but visa versa).

Apparent rules (CCP) are:

1) (x, y, z) can exchange with each other, or remove (Ψ_m, t)

2) (Ψ_m, t) can exchange with each other, or remove (x, y, z)

3) a center must be maintained

Therefore, (we have) exchanges within legitimacies (as from some perspective); removals otherwise.

So (for examples),

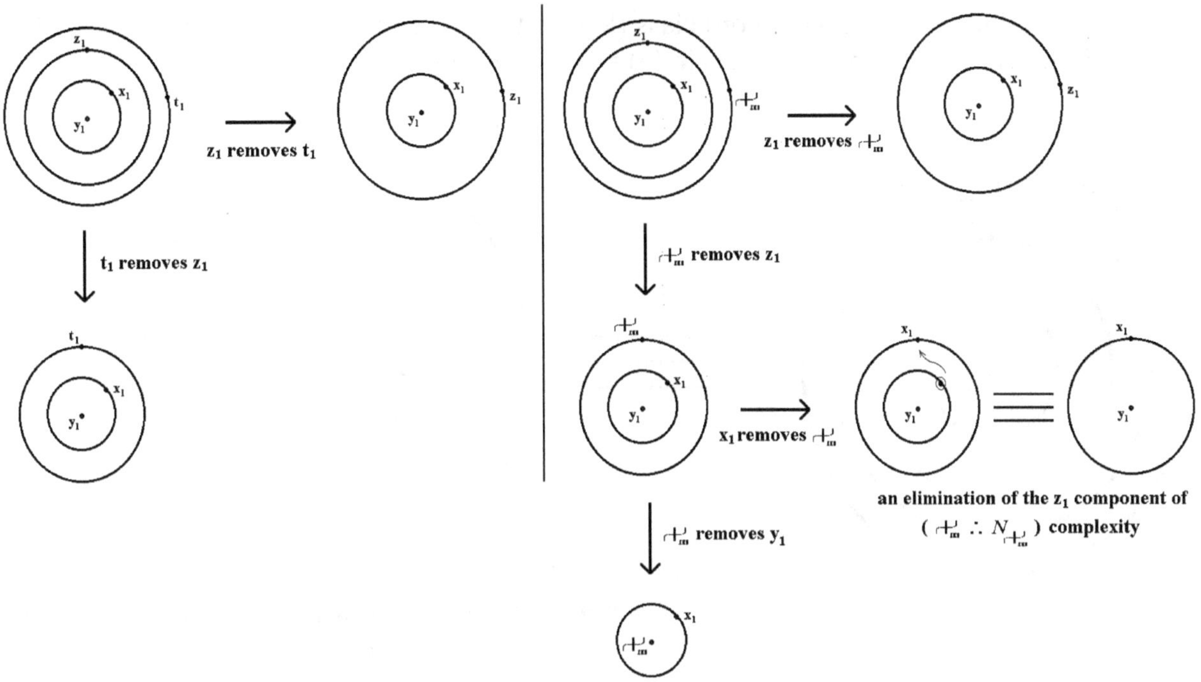

Visually, an operation such as:

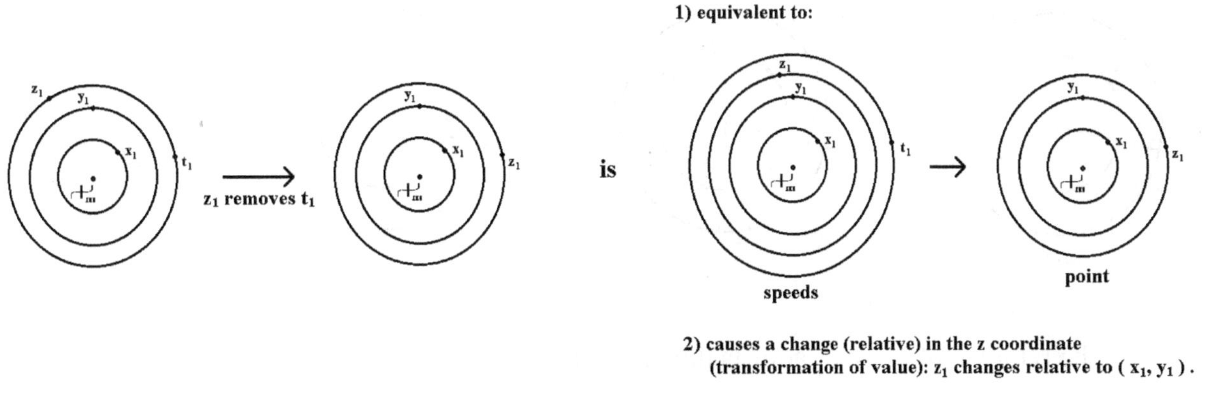

We can see that "mapping" of coordinates (to circles) is straightforward with CCP.

But this allows for some functional (or structural) restrictions for the CCP diagrams:

1) coordinates of like (common) class (legitimate, dis-legitimate) can be condensed of circles (i.e., placed on the same circle) == co-axial (for a common radius)

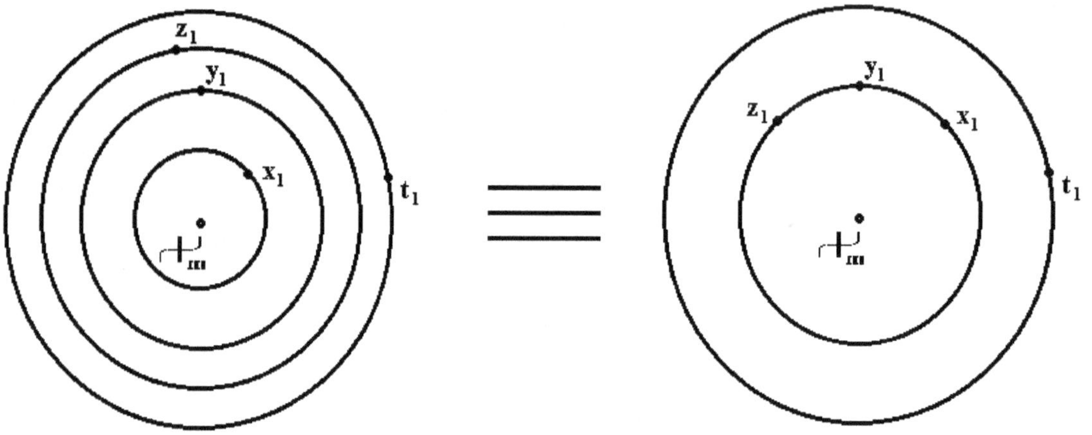

(distinguished from "mapping" uncommon coordinates)

2) lines (slopes) can be drawn between coordinates of common class

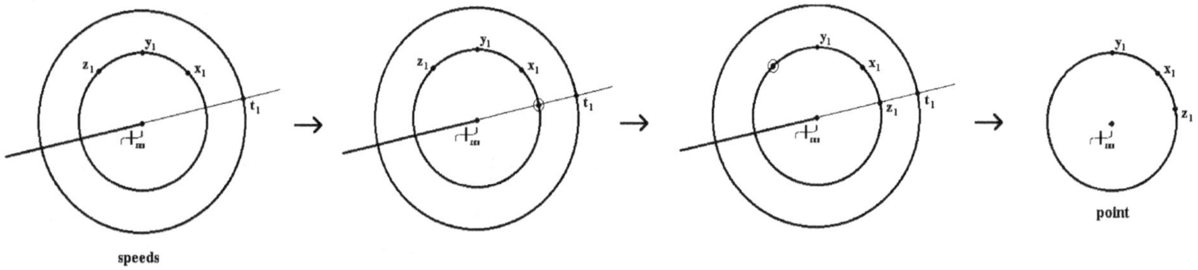

The removal of t_1 by z_1 adjusts the value (relative position) of z_1 .

This change in value (of z_1) is directly related to the (Ψ_m, t_1) line (and its slope).

Also, common class lines allow for exchanges (to further operations):

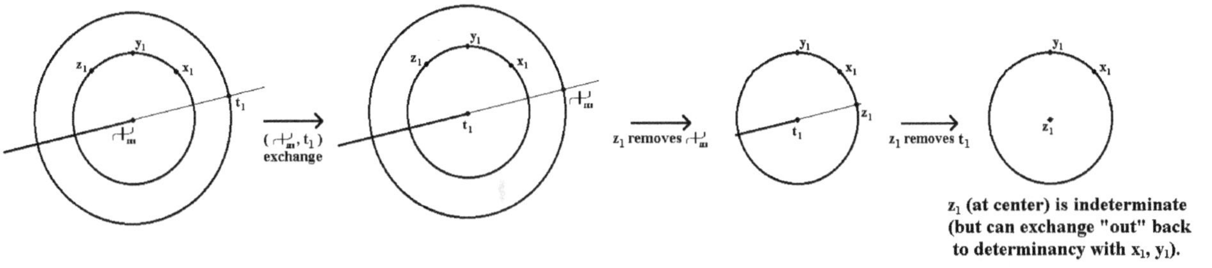

All of these operations occur on the same line (and slope). At the end (with z_1 in the center), the line (slope) is extinguished. This is why z_1 (relative to x_1, y_1) becomes indeterminate, the equivalent to converting a line of exchange to one (path) of removal.

Therefore, a diagram like

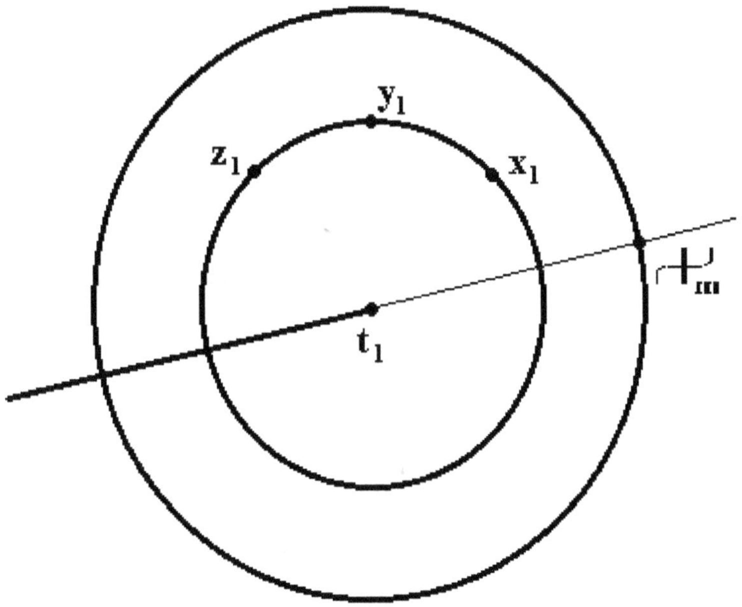

says much:

1) if (x, y, z) are legitimate, then (Ψ_m, t_1) must be dis-legitimate since a line (slope) is drawn (given)

2) (Ψ_m, t_1) may exchange, but not remove, each other; (x, y, z) can only remove

(Ψ_m, t_1) and visa versa

3) (x, y, z) is co-axial, therefore volumeric (volumetric)

4) as far as Ψ_m is concerned, t_1 is indeterminate (can have a value not determined)

5) to alter x_1, y_1, or z_1: a removal (e.g., through the center) of t_1 (or Ψ_m, after an exchange of t_1 for Ψ_m) must be involved, causing a dimension reduction (to areas or 'areal') and exchanges (among x, y, z)

Note the interesting divergences of fate (for these CCP operations):

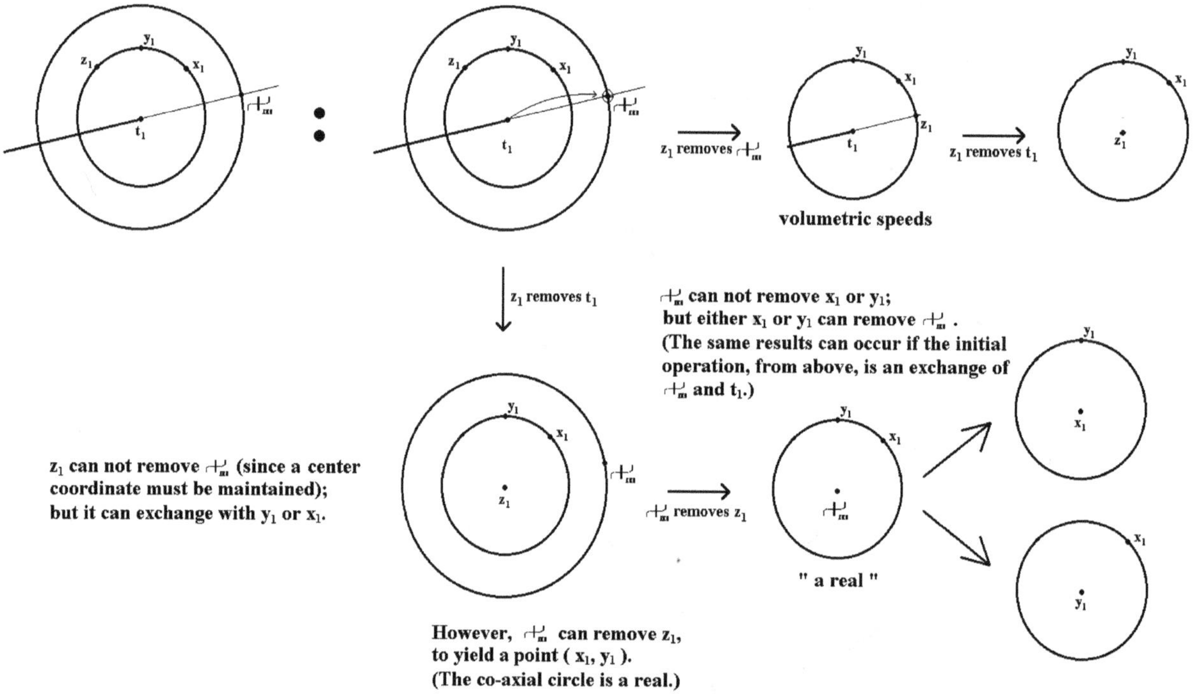

z_1 removes Ψ_m
z_1 removes t_1
volumetric speeds
z_1 removes t_1
Ψ_m can not remove x_1 or y_1;
but either x_1 or y_1 can remove Ψ_m .
(The same results can occur if the initial operation, from above, is an exchange of Ψ_m and t_1.)
z_1 can not remove Ψ_m (since a center coordinate must be maintained);
but it can exchange with y_1 or x_1.
Ψ_m removes z_1
" a real "
However, Ψ_m can remove z_1,
to yield a point (x_1, y_1).
(The co-axial circle is a real.)

Implausibles : part 2

The diagram can be very self-evident (of process).

Say, for example, that we have a speed x_1 / t_1, represented as:

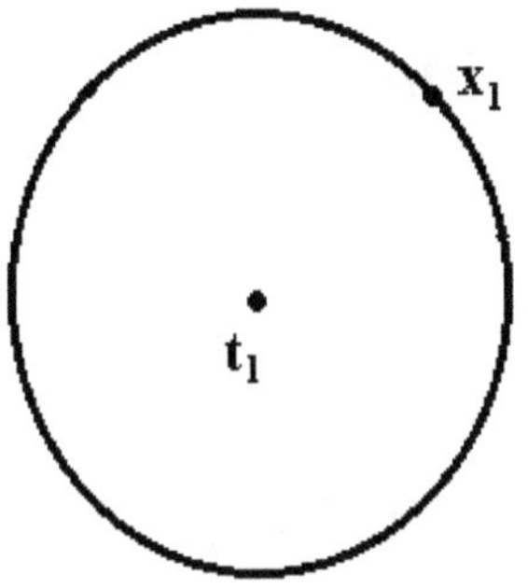

t_1 can not remove x_1, but x_1 can remove t_1 to reduce itself (and extinguish its relation with t_1) to a point coordinate:

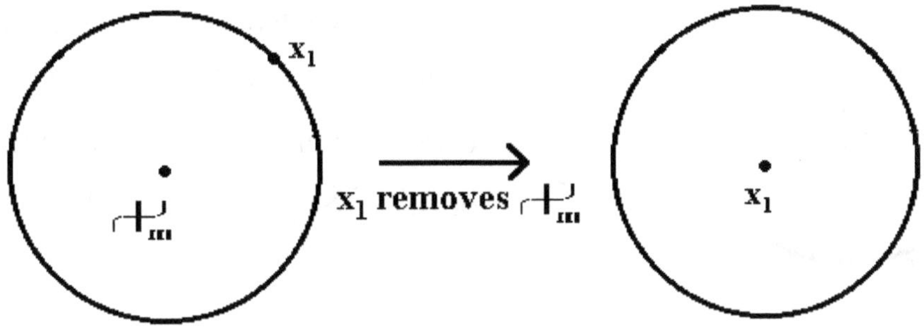

x_1 here is trapped (for eternity) as a coordinate and a result (i.e., it can not be removed by anything nor exchange with anything).

Interestingly, if complexity is defined with the coefficient(s):

$$(C_n / \Psi_m{}^n)(n + 1) = \boldsymbol{N}_{\Psi m}$$

then the complexity of x_1 can be reduced as:

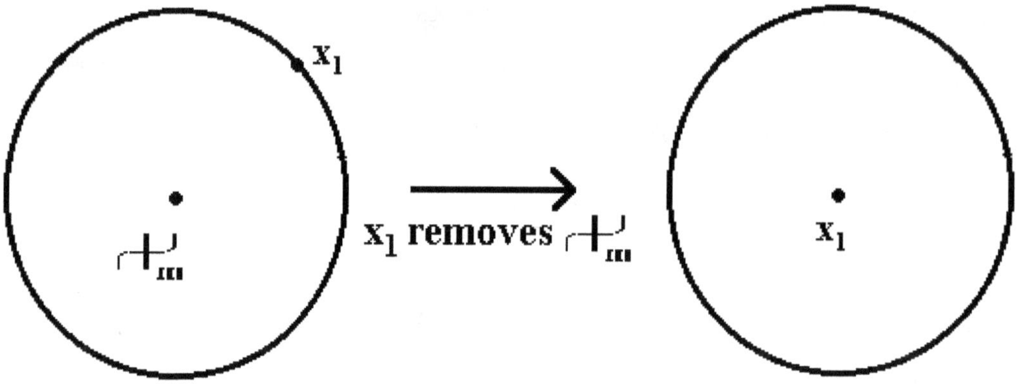

$(1 / \Psi_m^{\,1})(1 + 1)x_1 == (1 / 1.5231)(2)x_1 = 1.3131x_1$;

(likewise) using $N_\Psi \rightarrow 0.8732x_1$ ($N_e \rightarrow 0.7358x_1$)

Therefore, if

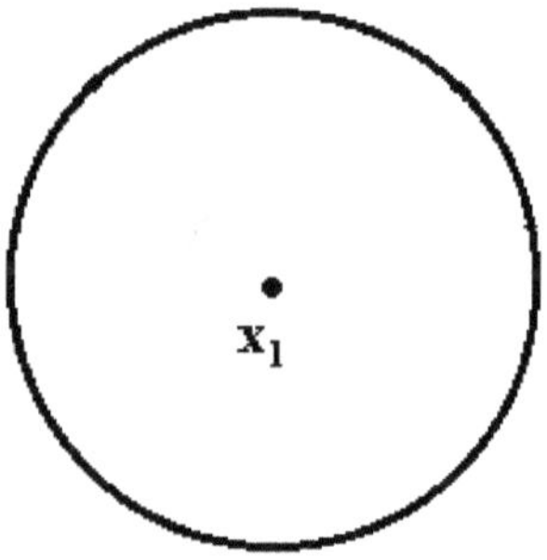

has a complexity of 1, as a coordinate or result, then in any relationship it starts out with a greater complexity using Ψ_m (i.e., $1.3131x_1$) or a lesser complexity using (base) Ψ or e (i.e., as $0.8732x_1$ or $0.7358x_1$ respectively);

i.e., there are inherent (Ψ_m, Ψ, e) complexities, or the inherent Ψ_m complexity of

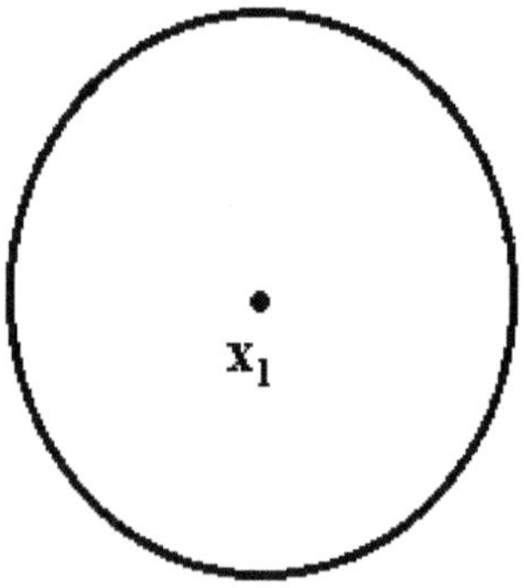

is $1.3131x_1$, etc.

How does x_1 / t_1 so comport (x legitimate; Ψ_m, t dis-legitimate)

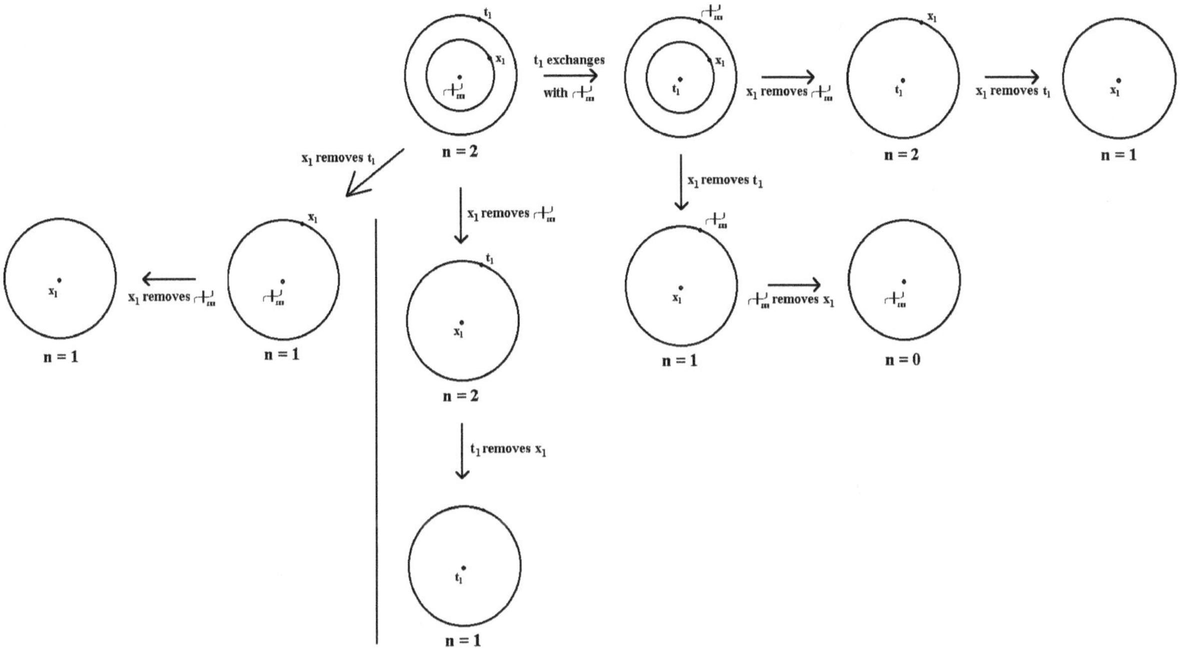

If we do strict dis-legitimacy of classes, i.e. (x, y, z), Ψ_m, t are separate classes, then Ψ_m and t can not exchange, but only remove (each other):

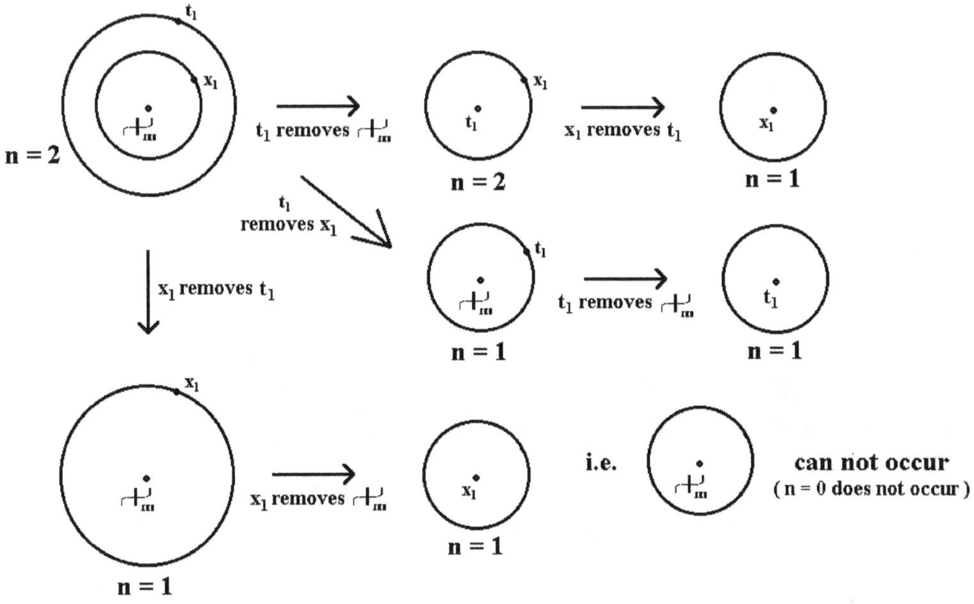

The species

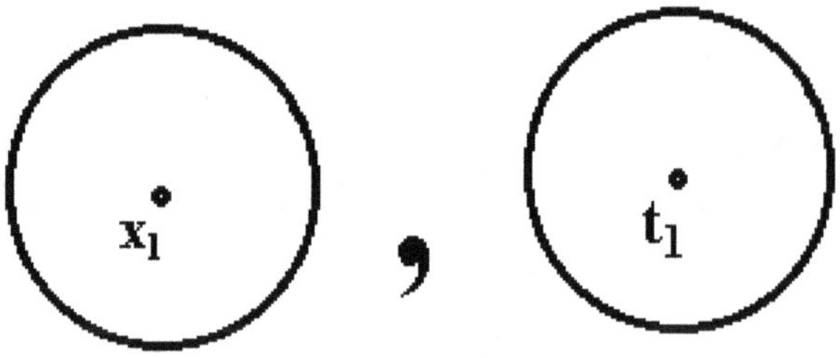

are without complexity (from Ψ_m), but also the speed

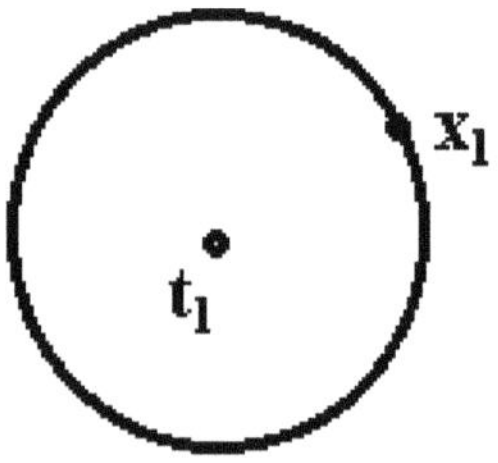

which can only devolve to

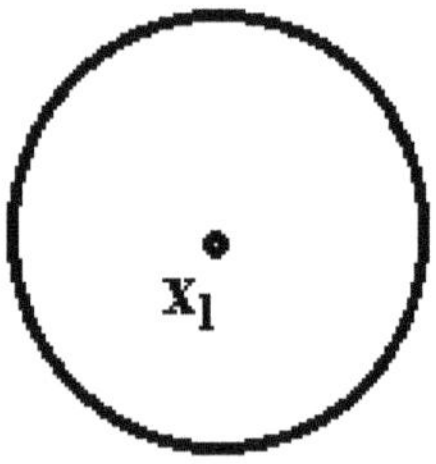

Therefore, isolated coordinates (points) and speeds (relations), involved with no other such points and relations, can be without complexity.

This is to say that the concept of complexity does not apply to species such as

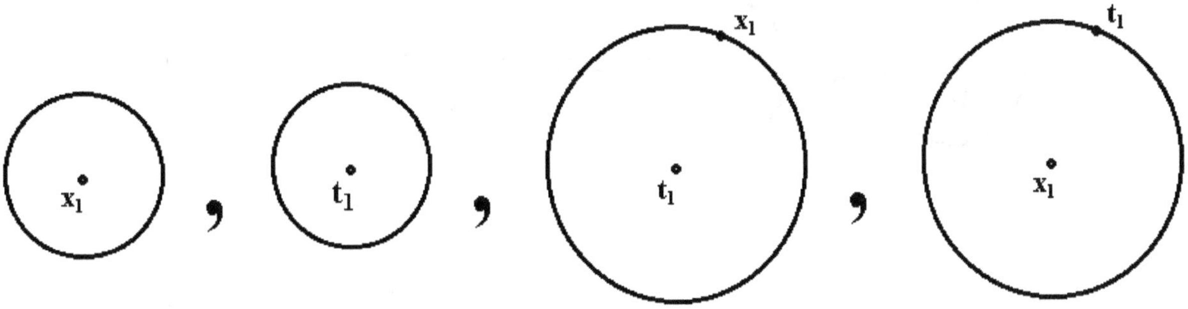

etc., aside from assigning to them (formally) the value 1 (as isolated elements).

[Note that we can not apply the suggestion (here) that Ψ_m exists such that $\Psi_m^{\;0} == 1$, forcing

complexities for 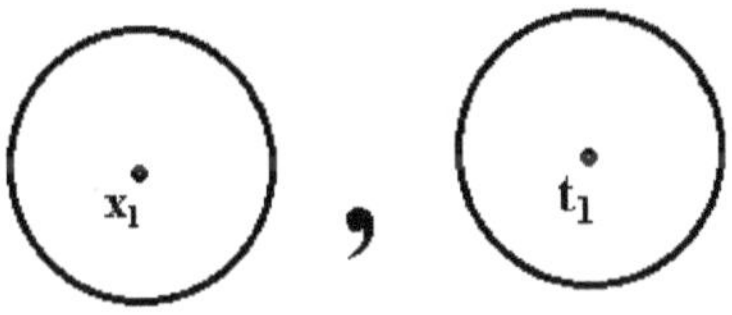

to be $\frac{1}{1}(1+1) = 2$

and

complexities for 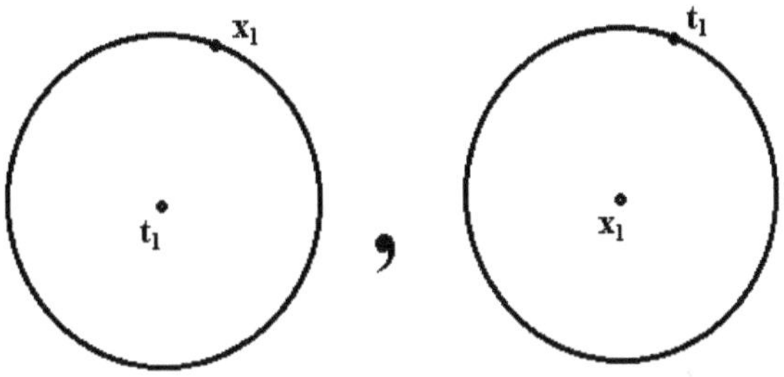

to be $\frac{3}{1}(2+1) = 9$

since n remains 1 and 2 (for above cases) such that $\Psi_m^{\;1}$ and $\Psi_m^{\;2}$ occur (i.e., $\Psi_m^{\;0}$ does not occur).]

Manners of Obtaining Complexity

What is the manner of elements obtaining complexity?

Apparently it is addition:

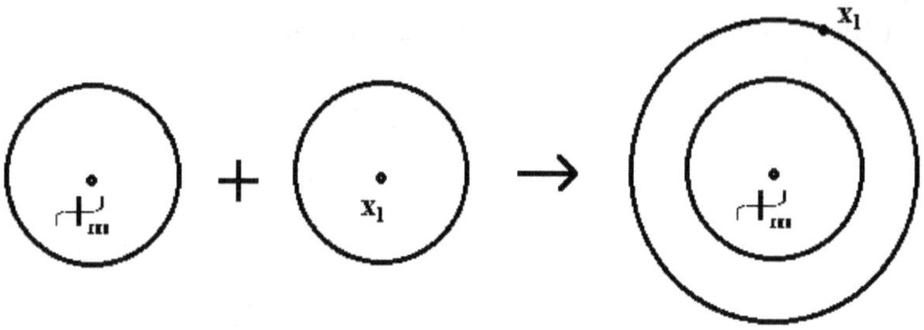

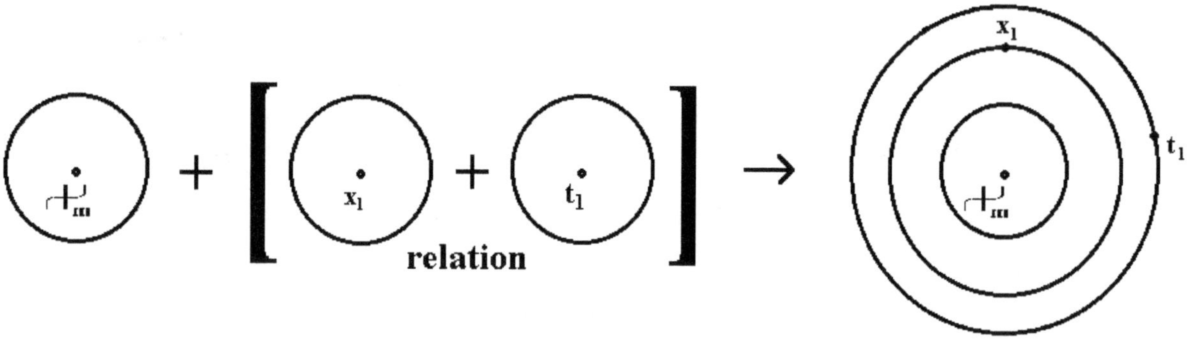

But what of the initial complexity?

i.e. that of

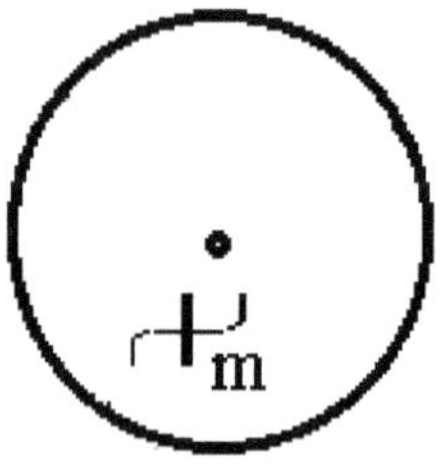

Again, since $n = 0$ does not occur, we can not allow for $\Psi_m^{\ 0}$ ($==1$) as that would lead to a complexity of: $(C_n / \Psi_m^{\ n})(n + 1) == (0 / 1)(0 + 1) = 0$.

Apparently, the complexity of

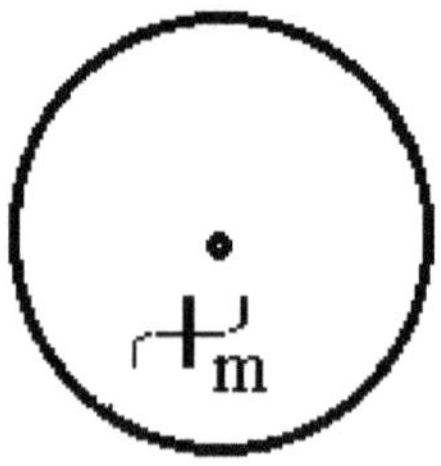

is found by subtraction, based on the formally charged complexity of an element (== 1) ;

therefore:

If complexity of

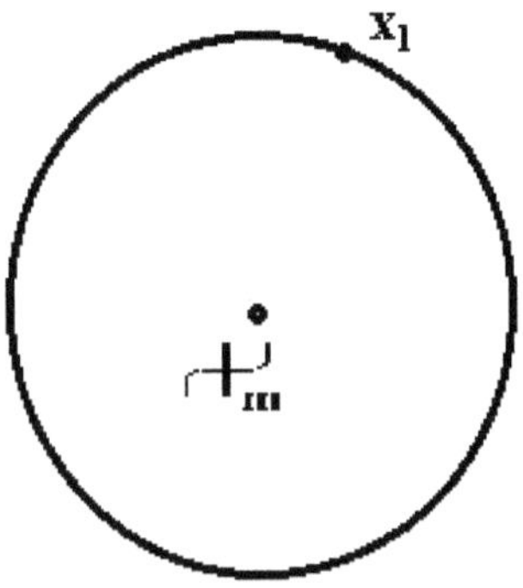

= 1.3131

and complexity of

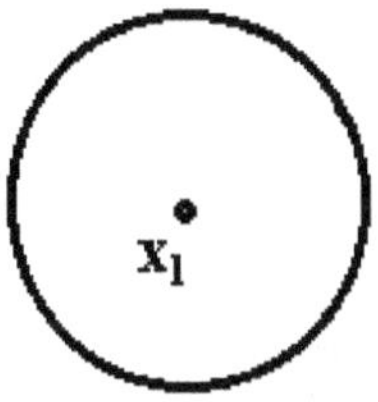

= 1

then complexity of

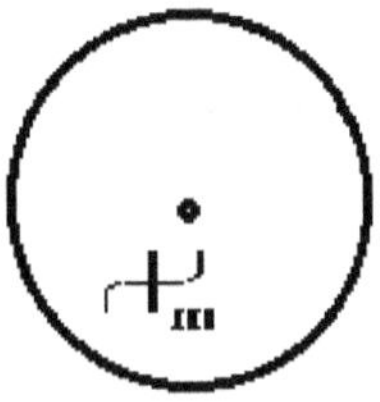

= 1.3131 - 1 = 0.3131

For Ψ, complexity of

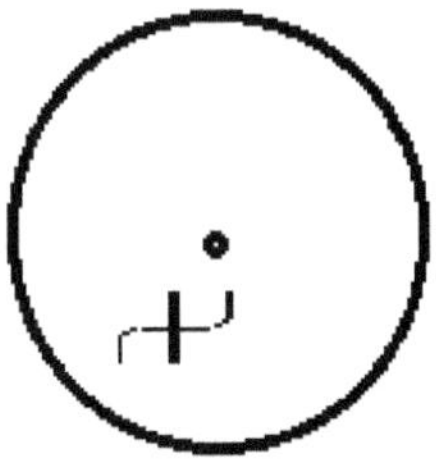

== -1 + 0.8732 = -0.1268 .

For e , complexity of

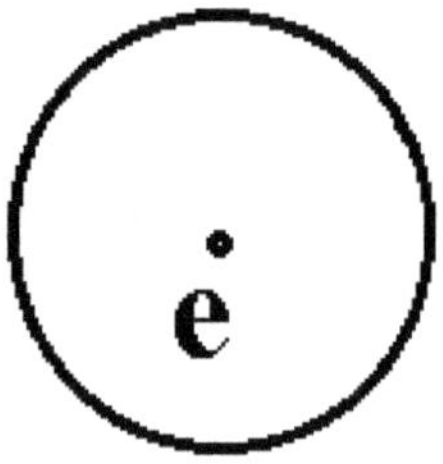

== -1 + 0.7358 = -0.2642 .

Continuing in this additive mode (of functionality):

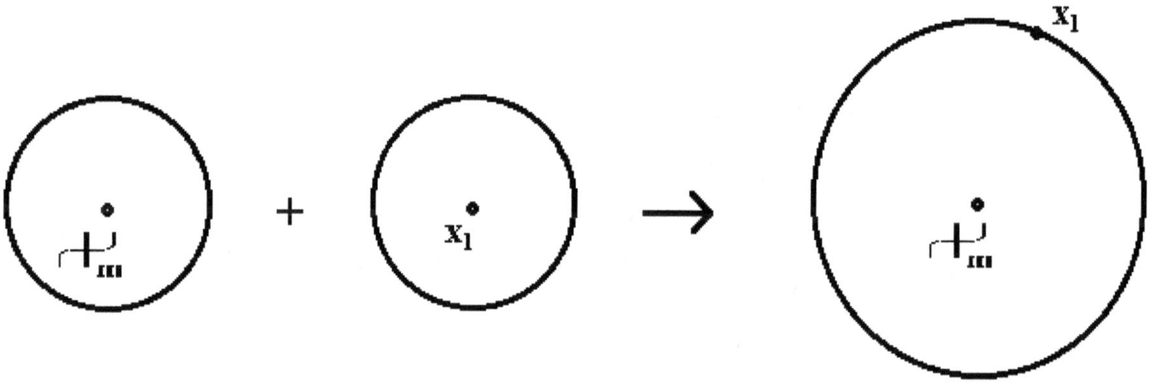

0.3131 **1** **1.3131**

(complexity)

but

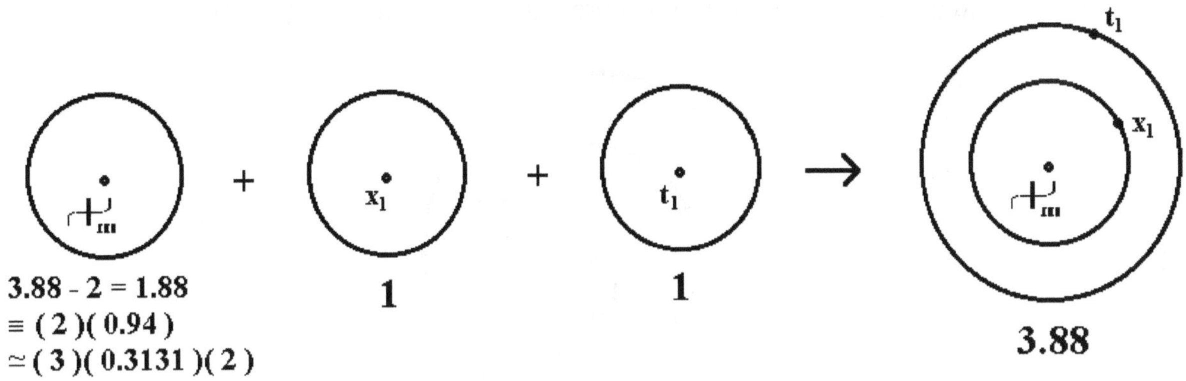

or

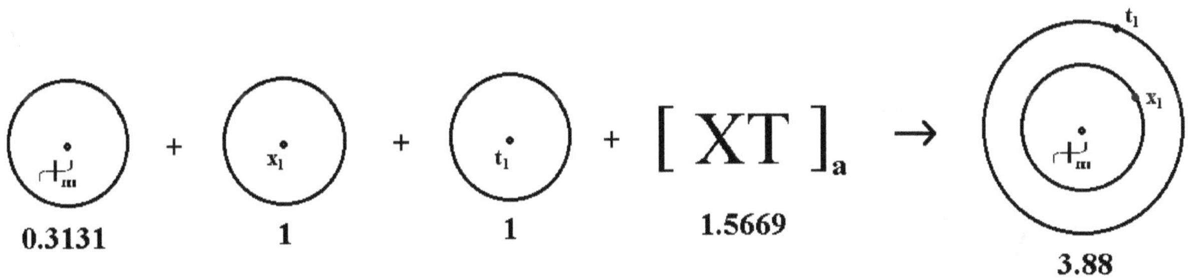

where [XT] represents a contribution to (a resultant) complexity due to a relationship between x and t; therefore, one might view

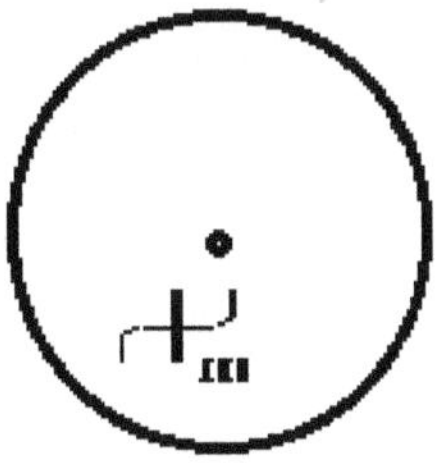

degenerate of complexity (depending on n).

Interestingly, one can partition $[XT]_a$ among the 3 species equally as (1.5669 / 3) = 0.5223 and one can partition $[XT]_b$ among the 4 species equally as (1.2538 / 4) = 0.31345 (or among the 2 species as (1.2538 / 2) = 0.6269).

So, we find at least 3 modes (or procedures) of creating (the complexity of)

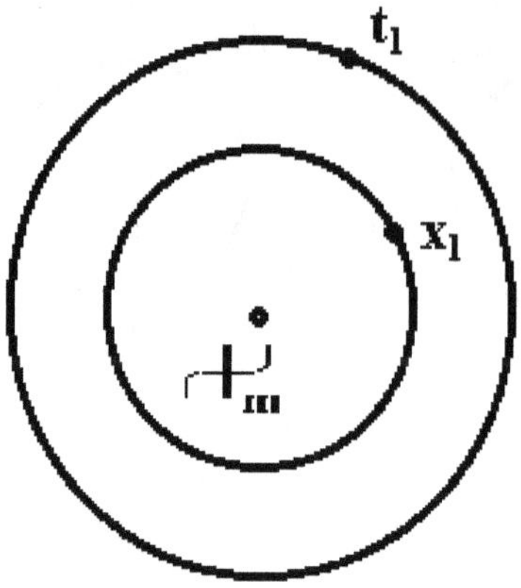

Therefore, for non-degenerate

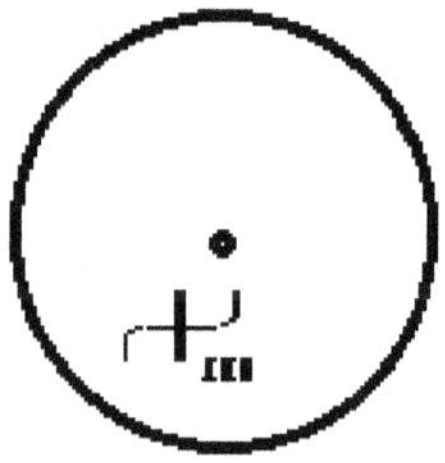

and judging from [XT] values,

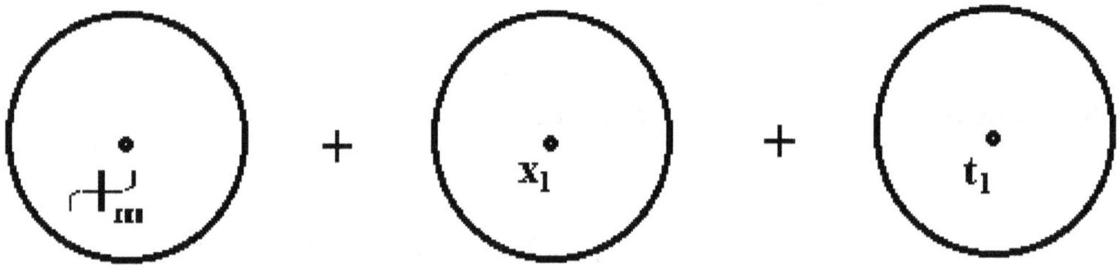

is more complex (of a procedure) than

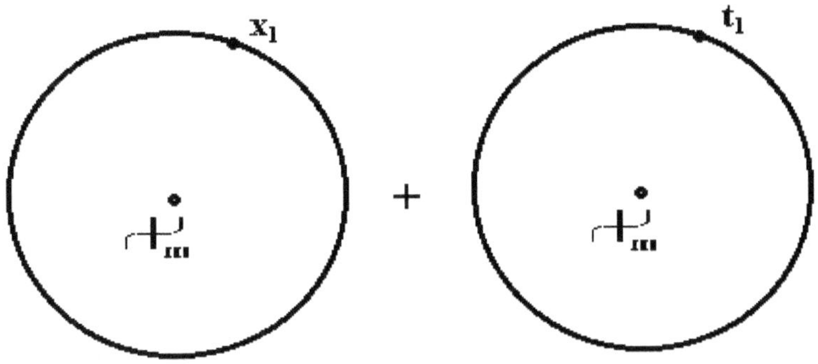

i.e., it (the latter) might be a preferred mode.

Still, a map to the complexity for speed (x_1 / t_1) presents itself:

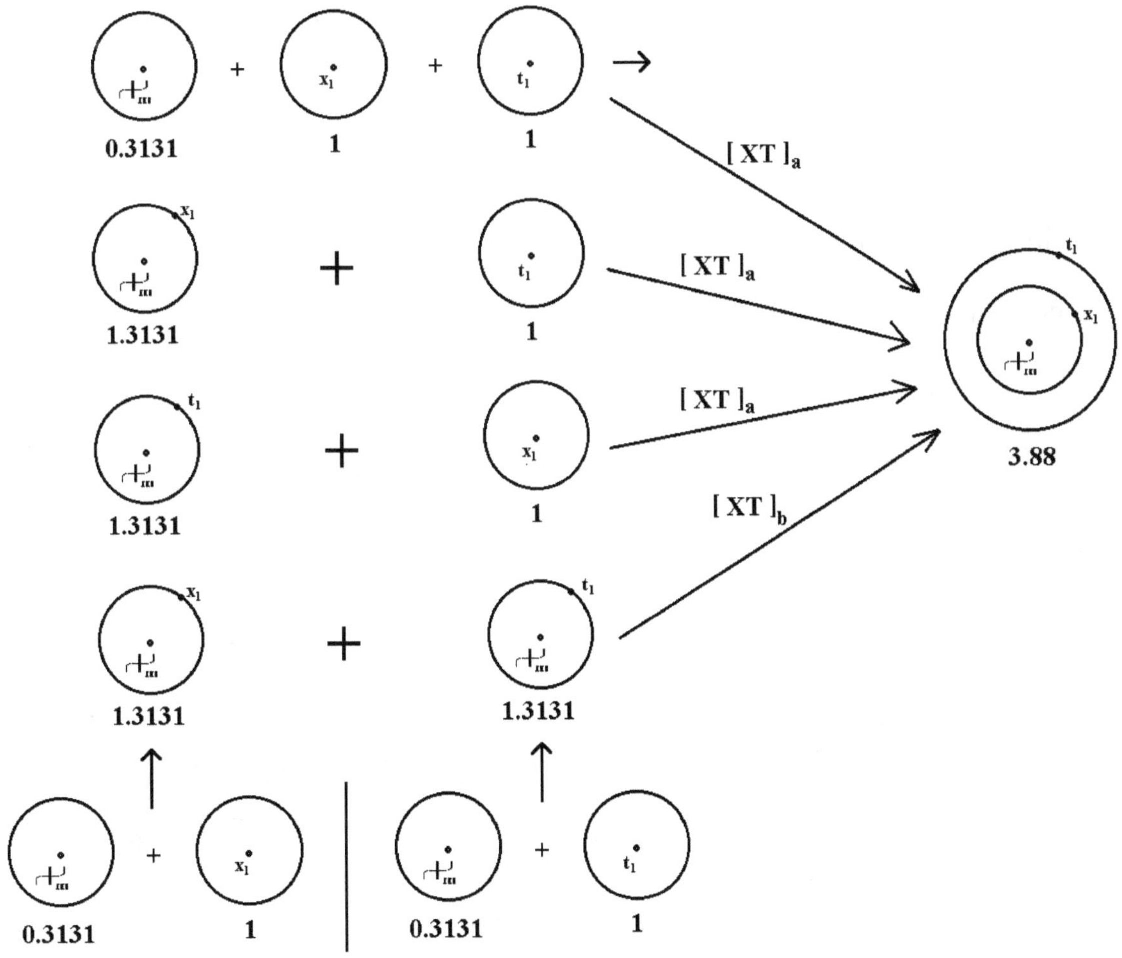

If one draws a line (slope or radius) between (connecting) elements of a common class, as distinguished from another class (such as can be considered legitimate), then that line establishes the positioning of the other class (i.e. legitimate) elements:

e.g., (x, y, z) legitimate and (Ψ_m, t) dis-legitimate (to x, y, z)

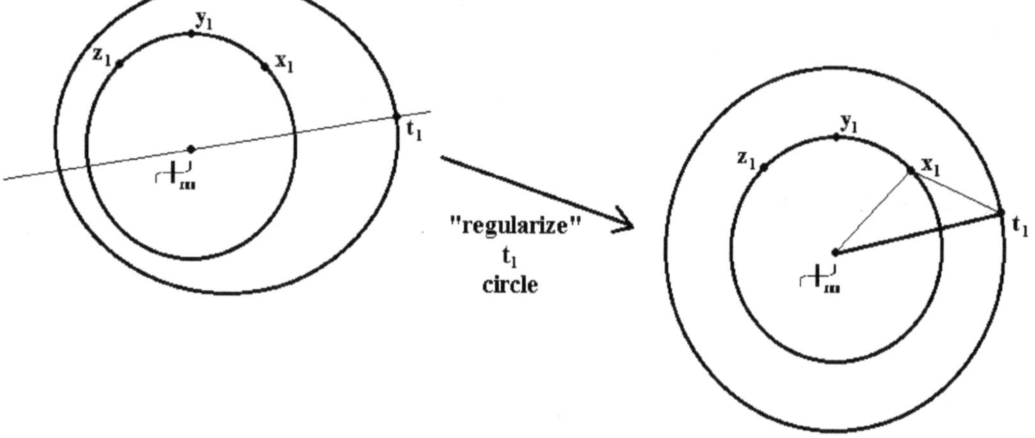

We can compare maps of (coefficients of) complexity thereby:

$$\text{line } x_1\Psi_m = \text{line } t_1\Psi_m == 1.3131\ ;$$

therefore, $\text{line } x_1t_1 = [XT]_b$;

therefore, $\Delta x_1t_1\Psi_m == \text{line } x_1\Psi_m + \text{line } t_1\Psi_m + \text{line } x_1t_1 = 3.88$

Using same radii (or extent), this can be re-diagrammed as:

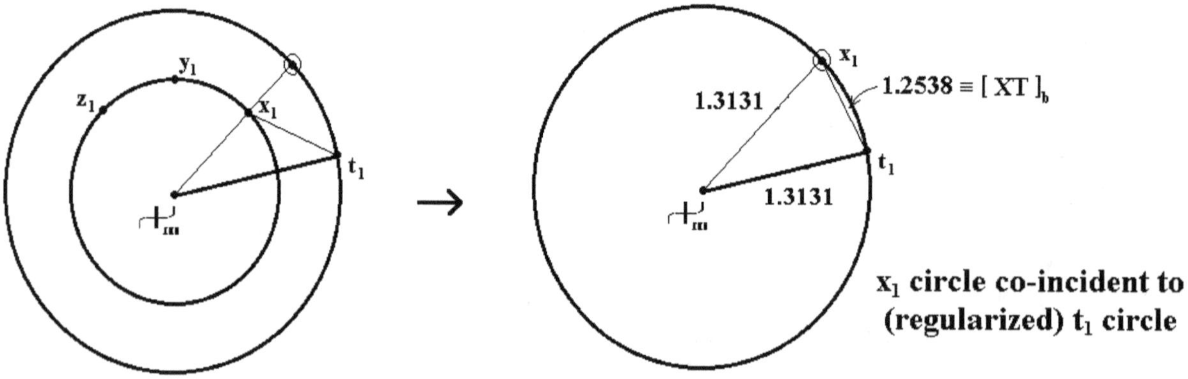

Based on the CCP diagramming, since each of x_1 / t_1, y_1 / t_1, z_1 / t_1 would have the same $N_{\Psi m}$ (or complexing coefficient), the functional range of (x, y, z) would seem to be limited (pictorially) as:

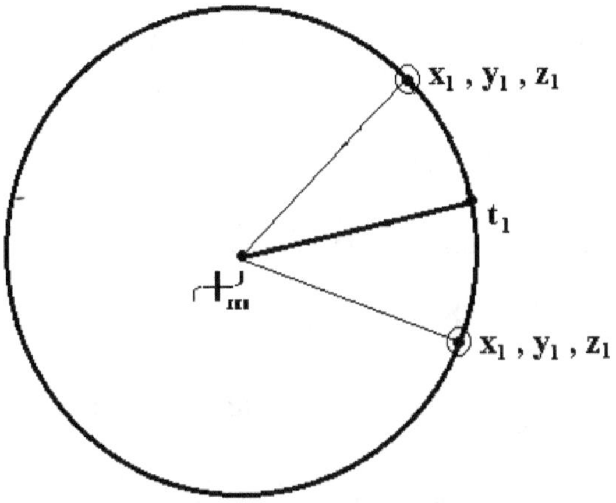

which imposes that two of the (x, y, or z) coordinates must superimpose since there are only two positions that allow for $[XT]_b$ while there are 3 (x, y, or z) coordinates. This appears to be a restriction of the complexity coefficient(s). The actual complexities depend on the values of x_1, y_1, z_1, and t_1.

The condensing (coincidence) [(x, y), (z, y), (x, z)] simply forces a volume into a planar representation (as stated earlier). But this could be equivalent to converting the circle into a sphere of surface, so that the mask of the circle only implies coincidence (of coordinates) where they might not actually occur:

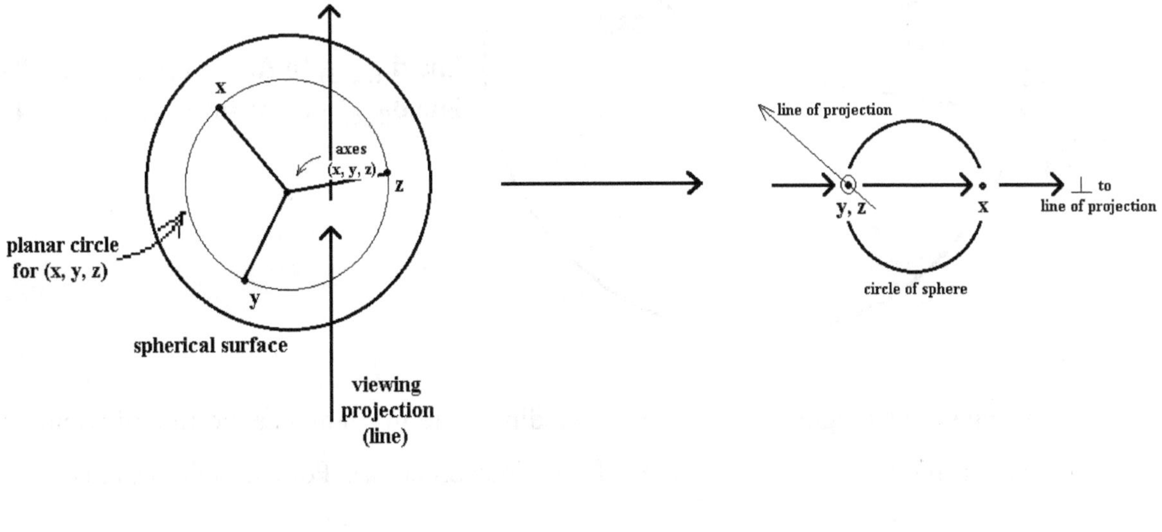

Note that (x, y, z) coordinates are constrained to a "planar" circle. The circle, here, is emphasized as being planar because it is formed from coordinates on the surface of a (volumetric) sphere. Note also that any two points on the circumference of a circle can be made to appear coincident when viewing the circle edge on as if a line.

For more sophisticated (i.e., among points) and higher complexities ($n > 2$), the (Ψ_m, t) radii will be larger (i.e., larger $N_{\Psi m}$), and the (CCP) plotting will be even more conceptualized.

Higher, more sophisticated CCP:

e.g., for points (1, 2, 3, 4) and (5, 6, 7, 8) of (x, y, z, t)

$t_1 = 4$, $t_2 = 8$; thus, $\Delta t = 8 - 4 = 4$

$d_{(1, 2, 3)} = 3.74$, $d_{(5, 6, 7)} = 10.49$ [see earlier calculations]

$n = 4$, $N_{\Psi m} = 23.23$, Complexity for (1, 2, 3, 4) = 21.84

and complexity for (5, 6, 7, 8) = 30.43

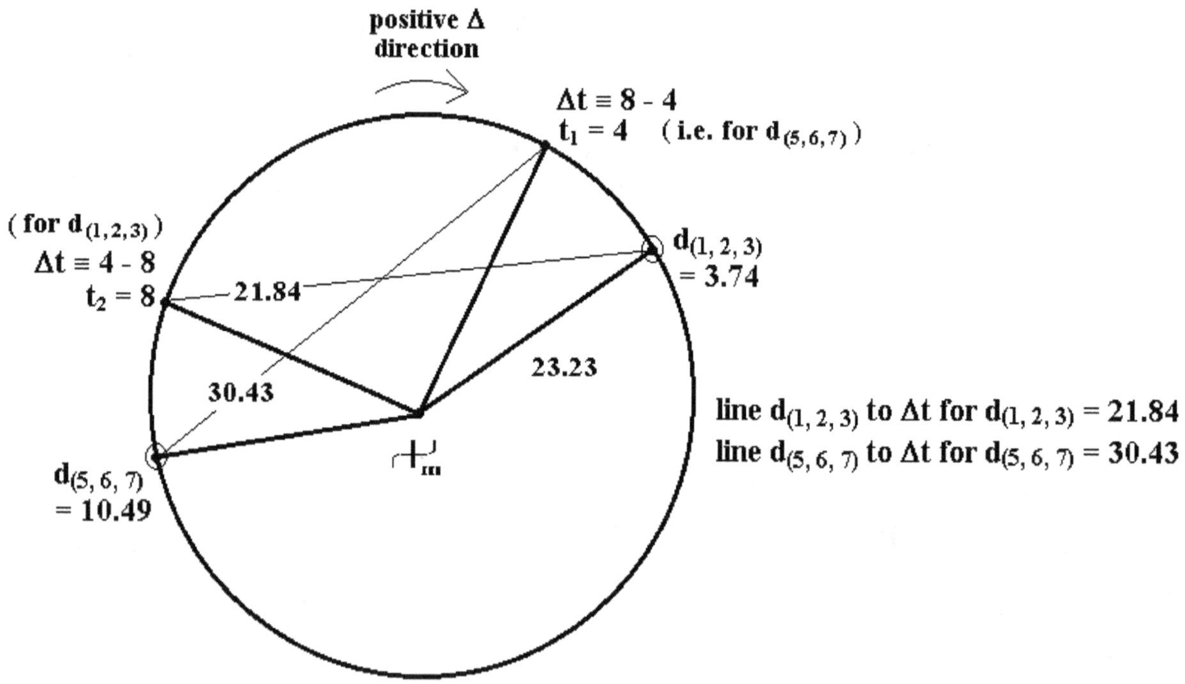

What can such a diagram elicit of understanding, in terms of an expenditure of complexity? We can propose "complexity charges" for various processes. For example, what is the complexity expenditure for converting, say, t_1 from 4 to 8 (in value)?

From the precepts of CCP manipulations, there are several manners (here) that can be followed:

e.g., t_1 exchanges with t_2 → complexity of +a

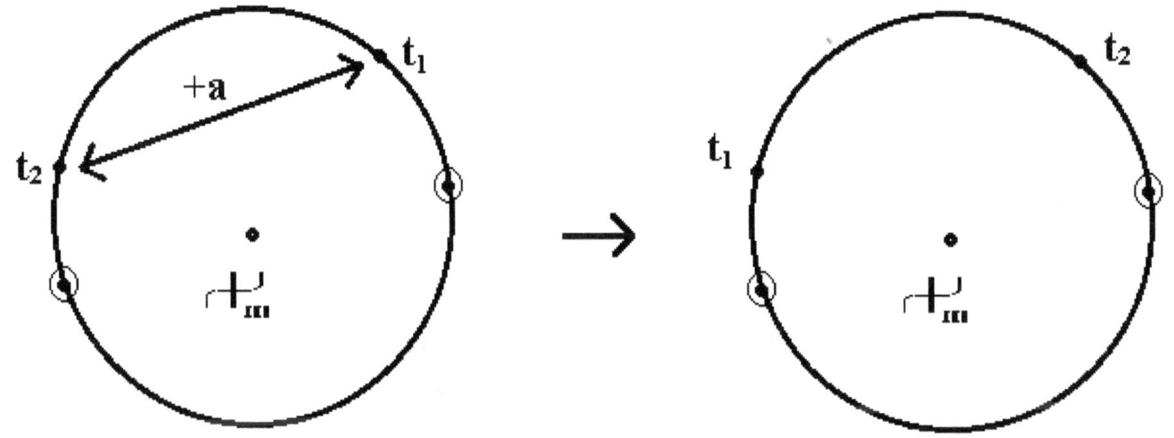

e.g., t_1 removes d(1, 2, 3) → complexity of -b

t_1 exchanges with t_2 → complexity of +c ;

thus, -b+c

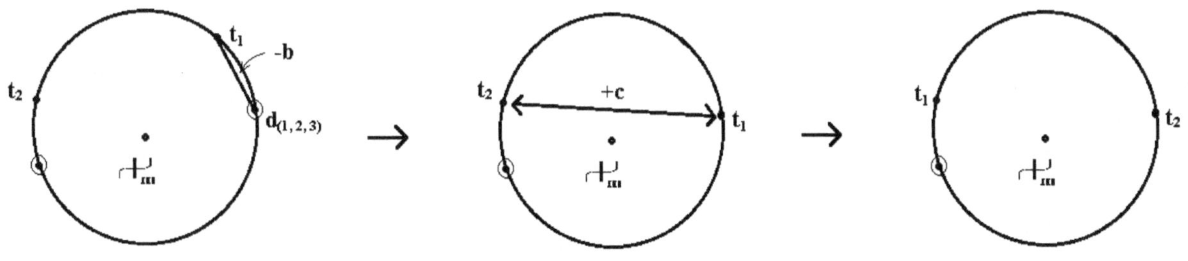

e.g., t_1 removes d(5, 6, 7) → complexity of -d

t_1 exchanges with t_2 → complexity of +e ;

thus, -d+e

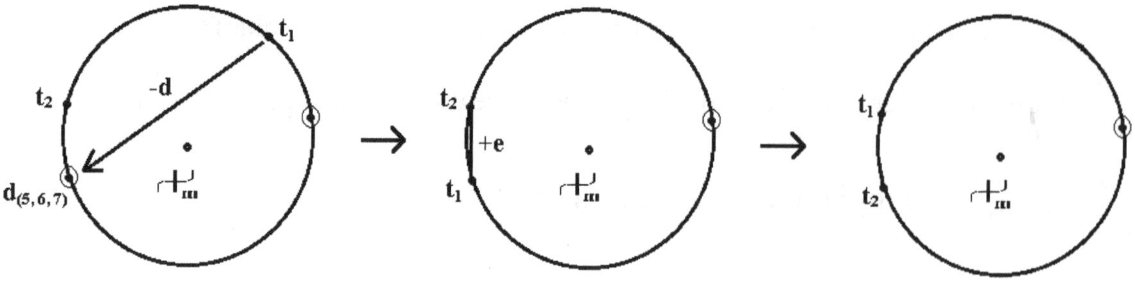

etc.,

or, in (CCP) summary for the 3 given methods:

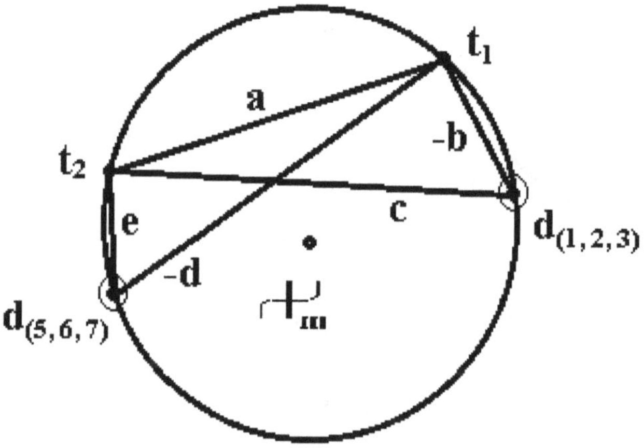

Of course, in the same manner, a $t_1 = 4 \rightarrow t_1 = 8$ conversion can be prevented:

e.g. $d_{(5, 6, 7)}$ removes t_1

e.g. $d_{(1, 2, 3)}$ removes t_1

e.g. $d_{(5, 6, 7)}$ removes t_2, followed by t_1 exchanges with Ψ_m or, using strict dis-legitimacies, t_1 removes Ψ_m (thus t_1 can no longer exchange, but can only be removed)

etc.

It is not necessary for t_2 to exist, to allow for the $t_1 = 4 \rightarrow t_1 = 8$ conversion:

e.g. $d_{(5,6,7)}$ removes t_2 (t_2 no longer exists)

t_1 removes $d_{(5,6,7)}$ ($d_{(5,6,7)}$ no longer exists)

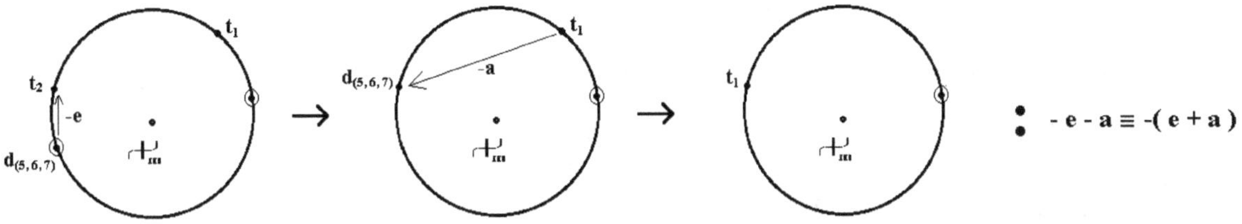

Note that, as vectorially, the order (exchange or removal) and sequence of the steps of a procedure are important (for the outcome). [The sign of an exchange is independent of direction, but the sign of a removal (replacement) is directional.]

Exchanges are positive; removals are negative (of complexity). Examples of complexity charges (here) are: a, b, c, d, e ; they are of the same units as the (CCP) radius (here, 23.23). A removal causes a reduction in complexity; exchanges are accompanied by increases in complexity.

We note that certain symbolic associations may be occasioned by the (allowed) CCP manipulations. For the above example, the (distal) relationship of t_1 to $d_{(1,2,3)}$ yields (essentially) a conversion of $b \rightarrow c$.

Thus, a notation such as, for a convolution of (e, a) to (b, c) :

$$b[\, -e-a \,] \rightarrow c == b[\, -(e+a) \,] = c$$

may be approached.

Therefore, $\{ \, (d_{(8,6,7)} \; r \; t_2), (t_1 \; r \; d_{(5,6,7)}) \, \} \rightarrow b[\, -(e+a) \,] = c$,

where r == “removes”

or

$d_{(5,6,7)} \; r \; t_2$	-e
$t_1 \; r \, d_{(5,6,7)}$	-a
$t_1 \; r \; (d_{(5,6,7)} \; r \; t_2) \rightarrow$	$b[\, -(e+a) \,] = c$

where the parentheses ( ) shows priority of operations(s).

To avoid confusion with complexity charges symbols, we might allow:

re == removes, ex == exchanges (with) ;

therefore (here), t_1 re ($d_{(5, 6, 7)}$ re t_2) → b[-(e + a)] = c

points or coordinates charges (distances)

Note that the distance of any coordinate to the center always remains the value of the radius ($N_{\Psi m}$).

Likewise: t_1 ex t_2 → b[a] = c

or (t_2 ex t_1) e[a] = d

(b, e)[a] = c + d

positive (exchanges) sign == order independent

negative (removals) sign == order dependent

Note: -e -a == -(e + a) =/= -(a + e) == -a -e

thus, (b, e)[a] == (b + e)[a] = c + d ;

thus, (b + e)[a] = b[-(e + a)] + d = b[-(e + a)] + e[a] = b[a] + e[a] ;

thus, b[-(e + a)] = b[a] , while -(e + a) does not necessarily equal a (which would yield: a = -e / 2) .

Coordinates of CCP Diagrams

How, then, can we directly relate the coordinates to the charges (distances), in a CCP diagram?

Each coordinate has the same radius from the center, as N. And each charge is bounded by two coordinates, to make an isosceles triangle with the center.

A removal or exchange involves two coordinates, and is therefore defined (or characterized) by such a triangle for inherent charge involvement (or participation).

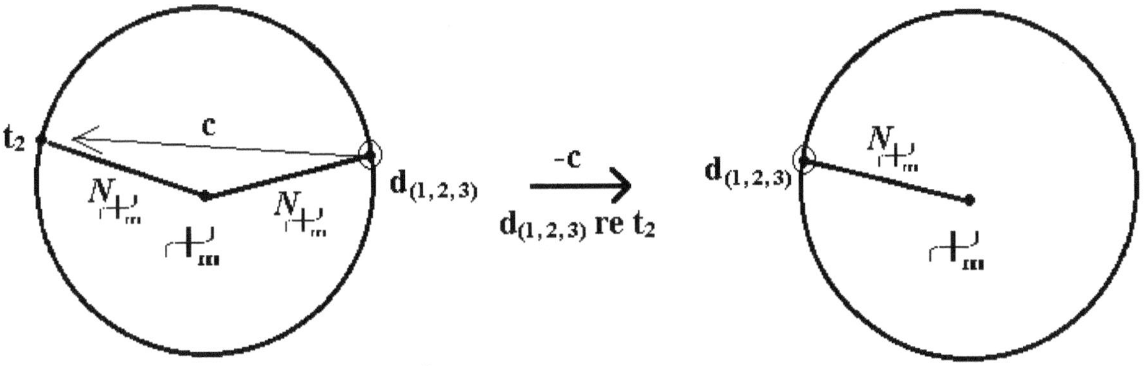

$d_{(1, 2, 3)}$ re $t_2 \rightarrow$ -c

For any removal, then, the perimeter of the triangle (an area conducing) is reduced from $2N$ + C(harge) $\longrightarrow$ N (a line), while for an exchange the perimeter is maintained ($2N$ + C). A route's distance from the center to any coordinate involved in an operation (removal or exchange) can be either N or (N + C) as taken.

If $d_{(1, 2, 3)}$ is represented by N and t_2 is represented by (N + C) in a path that includes $d_{(1, 2, 3)}$, then the operation:

$d_{(1, 2, 3)}$ re t_2

is equivalent (symbolically) to:

$$N - (N + C) = -C$$

Obviously, N as $d_{(1, 2, 3)}$ is a resulting definition (for $d_{(1, 2, 3)}$) by the operation (carried out), and (N + C) as t_2 is a definition before the operation (pre-operation).

For any exchange (of coordinates), you have:

$$N + (N + C) = 2N + C$$

[Note that the positioning of the coordinates on the circumference of the circle is relative, since they and (their) coordination and manufacture of the internal charges (their distance machinery and constructions) can all be rotated (in unison) about the center.]

Clearly, in our example, $d_{(1, 2, 3)}$ changes value: to that which was t_2 (of value). The change (in value) is the arc distance from $d_{(1, 2, 3)}$ to t_2; but this value distance is not necessarily related to the charge distances (unit-wise nor extent-wise); i.e., $2\pi N_{\Psi m}$ need not yield the circumference extent or size. The internal and circumferential distances are independent matters. This is because the charges are based on complexities, while the coordinates are not. The direct relation between the coordinates and the charges is due to the relation between the coordinates and the complexity:

d / t ,	$N_{\Psi m2}$,	$N_{\Psi m2} \bullet (d/t) \quad (n = 2)$
relation (of coordinates)	complexity coefficient (charge)	complexity

Note that for our example: $d_{(1, 2, 3)}$ re $t_2 \rightarrow$ -C

the complexity changes as if n goes from $n = 2$ to $n = 1$ (i.e., loss of t_2) .

As n goes from 2 to 1, so correspondingly must N decrease, and therefore the CCP radius (decreases). Thus, the CCP diagram for a removal more accurately depicts a process as such:

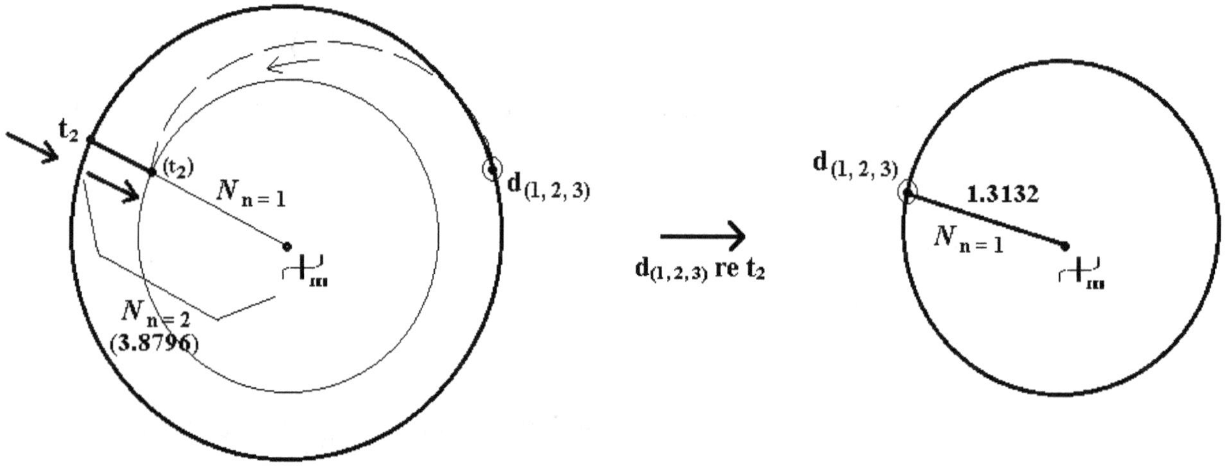

more complexity, n = 2

N = 3.8796

less complexity, n = 1

N = 1.3132

But, the actual complexity changes depend on the relationships of the coordinates (here t and d). If the relationship is d / t and $t > 1$, then the complexity change for $n = 2 \rightarrow 1$ can increase:

say $d = 2, t = 8$;

therefore, $N_{\Psi m2}(d/t) = (3.8796)(2/8) = 0.9699$

$$N_{\Psi m1} d = (1.3122)(2) = 2.6264$$

$d_{(1, 2, 3)}$ re t_2 (i.e., d re t) produces a complexity increase of ~ ~ 2.71 fold

(== 2.6264 / 0.9699)

Therefore, a decrease in complexity coefficient does not necessarily lead to a smaller complexity (for the result of a process).

Actually, for this example, as plotted, d increases from 2 to 8 (as a value of coordinate; i.e., d "carries" to t in value).

Therefore, $N_{\Psi m1}d == (1.3132)(8) = 10.5056$, and the complexity increase is 10.5056/0.9699 ~ ~ 10.83 fold.

Therefore, (here) two types of processes are suggested:

1) where one coordinate "carries" to the other (to take the other coordinate value; i.e., d carries to t, thus d becomes t in coordinate value)

2) one coordinate does not "carry" to the other; i.e., one coordinate remains constant, while the other (perhaps while able to be variable) is lost (or exchanged); therefore, d does not carry to t, and retains its d-value

Therefore, a more correct symbolic expression, for $d_{(1, 2, 3)}$ re t_2 or d re t, might be:

$$\{ N_{n=2} \rightarrow N_{n=1} ; N_{n=1} - (N_{n=1} + C) = -C \}$$

where we assume C retains its charge (distance). [The " ; " imposes simultaneity: i.e., " - " (as removal) implies " → " (as decrease in n) .]

If C retains its charge, the (d re t) reduction (replacement) might be something like:

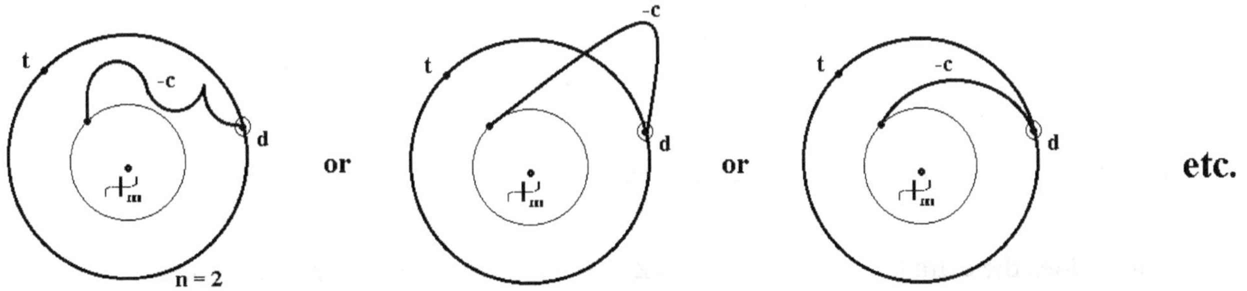

We can prescribe, then, a process (for our example):

$$N_{\Psi m2}(d / t) \rightarrow N_{\psi m1}d$$

where the complexity rises from 0.9699 to 10.5056, a 10.83 fold increase (with carry). Thereby, the value of t is delivered from $N_{\Psi m1}d = 10.5056$ (i.e., by solving for d), and the value of d (at $n = 2$) from $N_{\Psi m2}(d / t) = 0.9699$ (i.e., by solving for d with the delivered t value).

Of perhaps more pertinence would be to determine the value of C, which would simply be (in distance) $t - d$ along the circumference (i.e., of coordinate units); but this is not the complexity distance. The complexity distance for C would appear to be ($10.5056 - 0.9699 =$) 9.5357, which is ~ ~ 2.46 times the $n = 2$ radius and ~ ~ 7.26 times the $n = 1$ radius.

Now we have an interesting conundrum, since (for our example) the longest (straight) line within the $n = 2$ circle, and therefore able to directly connect two coordinate points, would have to be the diameter of the circle, which would be (2)(3.8796), while the estimated complexity charge between t and d is considerably larger (at 9.5357) when going from $n = 2$ to $n = 1$; i.e., $2N_{\Psi m2} < 9.5357$

Therefore, the complexity process (or processing of complexities) must extend the charge between t and d (in our case) when going from $n = 2$ to $n = 1$ (i.e., a removal); there is a distortion of the line (distance) for C from straight to (perhaps) curved (or vertex-ed, etc.).

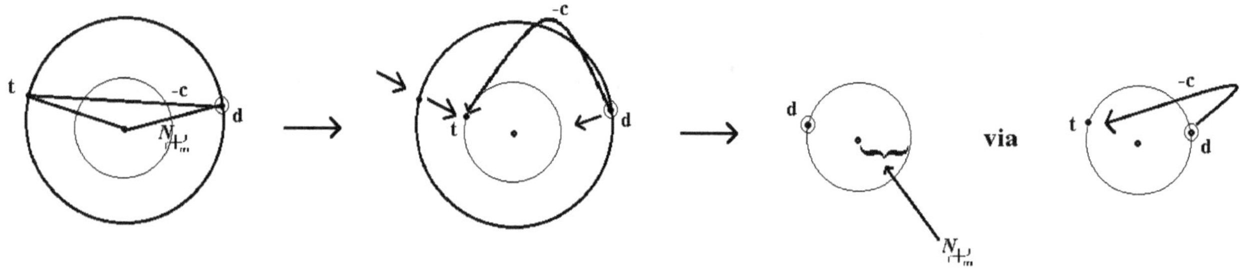

This complexity distortion (of charges) is, of course, dependent on the coordinate values involved.

Where does the extra ($9.5357 - 7.7592 =$) 1.7765 complexity come from?

Presumably it is due to the contraction of the (CCP) circle (on n reduction of N, as the radius) and the contortion of the C charge, as compared to the (maximal) complexity of a "diameter-ed" charge. The 8.2225 in extra complexity for the $n = 1$ circle resultant

(i.e., 9.5357 - 1.3132, where 1.3132 is for a "normal" $n = 1$ circle) might be due to the d-carry, with the (d re t) removal.

The numerical minutia suggests that even for an exchange (with the C charge being unaltered) an expenditure of complexity is required. This is apparent when there is "with carry" since the (here) t and d values are exchanged during their relation. "Without carry" may indicate something less obvious, perhaps the rotation (from the center) of the (CCP) circle's (family) sphere (the observed coordinates lying on the surface of that sphere), where another (smaller) circle including those (two) coordinates is produced:

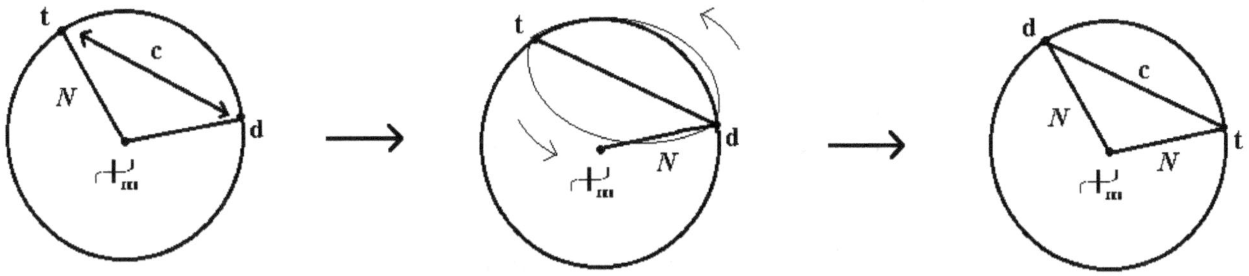

Since the entire sphere is rotating (with all coordinates relative to its surface) there is " no carry" for t and d. And as an exchange, C is unaltered, nor ***N*** radius. But what is altered is (our) perspective of t and d (e.g., now, t re d → -C).

The term "charge" is used here as a sort of **burden** or **obligation** between two points or coordinates, i.e., the distance separating them within the CCP circle as an internal or supra-internal (external but as internal) construction more or less mechanically analyzed. The meaning may be extended towards a (profitable) tendency for exchange or removal of points. Of course, all (circumferential) points are obligated to the central base (here Ψ_m), establishing the ***N*** radii for utility of computations and path descriptions of (re, ex) procedures and operations. It is assumed that all operations require an expenditure of complexity, including the formation - observation - and existence of coordinates themselves. That is, any relation has complexities associated with it, even if it is merely a variable to be coordinated. The formation (and utilization) of a circle, or a sphere and its surface, of coordinates requires a consumption of complexity, if only for orientation purposes. Therefore, all measurements result from manipulations of complexity.

CCP Proportional Rotation

There is one other obvious operation for (normalized → common unitage) coordinates on the (CCP) circle circumference: that of (necessarily) proportional rotation of coordinates about the center.

e.g.

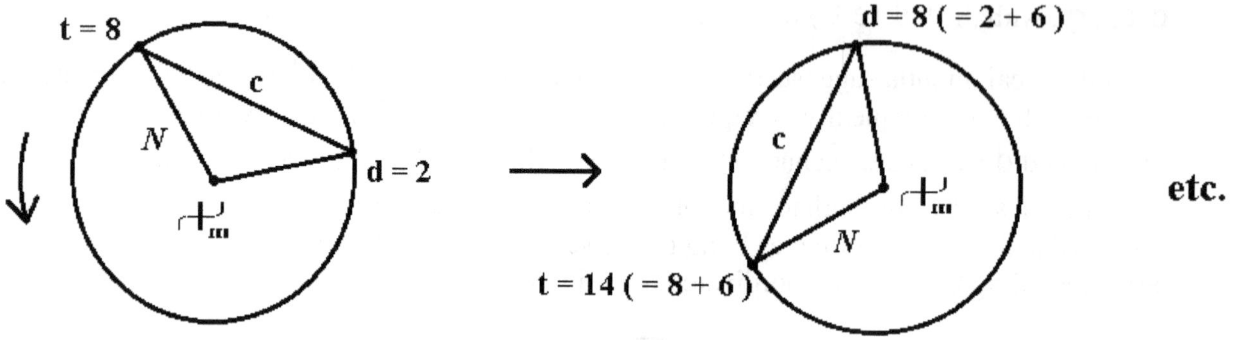

The (c) charges are unaltered in magnitude. But the relation between d and t can change of value: (2 / 8) =/= (8 / 14) for d / t . This rotation is positional, and so the coordinate change is additive rather than factorial (proportional).

A proportional rotation would be positional, but with a "resistance" proportionally affecting each coordinate (as a factor). This resistance varies as the rotation (directly or indirectly) of its extent (or effect).

Therefore, as $8 = 2 + 6$ (additive), so can $8 = (2)(4)$ as proportional (or factorial).

Yet, for a larger coordinate (than 2), the "resistance" yields a greater value

(here, 8 times $4 = 32$) and therefore requires a greater effort (for the rotation);

i.e., it is easier to raise 2 to 8 than to raise 8 to 32 .

The "resistance" or load or twist or burden of an effort on the center, in terms of the complexity expenditure, can be related to the center's base (here Ψ_m).

For example: to go from $(2, 8) \rightarrow (8, 32)$ for $(d, \rightarrow t)$ we might use, as a coordinate addendum (residual)

$$(\text{factor})\Psi_m^{(\text{d considered} / d)}$$

[Note that this requires that no coordinate "falls" to zero. The addition is a change (of coordinate).]

For coordinate (d =) 2 , with a factor (proportional) of 4 :

$$(4)\Psi_m^{2/2} = 4\Psi_m^{1} \sim\sim 6.0924 \quad (\Psi_m \sim\sim 1.5231)$$

Therefore, $2 \rightarrow 2 + 6.0924 = 8.0924$ (~ ~ 8) for coordinate (t =) 8 ;

thus, d considered = 8 (d = 2)

It is already factored (by consideration of d); thus,

$(4)\Psi_{m}^{8/2} = (4)\Psi_{m}^{4} \sim\sim (4)(5.3816) = 21.5265$; thus,

$8 \rightarrow 8 + 21.5265 = 29.5265 (\sim\sim 32)$

This is for the complexity for the process of d going to the t coordinate with a proportional rotation.

The (d, t) coordinate resistance (resistance factor), for a proportionality factor of 4, with (d, → t), for (the "point") (2, 8) is suggested to be 21.5265 / 6.0924 = 3.5333 (relative, in terms of Ψ_{m} complexity).

The "complexity factor" would then be 29.5265 / 8.0924 = 3.6487 , and the actual factor would be 8 / 2 (as d) = 32 / 8 (as t) = 4 .

So now, we have processes of CCP circle rotation (with no change of coordinates), coordinate rotation (positional) about the center (additive change of coordinates), and proportional rotation (positional with resistance, for change of coordinates).

The (d considered / d) ratio is, of course, an idealism that can be used to coordinate different coordinates (e.g., d, t). For the comparison (here of d to t), the (d, → t) operation suggests the resistance is:

$(t / d)\Psi_{m}^{(t / d)}$;

therefore, let (t / d) == r == factor of comparison.

Thus, resistance is $r\Psi_{m}^{r}$ for (d, → t) of relation (d / t).

The exponent for Ψ_{m} appears actually to run from r == (1) for d = d, to r == r for d = t (although the exponent can be any number).

Therefore, for practicality, let the exponent be (1 → r) ; thus, resistance is $r\Psi_{m}^{(1 \rightarrow r)}$.

This says that each coordinate starts out with at least $r\Psi_{m}^{1}$ of resistance complexity (for a proportional rotation). Therefore, the resistance factor is $(r\Psi_{m}^{r}) / (r\Psi_{m}^{1}) = \Psi_{m}^{(r - 1)}$; the complexity factor is $(a_{r=r} + r\Psi_{m}^{r}) / (a_{r=1} + r\Psi_{m}^{1})$, where a == coordinate; the actual (or proportional) factor is $a_{r=r} / a_{r=1}$.

(For our proposal) It is clear that for a proportional rotation, or proportional adjustment of values, there can be no coordinates at zero. Therefore, any (rotational) motion at all of the CCP circle eliminates any zero coordinates (proportionally considered). And since, in real nature, all coordinates are moving, then there are no zero coordinates; i.e., the fact that all relations of variables are factor-able (by r == 1 → r, or a factor of "1" occurs for each variable at least), and therefore proportional, says that zero coordinates do not exist. The CCP circle, then, is unbounded (as for any circle) and uninitiated (has no real starting point). The direction of its rotation determines positive and negative initiatives (signs), but always from non-zero points.

CCP Circles and Processes

A corollary for this proposal is that a Proportional Rotation (for a CCP) is always occurring, either in one direction or with oscillations. There is no "resting" of coordinates, although some constant values might seem to be measured (or measurable) due to "slow" rotations or regular oscillations (as observed).

The rotations, then, involve the physics (of a situation) as the coordinates lend (meaning) to the observations from measurements (or attempts of such). Of course, in this way, manner, or mode, one can envisage one CCP (circle) affecting or influencing another. In this case it is the resistance factor (rather than the complexity factor) which is signifying (or following) the "energy" of the compounded process, since "resistance" is directly related to rotational effort.

So, for example (just as illustration), one can have two (CCP) time curves affecting each other:

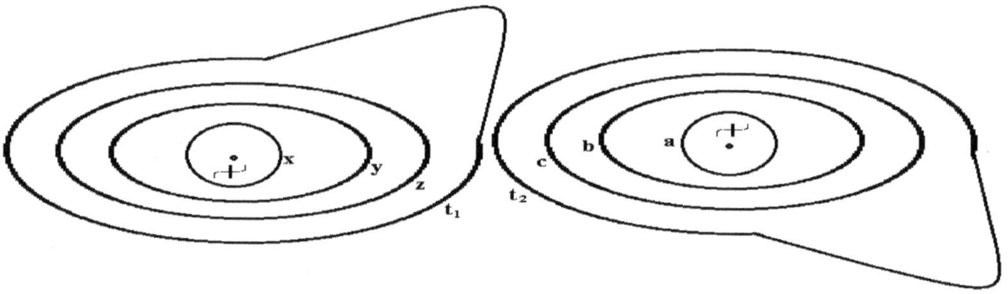

where there is a (supra-)relation between (x, y, z, t_1) and (a, b, c, t_2).

The overall resistance (factor) for the process could be (schematically)

$\Psi_{()}^{(rX-1)} \bullet \Psi_{()}^{(rA-1)}$ [as proportion's "and" probability], X == (x, y, z), A == (a, b, c)

Therefore, coupled

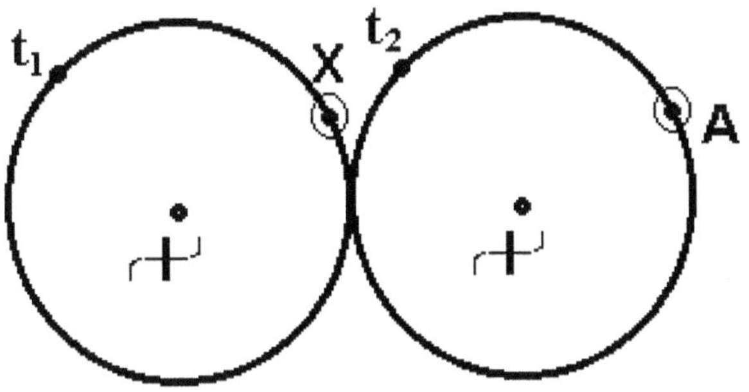

or uncoupled

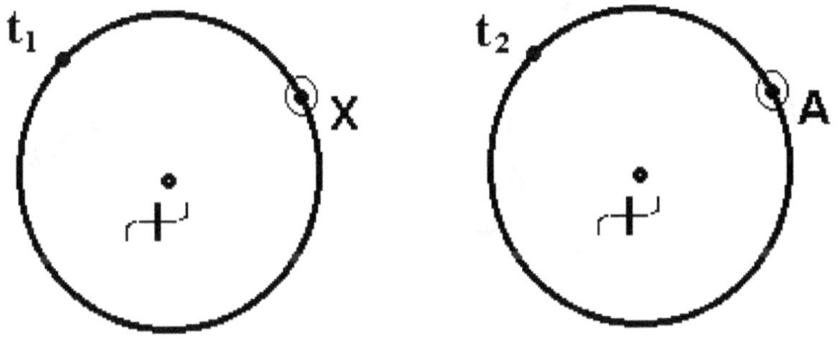

$\Psi_{()}^{(rX-1)} + \Psi_{()}^{(rA-1)}$ [as proportion's "or" probability] .

Note, then, how the (complexity) probabilities would differ for the process:

$(X \text{ re } t_1), (A \text{ ex } t_2)$

coupled $\rightarrow$ (X, A, t_2) or $(X; t_2, A)$ or $(X (t_2, A))$

uncoupled $\rightarrow$ $(X) + (t_2, A)$

For coupled, (X, A, t_2) would be a formal notation (with time coordinate last)

while $(X; t_2, A)$ or $(X (t_2, A))$ could essentially mean

$(X, (t_2, A))$ or $((X, t_2) , (X, A))$

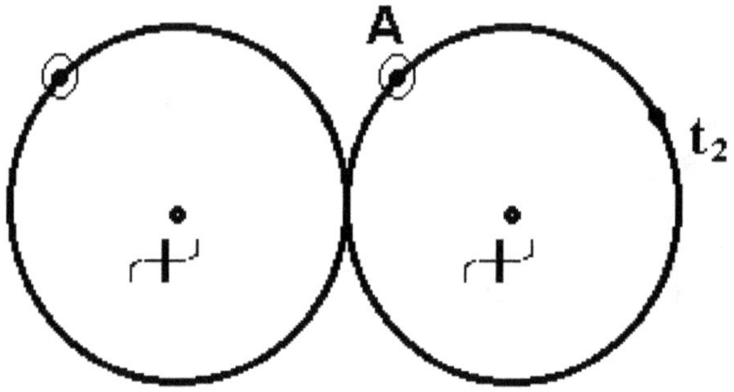

For uncoupled, you have

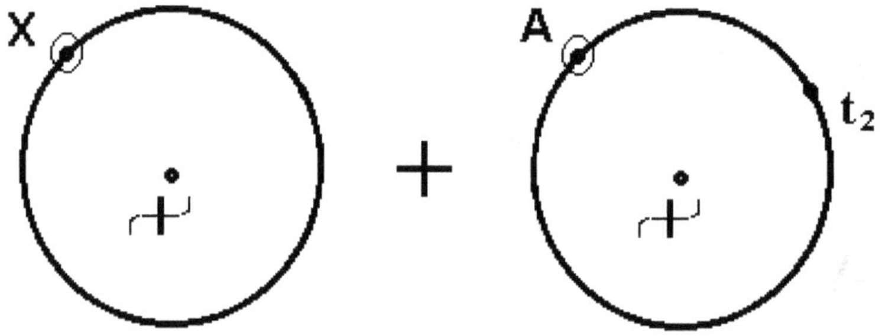

$\therefore\ (X), (t_2, A)$

For an exchange (of coordinates, here (A, t_2)) there is no (proportional) rotation about the (homed) CCP circle circumference, but there is a rotation about a (minor) circle intrinsic to the two coordinates:

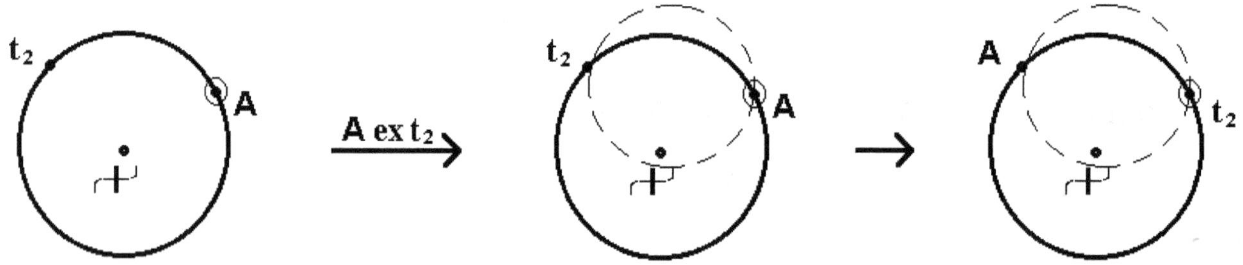

(Clearly,)This minor circle (or coordinate intrinsic circle) rotation is proportional and therefore with resistance if the charge connecting A and t_2 (i.e., the two coordinates) is not the diameter of the circle; it is (only) positional and with no resistance if the charge connecting the two coordinates is the diameter of the circle (or still proportional with double resistance).

Then, as a corollary, the proportional rotation resistance is (due to) the difference in arc distances for the two directions $A \rightarrow t_2$ and $t_2 \rightarrow A$. One direction would have greater resistance (due to length or distance) than the other. This suggests a deviation from the circle to ellipsoid (or other) circuit (or path), with the (directional) resistance being as (or proportional to) the ratio of one direction's arc length path to the other one's.

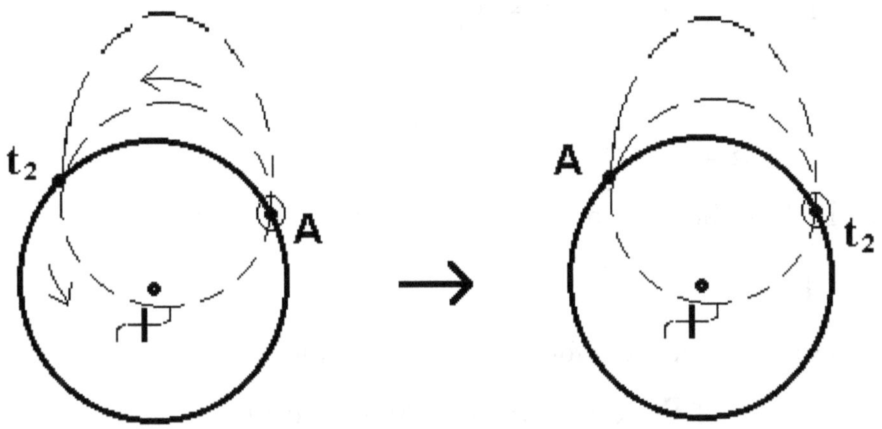

There is no observable resistance when this ratio is 1 ; i.e., $A \rightarrow t_2 = t_2 \rightarrow A$

Although, one could claim that the effective resistance is double ($== A \rightarrow t_2 + t_2 \rightarrow A$) for a consistent rotation to yield the coordinate exchange (proportionally).

CCP "Formuleic"s

Since, for an exchange, (in this case) $A \rightarrow t_2$ must occur with $t_2 \rightarrow A$, this is a probability's "and". Therefore, the (proportional) resistances are termed as (like):

$\Psi_{()}^{(rA \rightarrow t2 - 1)} \bullet \Psi_{()}^{(rt2 \rightarrow A - 1)}$ subject to the relation between A and t_2 .

Therefore, we have (for our example):

$$\Psi_{()}^{(rA - 1)} = \Psi_{()}^{(rA \rightarrow t2 - 1)} \bullet \Psi_{()}^{(rt2 \rightarrow A - 1)}$$

Therefore, for the entire process:

coupled $\Psi_{()}(r_X - 1) \bullet \Psi_{()}(r^{A \rightarrow t2}) \bullet \Psi_{()}^{(rt2 \rightarrow A - 1)}$

uncoupled $\Psi_{()}(r_X - 1) + \Psi_{()}^{(rA \rightarrow t2)} \bullet \Psi_{()}^{(rt2 \rightarrow A - 1)}$

[Since, as notated, A initiates the exchange with t_2, there might be (or is allowed) an extra "weight" for $r_{A \rightarrow t2}$ compared to $r_{t2 \rightarrow A}$; let this weight be w.]

Therefore, $\Psi_{()}^{(rA-1)} = \Psi_{()}(wrA \rightarrow t_2 - 1) \bullet \Psi_{()}^{(rt2 \rightarrow A-1)} == \Psi_{()}^{(wrA \rightarrow t2-1)+(rt2 \rightarrow A-1)}$

Therefore, a coupled process would yield

$\Psi_{()}^{(rX-1)} \bullet \Psi_{()}^{(wrA \rightarrow t2 + rt2 \rightarrow A-2)} == \Psi_{()}^{(rX + wrA \rightarrow t2 + rt2 \rightarrow A-3)}$

of resistance factor.

And if $r_{A \rightarrow t2} == r_{t2 \rightarrow A}$, the result is $\Psi_{()}^{[\,rX + (\,w+1)rA \rightarrow t2 - 3\,]}$

This is the (1, 2, 3) factorization of the Ψ exponent: $(\,1\,)r_X + (\,2\,)r_A - 3(\,1\,)$,

where $w == 1$;

which presents a "formuleic" (for the resistance factor) due to:

$(\,1\,)\,r_X$	of X re t_1	or a removal operation
$(\,2\,)\,r_A \rightarrow t_2$	of A ex t_2	or an exchange operation
$(\,3\,)1$	of X, A, t_2	or three coordinate initiatives

The termed relation between coordinates (e.g., (a•b), (a /b), etc.) is adopted to find the resistance factor using this (simple) formuleic (or formulaic progression or formalism). It is evident that, for CCP, the considered circles can be not only coordinates (of variables or constants), but also functions, with addition and subtraction of terms.

In order to retain the utility of our formuleic (i.e., to determine r terms for the Ψ exponent) we must find a convenient way to combine the terms of a function. One such method is to combine by subjecting each variable (or constant) of unitage to a common base, whereby a common factor may be derived, such as to show that addition (and subtraction) is actually a form of multiplication.

For example, if our relation were (instead of $d\,/\,t$) the function: $d + \alpha t$, where α is some constant, then we must combine $(\,d + \alpha t\,)$ to determine r whereby we have $r\Psi^r$ for (d, → t); thus, $d + \alpha t = a$ (say) and we determine for $r = a\,/\,d$.

Normally, (d, t, α) are all (numbers) of base 10, and so the combination (for: $d + \alpha t$) is straightforward.

But suppose α (via multiplication, here) does not convert t to a unitage of d. Then the fraction $a\,/\,d\,(\,= r\,)$ is not a proper factor. It is a proposed or relational factor.

Whereas for $t\,/\,d\,(\,= r\,)$ uses different unitages for t and d, the term is still a proper factor (of unitage $t \bullet d^{-1}$). This is required to allow for $r\Psi_{()}^{(\,1 \rightarrow r\,)}$ in calculation (of our formuleic). A unitage of (d + t) is too obscure of meaning or implacable to provide for a proper factor (i.e., it is not single or "term-like").

To yield "a" (for a proper factor) we might consider d a base, and convert α and t to that base.

Therefore, $\log_d d + (\log_{10}\alpha)(\log_{10}t)/\log_{10}d)^2 == 1 + (\log_{10}\alpha)(\log_{10}t)/(\log_{10}d)^2 = E$ (a "pure" number).

Therefore, $a = d^E$, $r = d^E / d$ (a unitage of d^{E-1}).

The same can be done for, such as, (x, $\rightarrow$ f(x)), if desired. This is, of course, to promote or utilize (the efficiency of) the formuleic.

So, we have a general formula for the resistance factor of a process:

$\Psi_{()}$[(Σ r operations coupled) - # coordinate initiatives]

where the CCP uses proportional rotation (to divine the complexity).

CCP Complexity Resistance and Coupling

The "resistance" refers (analogously) to the resistance or inhibition for an operation (re, ex) to occur as were the CCP circle's circumference to rotate. It represents a "complexity cost" for the operation. A complexity, of course, directly relates to the values of the coordinates (and variables or constants) achievable by them through (or during) their relationships with each other, therefore also to such special quantities as maximum and minimal values. The reason a CCP circle does not have a zero (start) point is that each (and any) coordinate (variable, constant, function, etc.) has an inherent (even initial) complexity (which can not, therefore, be zero) through its use in a relation (even singularity or alone) for its existence. Therefore, there is always a "resistance" associated with it, as the (CCP) circle is always in some motion (axial, circumferential, whole body, etc.). Perhaps even the center (base) of the circle can be subject to change, due to a motion or operation (e.g.: using strict dis-legitimacies, $n = 0$ can not occur).

So, we see, a character or element as like

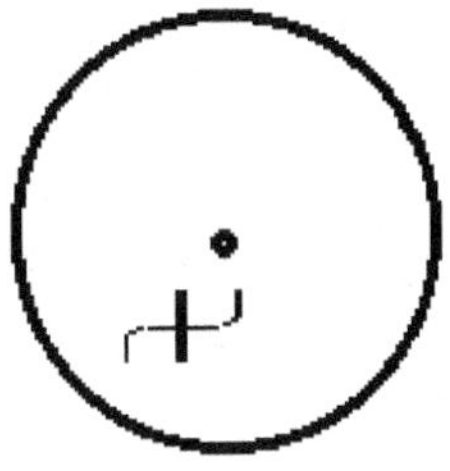

is really simply a utility of Ψ as a base (for some operation), and an element such as

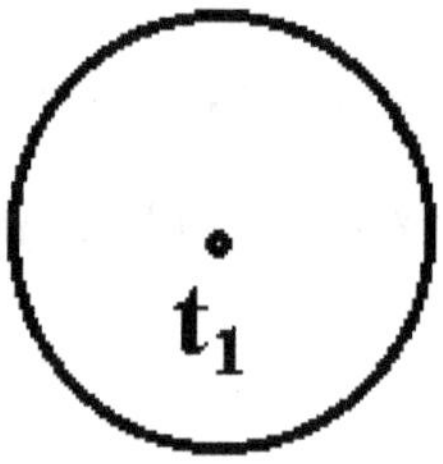

is the utility of a (particular) time coordinate for some operation. All elements, therefore, assume some inherent or initial complexity, in order to engage in an operation, and therefore come with (complexity) costs for the operation.

[These central or fundamental elements can of course be drawn without a circular circumference about them (since the circumference is empty), but they are easier to portray or "see" when so highlighted or "enhanced."]

We can assume, then, that "coordination of variables" is due to coupling, and that a coupling such as:

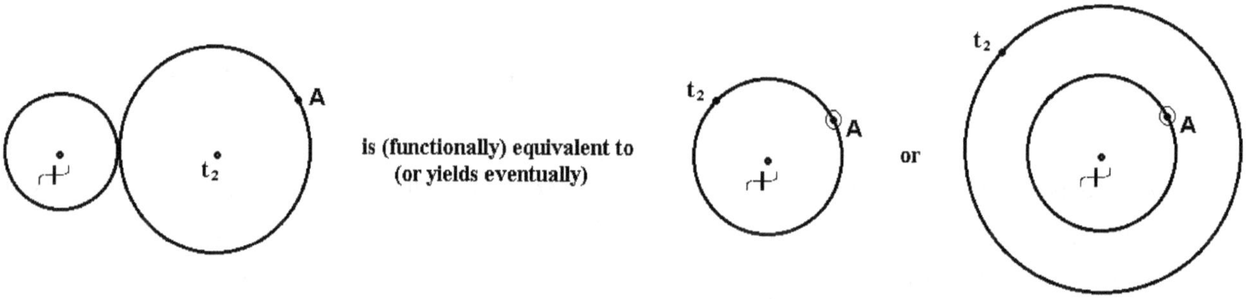

Note how coupling is distinguished from re and ex. It is more like a fusion of circles:

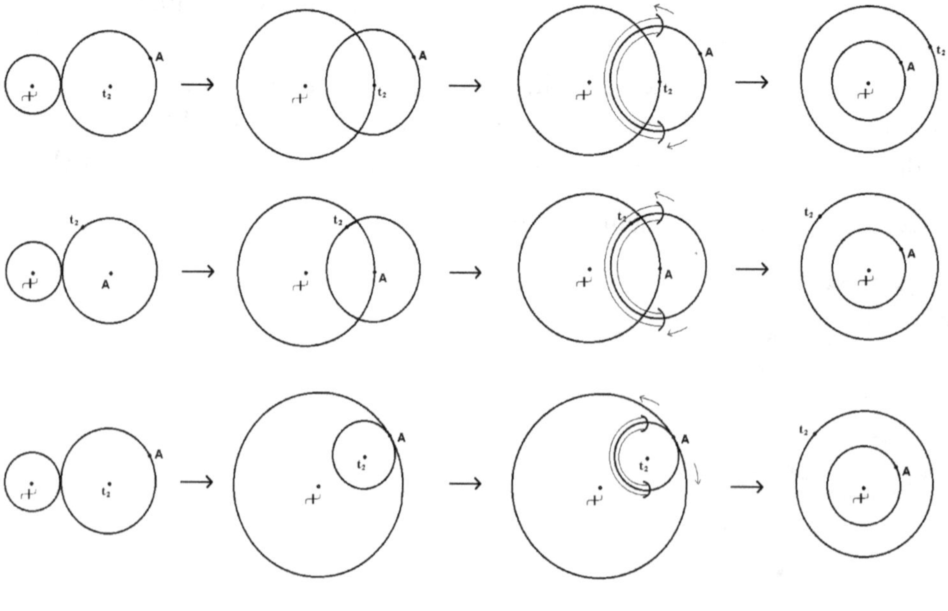

etc.

Coupling remains under the guidance (or proscriptions) of Proportional Rotation.

Perhaps the propensity (of elements, etc.) to couple [as "and" probabilities] is to increase the complexity (like entropy), to allow for a process (more efficiently achieved than with additive uncoupling: [as "or" probabilities]).

This, then, suggests a way (or manner) of "complexity basing" variables (here: Ψ to (A, t_2)).

This is dependent on the relation between the variables, and the process desired (type of CCP rotation, coupled or uncoupled, etc.).

Take, as an example:

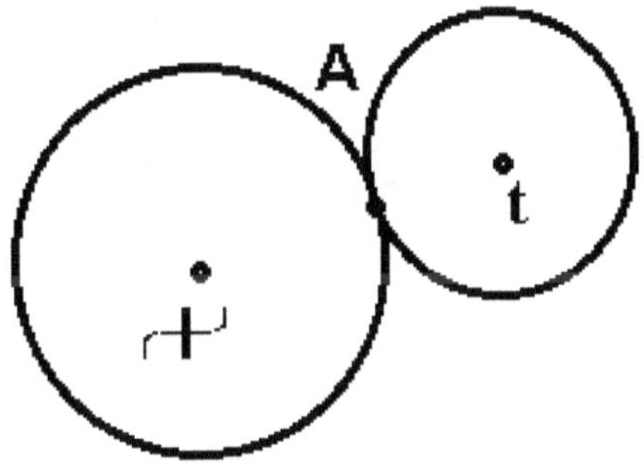

where (as diagrammed) A (a point, area, or volume) is "based" of complexity to Ψ, but t (time) is not. A is also "based" of complexity to t, and there is a coupling of the two spheres of base.

A can not exchange with t, but can remove t, and visa versa not. Therefore,

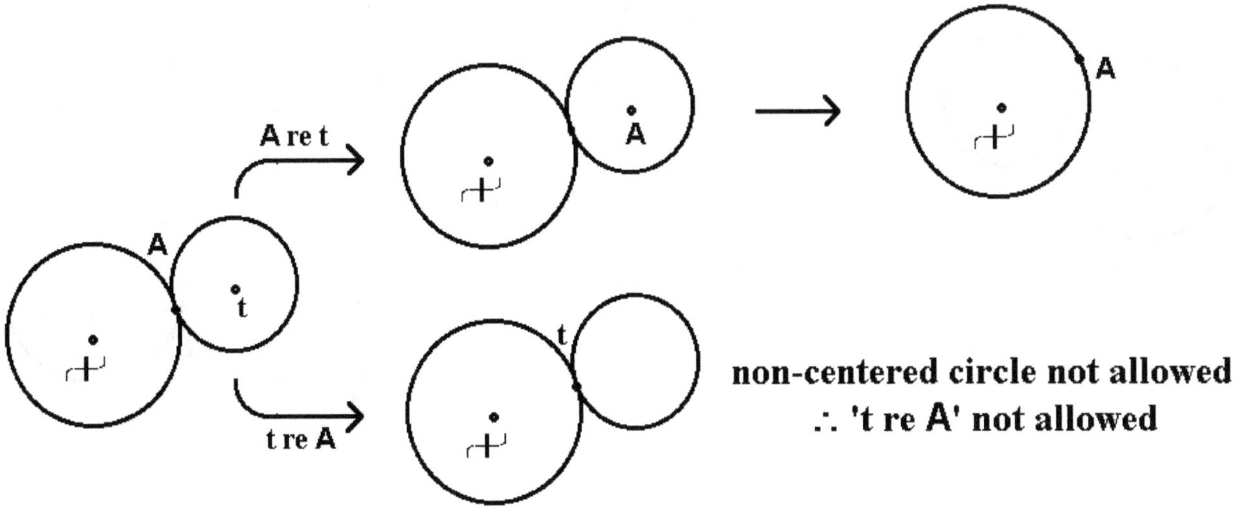

But we can also have (via CCP procedures):

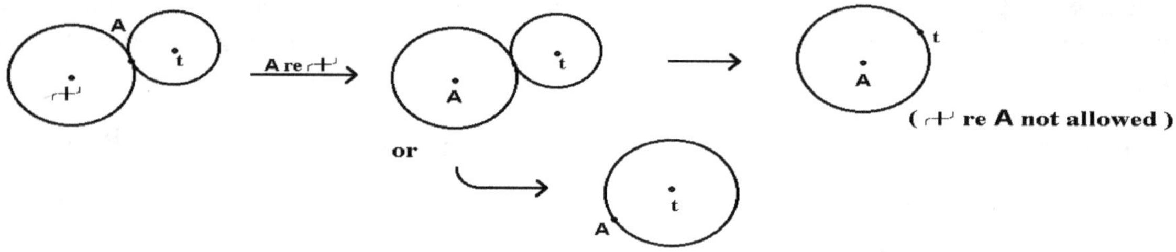

So, the 4 possibilities for

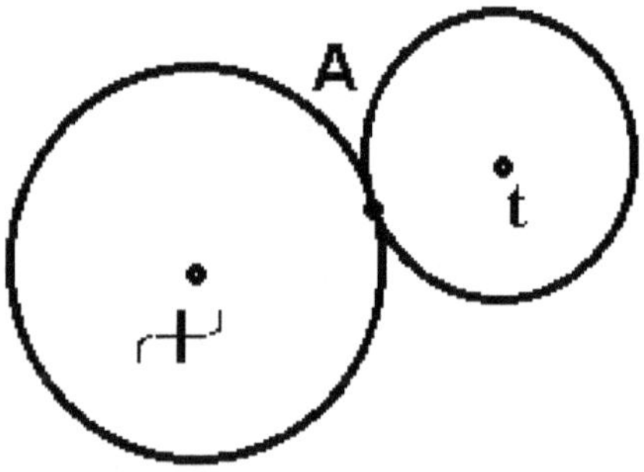

are

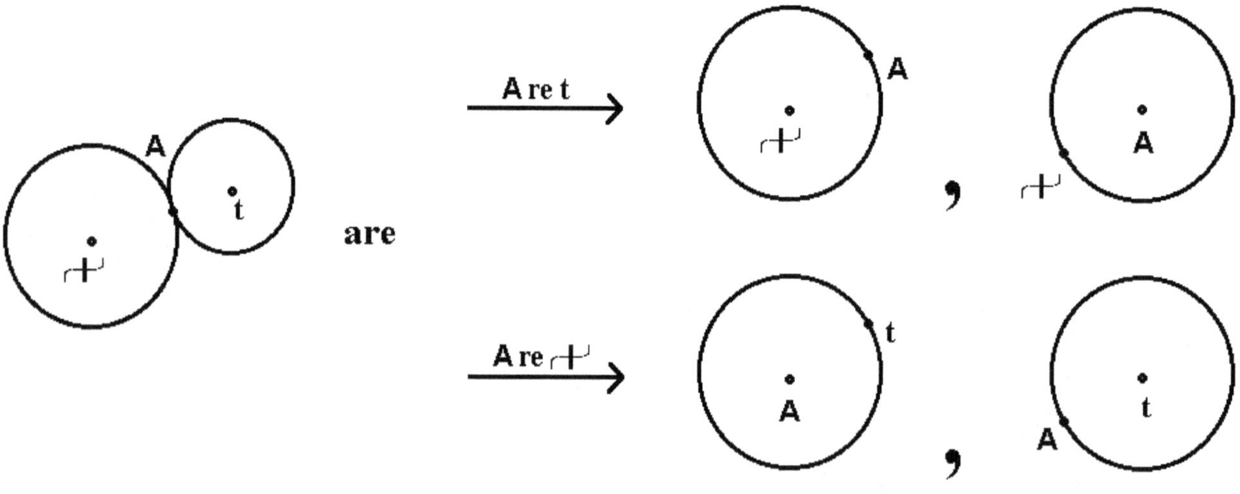

In terms of complexity, these (possibilities) can only be resolved by re operations:

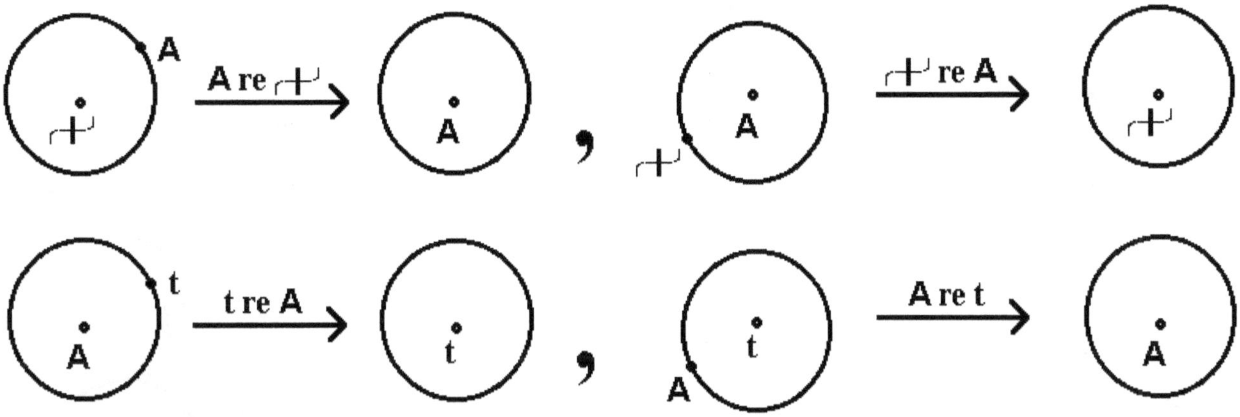

(and) therefore resolved down (eventually) to the three elements: Ψ, A, t .

[In each case, of course, the base can not result in zero, unless the base itself is zero, which is 'not of a procedure' and contradictory of kind since non-centered circles are not allowed. Some complexity must always occur for something to exist, such as a procedure or operation.]

Therefore, elements only "exist" when they are used for, or derived from, a procedure.

But, what does a non-$\Psi_{()}$ based element mean (or signify)?

Such elements must be coupled to

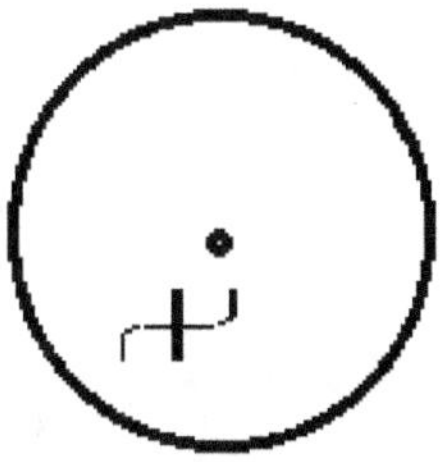

in order to become operable (of their complexities and modifications or changes).

Therefore, we can have:

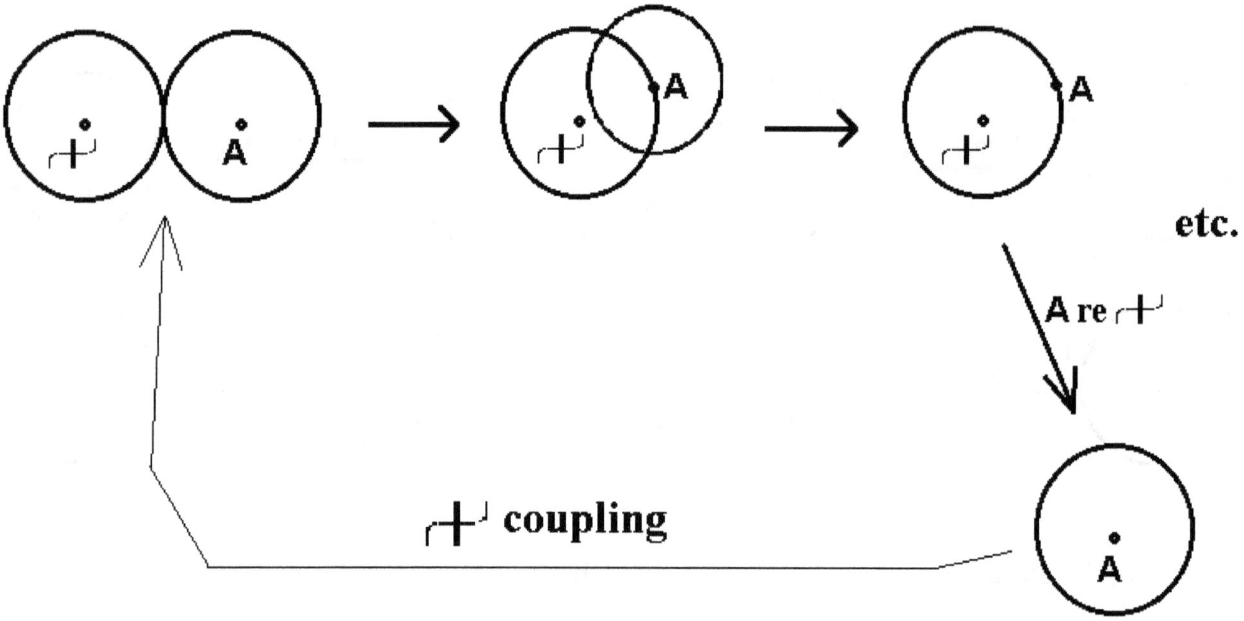

So, what compels (for) $\Psi_{()}$ coupling (of elements)?

→ a need for Proportional Rotation, during a (CCP) procedure;

i.e., in order to make A and t proportional (to each other), $\Psi_{()}$-coupling must occur.

A can have any (non-zero) value in regards to t $\left(\equiv \frac{t}{A} \right)$.

t can have any (non-zero) value in regards to A $\left(\equiv \frac{A}{t} \right)$.

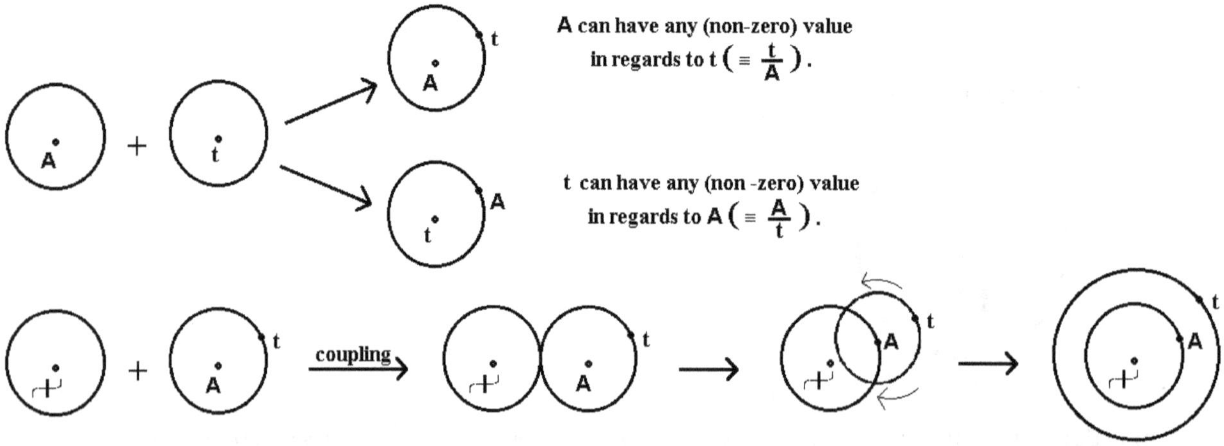

Ψ-coupling prescribes (here) to A a "determinable" value, in relation to t, for the operation t / A (apparently for the relation A / t).

Therefore, for (A, → t) of relation A / t , the resistance is $(t/A)\Psi_{()}(t/A)$,

with $r = t/A$, and the resistance factor is $\Psi_{()}^{[(t/A)-1]}$.

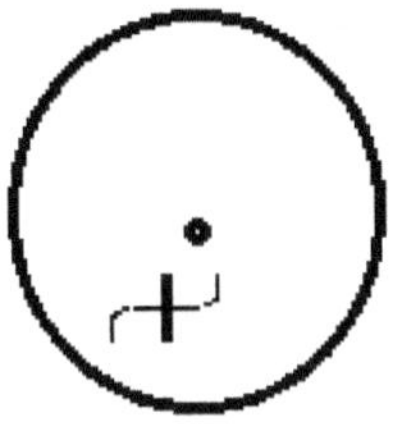

represents the Ψ base that can adapt to (or adopt) any of its exponential-ed values depending on the number of variables (elements) it couples to (i.e.: n). That adoption determines the complexity for the (or a) process involving its elements.

What compels non-Ψ-centered element coupling?

→ obviously the procedure or operation that uses them:

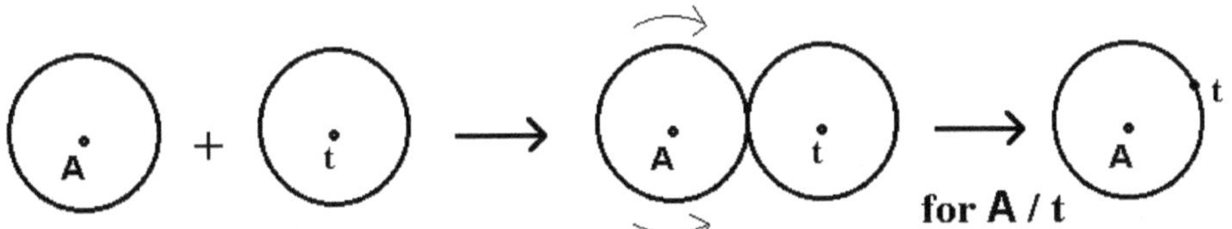

for A / t
[(t / A) operation's relation]

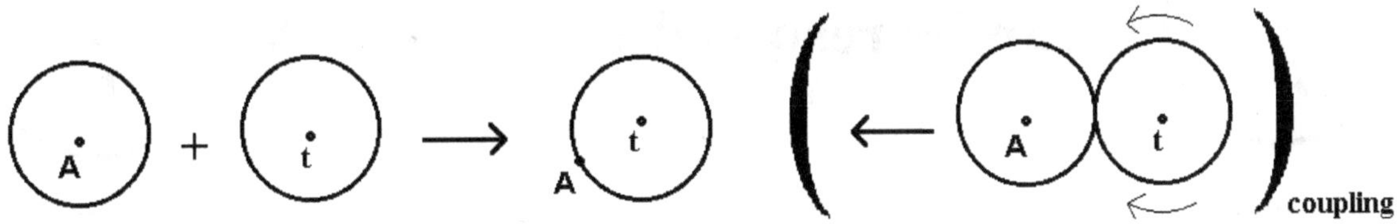

for t / A
[(A / t) operation's relation]

∴ for 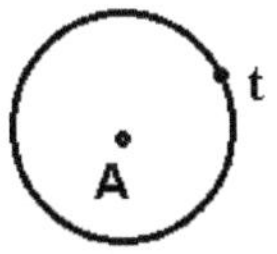

, t is "localized" (or is in regards to)
[If t is localized, it can form A / t relation.]

∴ for 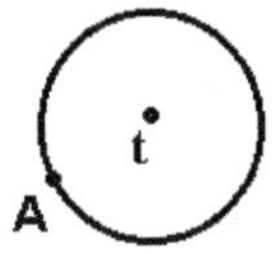

, A is "localized" (or is in regards to)
[If A is localized, it can form t / A relation.]

Therefore, the procedure (or operation) on variables (or elements) refers to the resistance (== r) employed during the Proportional Rotation; that is, the resistance towards complexity (complexity resistance); e.g., (A, $\rightarrow$ t) is the (or an) operation (which causes a resistance).

[Note: (A, $\leftarrow$ t) could still use the relation A / t, as distinguished from (t, $\rightarrow$ A) using the relation t / A .]

But are

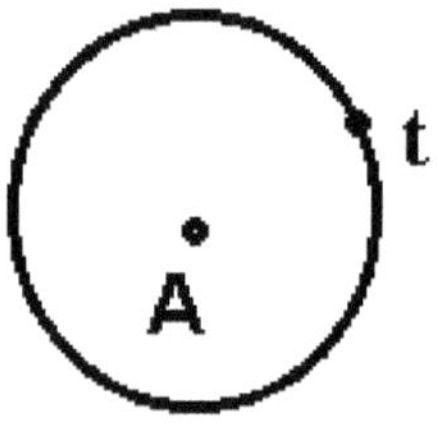

and

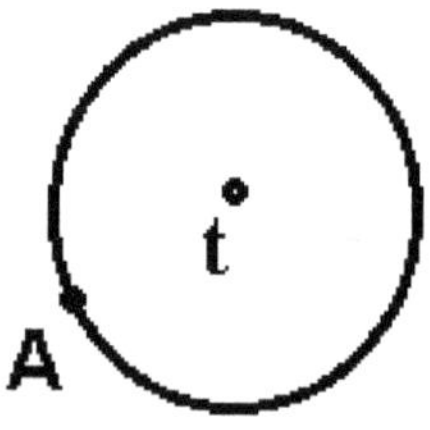

interchangeable?

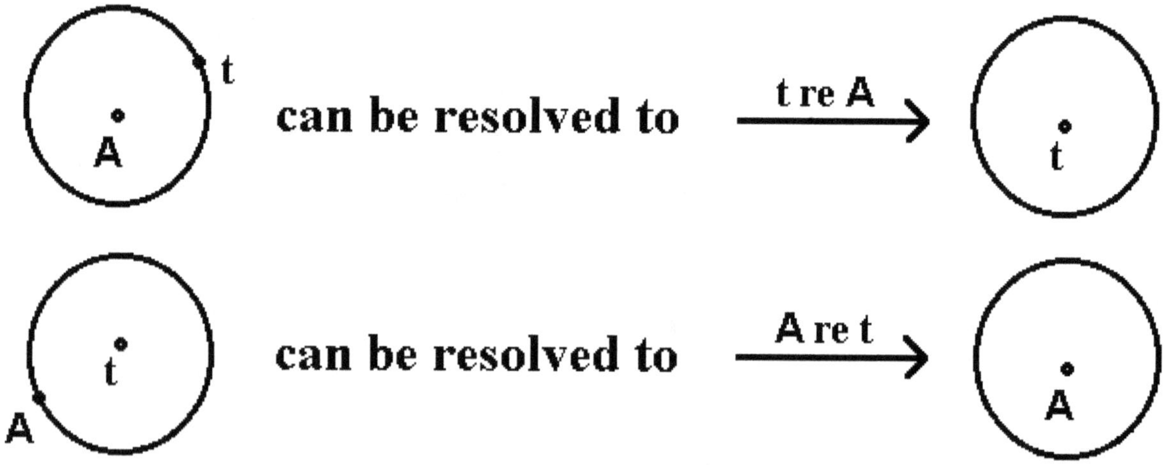

Therefore, they are not interchangeable (elementally).

By coupling, however:

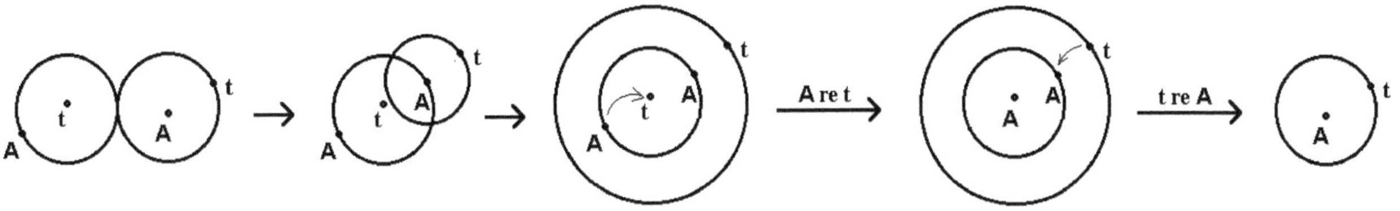

and likewise:

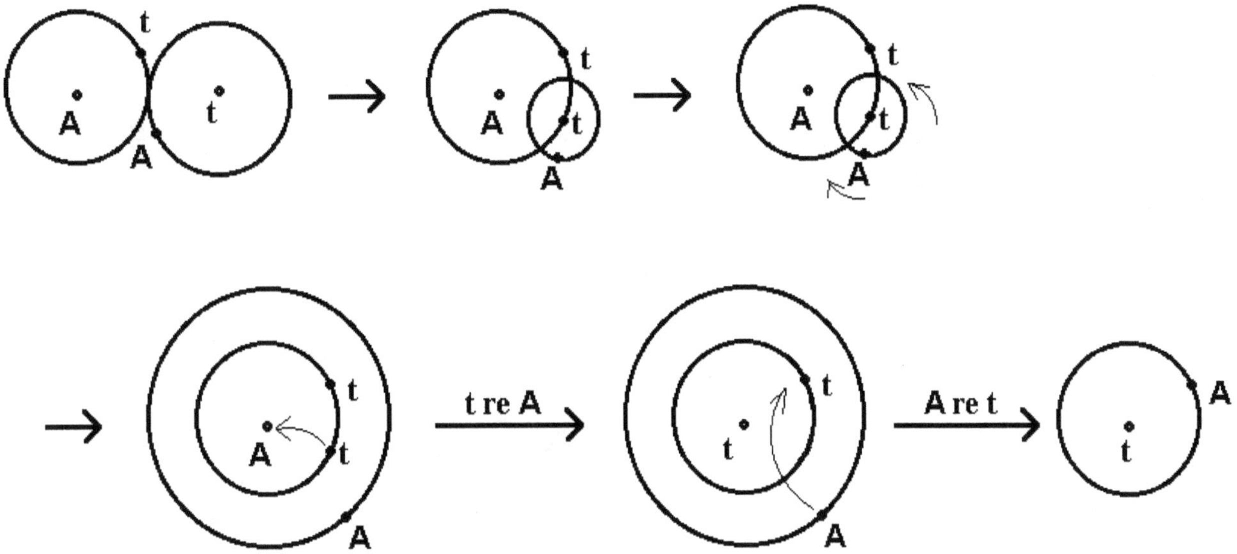

Therefore, through coupling of both, either one can result, of resolution, of kind.

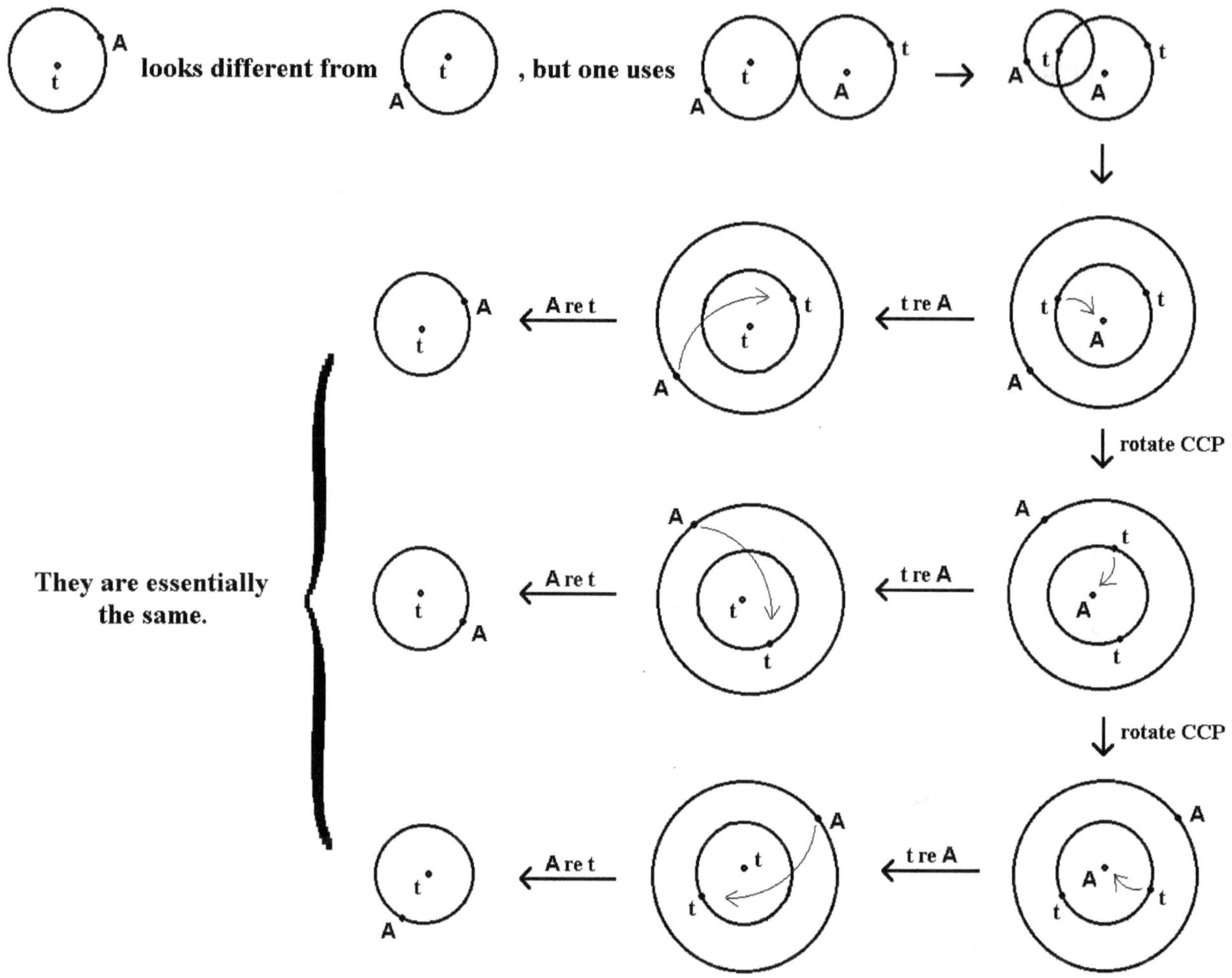

They are essentially the same (t can take any non-zero value, while A remains at its value). Still, it may be significant and utilizable that a CCP coupling diagram seems vectorial (directional), of resolution; e.g., let's couple 3 elements:

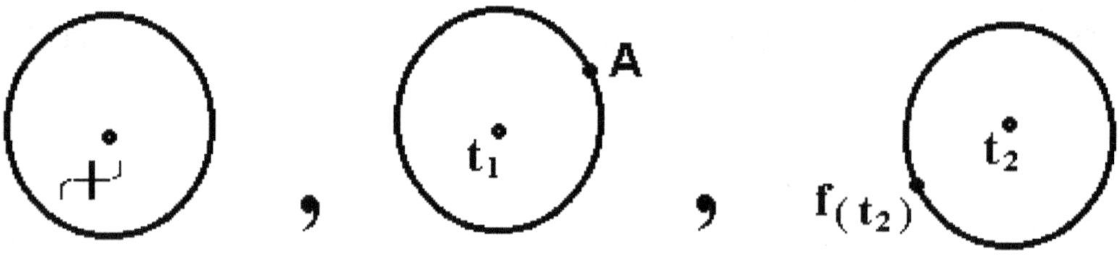

where $f(t_2)$ is a function of (A, t_2) and t_2 is the independent variable (A is the dependent, to t_2, variable).

Therefore,

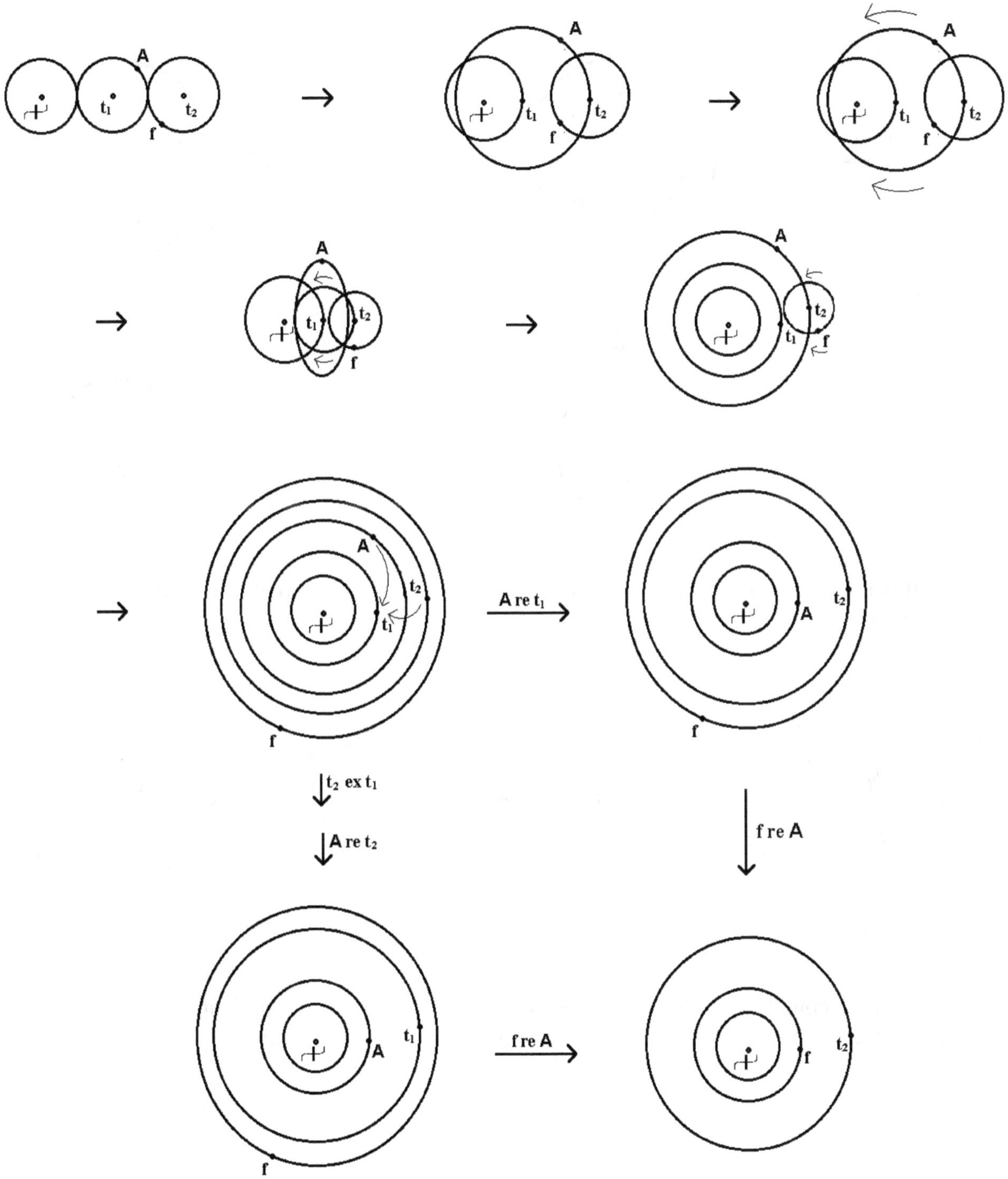

This can of course be resolved elementally (i.e., to the elements) via:

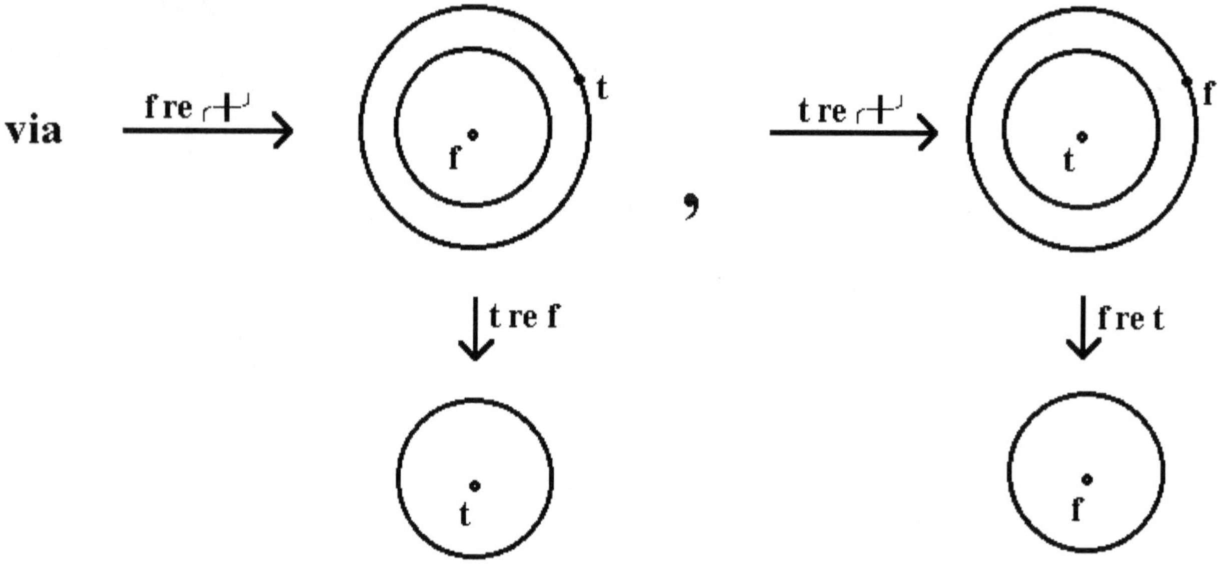

So, it is interesting that this coupling ultimately can lead to the function with its dependent variable(s):

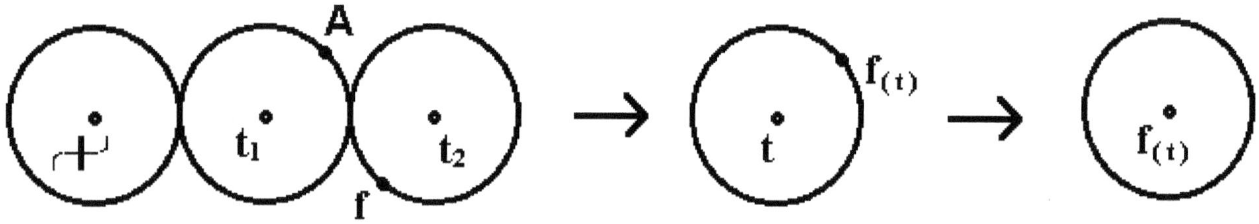

Therefore, a coupling can be used to define such functions, at least derivatively. This indicates (or illustrates) that there can be various complexity paths to a function.

In fact, from

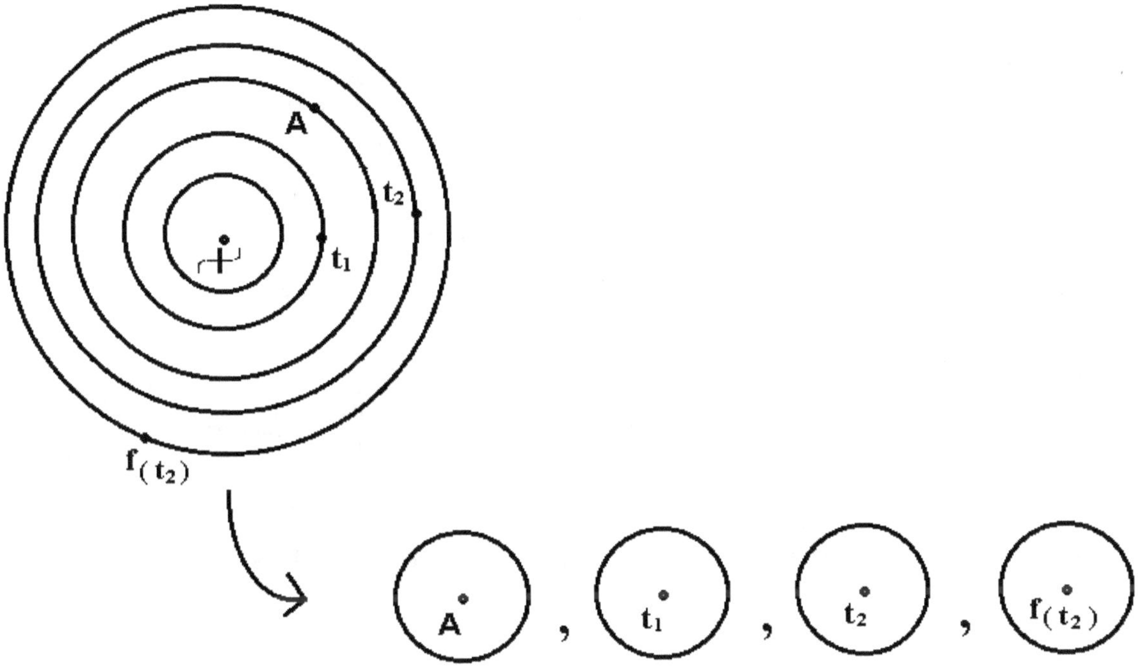

each of the elements can be derived, except for

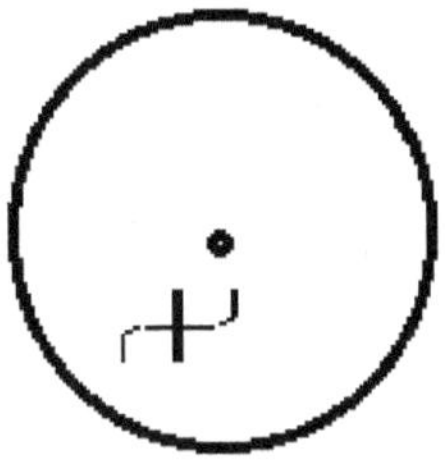

itself.

(All operations, [re, ex], using strict dis-legitimacy, must eventually eliminate Ψ.)

Therefore, in CCP, a unique (legitimacy-wise) central element can not itself be derived (i.e., be resolved to itself).

Expressing the CCP in a condensed form makes perhaps a clearer interpretation of the coupling result:

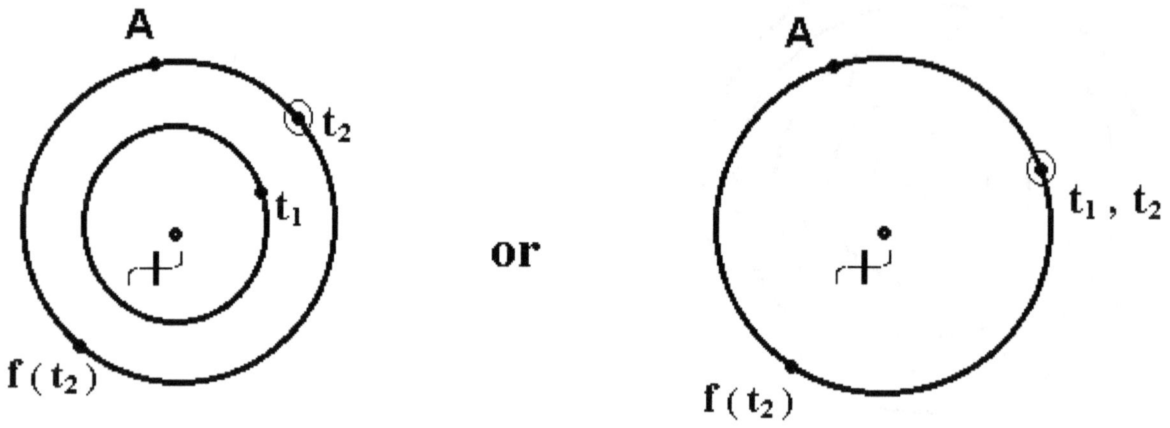

If we take $f(t_2)$ and A to be of the same class (as t_1 and t_2 are of the same class), using non-strict-dis-legitimacy, then they can only exchange with each other but may remove t or Ψ or be removed by t.

One can have the interesting phenomenon that $f(t_2)$ eventually becomes A:

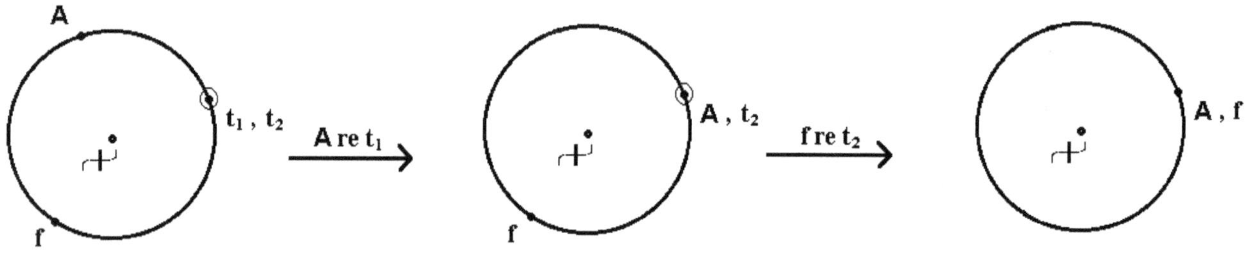

Likewise, t_1 and t_2 can be made distinguished from each other:

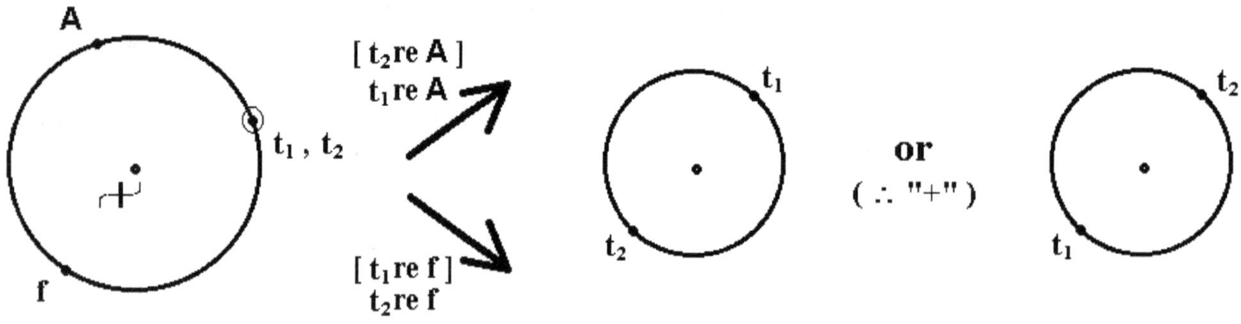

In the same manner, one can have ([A re t_1, t_2 re f], etc.):

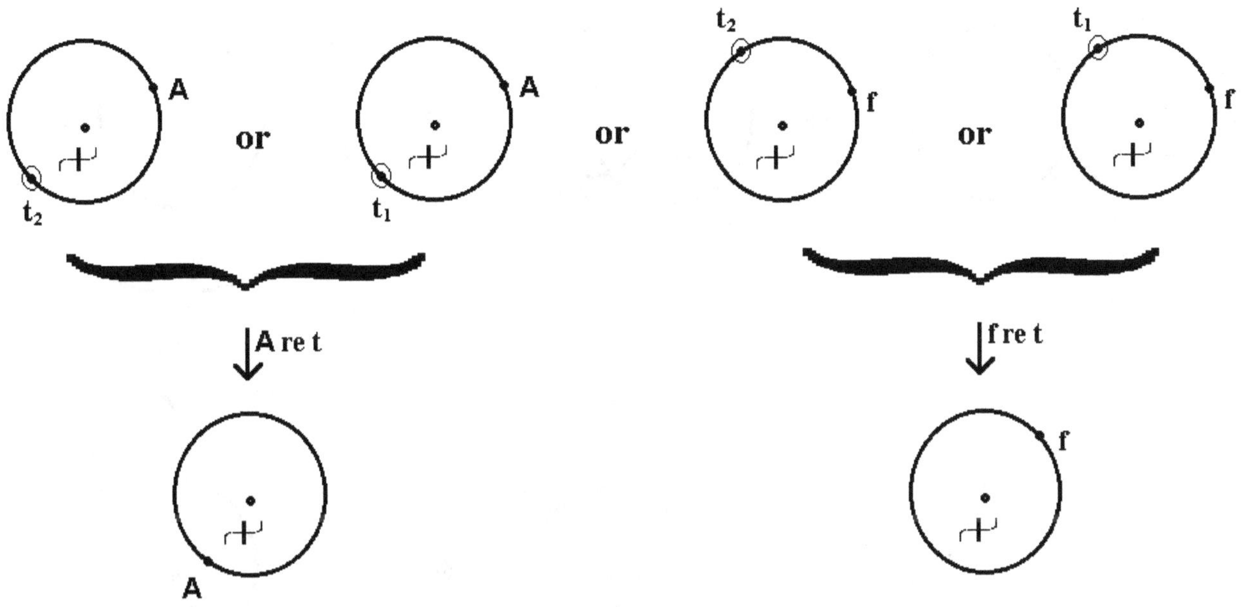

This suggests an alteration of (A, $f(t_2)$) values, which would be similar (but not quite the same) as an exchange (within a class).

The diagram also suggests a manner for common center coupling:

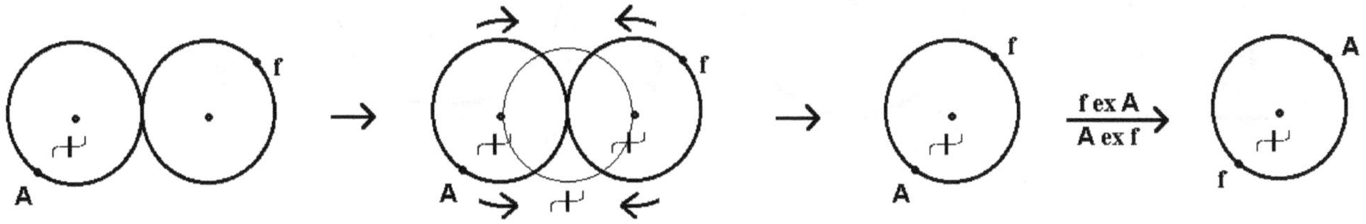

In a more strict sense, as done previously, this coupling should be something more like:

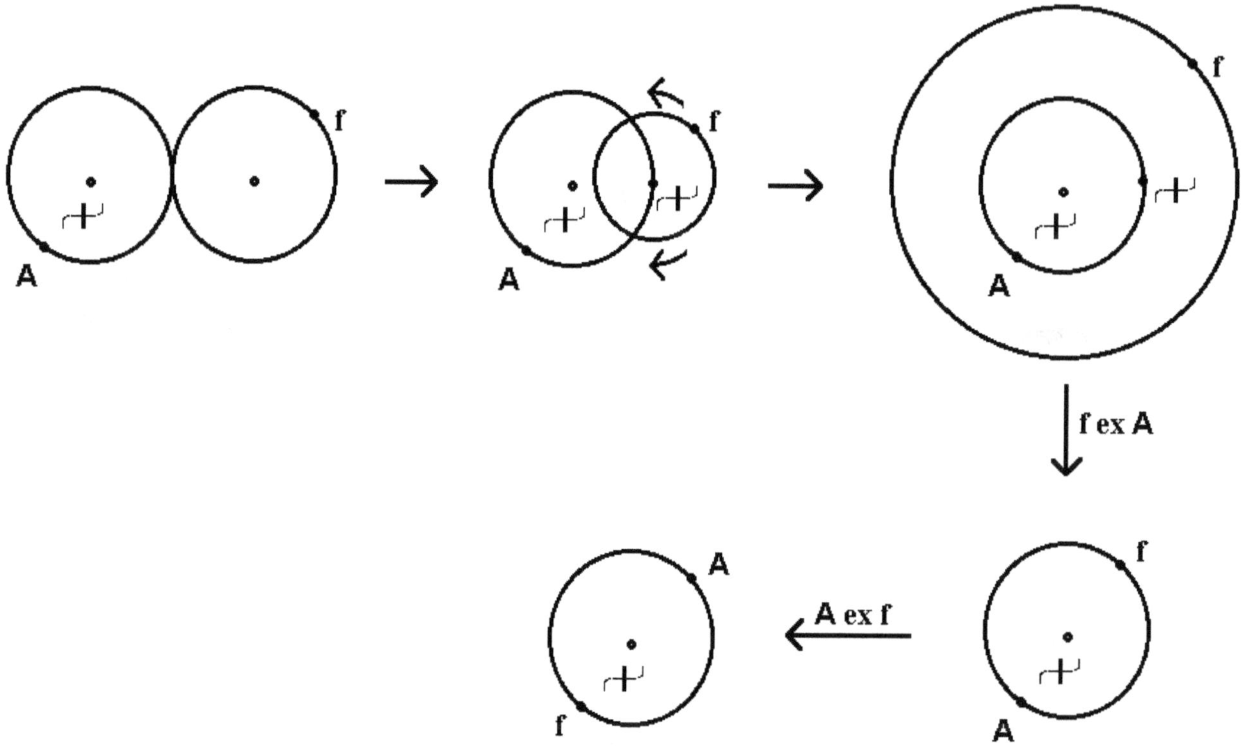

But, of course, there are other paths that can lead to:

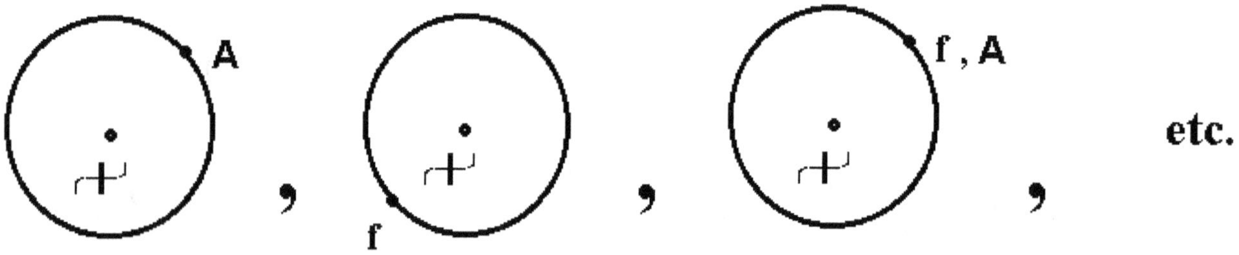

Therefore, the coupling operation is not merely a combining (or addition), but also a "coalescing" of elements (that allows for a furthering of procedure).

Notice that an interesting multiplication can result, in the form of the element

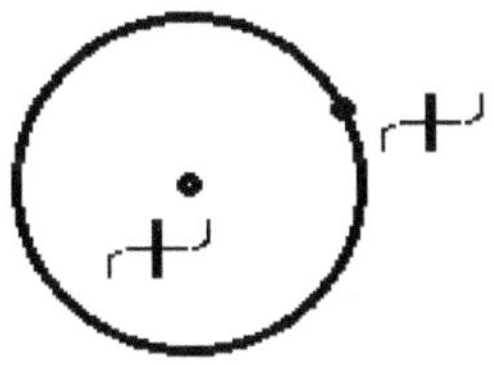

i.e,

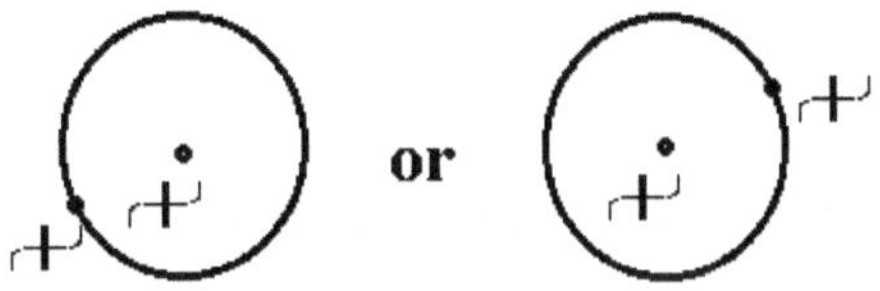

e.g.

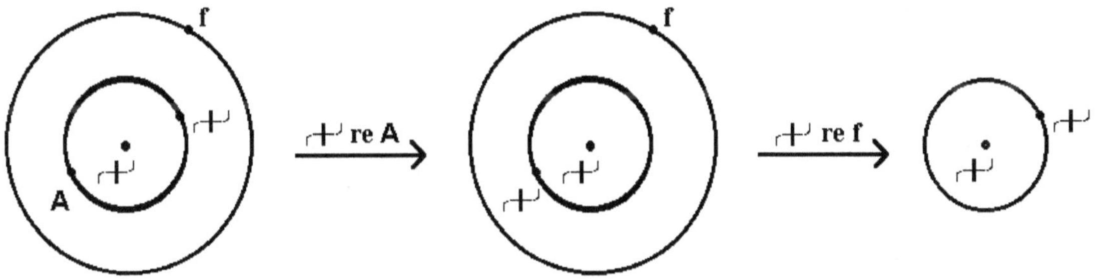

The (operational) value of such a multiplicative element would seem to depend on its utility during further processing (with other elements). The "multiplication" is of course complexity based (here).

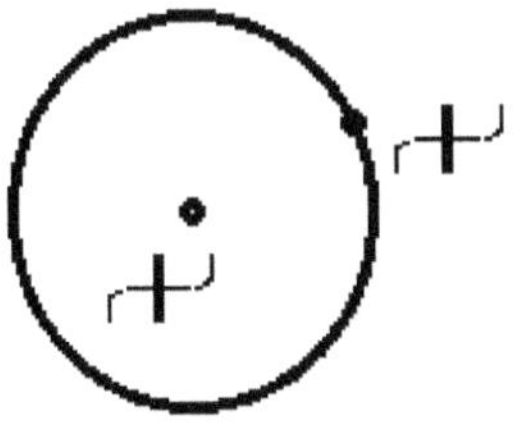

would be interpreted as (for) n = 1

$\therefore N_{n\,=\,1} \equiv$ **radius**

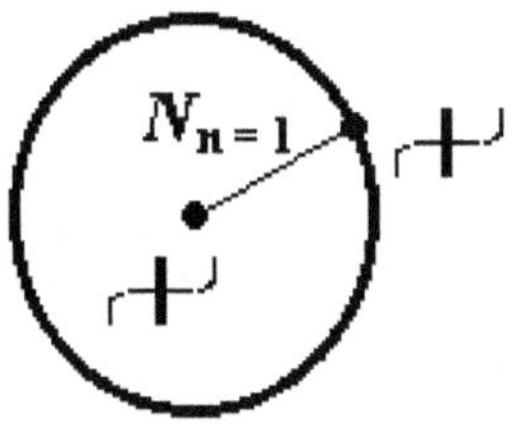

and its complexity would be:

$(N_{n=1})\Psi_{()}$; e.g., for Ψ_m at $n = 1$, $N_{n=1} \bullet \Psi_m == (1.3132)(1.5231) = 2.0001$;

i.e., $[(C_n / \Psi_m^{\ 1})(n + 1)] \bullet \Psi_m^{\ 1} = [(1 / \Psi_m^{\ 1})(1 + 1)] \bullet \Psi_m^{\ 1} = 2$

Therefore, (at $n = 1$), it simply represents $(n + 1)$; at $n > 1$ (or more generally) it represents $C_n \bullet (n + 1)$.

Complexities of Combinations and Clo Elements

So, we have modes (methods) of enhancing processes (coupling) and deriving elements from processes, using Proportional Rotation.

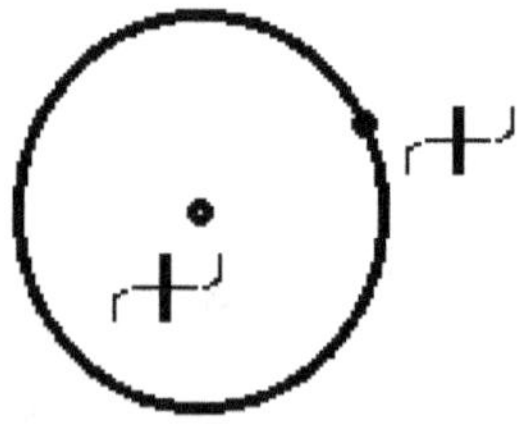

as such, is a particularly convenient element of utility since its complexity is simply $[C_n \bullet (n + 1)]$ and therefore is an enhanced (combination) complexity (by $n + 1$) or "Combination Complexity." Thus, the element may be called the (or a) combination complexity element, or shortened to 'combp' , which can be symbolized to: C|o or C |o . The combp element can be coupled to a process (asserting its complexity) and so (under CCP) be used to derive elements from the process (through the complexities allowed with Proportional Rotation).

For example, say we have a velocity $d / t == v$, and therefore a process

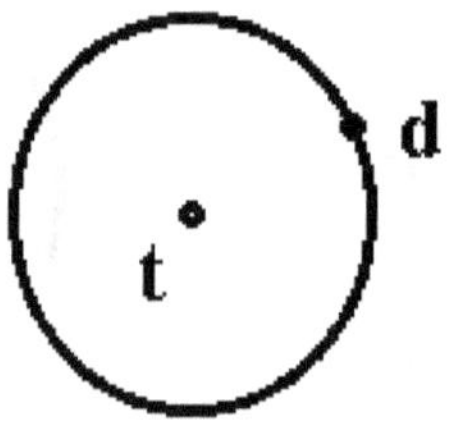

Thus, C |o indicates the coupling:

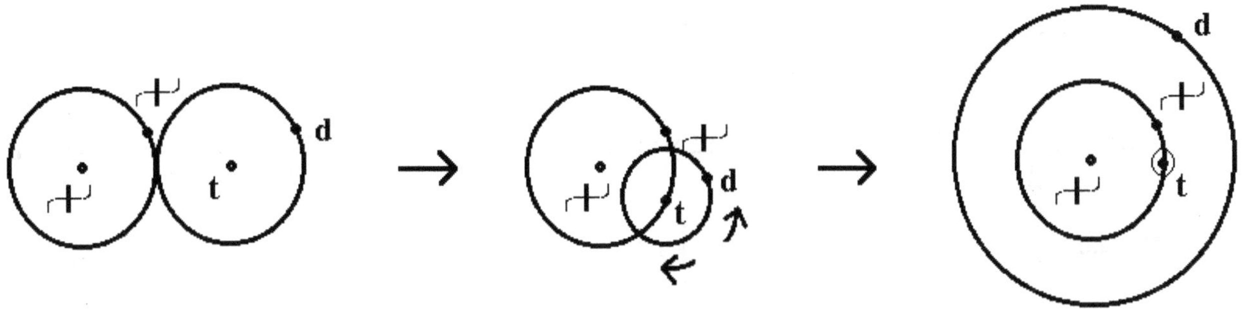

Now, we have (n = 2)

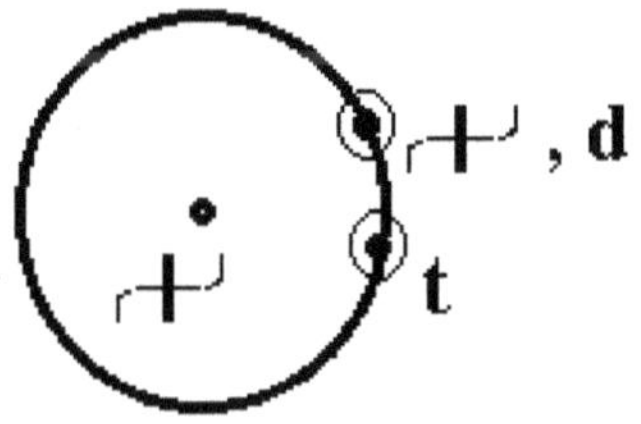

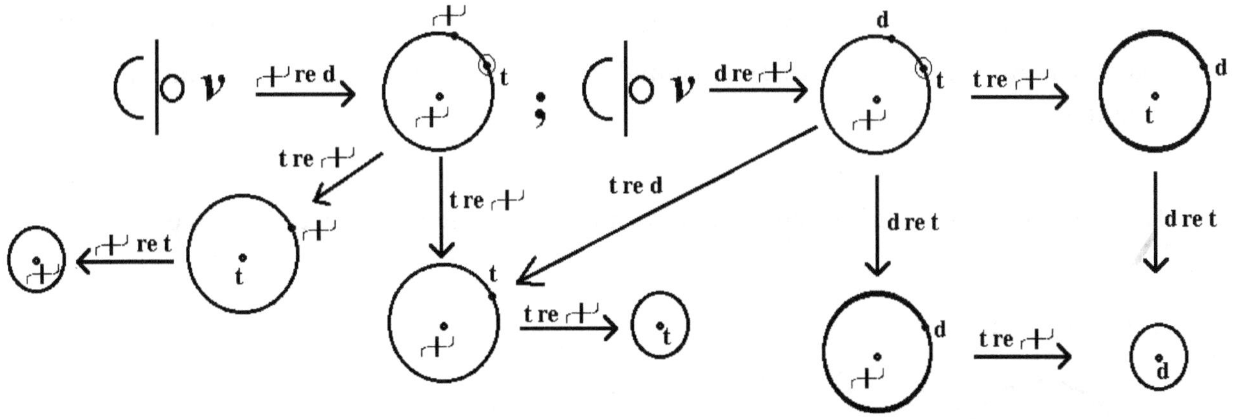

There are some interesting identities established here:

$C|o\ v \xrightarrow{d\,re\,⊬} \xrightarrow{t\,re\,⊬} v$; $C|o\ v \xrightarrow{d\,re\,⊬}$ **complexity of (d / t)**

$v \xrightarrow{d\,re\,t} d$; $C|o\ v \xrightarrow{d\,re\,⊬} \xrightarrow{d\,re\,t} \xrightarrow{d\,re\,⊬} d$; $C|o\ v \xrightarrow{d\,re\,⊬} \xrightarrow{t\,re\,⊬} \xrightarrow{d\,re\,t} d$

$C|o\ v \xrightarrow{⊬\,re\,d} \xrightarrow{2(t\,re\,⊬)} t$; $C|o\ v \xrightarrow{d\,re\,⊬} \xrightarrow{t\,re\,d} \xrightarrow{t\,re\,⊬} t$

$C|o\ v \xrightarrow{⊬\,re\,d} \xrightarrow{t\,re\,⊬} \xrightarrow{⊬\,re\,t}$ ⊬ **complexity (i.e., "base")**

$C|o\ v \xrightarrow{d\,re\,⊬} \xrightarrow{d\,re\,t}$ **complexity of d**

$C|o\ v \xrightarrow{⊬\,re\,d} \xrightarrow{t\,re\,⊬}$ **complexity of t**

} **resultant n = 1**

Because the notation ' C |o ' looks (visually) like "clov," we can call (or term) this family of manipulations (of velocity) "clov" manipulations, and C |o coupling in general as "clo" coupling.

Here, we are adopting a convention that a (or any) variable may function as its own base.

Therefore, e.g.,

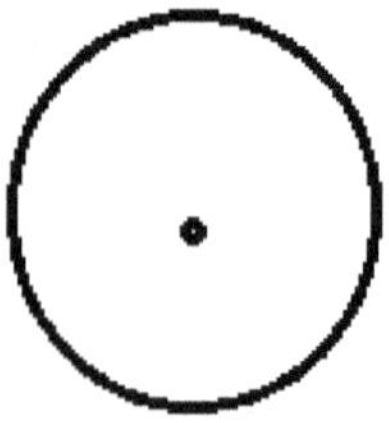

== •

or

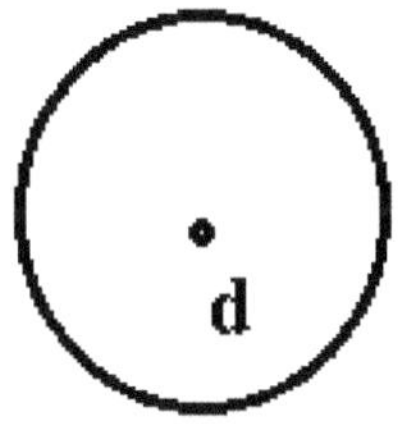

== d , etc.

Such functioning does not (automatically) provide a complexity for the variable.

It is evident that coupling enhances a process (provides further processing of variables), while CCP operations (re, ex) resolve to elements from a process. When only a variable is derived to its (CCP) center, then that element is a "basic" element (or "base element").

Note that (for generalization of clo elements):

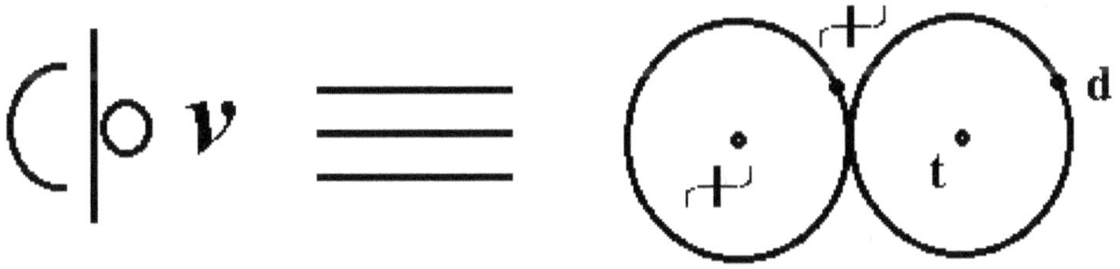

(ultimately) represents not a (formal) complexity, but rather an enhanced combination:

$[\, C_n(\, n + 1 \,)\,] \bullet (\, d / t \,)$; at $n = 1$ (for Ψ). $\therefore\ 2(\, d / t \,)$.

It is a special combination that (as shown) can be used to (directly) derive each of the considered base elements, including

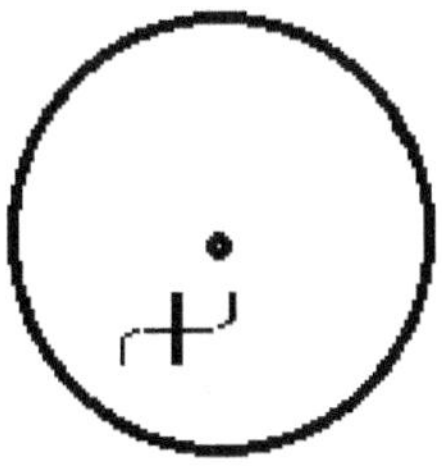

or the base Ψ.

The (formal) complexity of v is: $N_{\Psi()} \bullet (\, d / t \,) == [\, C_n \bullet (\, n + 1) / \Psi_{()}^{\ n}\,] \bullet (\, d / t \,)$.
(If d is one-dimensional, then $n = 2$ here.)

Therefore, there can be various clo elements:

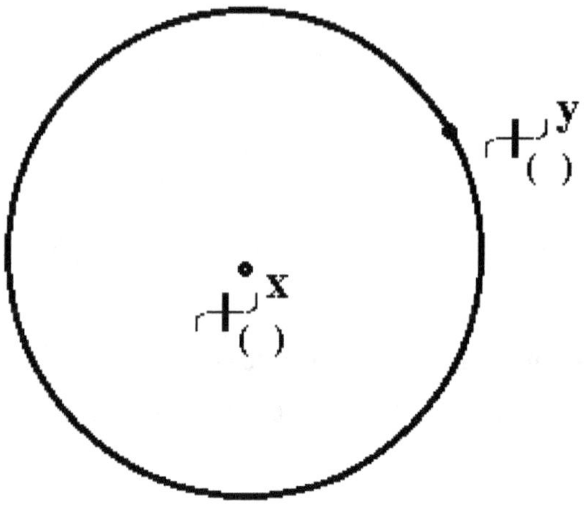

where x (nor n) need not equal y.

When x =/= y , C |o v does yield a (residual) complexity:

$$[\, C_n \bullet (n + 1)\,] \bullet (\Psi^{y} / \Psi^{x}).$$

Operationally, for the coupling, it seems advantageous to have x = y and this value to reflect the n for v (or other process), so that the resultant is a simple (easily predictable) multiple of v (or process). But this need not be so, and one can (just as simply) use the clo where x = y = 1, to yield a doubling of v.

[There may be more than one Ψ^{y} on the circumference of the CCP circle of the clo element, thereby increasing n from n = 1 to n = this number of Ψ^{y}s. Such circumferential Ψs can even have different exponents (from y).]

The circumferential Ψs, therefore, must be additive (for coupling) and (perhaps) yield to a resultant (or residual) complexity component:

e.g.,

$$\frac{C_n(n+1)}{\Psi^{x}}\left(\frac{\textcircled{1}}{\Psi^{y}} + \frac{\textcircled{2}}{\Psi^{z}} + \ldots\ldots + \frac{\textcircled{n}}{\Psi^{(\ )}}\right)$$

If [y = x, z = x, ...( ) = x], then the resultant is:

$$[\, C_n \bullet (n + 1)\,] \bullet n == C_n \bullet (n^2 + n)$$

Of course, (n, C_n, Ψ^n) remain pure numbers (in any case). So they contribute to the coefficients of a process' complexity, as a pure factor of the process (e.g., a factor of v).

It is clear that coupling approximates a form of multiplication, but we must be conscious of the coordinate types (circumferential and central or centered) and their appropriate associations; e.g, for the v process ($== d / t$) :

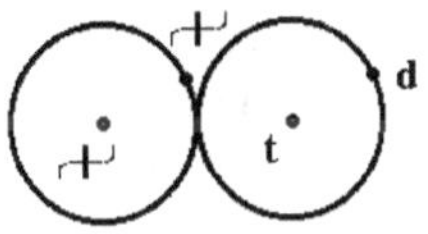

$$\equiv \frac{[C_n(n+1)]}{\Psi_{center}} \bullet \Psi_{cir} \bullet \left(\frac{d}{t}\right) \equiv \frac{[C_n(n+1)]}{\Psi_{center} \bullet t} \bullet \frac{\Psi_{cir} \bullet d}{1}$$

What if there is another process, say $u == d \bullet t$, with the same (type of) coupling?

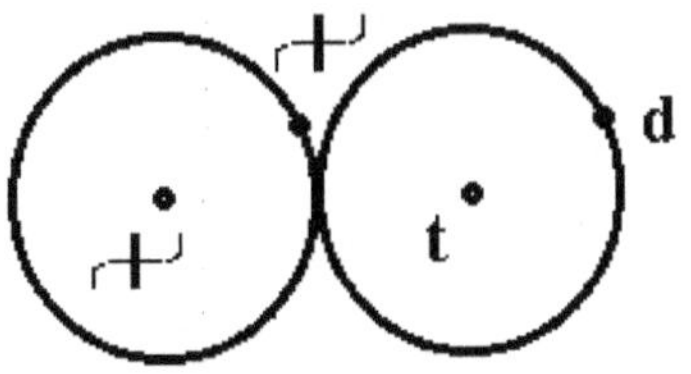

$$\equiv \frac{[C_n(n+1)]}{\Psi_{center}} \bullet \Psi_{cir} \bullet (d \bullet t)$$

but coordinate t is "centered" as is Ψ_{center}. Therefore,

$$([\, C_n(n + 1)\,] / \Psi_{center} \bullet t) \bullet \Psi_{cir} \bullet (d \bullet t) \rightarrow [\, C_n(n + 1)\,] \bullet (\Psi_{cir} / \Psi_{center}) \bullet d$$

suggesting a type of distance (d-like) as a result(ant) for u (this would appear to force "t" to become unit-less, or a pure number). Here, then, use of a clo element might be used to reveal (or elucidate) the nature of a coordinate (e.g.: t), in a process.

We can, thus, make (Proportional Rotation) distinctions between the processes:

$$\frac{[C_n(n+1)]}{r^n_{()}} \bullet (d \bullet t) \qquad \frac{[C_n(n+1)]}{r^n_{center} \bullet t} \bullet r^n_{cir} \bullet (d \bullet t) \qquad \frac{[C_n(n+1)]}{r^n \bullet t} \bullet (d \bullet t) \qquad \frac{(d \bullet t)}{t} \qquad \frac{(d/t)}{t}$$

$$[C_n(n+1)] \bullet \frac{r_{cir}}{r_{center}} \bullet d \qquad \frac{[C_n(n+1)]}{r^n} \bullet d \qquad d \qquad d/t^2$$

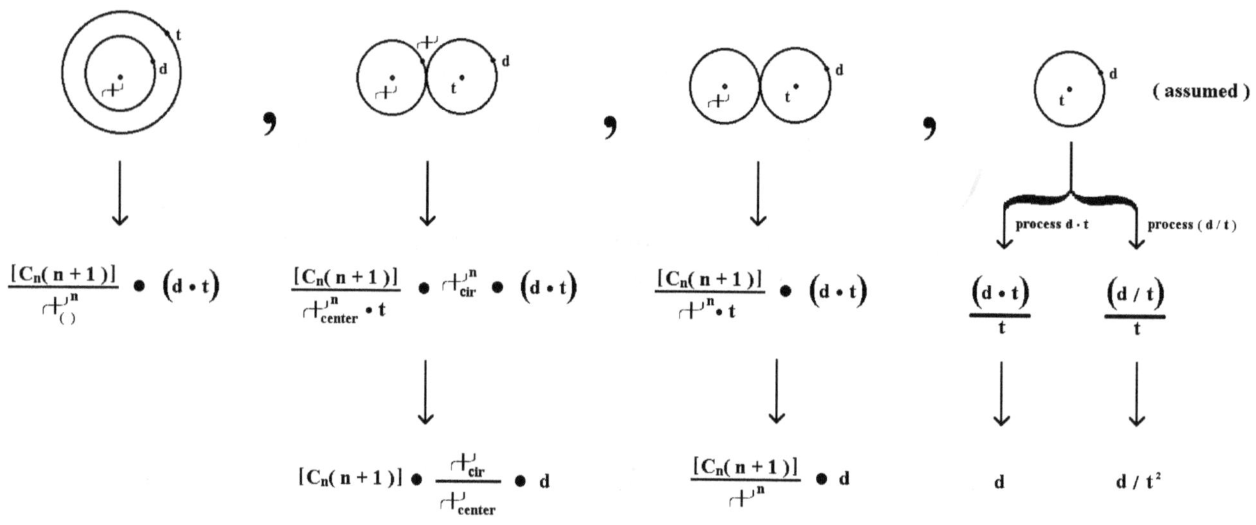

Note that

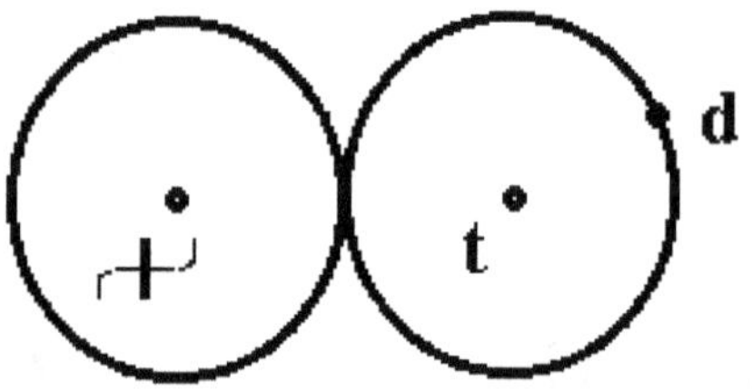

can be interpreted as the formal coupling:

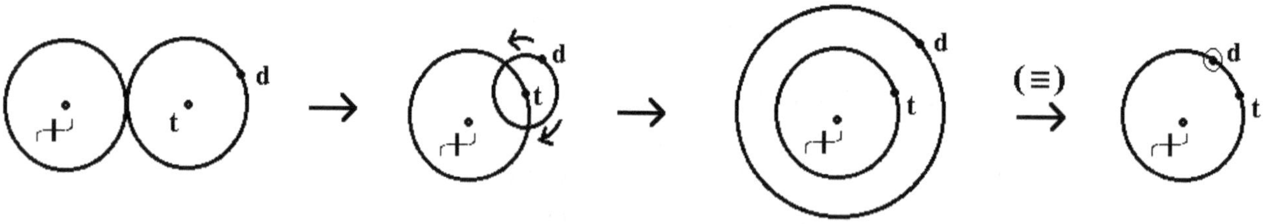

Multiplicatively:

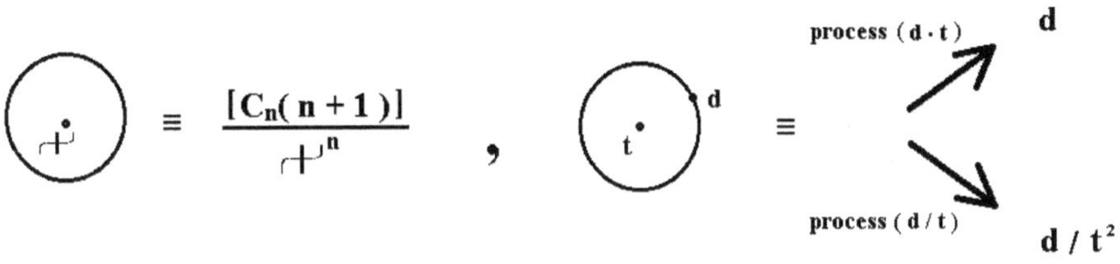

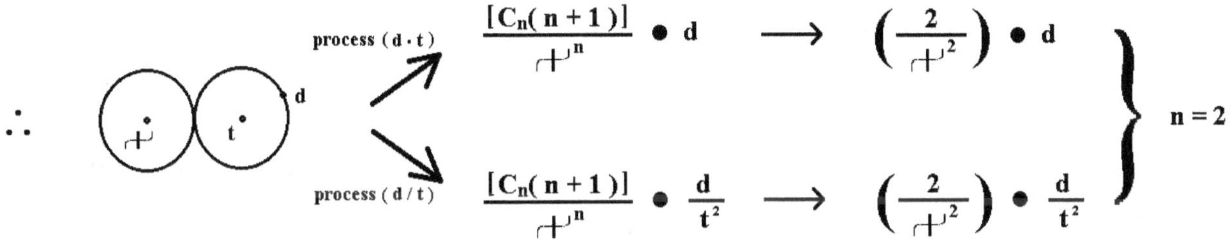

$$\left(\Psi\right) \equiv \frac{[C_n(n+1)]}{\Psi^n} \quad , \quad \left(t\right)^d \equiv \begin{array}{c} \text{process }(d \cdot t) \nearrow \quad d \\ \\ \searrow \\ \text{process }(d/t) \quad d/t^2 \end{array}$$

$$\therefore \quad \left(\Psi\right)\left(t\right)^d \begin{array}{c} \text{process }(d \cdot t) \nearrow \dfrac{[C_n(n+1)]}{\Psi^n} \bullet d \longrightarrow \left(\dfrac{2}{\Psi^2}\right) \bullet d \\ \\ \searrow \\ \text{process }(d/t) \quad \dfrac{[C_n(n+1)]}{\Psi^n} \bullet \dfrac{d}{t^2} \longrightarrow \left(\dfrac{2}{\Psi^2}\right) \bullet \dfrac{d}{t^2} \end{array} \Bigg\} \; n = 2$$

But

$$\left(\Psi\right)^{d}_{t} \begin{array}{c} \text{process }(d \cdot t) \nearrow \dfrac{[C_n(n+1)]}{\Psi^n} \bullet (d \bullet t) \longrightarrow \dfrac{2dt}{\Psi^2} \\ \\ \searrow \\ \text{process }(d/t) \quad \dfrac{[C_n(n+1)]}{\Psi^n} \bullet \left(\dfrac{d}{t}\right) \longrightarrow \dfrac{2d}{\Psi^2 t} \end{array} \Bigg\} \; n = 2$$

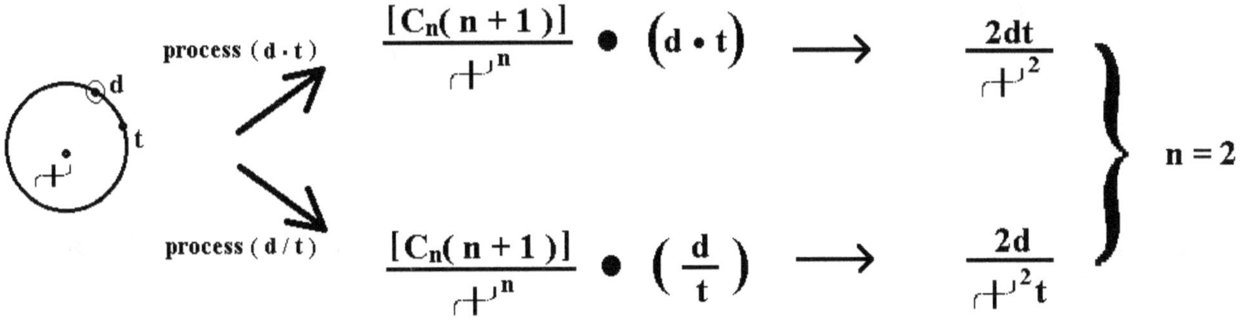

This would seem to force $t = 1$, as $2d / \Psi^2 = 2dt / \Psi^2$, $2d / (\Psi t)^2 = 2d / \Psi^2 t$

Therefore,

$$\therefore \qquad \bigcirc t \quad d \qquad \equiv \qquad \bigcirc 1 \quad d$$

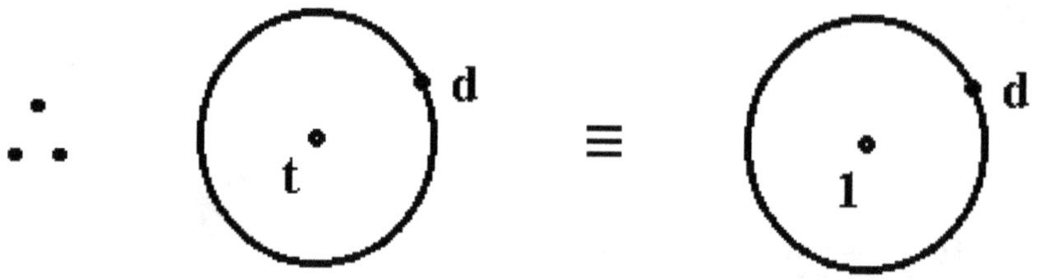

Since **t-centered** is allowed any value (in CCP), the apparent hysteresis (lag of t processing) comes from the observation that

is not equivalent to

process (d · t)

$$\frac{[C_n(n + 1)]}{\boxed{+}^n} \bullet \big(d \cdot t \big) \qquad\qquad \frac{[C_n(n + 1)]}{\boxed{+}^n} \bullet d$$

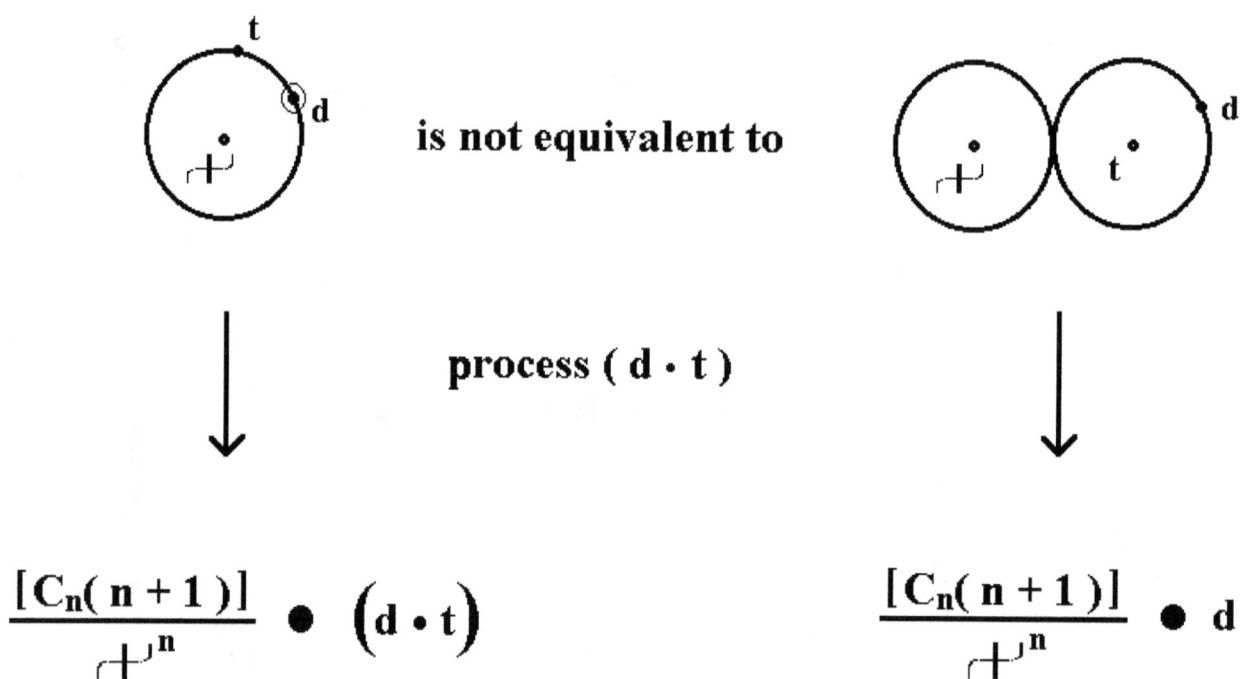

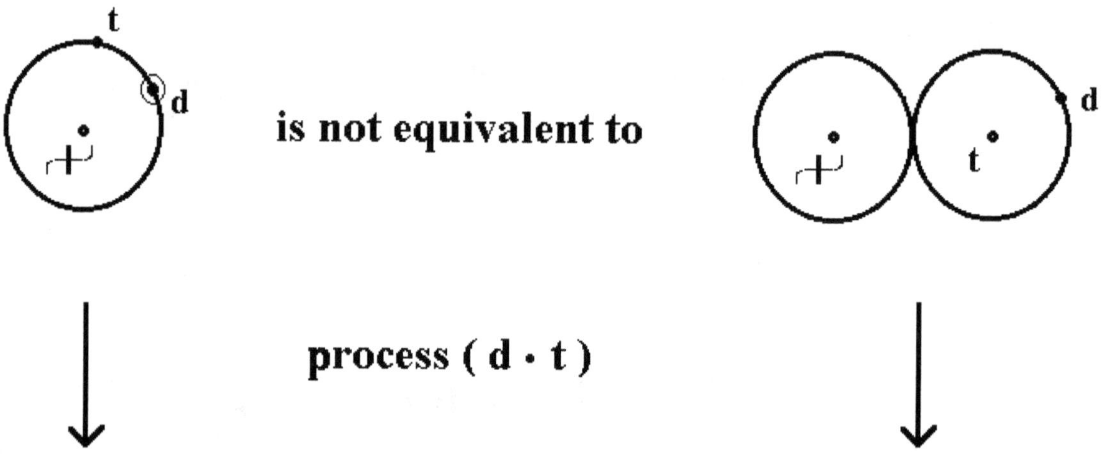

which can not be correct as postulated, except by t centering with Ψ (and thus a distinction between coupling and multiplication).

Obviously, then,

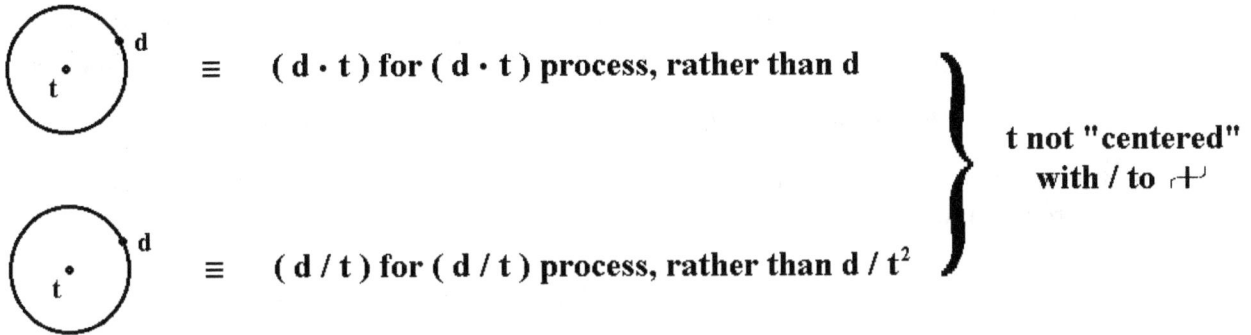

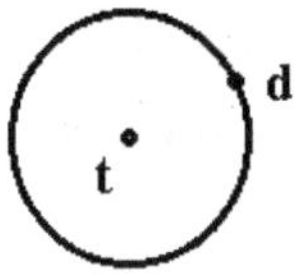

$\equiv$ (d $\cdot$ t) for (d $\cdot$ t) process, rather than d

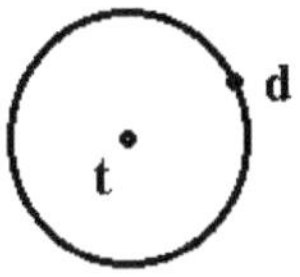

$\equiv$ (d / t) for (d / t) process, rather than d / t^2

$\Bigg\}$ **t not "centered"
with / to ⊢**

[This suggests an inductive proof for CCP.]

So, we have two interpretations of functioning (further processing) for:

t centering

coupling

process (d $\cdot$ t)

$$\frac{[C_n(n + 1)]}{⊢^n} \bullet d$$

$$\frac{[C_n(n + 1)]}{⊢^n} \bullet \left(d \bullet t\right)$$

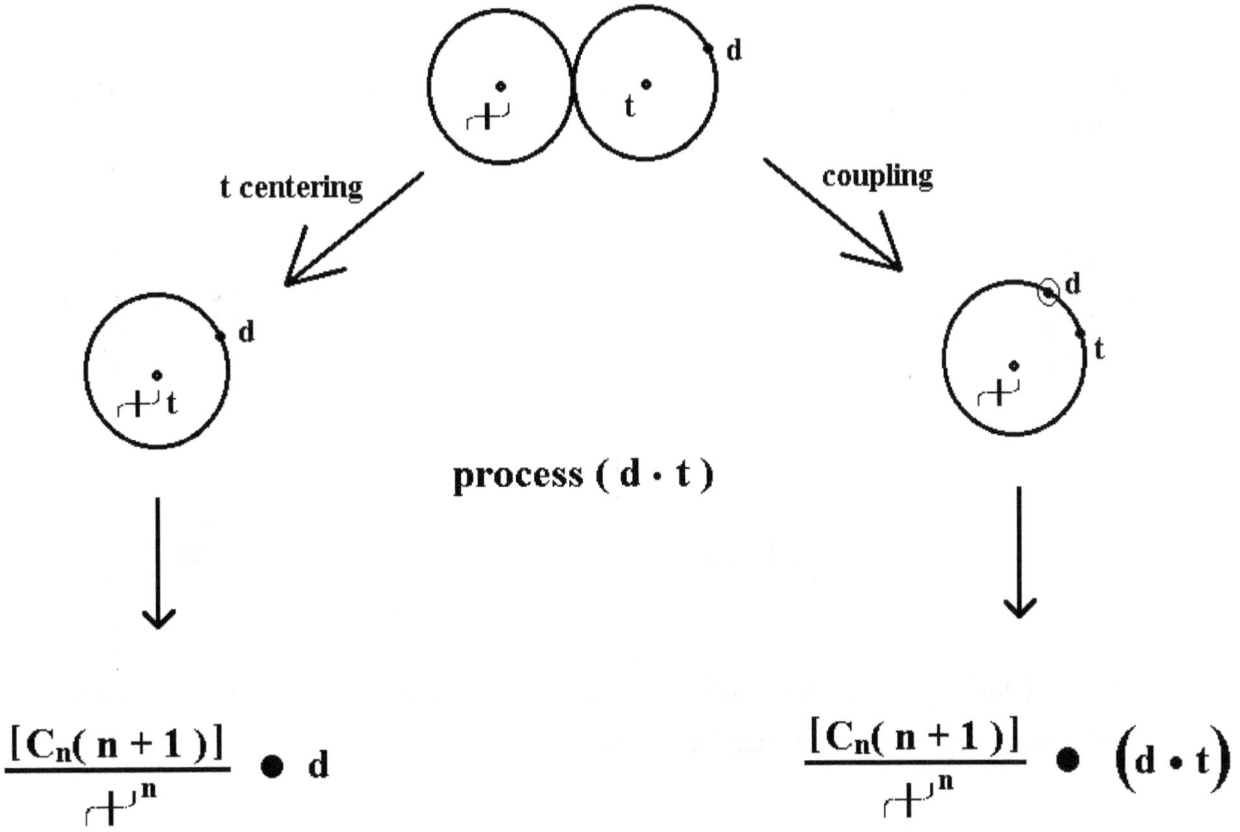

These "functionings" are clearly two different operations, the t "centering" being multiplicative and the coupling being associative (or de-centering t, essentially, to make it circumferential of coordinate). Both (methods) produce (or yield) formal complexities (i.e., via N coefficients).

We can note that coupling creates (apparently) equivalent products (or elements):

Say Ψ couples to d (rather than t), thus (the) process is independent of coordinate position (center or circumference)

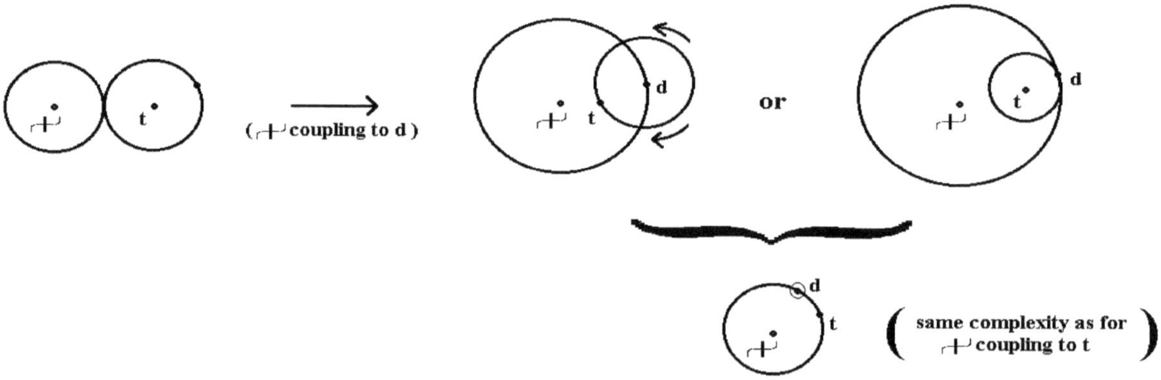

while "centering" yields different resultants based on positions of coordinates ("compounded" centers):

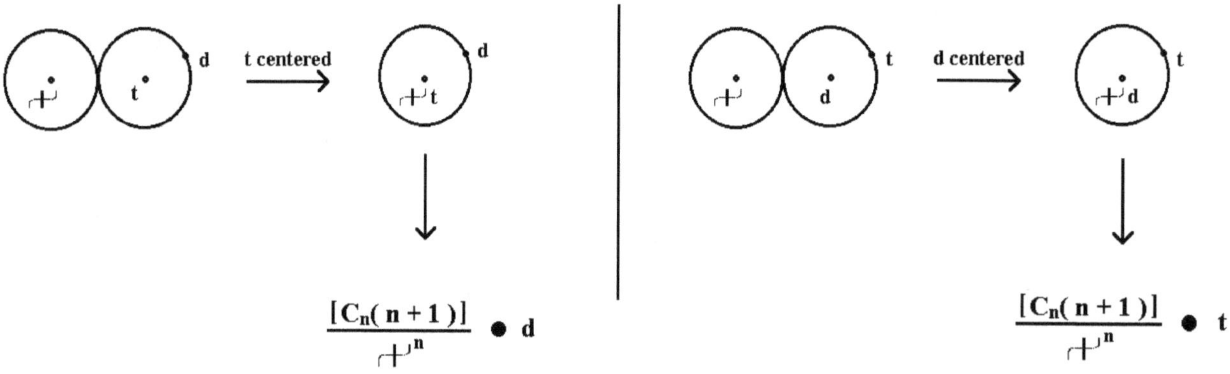

Therefore, coupling "circumferentializes" coordinates (to coalesce to a common center).

In both cases (methods) centers may be changed (or lost).

How do we analyze "compounded centers"?

They can be observed of factors, so that the CCP operations (as of re, ex) can or might not be readily made (or established):

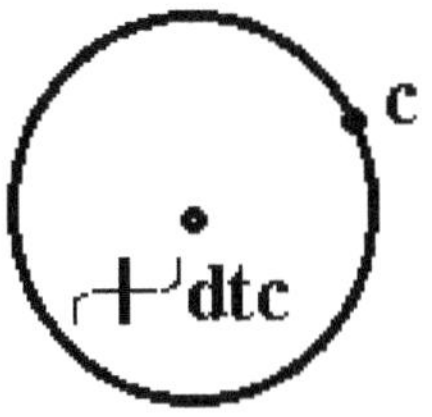

center is ┌+┘ · d · t · c ≡ X ∴

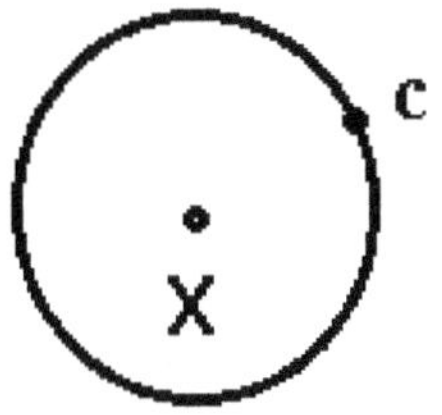

If only X is known (or observed), one can not make apparent operations such as (c re d), or (c ex c), etc.

Here, one might try "presumptive" operations. One can assume that (at least) $\Psi_{()}$ is centered, and that a greater compounding of the center could imply a coordinate (circumferential) may be applicable for a re operation, and perhaps an ex exchange, if it is advantageous to some analysis (of process resolution). The resultants would have to be examined as potential for utility.

So, for example:

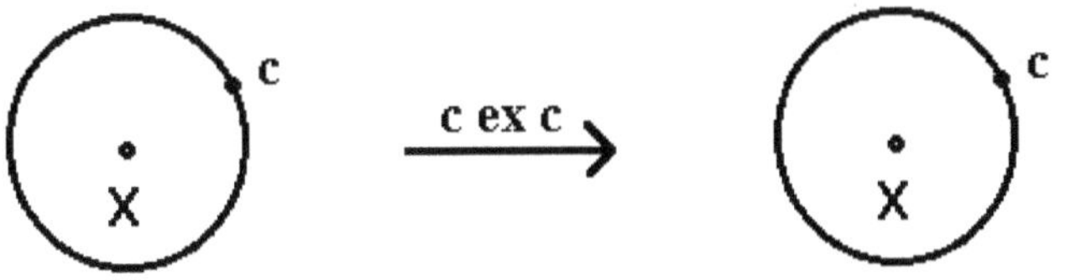

suggests c is potentially centered (for the element)

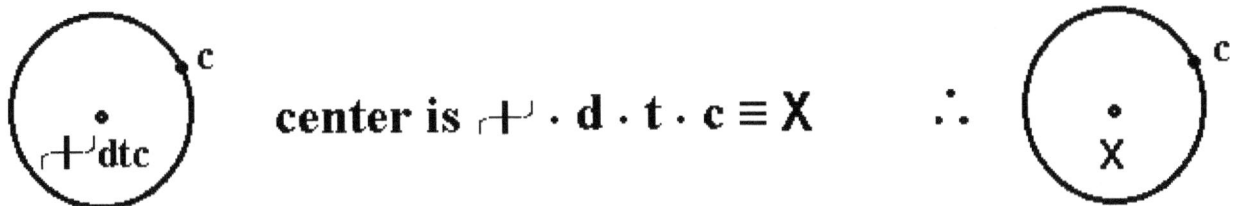

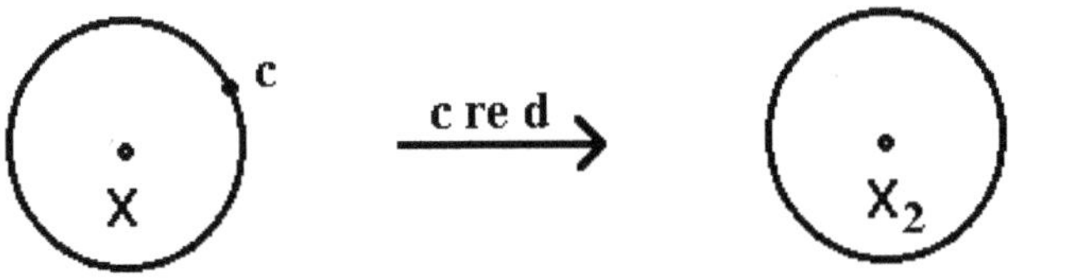

suggests d was centered for X and c replaces d for X_2

etc.

To analyze (study) such an element (of X center), one can center with a test element, such as a clo:

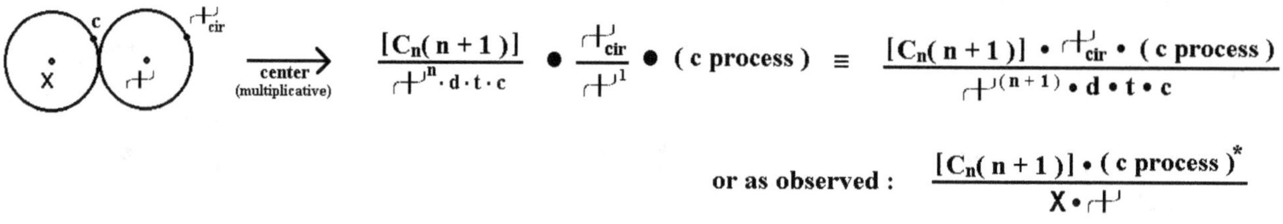

where (c process)* $==\Psi_{cir}\bullet$ (c process), and compare this result to formal coupling (with its various resolutions):

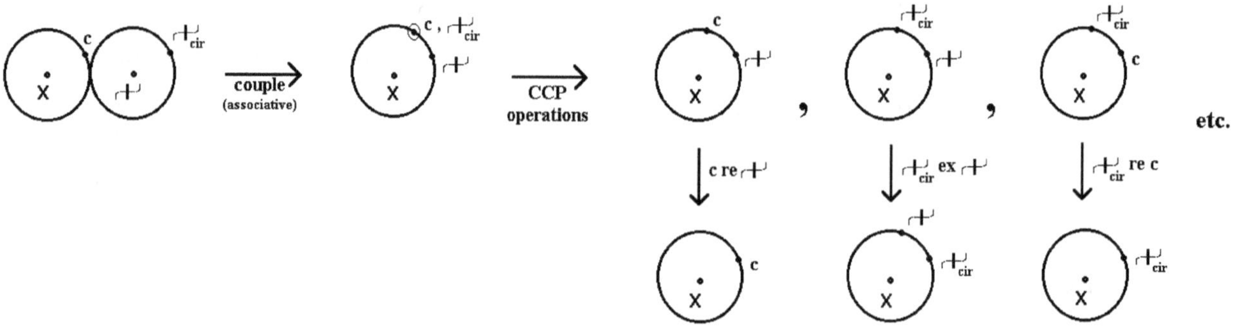

Therefore, the circumferential coordinates can be used as probes of the center.

We note, then, the distinction between "centering" and coupling (in this case):

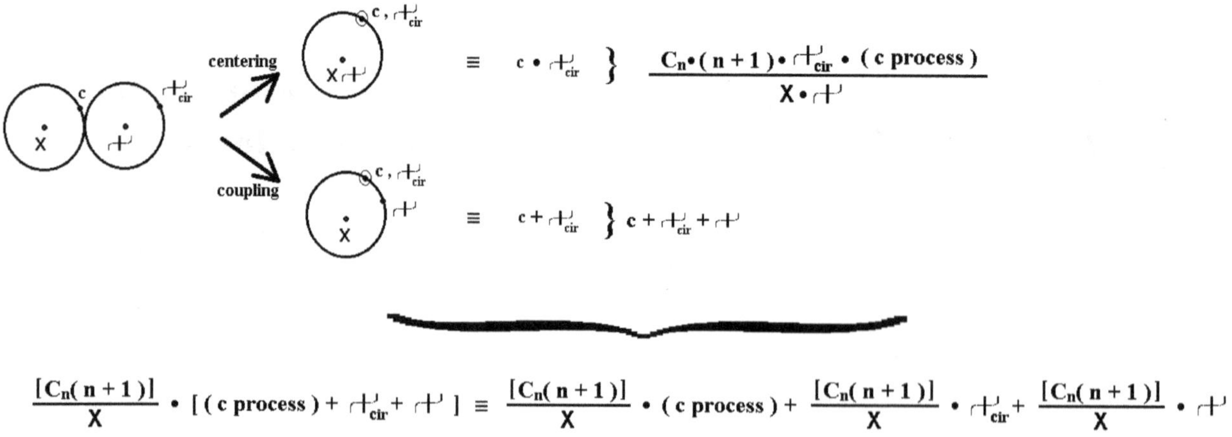

For the coupling, if $\Psi_{cir} == \Psi$, and X includes Ψ as a factor, then the resultant is:

$$\frac{[C_n(n+1)]}{X} \bullet (\text{c process}) + \frac{2[C_n(n+1)]}{{}^{r}X} \quad \text{where } {}^{r}X \equiv X \text{ "reduced"} = (X/\Psi)$$

$$= [C_n(n+1)] \bullet \left[\frac{(\text{c process})}{X} + \frac{2}{{}^{r}X}\right]$$

or

$$[C_n(n+1)] \bullet \left[\frac{(\text{c process})}{X} + \frac{2\Psi}{{}^{r}X}\right] = \frac{[C_n(n+1)]}{X} \bullet [(\text{c process}) + 2\Psi]$$

$$\therefore \quad \frac{[C_n(n+1)]}{{}^{r}X \bullet \Psi} \bullet [(\text{c process}) + 2\Psi]$$

The corresponding resultant for centering is: $([C_n(n+1)] / {}^{r}X \bullet \Psi) \bullet (\text{c process})$

Therefore, with this clo (as a test element):

coupling == centering $+ 2\Psi \bullet {}^{c}N$ (is possible)

${}^{c}N == [C_n(n+1)] / ({}^{r}X \bullet \Psi) == N$ from a compounded center $== N / {}^{r}X$

The 2Ψ of the $2\Psi \bullet {}^{c}N$ term represents the # of Ψ coordinates from the clo, used for the coupling.

Therefore, we can generalize the term $2\Psi \bullet {}^{c}N$ to allow for the following relation:

coupling == centering + (each coupled test element coordinate) $\bullet N / {}^{r}X$

This provides a very convenient posit for explaining the nature of a coupling (i.e., by simply looking at the coupling diagram):

e.g.,

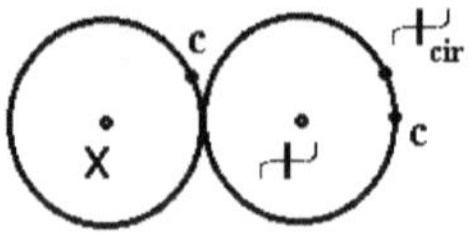

$$\xrightarrow{\text{coupling}} \frac{[C_n(n+1)]}{X\Psi} \bullet \Psi_{cir} \bullet (\text{c process}) + \frac{[C_n(n+1)]}{X\Psi} \bullet c \bullet (\text{c process})$$

$$+ (\Psi_{cir} + c) \bullet \frac{N\Psi}{X} + \frac{N\Psi^2}{X}$$

$$\equiv \frac{[C_n(n+1)]}{X\Psi} \bullet (\text{c process}) \bullet (\Psi_{cir} + c) + (\Psi_{cir} + c) \bullet \frac{N\Psi}{X} + \frac{N\Psi^2}{X}$$

$$\equiv (\Psi_{cir} + c) \left[\frac{[C_n(n+1)]}{X\Psi} \bullet (\text{c process}) + \frac{N\Psi}{X} \right] + \frac{N\Psi^2}{X}$$

- $(\Psi_{cir} + c)$ ↑ **coupled coordinates (cir. of test element)**
- first brace: $\overset{c}{N}$
- (c process) ↑ **fundamental process considered (or "coupling" process)**
- second brace: $\frac{N}{{}^{r}X}$
- $\frac{N\Psi^2}{X}$ ↑ **"centered" contribution (of test element)**

$$\equiv \frac{(\Psi_{cir} + c)}{X} \bullet \Psi \left[\frac{[C_n(n+1)]}{\Psi^2} \bullet (\text{c process}) + N \right] + \frac{N\Psi^2}{X}$$

$$\equiv \left(\frac{\Psi}{X}\right) \bullet (\Psi_{cir} + c) \bullet \Psi \left[\frac{[C_n(n+1)]}{\Psi^2} \bullet (\text{c process}) + N \right] + \left(\frac{\Psi}{X}\right) \bullet N\Psi$$

$$\equiv \left(\frac{\Psi}{X}\right) \bullet (\Psi_{cir} + c) \bullet \left[\frac{[C_n(n+1)]}{\Psi} \bullet (\text{c process}) + N\Psi \right] + \left(\frac{\Psi}{X}\right) \bullet N\Psi$$

[Note that here, (each coupled test element coordinate) == $(\Psi_{cir} + c + \Psi)$]

$$(\Psi / X) == \Psi / \Psi \bullet {}^{r}X = 1 / {}^{r}X$$

Therefore, we have

components of coupling

$$\underbrace{\frac{(\Psi_{cir} + c)}{{}^{r}X} \bullet \left[\frac{[C_n(n+1)]}{\Psi} \bullet (\text{c process}) + N\Psi \right]}_{\text{"circumferential" component}} + \underbrace{\frac{\Psi}{{}^{r}X} \bullet N}_{\text{"centered" component}}$$

A goal would be to find the constituent factors of ${}^{r}X$, assumed to be other than Ψ.

If we (autocratically) keep Ψ as equivalent to $\Psi_{n=1} = \Psi^1 = \Psi$, and then multiply the coupling components by ${}^{r}X$, they reduce to:

$$(\Psi_{cir} + c)\,[\,(C_n(n+1)/\Psi) \bullet (\text{c process}) + C_n(n+1)\,] + C_n(n+1),$$

where $\boldsymbol{N} == C_n(n+1)/\Psi$.

If we again say (or assume) $\Psi_{cir} == \Psi$, this comes to:

$$(\pitchfork + c)\left[\frac{C_n(n+1)}{\pitchfork} \bullet (c\ \text{process}) + C_n(n+1)\right] + C_n(n+1)$$

$$= \frac{(\pitchfork + c)C_n(n+1)}{\pitchfork} \bullet (c\ \text{process}) + (\pitchfork + c)C_n(n+1) + C_n(n+1)$$

$$= \frac{\pitchfork C_n(n+1) \bullet (c\ \text{process})}{\pitchfork} + \frac{(c)C_n(n+1) \bullet (c\ \text{process})}{\pitchfork} + (\pitchfork + c)C_n(n+1) + C_n(n+1)$$

$$= C_n(n+1) \bullet (c\ \text{process}) + \frac{(c)C_n(n+1) \bullet (c\ \text{process})}{\pitchfork} + (\pitchfork + c)C_n(n+1) + C_n(n+1)$$

$$= (c\ \text{process})\left[C_n(n+1) + \frac{(c)C_n(n+1)}{\pitchfork}\right] + (\pitchfork + c)C_n(n+1) + C_n(n+1)$$

$$= C_n(n+1) \bullet (c\ \text{process})\left[1 + \frac{c}{\pitchfork}\right] + (\pitchfork + c)C_n(n+1) + C_n(n+1)$$

$$= C_n(n+1) \bullet (c\ \text{process})\left[1 + \frac{c}{\pitchfork}\right] + C_n(n+1)\left[(\pitchfork + c) + 1\right]$$

$$= C_n(n+1)\left[(c\ \text{process})\left(1 + \frac{c}{\pitchfork}\right) + \left[(\pitchfork + c) + 1\right]\right]$$

$$= C_n(n+1)\left[(c\ \text{process})\left(1 + \frac{c}{\pitchfork}\right) + (\pitchfork + c + 1)\right]$$

$$= C_n(n+1)\left[(c\ \text{process})\left(1 + \frac{c}{\pitchfork}\right) + \pitchfork + c + 1\right]$$

Therefore, we can simply impose the ${}^{r}X$ (== as ${}^{r}X^{-1}$), to get:

$$(C_n(n + 1) / {}^{r}X)[(c \text{ process })(1 + c / \Psi) + \Psi + c + 1]$$

Probing Complexity Components

Now, we can consider the actual coupling elements:

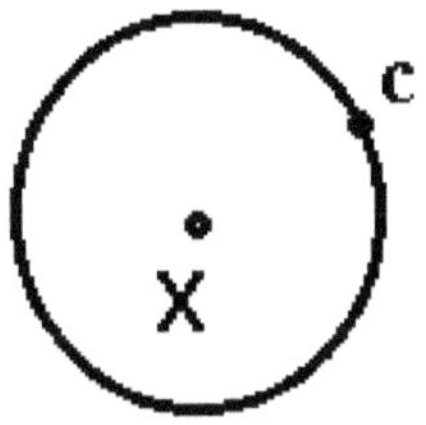

has n = ? ;

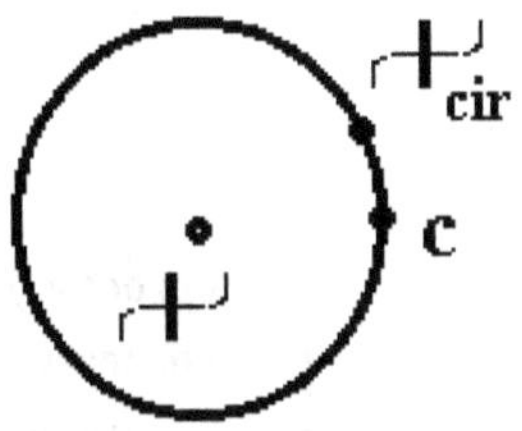

has n = 2 ; therefore, $\Psi == \Psi^2$.

So this modifies the coupling expression to:

$$(C_n(n + 1) / {}^{r}X)[(c \text{ process })(1 + c / \Psi^2) + \Psi^2 + c + 1]$$

There are 5 terms to this expression, hierarchically given as:

$$(C_n(n + 1) / {}^{r}X)[\Psi^4 c \Psi^{-6}(c \text{ process }) + \Psi^2 + (c \text{ process }) + c + 1] ,$$

$$\Psi^4 \bullet \Psi^{-6} == \Psi^{-2} = 1 / \Psi^2 .$$

The terms in the [....] essentially "probe" the ${}^{r}X$ factor (as $C_n(n + 1) \bullet {}^{r}X^{-1}$):

$$(C_n(n + 1) / {}^{r}X)[\Psi^4 \bullet (\Psi^{-2})^3 \bullet (c \text{ process }) \bullet c + \Psi^2 + (c \text{ process }) + c + 1]$$

The probing can be very apparent.

For example, if ${}^{r}X == \Psi^2$, then $(C_n(n + 1) / {}^{r}X) \bullet \Psi^2 \rightarrow C_n(n + 1)$, and similarly for the terms (c process), c, 1, and the compounded term.

For within the compounded term, it is more obscure:

e.g., if ${}^{r}X == (c \text{ process }) \bullet c$, then

$$(C_n(n + 1) / {}^{r}X) \bullet (c \text{ process }) \bullet c \rightarrow C_n(n + 1) \bullet \Psi^{-2} ;$$

but if ${}^{r}X == \Psi^4$, then

$$(C_n(n + 1) / {}^{r}X) \bullet \Psi^4 \rightarrow C_n(n + 1) \bullet (\Psi^{-2})^3 \bullet (c \text{ process }) \bullet c \text{ , etc.}$$

A compounded center is, of course, due to centering:

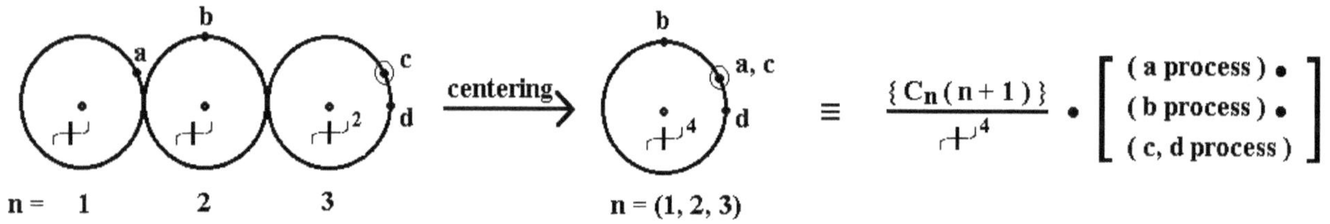

Note that here:

$$\{ C_n(n + 1) \} == [C_1(1 + 1)] \bullet [C_1(1 + 1)] \bullet [C_2(2 + 1)]$$

$$= 4 \bullet [C_2(2 + 1)] \text{ rather than } C_4(4 + 1) \text{ ; i.e., } n = (1, 1, 2) =/= 4$$

[For a coupling, $n = 4$ with $(a + b + c + d)N / {}^{r}X$; therefore,

$$\{ {}^{r}X == X / \Psi = \Psi^4 / \Psi = \Psi^3 , N == C_4(4 + 1) / \Psi^4 \} \rightarrow N / {}^{r}X$$

$$== [C_4(4 + 1) / \Psi^4] \bullet \Psi^{-3} = C_4(4 + 1) / \Psi^7]$$

The

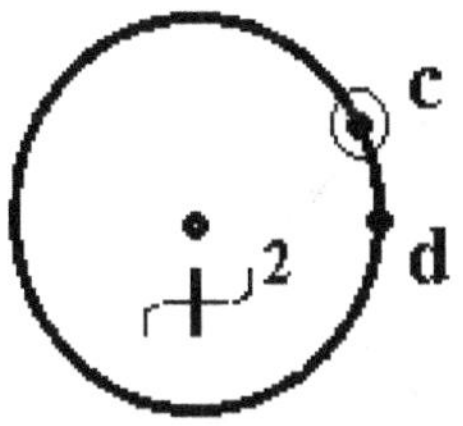

element is itself compounded of center:

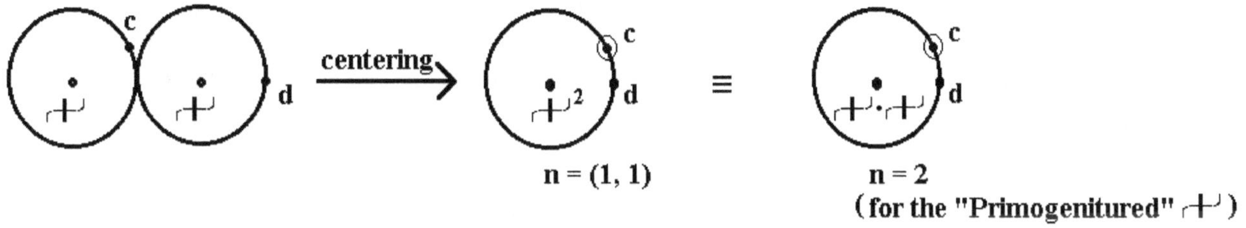

This is distinct from a coupling:

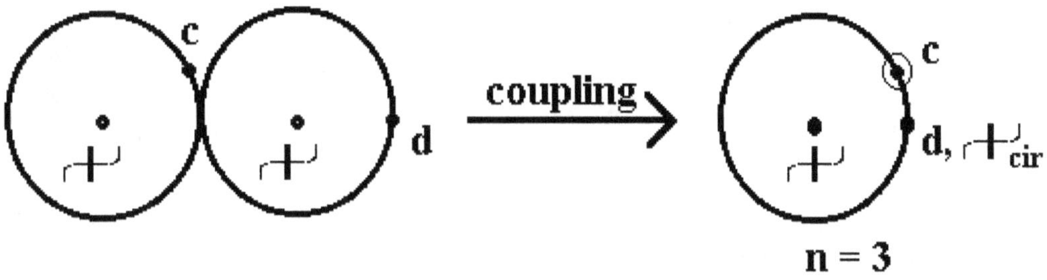

Therefore, the

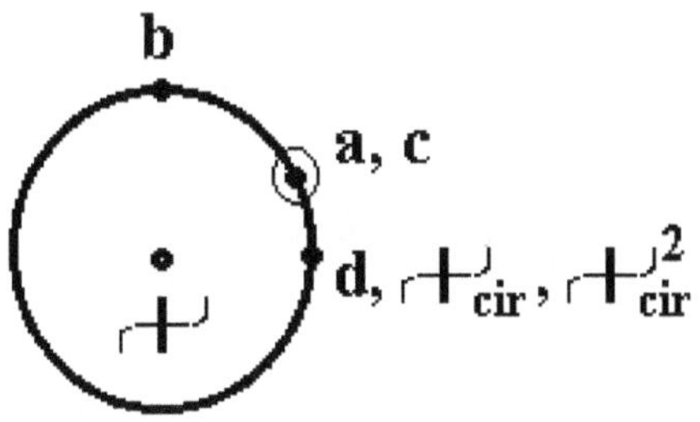

(element) would have $n = 6$ of coupling:

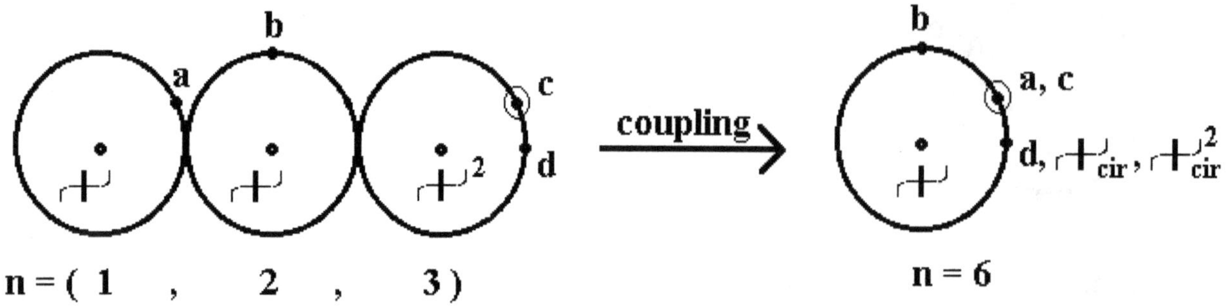

For a proper analysis, the process would have to be considered (of coordinates) as: $(a, b, c, d, \Psi_{cir}, \Psi_{cir}^{2})$.

For an indirect (improper or peripheral) analysis, one might consider a (a, b, c, d process) with $n = 4$ and Ψ_{cir} and Ψ_{cir}^{2} (perhaps circularly) exchanging with Ψ, but this could exclude possible re targets for the other coordinates.

[i.e.:

1) ignore Ψ_{cir}, Ψ_{cir}^{2};

2) Ψ_{cir} ex Ψ; ignore Ψ, Ψ_{cir}^{2};

3) Ψ_{cir}^{2} ex Ψ_{cir}; ignore Ψ, Ψ_{cir}; etc.]

The order of the elements applied to centering or coupling does not seem to matter, but the arrangement of circumferential coordinates is directional (for each element).

Formalism of N Complexity Coefficient

The formalism of $\boldsymbol{N}$ as (defined):

$$N \equiv \frac{C_n(n+1)}{\text{r+}^{\jmath\, n=1}} = \frac{C_n(n+1)}{\sum_{n=1}^{\infty} \frac{n}{C_n}} \equiv \frac{C_n(n+1)}{\sum_{k=1}^{n} \frac{k}{C_k}}$$

is essentially as $(n = 1)$ of a character (nature):

$$C_n(n+1) / (n / C_n) = C_n(n+1)C_n / n \sim\sim C_n^{\,2} \text{ (or } 2C_n^{\,2} \text{ for } n = 1)$$

or more generally (for large n):

$$C_n(n+1) / (n / C_n)^n = C_n(n+1) / (n^n / C_n^{\,n}) = C_n(n+1)C_n^{\,n} / n^n$$
$$\sim\sim C_n^{\,(n+1)} / n^{(n-1)}.$$

We can define a "marginalized" $\boldsymbol{N}$ by dividing $\boldsymbol{N}$ by this generality:

$$\frac{\left(\dfrac{C_n(n+1)}{\sum_{k=1}^{n} \frac{k}{C_k}}\right)}{\left(\dfrac{C_n(n+1)C_n^n}{n^n}\right)} = \frac{\left(\dfrac{1}{\sum_{k=1}^{n} \frac{k}{C_k}}\right)}{\left(\dfrac{C_n^n}{n^n}\right)} = \frac{n^n}{C_n \sum_{k=1}^{n} \frac{k}{C_k}} = \left(\frac{n}{C_n}\right) \cdot \left(\frac{1}{\sum_{k=1}^{n} \frac{k}{C_k}}\right)$$

$$\text{r+}^{\jmath\, n} \equiv \left(\sum_{k=1}^{n} \frac{k}{C_k}\right)^n$$

Therefore, we have $(n / C_n)^n \bullet \Psi^{-n}$ for Ψ raised by n .

$\therefore\ {}_m\boldsymbol{N} == (n / C_n\Psi)^n$, which comes down to $(1 / 1\Psi)^1 = \Psi^{-1}$

$\sim = 1 / 2.2905$ when $n = 1$.

So, the marginalized nature would be to divide a (process) by the n exponential of Ψ:

$((\text{process}) / \Psi^n) \bullet (n / C_n)^n == {}_m\boldsymbol{N} \bullet (\text{process})$

Note that this is a variant of changing (process) from base 10 to a Ψ-like base:

$\log_{\Psi n}(\text{process}) = \log_{10}(\text{process}) / \log_{10}\Psi^n = \log_{10}(\text{process}) / n\log_{10}\Psi$.

Therefore, let $(\Psi^n)^x = (\text{process})$, $\log_{10}(\Psi^n)^x = \log_{10}(\text{process})$

$\log_{10}(\Psi^{nx}) = nx\log_{10}\Psi = \log_{10}(\text{process})$

Thus, $x = \log_{10}(\text{process}) / n\log_{10}\Psi$; i.e., $\Psi^{nx} = (\text{process}) = (\Psi^n)^x$

[i.e., x is in base Ψ^n; thus $x = (\text{process})$ in base Ψ^n]

$\therefore\ [(\text{process}) / (\Psi^n)^x] \bullet (n / C_n)^n == (1 / 1) \bullet (n / C_n)^n = (n / C_n)^n$;

i.e., (as for any division or ratio) we can ask: How can Ψ^n be made to equal (process) ?

We can also define the "attenuated" Ψ, where the calculation of Ψ is limited to the n of C_n :

$$\Psi_{att} = \sum_{k=1}^{n} \frac{k}{C_k} \qquad \textbf{using n from } \mathbf{C_n(n + 1)} \textbf{ term}$$

Therefore, $\boldsymbol{N}_{att}$ or ${}_{att}\boldsymbol{N} == C_n(n + 1) / \Psi_{att}$ (rather than $C_n(n + 1) / \Psi^n$).

Then ${}_m\boldsymbol{N}_{att} == (n / C_n)^n \bullet \Psi_{att}^{\text{(here, n == 1)}}$, mirroring $[(\text{process}) / \Psi^n] \bullet (n / C_n)^n == {}_m\boldsymbol{N} \bullet (\text{process})$.

Here, when $n = 1$: $\Psi_{att} = 1$, $\Psi_{att}^{-1} = 1$, ${}_m\boldsymbol{N}_{att} = 1$.

And, of course, we can have the modulated Ψ_{att} :

$(\, ^{mod}\Psi_{att} =) \, \Psi_{mod} == (\, \Psi_{att} \,)^n$, using the n used to calculate Ψ_{att} .

$\therefore \, ^{mod}_{\ m}N_{att} == (\, n \, / \, C_n \Psi_{att} \,)^n$. Again, at $n = 1$: $\Psi_{att} = 1$, $\Psi_{mod} = 1$, $^{mod}_{\ m}N_{att} = 1$.

Thus, **Table of Coefficients** :

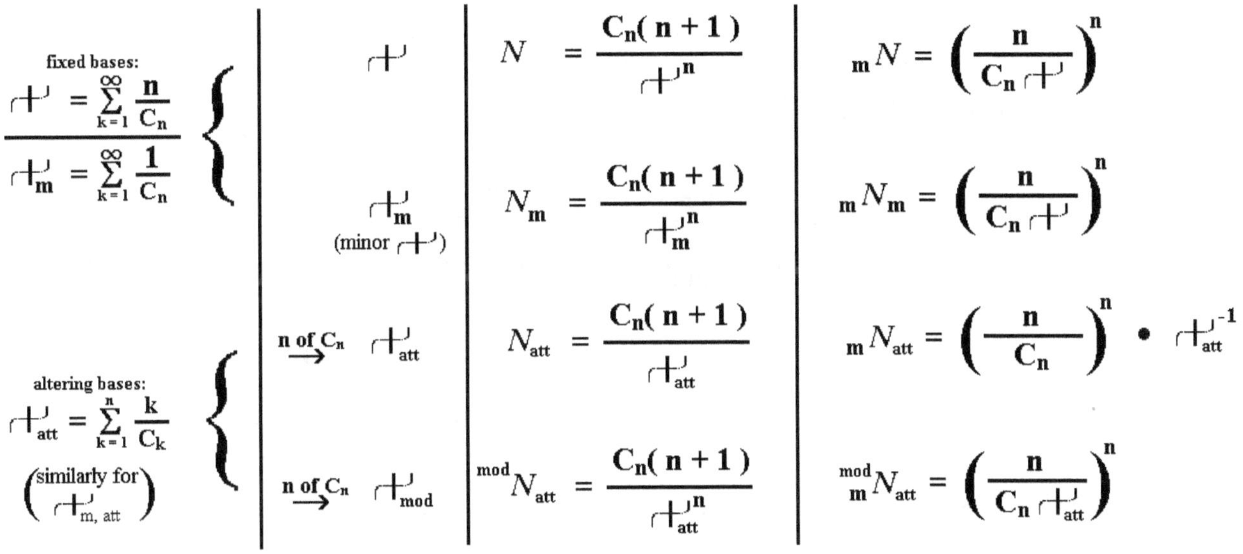

$$
\begin{array}{l}
\text{fixed bases:} \\[2pt]
\Psi = \sum_{k=1}^{\infty} \dfrac{n}{C_n} \\[6pt]
\hline
\Psi_m = \sum_{k=1}^{\infty} \dfrac{1}{C_n}
\end{array}
\left\{
\begin{array}{l}
\\[8pt]
\\[8pt]
\end{array}
\right.
\qquad
\begin{array}{l}
\text{altering bases:} \\[2pt]
\Psi_{att} = \sum_{k=1}^{n} \dfrac{k}{C_k} \\[6pt]
\left(\substack{\text{similarly for} \\ \Psi_{m,\,att}} \right)
\end{array}
\left\{
\begin{array}{l}
\\[8pt]
\\[8pt]
\end{array}
\right.
$$

	Ψ	$N = \dfrac{C_n(\,n+1\,)}{\Psi^n}$	$_mN = \left(\dfrac{n}{C_n\,\Psi} \right)^n$
	Ψ_m (minor Ψ)	$N_m = \dfrac{C_n(\,n+1\,)}{\Psi_m^{\,n}}$	$_mN_m = \left(\dfrac{n}{C_n\,\Psi} \right)^n$
$\xrightarrow{n \text{ of } C_n} \Psi_{att}$		$N_{att} = \dfrac{C_n(\,n+1\,)}{\Psi_{att}}$	$_mN_{att} = \left(\dfrac{n}{C_n} \right)^n \bullet \Psi_{att}^{\,-1}$
$\xrightarrow{n \text{ of } C_n} \Psi_{mod}$		$^{mod}N_{att} = \dfrac{C_n(\,n+1\,)}{\Psi_{att}^{\,n}}$	$^{mod}_{\ m}N_{att} = \left(\dfrac{n}{C_n\,\Psi_{att}} \right)^n$

$\therefore$ 12 complexity coefficients:

N , N_m ; N_{att} ; $^{mod}N_{att}$; $N_{m,\,att}$; $^{mod}N_{m,\,att}$;

$_mN$; $_mN_m$; $_mN_{att}$; $^{mod}_{\ m}N_{att}$; $_mN_{m,\,att}$; $^{mod}_{\ m}N_{m,\,att}$.

$N \bullet \, _mN == (\, C_n(n+1) \, / \, \Psi^n \,) \bullet (\, n \, / \, C_n\Psi \,)^n = (\, C_n(n+1) \, / \, \Psi^n \,)(\, n^n \, / \, C_n^{\,n}\Psi^n \,)$

$= (\, C_n \, / \, C_n^{\,n} \,) \bullet (\, 1 \, / \, \Psi^n \bullet \Psi^n) \bullet (\, (n+1)n^n \, / \, 1 \,)$

$= (\, n+1 \,)n^n \, / \, (\, C_n^{\,n-1} \bullet \Psi^{n+n} \,) = (\, n^{n+1} + n^n \,) \, / \, (\, C_n^{\,n-1} \bullet \Psi^{2n} \,)$

$N \, / \, _mN == (\, C_n(n+1) \, / \, \Psi^n \,) \bullet (\, C_n^{\,n}\Psi^n \, / \, n^n \,) = C_n(n+1) \bullet C_n^{\,n} \, / \, n^n = C_n^{\,n+}$

$^1(n+1) \, / \, n^n$ (as Ψ independent).

$N \bullet {}_m N == (n^{n+1} + n^n) / (C_n^{\,n-1} \bullet \Psi^{2n}) = (n + 1)n^n / (C_n^{\,n-1} \bullet \Psi^{2n})$.

$N + {}_m N == (C_n(n + 1)n^n + C_n^{\,n}\Psi^{2n}) / (\Psi \bullet n)^n = C_n[n^n(n + 1)$

$= C_n^{\,n-1}\Psi^{2n}] / (\Psi \bullet n)^n$.

Therefore, for large n (or where $n + 1 \sim\sim n$) :

$n / {}_m N \sim\sim n \bullet (C_n / n)^n = C_n(C_n / n)^{n-1}$

which appropriately looks like (or is akin to) the inverse of a marginalization.

Making the ratio Ψ dependent:

$(N / {}_m N) \bullet \Psi_{att} \sim\sim n \bullet (C_n / n)^n \bullet \Psi_{att} = n / {}_m N_{att}$

$\therefore N\Psi_{att} / {}_m N \sim\sim n({}_m N_{att})^{-1}$, $n \sim\sim N \bullet \Psi_{att} \bullet {}_m N_{att} / {}_m N$

[Can $N\Psi^n / {}_m N \sim\sim n({}_m N)^{-1}$, which implies $N \bullet \Psi^n \sim\sim n$?

But $N == C_n(n + 1) / \Psi^n$; therefore, $N \bullet \Psi^n = C_n(n + 1)$]

$(N / {}_m N) \bullet \Psi^n \sim\sim n(C_n / n)^n \bullet \Psi^n = n \bullet (C_n\Psi / n)^n == n \bullet ({}_m N)^{-n}$

$\therefore N\Psi^n / {}_m N \sim\sim n / ({}_m N)^n \rightarrow n \sim\sim N \bullet \Psi^n \bullet ({}_m N)^{n-1}$

Therefore, for large n:

$n \sim\sim N \bullet \Psi_{att} \bullet {}_m N_{att} / {}_m N \sim N \bullet \Psi^n \bullet ({}_m N)^{n-1}$.

If we generalize $N / {}_m N$ even more (by making the exponent = 1), we are left with simply:

$n(C_n / n)^1 = C_n$

Therefore, we have direct approximations to both n and C_n (by the above).

Likewise: $N \bullet {}_m N == (C_n(n + 1) / \Psi^n) \bullet (n / C_n\Psi)^n = (C_n(n + 1) / \Psi^n)$
$\bullet (n^n / C_n^{\,n}\Psi^n) = (n + 1)n^n / (\Psi^{2n} \bullet C_n^{\,n-1})$;

or, for large $n \sim\sim (n / (\Psi^2 \bullet C_n))^n$, with $(n + 1) \sim (n - 1) \sim= n$.

And if we generalize for $n = 1$, this comes to:

$(1+1)1^1 / (\Psi^2 \bullet 1) = 2/\Psi^2$, or for large $n \sim\sim 1/\Psi^2 = \Psi^{-2}$;

to compare to (at $n = 1$) $N / {}_mN == C_n{}^2(2)/1 = 2C_n{}^2$

Therefore, (at $n = 1$) $N \bullet {}_mN \sim= 0.38$, $N / {}_mN = 2$

Thus, $(N \bullet {}_mN)(N / {}_mN) == N^2 / 1 \sim= (0.38)2 \sim= 0.76$

$N == C_n(n+1)/\Psi^n \sim= 1(1+1)/2.2905 \sim= 0.87$

$N^2 \sim [1(1+1)/2.2905]2 \sim= 0.76$

${}_mN \sim== (1/(1 \bullet 2.2905))^1 = 1/2.2905 =\sim= 0.44$ (i.e., $N \sim 2\,{}_mN$).

N relation to the Golden Ratio

Is it possible, then, that $2\Psi^{-2}$ is $\sim\sim$ golden ratio's, of $(1-\tau)^2$, fraction squared $(0.61803...)^2$, since $N \bullet {}_mN \sim= 0.3812$,

which is similar to $(1-\tau)^2 = 0.3820$ ($\tau = 1.61803...$) ?

[**Golden Ratio** occurs when $(a+b)/a = a/b = 1.61803...$, and must be for $a > b$]

$1 - (a+b)/a = 1 - a/b$, $(a/a) - (a+b)/a = (b/b) - (a/b)$

$\therefore [a - (a+b)]/a = (b-a)/b$, $-b/a = (b-a)/b$

$\therefore (1-\tau)$ occurs when $(b-a)/b = -b/a = -0.612803...$,

$(1 - 1.61803)^2 = (-0.61803)^2 = 0.3820$

τ occurs when $(a+b)/a = a/b = 1.61803...$, $1 - 1.61803 = -0.61803$

$[(1-\tau)^2]^{1/2} = 0.61803$, $(2\Psi^{-2})^{1/2} = 0.61742$

If (we are) accepting the equivalence, then

$\Psi \sim\sim [(2)^{1/2} / (1-\tau)]_{\text{absolute value}} \sim= 2.2883$

As a base, Ψ must be positive: $\Psi \sim= 2.2905$

$2 / \Psi^2 \sim= (1 - \tau)^2 , (2)^{1/2} / \Psi \sim= (1 - \tau)$; thus,

Ψ is roughly $(1.00096)[(2)^{1/2} / (1 - \tau)]_{\text{absolute value}}$.

With $\Psi = [(2)^{1/2} / (1 - \tau)]_{\text{absolute value}} \bullet (1.00096)$, i.e. less than (1 / 1000th) off the relation, this would, of course, make the golden ratio:

$| 1 - \tau | = ((2)^{1/2} / \Psi)(1.00096)$

$\therefore \tau = 1 + ((2)^{1/2} / \Psi)(1.00096) \sim (\Psi + (2)^{1/2}) / \Psi \sim= 1.61742$

$\tau / ((\Psi + (2)^{1/2}) / \Psi) \sim\sim 1.61803 / 1.61742 = 1.00038$

Therefore, we can approximate "golden ratios" by n :

$\tau_n \sim\sim (\Psi + (n + 1)^{1/2}) / \Psi$, where $t_1 \sim == \tau$

e.g. (the following table):

n	τ_n	||	n	τ_n	||	n	τ_n
1	**1.61742**	||	**6**	**2.15510**	||	**100**	**5.38763**
2	**1.75619**	||	**7**	**2.23485**	||	**1000**	**14.8130**
3	**1.87317**	||	**8**	**2.30976**	||	**10000**	**44.6608**
4	**1.97624**	||	**9**	**2.38061**	||	$\mathbf{10^5}$	**139.061**
5	**2.06941**	||	**10**	**2.44799**	||	$\mathbf{10^6}$	**437.586**

This suggests that such ratios are related (or can be related) to (C_n, Ψ^n) complexities (under the following progression):

$n \rightarrow$ index of terms

$C_n \rightarrow$ combination of terms ($C_\#$)

$\Psi^n \rightarrow$ base powers

$C_n(n + 1) / \Psi^n \rightarrow$ complexity coefficient

$(\Psi + (n + 1)^{1/2}) / \Psi \rightarrow$ "golden ratios" (Grs)

$(\Psi^n + (n + 1)^{1/2}) / \Psi^n \rightarrow$ GR Powers ratios

An obvious implication is an "adapted" complexity coefficient:

$$C_n(n + 1) / \Psi^n == C_n(n + 1) / [(n + 1)^{1/2} / (1 - \tau_n)]_{abs.\ value}^{n}$$

$$= C_n(n + 1) / [(n + 1)^{n/2} / (1 - \tau_n)^n]_{abs.\ value}$$

$$= C_n(n + 1)| (1 - \tau_n)^n / (n + 1)^{n/2} == C_n| 1 - \tau_n |^n / (n + 1)^{((n/2)-1)} == a\boldsymbol{N};$$

at $n = 1$, $a\boldsymbol{N} \sim = 1| 1 - 1.61742 |^1 / (1 + 1)^{-1/2} = 0.87316$

while $\boldsymbol{N} \sim = 1(1 + 1) / 2.2905^1 = 0.87317$

[Past $n = 1$, i.e. $n > 1$, $\boldsymbol{N}$ and $a\boldsymbol{N}$ must diverge since $\tau_{n > 1}$ is not $\sim \sim t_1$ nor t .]

An interesting observation is the combination times the GR Powers ratio:

$$C_n \bullet [(\Psi^n + (n + 1)^{1/2}) / \Psi^n] = [C_n\Psi^n + C_n(n + 1)^{1/2}] / \Psi^n$$

$$= (C_n\Psi^n / \Psi^n) + C_n(n + 1)^{1/2} / \Psi^n = C_n + C_n(n + 1)^{1/2} / \Psi^n$$

which is somewhat similar to the complexity coefficient: $C_n(n + 1) / \Psi^n$.

The combination times the Grs ratio would be:

$$C_n \bullet ((\Psi + (n + 1)^{1/2}) / \Psi) = C_n + C_n(n + 1)^{1/2} / \Psi == C_n \bullet \tau_n .$$

For $n = 1$, this is: $1 + 1(1 + 1)1/2 / 2.2905 = 1.61743 \sim \sim t$

(i.e. at $n = 1$ this is τ_1; but it diverges from τ_n for $n > 1$).

The corresponding (combination) (GR Powers ratio) product, for $n > 1$, would be less than $C_n \bullet \tau_n$ (due to the term with division by Ψ^n, Ψ^n being $> \Psi^1$ for $n > 1$).

Therefore, we can divine (and define) a τ_n complexity coefficient:

$$\boldsymbol{N}_{\tau n} == C_n\tau_n / \Psi^n = (C_n / \Psi^n) + C_n(n + 1)^{1/2} / \Psi^{n+1} .$$

The corresponding coefficient using a GR Powers ratio is:

$$_{powers}\boldsymbol{N}_{\tau n} == (C_n / \Psi^n) + C_n(n + 1)^{1/2} / \Psi^{2n} == (C_n + (C_n(n + 1)^{1/2} / \Psi^n)) / \Psi^n .$$

Therefore, we have coefficients (of complexity): $\boldsymbol{N}, \boldsymbol{N}_{\tau n}, \boldsymbol{N}_{\tau n,\ powers}$;

so we have (what looks like) an evolving series (of coefficients):

$$C_n(n+1)/\Psi^n,\ (C_n/\Psi^n) + C_n(n+1)^{1/2}/\Psi^{n+1},\ (C_n/\Psi^n) + C_n(n+1)^{1/2}/\Psi^{2n}.$$

For $n = 1$, this (series) is (with $C_n = C_1 = 1$):

$1(2)/2.2905,\ (1/2.2905) + 1(2)^{1/2}/(2.2905)^2,$

$(1/2.2905) + 1(2)^{1/2}/(2.2905)^2 == 0.87317, 0.70615, 0.70615$

For $n = 2$, this is (with $C_n = C_2 = 3$):

$3(3)/(2.2905)^2 = 1.7155$;

$(3/(2.2905)^2) + 3(3)^{1/2}/(2.2905)^3 = 0.43241 + 0.57182$;

$(3/(2.2905)^2) + 3(3)^{1/2}/(2.2905)^4 = 0.18878 + 0.57182$

$== 1.7155, 1.00423, 0.76060$

The "divergence" of from $n = 1$ to $n = 2$ is:

$1.7155/0.87317,\ 1.00423/0.70615,\ 0.76060/0.70615$

$== 1.9647, 1.4221, 1.07711\ (\sim 2.0, 1.4, 1.1)$

Clearly, the complexity decreases across the series, from $\boldsymbol{N}$ to GR to GR Powers. The complexity increases with n, but the relative complexity (divergence) decreases (across the series).

Utilization of Complexity

How is this information (to be) utilized?

Say, we have a velocity: $v = d/t$

v has two components (variables): d, t

The complexity of each component is (for itself, $n = 1$) : $\boldsymbol{N}d \sim== 0.87d$; $\boldsymbol{N}t \sim== 0.87t$

The complexity of the velocity is (for itself, $n = 2$): $\boldsymbol{N}v \sim== 1.7v$

The τ_n complexity would be close to the velocity itself: $\boldsymbol{N}_{\tau 2} \sim== 1.0v$, while the $\tau_{n,\ powers}$ complexity would be less: $\boldsymbol{N}_{\tau 2,\ powers} v \sim== 0.76v$

Is there significance to a τ_n coefficient of $n = 2$ causing the complexity to be essentially the relation's value itself? It might suggest that a relation involves basically (or fundamentally) two variables at least, only one being a modification of the original (e.g. $v_2 = v_1 + c$, $c == a$ *constant*). Therefore, $v_2 = v_1 + x$; x as a variable would force v_1 to be a variable also (for $n = 2$), or a constant (for $n = 1$), and so on (e.g., $v_2 = v_1 x$, etc.).

If v_1 and v_2 are of like kind, then $n = 1 \rightarrow$ modification (of extent or v quality, magnitude), and $n = 2 \rightarrow$ change of kind or proportion:

e.g., $v_2 = v_1 + x$, where all are of same kind, but

x is a variable, v_1 is a constant $\rightarrow n = 1$;

x is a variable, v_1 is a variable $\rightarrow n = 2$ and there is a change of proportion (for v_2) ;

or, x is a variable of different kind from $v_1 \rightarrow n = 2$ whether v_1 is a variable or not, therefore forcing v_2 to be a change of kind (from v_1). [This, of course, descends into a sort of logic or rationality.]

Therefore, we may term $N_{\tau 2}$ as a logical (rational) discriminant. Thus as we see complexity coefficients to be (in general) discriminants of a nature (towards n). That is, they restrict or guide the application of n; for a relation. "n" is, of course, defined by the relation (or established as such);

e.g., $d = a$, $n = 1$, complexity $== {}_nN_{c1} \bullet a \sim= 0.87a$

(say) $d = ab$ or $a + b$, $n = 2$, complexity $== N_2(ab, a + b) \sim= 1.7(ab, a + b)$

$d = abc$ or $a + b + c$ (etc.) , $n = 3$, complexity $== N_3(abc, a + b + c, \text{etc.})$

$\sim= 2.3$ (abc, a + b + c, etc.) [etc. $==$ ab + c, a + bc, ac +b,

and higher combinations (e.g., $a^2b + c^3$)]

Initially, the coefficients somewhat mirror n (in progression):

n	N_n	N_n / N_1	$N_{\tau n} / N_{\tau 1}$	$N_{\tau n}$	$N_{\tau n,P} / N_{\tau 1,P}$	$N_{\tau n,}P$
1	0.87	1	1	0.71	1	0.71
2	1.7	1.95 ~ ~ 2	1.4	1.0	1.1	0.76
3	2.3	2.64 ~ ~ 3	1.55	1.1	0.96	0.68

Therefore, the GR Powers coefficients contract by n = 3 , suggesting the complexity, from base (Ψ) powers, decreases after more than 2 variables. But this is only a dipping point, since the coefficient recovers for n = 4 ($N_{\tau 4,\text{ Powers}}$ ~ == 0.98) :

n	C_n	N_n	N_n/N_1	$N_{\tau n}$	$N_{\tau n}/N_{\tau 1}$	$N_{\tau n,P}$	$N_{\tau n,P}/N_{\tau 1,P}$
1	1	0.87	1	0.71	1	0.71	1
2	3	1.7	1.95	1.01	1.4	0.76	1.1
3	7	2.3	2.64	1.1	1.5(5)	0.68	0.96
4	25	4.54	5.22	1.80	2.54	0.98	1.38
5	161	15.32	17.61	5.30	7.46	2.65	3.73

In general, then, the (3 types of complexity) coefficients increase with n.

One casually observing the "dip" for the progression of the powers coefficients (with increasing n) might conclude it to be a retardation of increase as of a hysteresis (compared to the other types of N coefficients), for whatever phenomena of variables the coefficients are associated with. But we (here) see it is a "natural" arithmetic enlistment of the series.

The "robustness" of variable (or relation) association is:

$N_{\tau n,\text{ Powers}} <= N_{\tau n} <= N_n$ (in coefficient value) .

But, in each case it is clear that there is a "shallowing" of slope (N vs n) between n = 2 and n = 3 .

Is this because with n = 3 another type (or phase) of complexity (different from n = 1 $\rightarrow$ 2) ensues (i.e., a more complex form, or more sophisticated)? Perhaps this enlists an alternative mode of complexity dependence (upon possibility of relation):

n = 1	**r = a**	**r = a**
n = 2	**r = a + b**	**r = ab**
n = 3	**r = a + b + c**	**r = abc**
	→ a + (b + c) , (a + b) + c , (a + c) + b	**→ a(bc) , (ab)c , (ac)b**

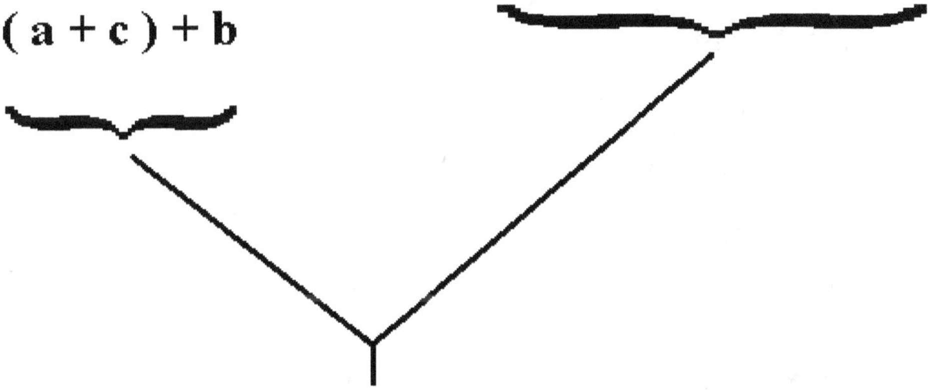

look like (a new)
n = 2 mode

i.e., all modes past n = 2 can be made to look like (or consider as) n = 2 modes. Therefore, n = 2 → n = 3 can represent a transition of complexity.

Note that there is only one way to group:

a, n = 1

[a + b == b + a , ab == ba] → n = 2

But at n = 3, suddenly 3 groupings are formed (found). The "jump" from 1 (grouping) to 3 (i.e. skipping 2 groupings) essentially established a new (or another) complexity curve. The 3 groupings (and all further groupings, presumably) emulate a "pseudo" n = 2 mode.

One can then assume that the (various kinds of) $\boldsymbol{N}$ coefficients demarcate the "difficulties" of change of kind for a relation. The greater the change of kind, the greater the inherent complexity. Apparently, $\boldsymbol{N}_{\tau n,\ \text{Powers}}$ provides for a less stringent (or slower) increase of complexity (for a relation) than $\boldsymbol{N}_{\tau n}$ and $\boldsymbol{N}_{n}$.

Complexity's Numerical Relation to Infinity

What is the utility of the (a) complexity coefficient?

Clearly, as towards a relation, it can be used as a variable (to be determined or discovered); e.g., (complexity) relation $r_c = \mathcal{N}X_c$, where $X == d / t = v$.

This suggests a set of values (for r_c) whereby the variables d and t may be accessed:

$r_1 = \mathcal{N}X_1$, $r_2 = \mathcal{N}X_2$, etc.

If either d or t is held constant, $n = 1$.

If both are held constant, r_c is a constant.

If neither are held constant, $n = 2$ (both are variable).

By presumption, then, $\mathcal{N} = r_c / X == r_c t / d$.

But $\mathcal{N}$ is confined to definite (discreet) values, which must restrict $r_c t / d$ (and thus t and/or d).

Therefore, only certain values of r_c can be obtained (measured). This discreetness, however, is (ably) obscured by v. For a simple relation, i.e. one without a complexity component, is commanded by the falsehood of (the concept of) infinity, with an infinity of values to relate to (its variables). This is an (or due to an) incorrect logic. Let N_1 be a number, and say that it is infinite. Now, let N_2 be a number and say that it can have any value except a particular one. The conclusion is that both N_1 and N_2 are infinite. Yet it is clear that N_1 =/= N_2 (for at least one particular case). Yet, infinity is only of one sort. So the logic is faulty. There is no legitimate concept as infinity (for a physical reality and a mathematical conspicuity). It is, again, like finding the center (point) of an infinite number-line.

One can perhaps view the data (presented by the relation) as like an infinite number-line with either gaps or markers similar to frequency spikes for an energy band. This number-line (energy band), however, would represent a curve not necessarily straight:

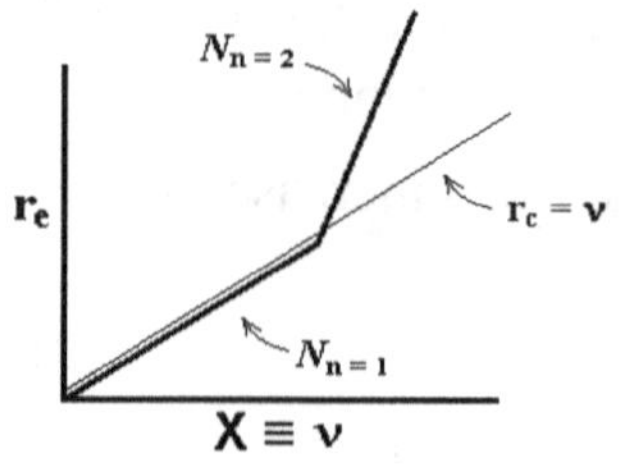

,

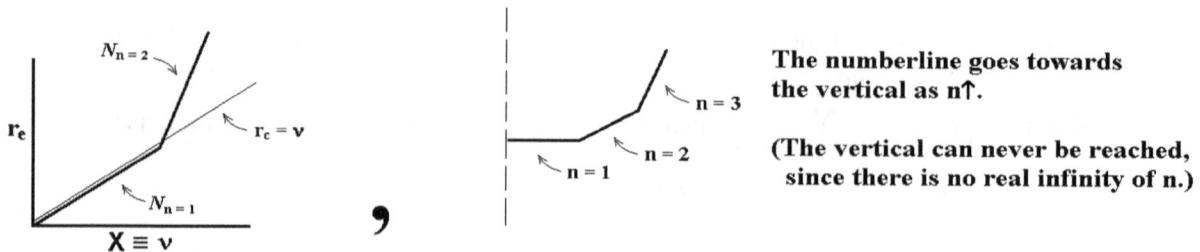

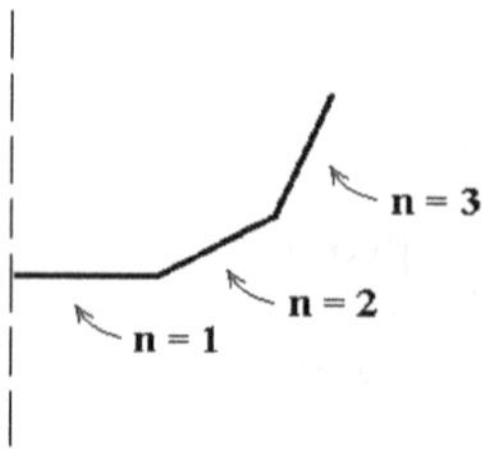

The numberline goes towards the vertical as n↑.

(The vertical can never be reached, since there is no real infinity of n.)

The steepness of the curves (i.e. N) emphasizes the rapid increases in complexity with rising n, as well as the increasing lengths of the "n" segments (to encompass both r_c and v). In this sense it is reminiscent of the (Ψ^n, C_n) plots. There is a tremendous rise in r_c with just a slight rise in v, as n increases. Therefore,

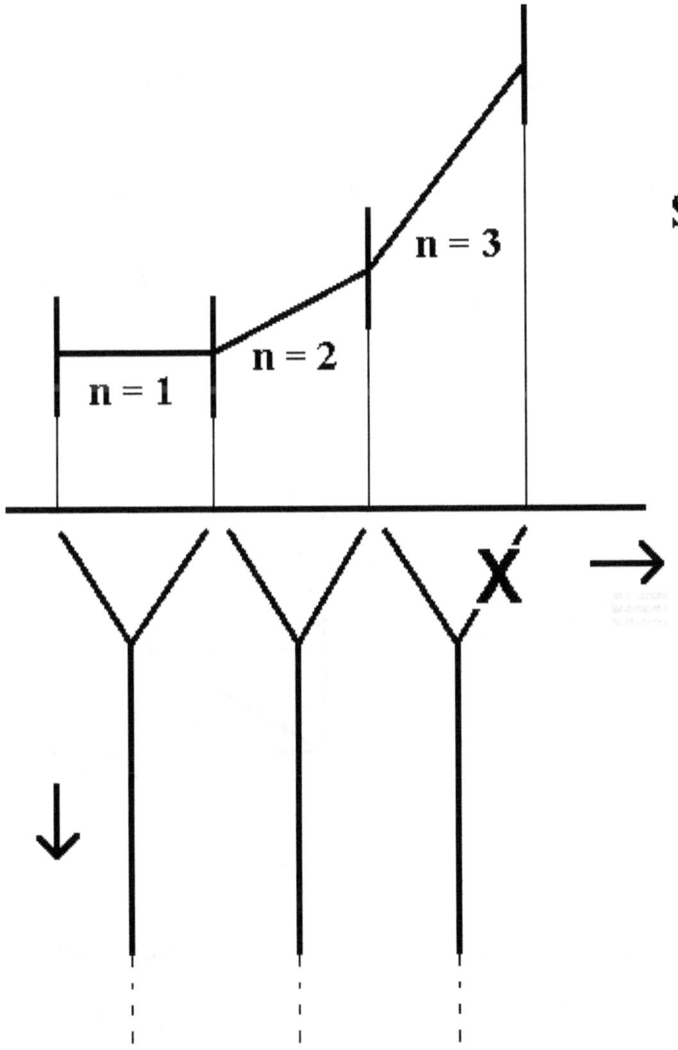

slope and length for n increases

Projection onto X: suggests domains of applicable X. But each domain (projection) is itself as an infinite number-line (for the variables). Therefore, the (variable) values must be (de-)limited (upon their n number-line) by r_c.

[Each 'domain of n' should probably be of equal length of X projection.]

Then, how can $\infty = \infty - 0$? or $\infty = \infty - n$, where $n ==$ any number?

Obviously, "∞" does not relate to the "=" operation properly (or typically), since $\infty + n = \infty$.

Condensation of Coordinates and Con-axial Locations

Clearly, then, such considerations require a condensation of coordinates.

For example, the (X vs r_c) plot is more properly con-axial:

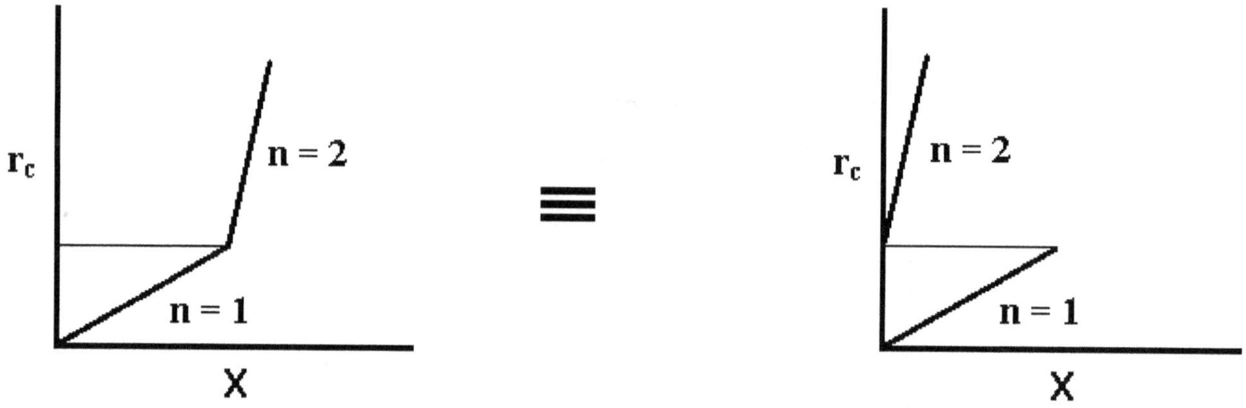

Note that the 'speed' with which X increases slows as n and r_c rise. The complexity increases faster than the variables, as n increases. So, a small change in variables causes greater complexity as the number of variables increase.

In particular, it takes much longer (of X axis) to obtain a value for X as $n\uparrow$. This is a sign that the complexity increases (and r_c is climbing higher more rapidly, for the relation).

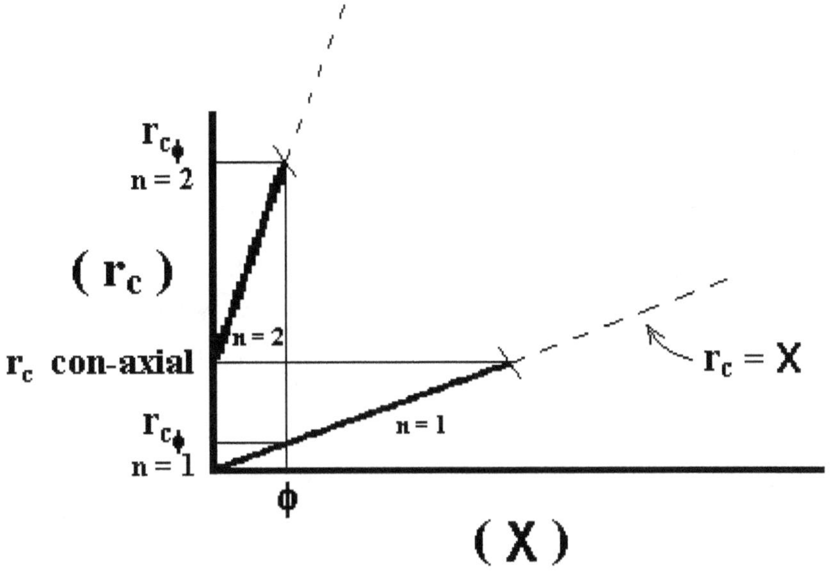

If the 'length' of an 'n-step' is constant (for each n), then (here),

to reach X_ϕ takes a whole $n = 2$ step but only a fraction of the $n = 1$ step.

$r_{c\phi,\, n=2} > r_{c\phi,\, n=1}$. $r_{c\phi,\, n=2}$ falls on a con-axial for r_c ($r_{c,\, n=2}$),

as $r_{c,\, n=1}$ falls on a con-axial for r_c.

Of course, when $n = 1$, $r_c \sim= 0.87X$ ($N \sim= 0.87$)

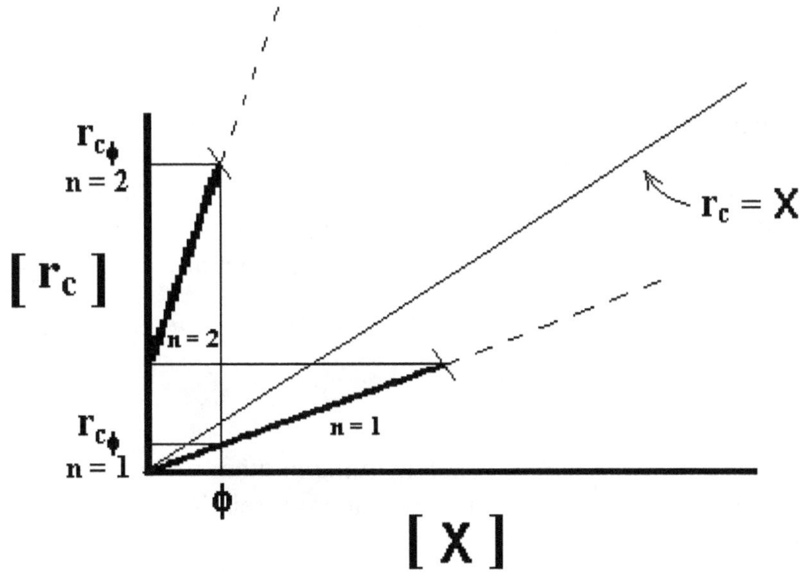

(etc. for $\boldsymbol{N}_{\tau n}$, $\mathbf{N}_{\tau n,\ \text{Powers}}$)

What delineates, then, a con-axial (location), or the functional extent of an n-step?

This is when a constant becomes a variable (and visa versa).

The plot presents an interesting graphical/geometric correlation for conversions of n (i.e. numbers of variables): con-axial motivation of complexity. When a con-axial is established, for considering a particular X (as ϕ), then too is determined a difference ($\phi_{\text{con-axial}} - \phi$) along X and a difference ($r_{c,\,\phi} - r_{c,\ \text{con-axial}}$) along r_c.

Therefore, a triangle can be proposed (with the above differences as legs) that is right-angled, to form an hypotenuse that leads to a 'reduction' in X as r_c increases; i.e., the n conversion accompanies greater complexity simultaneously with decreasing X values (of the relation).

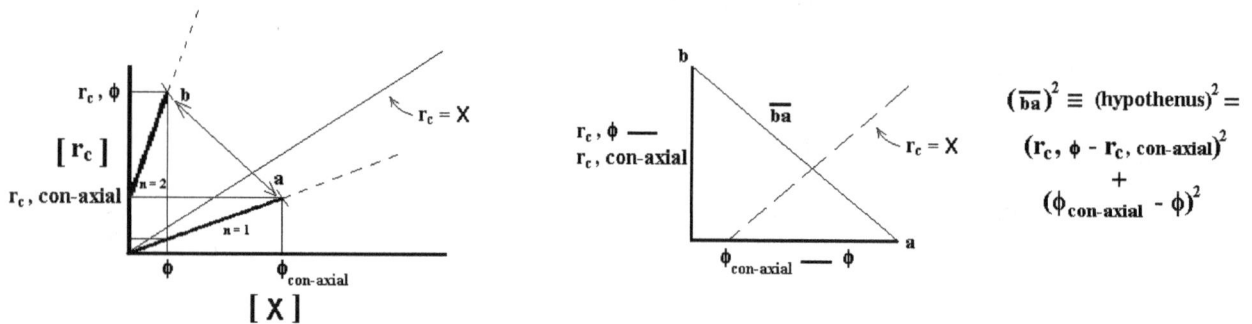

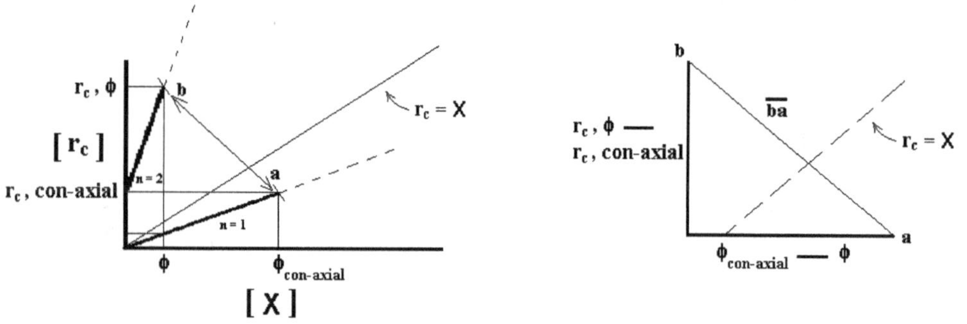

$$(\overline{ba})^2 \equiv (\text{hypothenus})^2 = (r_{c,\,\phi} - r_{c,\ \text{con-axial}})^2 + (\phi_{\text{con-axial}} - \phi)^2$$

The path from a to b need not be linear (i.e. as an hypotenuse), but is propitiously described as such, for consideration of a particular ϕ (== X value). This directly establishes, for example, that (here) $\phi_{n=2}$ is more complex than $\phi_{n=1}$ (though they share the same X value).

Of course, the operative path from a to b is to go from a along the con-axial to the r_c axis ($r_{c,\ \text{con-axial}}$), and then up along $n = 2$ to b. But there are many other paths, varying with ϕ. The shortest path would be from $\phi_{\text{con-axial}}$ (== a) to b (i.e. hypotenuse line ba):

thus, $\phi = \phi_{\text{conaxial}}$.

As a con-axial applies when there is a change in the number of variables, this automatically defines an alteration in complexity for the (considered) relation. In particular, coefficients, once considered as constants, can (or might) eventually be partitioned (multiplicatively, etc.) into (various) variables, increasing the complexities of the relation. Therefore, such variation (for coefficients) can be decided for when choosing (among) $\boldsymbol{N}$, $\boldsymbol{N}_{\tau n}$, $\boldsymbol{N}_{\tau n,\text{ Powers}}$ (or other), as well as more fundamentally (in, for example, n and base).

Perception of Complexity

Human perception is more perceptive (acute) or sensitive (and responsive) to the regularities of lines than the irregularities on (non-straight) curves, but it is within these hidden (from our normal senses) irregularities that complexity abides. We choose the geometrical regularity of shapes to observe over the complexities of matter condoning these shapes, and define forthwith their properties (often) in abeyance to their complex(-ional) foundations. Consideration of r_c helps to temper this (mental or intellectual) impedance (for a relation).

The apparent irregularity (of the resultant curves) is due to the physical adoption (and need of this) to vary the number of variables (for a relation or realistic outcome of a process).

Obviously, the variation in (complexity) coefficients is due to n (as an unknown). In particular, a set (to series) of coefficients can be derived by changing the addend "1" to "i" (as a variable) for the $(n+1)$ term ($\rightarrow (n+i)$), etc. Therefore,

$${}^{i}\boldsymbol{N} == C_n(n+i)/\Psi^{n}; \qquad {}^{i,i}\boldsymbol{N} == C_n(n+i)/\Psi^{in}.$$

$${}^{i}\boldsymbol{N}_{\tau n} == (C_n/\Psi^{n}) + (n+i)^{1/2}/\Psi^{n+I}; \quad {}^{i,i}\boldsymbol{N}_{\tau n} == (C_n/\Psi^{n}) + (n+i)^{1/2}C_n/\Psi^{n+i}$$

$${}^{i}\boldsymbol{N}_{\tau n,\text{ Powers}} == (C_n/\Psi^{n}) + C_n(n+i)^{1/2}/\Psi^{2n}; \qquad {}^{i,i}\boldsymbol{N}_{\tau n,\text{ Powers}} == (C_n/\Psi^{n}) + C_n(n+i)^{1/2}/\Psi^{2in}.$$

[Note that the "C_n" term is not affected by "i" use.]

The coefficients can, thus, be further generalized by marginalization, attenuation, and modulation: ${}_{(m)}{}^{(i)}\boldsymbol{N}$, ${}_{(m)}{}^{(i)}\boldsymbol{N}_{m}$, ${}_{(m)}{}^{(i)}\boldsymbol{N}_{att}$, ${}_{(m)}{}^{(i),mod}\boldsymbol{N}_{att}$; etc.

The "i" can conceivably be replaced by its own relation, to characterize a complexity even farther (or more indelibly). For example, a recursion can be produced:

$$\boldsymbol{N} == C_n(n+\boldsymbol{N})/\Psi^{n}, \text{ etc.}$$

Note that if $(n+\boldsymbol{N})$ is index-able, then the sum will only (effectively) occur when $\boldsymbol{N}$ is a whole number (i.e. has no fractional parts). This can lead to "stationary" (non-moving or statistical) coefficients of the type:

$$\boldsymbol{N} == C_n(n+\boldsymbol{N})/\Psi^{n} \rightarrow C_n(n)/\Psi^{n}, \ \boldsymbol{N} \text{ with fraction.}$$

The recursion can be minimized by simply taking the whole number part of $\boldsymbol{N}$ for the indexation. This means, of course, that $\boldsymbol{N}_{indexed}$ is estimated (beforehand).

This proposes, then, how a "moving" recursion is accomplished:

$$\boldsymbol{N}_{stationary} = C_n(n) / \Psi_n , \boldsymbol{N}_{indexed} == \boldsymbol{N}_{stationary} - (\text{ its fractional part})$$

$$\therefore \boldsymbol{N}_{(moving)} == C_n(n + \boldsymbol{N}_{indexed}) / \Psi^n = \boldsymbol{N}_{moving} .$$

There is no particular reason, though, why a factor for C_n need be only a whole number, except through the physical illustration that derives a (n + i) increment, or (n + 1) increment in the original construction (of motion). If a model calls for a complete cycle of motion, then presumably a whole number increment, of $C_n(n + i)$, is required. This suggests the denominator Ψ^n imposes a fractional*ization* to the complexity (coefficient), as a nature of division as well as base.

Complexities with Fractional Components

Why would a complexity (coefficient) have a fractional component?

We can partition (it) into:

an obligate (or base) fraction → denominator Ψ^n

an obligate (or index-able) whole number → numerator $C_n(n + 1)$

Conceptually, then, the relation (to which the coefficient is multiplied or attached to) should be (and generally is) fractional, so that (the) fractional properties cancel out.

Therefore, using C, N, B, R as whole numbers and b, r as fractional numbers, the complexity coefficient can be expressed as:

$$(C (N) / (B + b)) \bullet (R + r) = L + l$$

where L is whole number and l its fractional addition.

$$\therefore CNR + CNr = (L + l)(B + b) = BL + Bl + bL + bl$$

$\therefore$ (a whole number) $CNR – BL = Bl + bL + bl - CNr$ (which must also be a whole number).

So, a complexity expression can be converted into a whole number essential (value):

CNR – BL , as well as into fractional (total) components: Bl + bL + bl – CNr .

Note that the equation:

(C(N) / (B + b)) • (R + r) = (L + l)

shows:

1. that division itself is fractional (thus counts as a fractional)
2. 3 explicit fractionals are also involved, i.e., the (...+...) terms.

∴ There are 4 fractionals, which must cancel each other out to lead to a whole number result (or essential value).

Therefore, perhaps it is this "essential value" that is most important in considering the complexity of a relation (e.g., for comparisons).

For example, using velocities: d / t

case 1

n = 1 , d = 2.5 units, t = 3 time units

let d be a constant (→ n = 1)

C = 1 , N = 2 , (R + r) = 2.5 / 3.0 = 0.8333 == R = 0 + r = .8333

B = 2 , b = 0.2905 , (B + b = Ψ^1)

(L + l) = [(1)(2) / 2.2905](.8 1/3) = 0.7276 == 0 + 0.7276

Thus, L = 0 , l = 0.7276 ; CNR – BL == (1)(2)(0) - (2)(0) = 0

case 2

case 1, but with d also as a variable: thus n = 2

C = 3 , N = 3 , R = 0 , r = 0.8333 == .8 1/3

B = 2 , b = 0.2905

(L + l) = ((3)(3) / 2.2905)h(.8 1/3) = 9(0.8333)/2.2905 = 3.274

Thus, L = 3 , l = 0.274 ; CNR – BL == (3)(3)(0) - (2)(3) = -6.0

Therefore, case 2 is more complex; it is presumably because its absolute value is larger than that of case 1 . But what does the negative sign (for the essential value, in case 2) signify?

We can consider this arithmetically: if we subtract from the (whole number) term CNR all of the fractional(ized) components (the sum of which is a whole number), then we are left with the (positively signed) BL term.

CNR – BL = (Bl + bL + bl – CNr) ; ∴ CNR - (Bl + bL + bl – CNr) = (+)BL

Note, then, that

CNR + BL is also a whole number [== 2CNR - (Bl + bL + bl – CNr)]

which would be:

CNR + BL = 0 for case 1 (n = 1) and +6 for case 2 (n + 2).

Note also that if we combine related terms:

for CNR – BL = Bl + bL + bl – CNr

then CNR + CNr = Bl + bL + bl + BL (as shown earlier).

(CNR + BL) might serve as a useful (and objective) measure (metric) for comparisons of complexities for relations, as well as the "essential value": (CNR – BL). Clearly, for a relation, the complexities may change as the variable values change: e.g.,

as in case 1 earlier, let d go from 2.5 to 6.5 in steps of 1 (dist. unit); t = 3.0 (time units); B = 2

(R + r)	d	R	L	(CNR + BL)	(CNR – BL)	(L + 1)
0.8333	**2.5**	**0**	**0**	**0**	**0**	**0.7276**
1.1667	**3.5**	**1**	**1**	**2 + 2 = 4**	**0**	**1.0187**
1.5000	**4.5**	**1**	**1**	**2 + 2 = 4**	**0**	**1.3098**
1.8333	**5.5**	**1**	**1**	**2 + 2 = 4**	**0**	**1.6008**
2.1667	**6.5**	**2**	**1**	**4 + 2 = 6**	**2**	**1.8919**

So, we have:

{ (CNR + BL) | 0, 4, 4, 4, 6 }
{ (CNR – BL) | 0, 0, 0, 0, 2 }
{ L | 0, 1, 1, 1, 1 }
{ R | 0, 1, 1, 1, 2 }
(d | 2.5, 3.5, 4.5, 5.5, 6.5 }
[all at: n = 1 , t = 3.0 , B = 2 , CN = 2]

Note : { CNR | 0, 2, 2, 2, 4 }

Yet : { (CNR + BL) | 0, 4, 4, 4, 6 }

\+ { (CNR – BL) | 0, 0, 0, 0, 2 }

{ (CNR + BL) + (CNR – BL) | 0, 4, 4, 4, 8 } == { 2(CNR) | 0, 4, 4, 4, 8 }

It is interesting that (here) BL complements CNR – BL :

{ CNR – BL | 0, 0, 0, 0, 2 }

\+ { BL | 0, 2, 2, 2, 2 }

{ CNR | 0, 2, 2, 2, 4) (as it should).

The CNR component is not "base-d" (upon Ψ).

The BL component is quasi-unbased (upon Ψ), since BL ~ ~ (CNR / B) • B = CNR (but the "quasi" implies that it is so based).

Obviously, a base may change for the BL component ($\Psi_{()}{}^{1}$, e^{1}, etc.), while the CNR component is thoroughly independent of this. But the "C" derives (or is used to derive) $\Psi_{()}$ definitions (or constructions). So there is correlation here (between C and $\Psi_{()}$). Therefore, both components are dependent on (the definition of) "C" when base $\Psi_{()}$ is employed.

Since (Bl + bL + bl – CNr) / (CNR – BL) = 1

the characteristic, or descriptive, fractional components are:

BL / (CNR – BL) , bL / (CNR – BL) , bl / (CNR – BL) , CNr / (CNR – BL) .

Here, we note, arithmetically, that division by zero may occur.

So, we substitute for CNR – BL → W (for "whole" number).

Therefore, the fractions are: Bl / W , bL / W , bl / W , CNr / W ,

and the actual fractions are: $(\mathrm{Bl})W^{-1}$, $(\mathrm{bL})W^{-1}$, $(\mathrm{bl})W^{-1}$, $(\mathrm{CNr})W^{-1}$;

(e.g., for case 1 (n = 1) at d = 2.5, the values are :

$[(2)(0.7276) + (0.2905)(0) + (0.2905)(0.7276) – (1)(2)(0.8333)]W^{-1}$

$= (1.4552 + 0 + 0.2114 – 1.6666)W^{-1} = (0)W^{-1} = 0$;

Bl = 1.4552, bL = 0.0, bl = 0.2114, CNr = 1.6666; CNR – BL = 0 == W .

[Note the inconsistency that $0/0 == 0$ instead of 1, nor ∞ : multiplication takes precedence over division.]

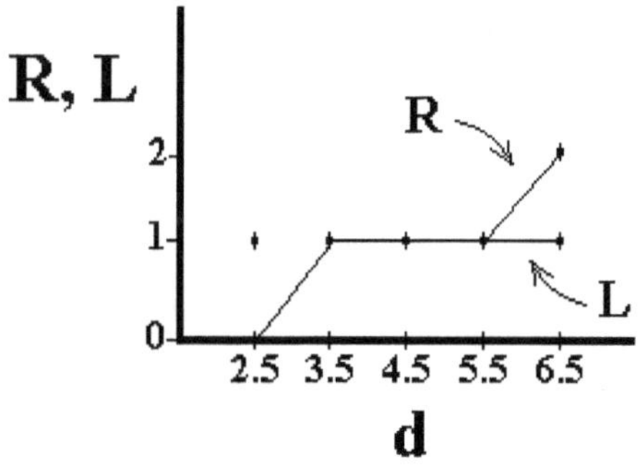

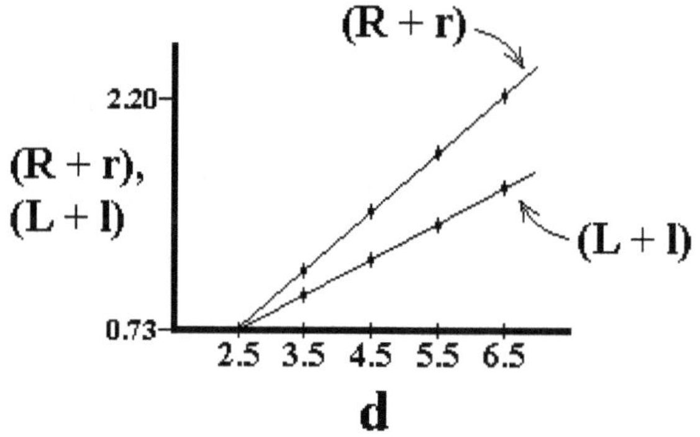

The distinction between the Whole Number analysis and the relational (or, here, linear) analysis is clear.

The corresponding fractional (number) analysis appears rather cryptic: e.g.,

d	r	l	(Bl + bL + bl – CNr)	r / l
2.5	**0.8333**	**0.7276**	**0**	**1.1453**
3.5	**0.1667**	**0.0187**	**0**	**8.9144**
4.5	**0.5000**	**0.3098**	**0**	**1.6139**
5.5	**0.8333**	**0.6008**	**0**	**1.3870**
6.5	**0.1667**	**0.8919**	**2**	**0.1869**

r (here) acts almost like a sinusoidal carrier for l (due to the regularity of the d step):

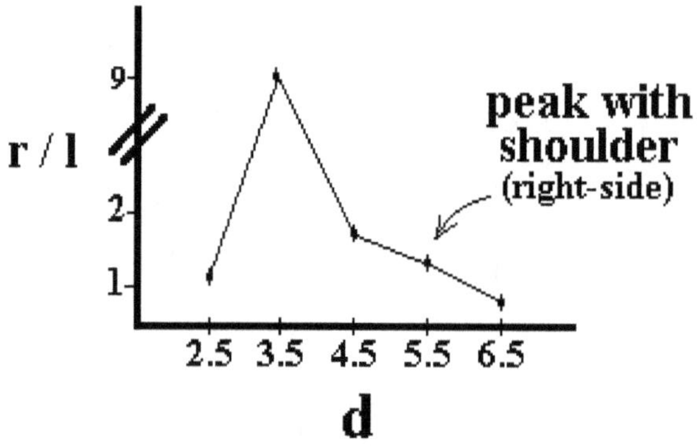

We might call this a “residuals” trace, but it could be emblematic (in some fundamental way) of the complexity change in the relation (via: L + l). The d change (from $2.5 \rightarrow 3.5$) for this relation (with t = 3.0) represents the first whole number conversion for this limited (or set) series (i.e., 3.5 > 3.0). It says that the complexity (L + l) is close to being a whole number (having a small l fractional component), at this juncture (d = 3.5). If effective complexities (for a relation) are limited to (approximately) whole number values, then such a (r / l) trace could be indicative (and illustrative) of these (such) occurrences. The limitation is in terms of observation (or measurement). [l alone can probably not be used for this diagnostic, but must be compared to the regularization of r as a data carrier ('correlate-or').]

Note that the only relevant "fraction" from the essential value (CNR – BL) with regards to (r / l) is: -CNr , for we can divide this by l, to yield : -CNr / l == (-CN) • (r / l) .

The (prescribed) unknown, for a complexity consideration, is of course n, since this determines C_n, $\Psi_{()}^{n}$, N (= n + 1), and ultimately (L + l), knowing (R + r).

If we plot (r vs l), we see a cyclic (or relational) behavior based on r, and a complexity behavior based on l:

d		r	l [determined from (L + l)]		
(0.5)		**(0.1667)**	**(0.1455)**	|	
(1.5)		**(0.5000)**	**(0.4366)**	|	**slope == 0.873**
2.5		**0.8333**	**0.7276**	|	
3.5	[	**0.1667**	**0.0187**	]	**slope of line**
4.5	[	**0.5000**	**0.3098**	]	**== (0.6 – 0.02) / (0.8 – 0.2)**
5.5	[	**0.8333**	**0.6008**	]	**~= 0.87(324)**
6.5		**0.1667**	**0.8919**	|	
(7.5)		**(0.5000)**	**(0.1829)**	|	
(8.5)		**(0.8333)**	**(0.4740)**	|	**slope == 0.873**

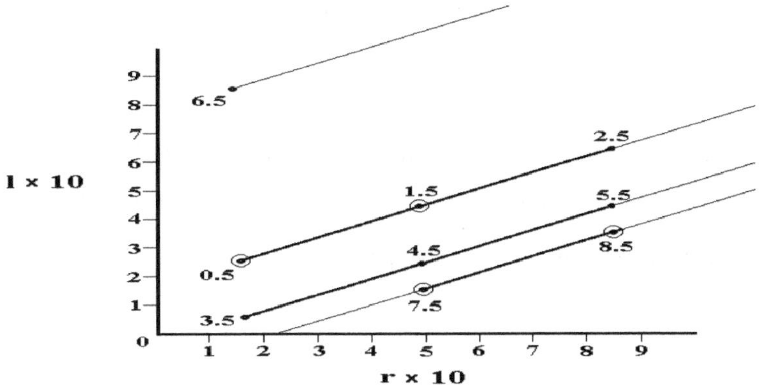

Parallel lines == equal slopes (dashed lines presumed)

N == (1)(1 + 1) / 2.2905 = 0.8732

This strongly suggests the relationship: slope (of 'r vs l') • Ψ^n = CN

If we let Ψ^n == Ψ base, then :

((0.6008 – 0.0187) / (0.8333 – 0.1667)) • (2.2905) = 2.0002 == 2

Therefore,

N = 2 = n + 1   (C = 1 for N = 2)

n = 1 , C_n == C_1 = 1

So, the argument is:

$\Psi_{()}^{\ n} = \int n\Psi_{()}^{\ n-1}\, d\Psi$ = CN(Δr / Δl) == CN(r / 1)$_\Delta$.

d = 3.5 is clearly a start (point) for the (r) cycle (so are d = 0.5, d = 6.5, etc.). The period (of the cycle) is 3.0, but there is a clear complexity (i.e. l based) discontinuity (of plot) between d = 6.5 and (d = 7.5, d = 8.5). This is because the (d = 7.5, 8.5) line would extend below the r(x 10) axis, to initiate (for d = 6.5) at presumably l = -0.1081

(== -[1.0 – 0.8919] = 0.8919 – 1.0) ;

i.e., (l – 0.4790) / (0.1667 – 0.8333) = 0.8732 → l = -0.1081 .

Therefore, we see that the (plot) lines (of period 3, or 3 points per line) would tend to run negatively (i.e. pertaining to the l axis) with increasing d, (for) with each initiation of cycle (0.5, 3.5, 6.5, 9.5, etc.). This may pertain to the CNr fraction (of the essential value) being assigned a negative (i.e. being as: -CNr). For our constructions (of complexity), l is restricted to positive values; -CNr is the base independent component of the essential value.

Suppose, then, that both d and t are variables, so that n = 2; and further that both increase by 1 unit (distance and time units) sequentially:

C_2 = 3, N = 3, Ψ^2 ~= 5.2464, thus N ~= 1.7155

d	t	R + r	L + l	r / l
2.5	**3.0**	**0.83333**	**1.42958**	**1.9397**
3.5	**4.0**	**0.88750**	**1.52251**	**1.6986**
4.5	**5.0**	**0.90000**	**1.54400**	**1.6544**
5.5	**6.0**	**0.91670**	**1.57260**	**1.6009**
6.5	**7.0**	**0.92860**	**1.59300**	**1.5659**

Slope (r vs l) from d = 2.5 to d = 3.5 is:

$\Delta l / \Delta r = (0.5930 - 0.4296) / (0.8875 - 0.8333) = 1.7140$

Slope (r vs l) from 2.5 to d = 6.5 is:

$\Delta l / \Delta r = (0.5930 - 0.4296) / (0.9286 - 0.8333) = 1.7146$

The (two) slopes are roughly equivalent, and seem similar to $\boldsymbol{N}$

$\boldsymbol{N} == C_n(n + 1) / \Psi^n = C_n N / \Psi^n$.

Does $\Psi^n = CN(r / l)_\Delta == CN \bullet \text{slope} (r \text{ vs } l)^{-1}$?

$\Psi^2 = 5.2464 \ \ (\Psi \sim 2.2905)$

$CN \bullet \text{slope}^{-1} = (3)(3)(1.7146)^{-1} = (3)(3) / 1.7146 = 5.2490$

Therefore, Ψ^2 and $CN \bullet \text{slope}^{-1}$ are roughly equivalent;

thus from the slope (r vs l), n = 2 .

Note that there is a distinction between: $\Delta(r / l) == (r_2 / l_2) - (r_1 / l_1)$

and $(r / l)_\Delta == (\Delta r / \Delta l) == (r_2 - r_1) / (l_2 - l_1) == \text{slope} (r \text{ vs } l)^{-1}$.

The relationship $\Psi^n = CN(r / l)_\Delta$ is validated due to the fact (or occurrence) that $\boldsymbol{N}$ is a (constant) coefficient and a steady slope [as $(r / l)_\Delta^{-1}$] is obtainable.

This, then, suggests (there are) valid comparisons between $(r / l)_\Delta$ points.

The real query, though, is whether (L + l) can be measured or predicted, and to what significance (physical or relational, to R + r).

A changing or inconsistent slope (r vs l) would allow for varying values of n between (2) points, or thereby modifying the (value of the) base for a complexity based on an established (or fundamentally recognized) n. This, then, would seem to demand that a relationship (for R + r) be functionally consistent (in a physical process).

Essentially, then, N (= n + 1) is the "complex-ed" slope of a plot of (r vs l):

$\Psi^n = CN(r/l)_\Delta$; thus $N = (\Psi^n / C) \bullet (r/l)_\Delta^{-1}$.

$\mathcal{N} = CN / \Psi^n$; thus $N = \mathcal{N}\Psi^n / C$.

$\therefore \mathcal{N}\Psi^n / C = (\Psi^n / C) \bullet (r/l)_\Delta^{-1}$, $\mathcal{N} = (r/l)_\Delta^{-1}$ = slope (r vs l) ;

i.e. "complex-ation" multiplies a slope by (Ψ^n / C_n) .

This fundamental "relief" (as for a sculpture) can now be applied to (the variables) of any physical change (i.e., as of "raising" slopes, or raising resultants above their relational slopes). Obviously, for any $n > 1$, this "raising" actually lowers the slope (to bring the original slope into relief), because C_n increases faster than Ψ^n with growing n.

[For :

Ψ_m, $n > 1$

Ψ , $n > 3$

e , $n > 4$;

i.e., C_n becomes greater than $\Psi_{()}^n$ or n-powered base.]

It might not be surprising that most physical processes we are (or allow ourselves to be) conscious of involve (from our perspectives) only a few variables (i.e., low n value).

So, we have $N = n + 1 = (\Psi^n / C_n)(\Delta l / \Delta r)$;

thus $n = (\Psi^n / C_n) \bullet (\Delta l / \Delta r) - 1$, or $(n + 1) \bullet \Delta r = (\Psi^n / C_n) \bullet \Delta l$.

The calculus would yield to (for infinitesimal changes in r and l):

$$\int (n+1)\,dr = \int \left(\frac{\Psi^n}{C_n}\right) dl \quad , \quad (n+1)\int dr = \frac{\Psi^n}{C_n}\int dl$$

$$\therefore \quad (n+1)r = \left(\frac{\Psi^n}{C_n}\right) l$$

or, for definite integrals:

$$\left.\begin{array}{l} \int_{r_1}^{r_2} (n+1)\,dr = (n+1)(r_2 - r_1) + {}_rC \\[2ex] \int_{l_1}^{l_2} \left(\frac{\Psi^n}{C_n}\right) dl = \left(\frac{\Psi^n}{C_n}\right)(l_2 - l_1) + {}_lC \end{array}\right\}$$

$[\,{}_rC \equiv {}_{r_2}C - {}_{r_1}C = {}_{l_2}C - {}_{l_1}C \equiv {}_lC\,]$

${}_rC$, ${}_lC$ are constants of integration

(for these definite integrals: ${}_rC = {}_lC$)

$$\therefore \quad (n+1)(r_2 - r_1) = \left(\frac{\Psi^n}{C_n}\right)(l_2 - l_1) + {}_lC - {}_rC$$

Thus, as expected, the relation (i.e. R + r) is directly relational to n (its variables; i.e., n + 1), while the complexity (i.e. L + l) is directly relational to (C_n, Ψ^n).

But the calculus assumes, for infinitesimal changes, a steady slope; therefore, by this, one could replace r with (R + r) and l with (L + l), without revealing the (important or suggestive) residual natures of r and l. It is interesting, though, how the coefficients (from or desired from *N*) become separated and properly associated (prior to the calculus).

The simplest complex-ation, then, using Ψ base, would simply be to multiply a (or the) slope by ~ 2.2905, at n = 1.

$\Psi^1 / C_1 == \Psi^n / C_n$, n = 1 ; $\Psi^1 / 1 == \sim 2.2905 / 1 = 2.2905$

Here, the residuals are being used as surrogate variables. So we have a demonstration of results being variables. But opposed to a limit leading to (but avoiding) an improper (for the calculus: division by zero), a residual is a definite number to be signatory to a relation. It deals not with a rate of change (as slope) but rather a rate of occurrence (for r and l). And, instead of the scaffolding relation being

$$\lim_{\Delta x \to 0} \frac{f(x + \Delta x) - f(x)}{\Delta x}$$

it is: CNR – BL = Bl + bL + bl – CNr .

Notably, were the base = 1, $B^n = 1$, $(B + b)^n = 1$, and the relation reduces to:

CNR – L = l + bL + bl – CNr

and then to: CNR – L = l – CNr .

Note that the essential value (and its relation) avoids divisions.

The fundamental residual (r, l) occurrence slope, then, is found to be, from a base 1 perspective:

l / r = (CNR – L + CNr) / r = ((CNR – L) / r) + CN , (B + b) = 1 ; thus,

$(l / r)_{base = 1}$ == $(l / r)_1$ = ((CR(n + 1) - L) / r) + C(n + 1)

= ((CRn + CR – L) / r) + C_n + C

Therefore, $(l / r)_1 = C_n(R / r) + C(R / r) - (L / r) + C_n + C$;

and, if n = 1, then $C_n = 1$:

$(l / r)_1$ = (R / r) + (R / r) - (L / r) + 1 + 1 = (2R / r) - (L / r) + 2 ,

or: $[\, l = 2R - L + 2r \,]_{base = 1,\ n = 1}$.

This, then, gives a relative definition for base 1: L + l = 2(R + r) ; i.e., the complexity is twice the relation.

This seems merely equivalent to (n + 1) = 2 , for n = 1 .

It is interesting that base Ψ approaches the complexity of base 1, but with a residual of ~ 0.2905 (i.e., it approaches 2).

If we treat Ψ as (R + r) :

at n = 1, $N = (1)(1 + 1) / \Psi^1$ ~ ~ 2 / 2.2905 ~ = 0.8732 == $(L + l)_N$.

Therefore, $1 / r = 0.8732 / 0.2905 \sim= 3.006 == (1 / r)_{N,\, n=1}$.

Of course, $\boldsymbol{N}$ for Ψ as $(R + r)$ is (at $n = 1$) :

$$(C_n (n + 1) / \Psi^n) \bullet \Psi = ((1)(2) / 2.2905) \bullet 2.2905 = 2.0 == (L + 1)_{\Psi} .$$

Therefore, $(1 / r)_{\Psi,\, n=1} = 0.000 / 0.2905 = 0$; thus, no complexity at $n = 1$ (to/for a measurable slope).

Yet, at $n = 1$, there is always a residual complexity of ~ 0.8732, for a relation. Such a residual (as a factor) reduces $(R + r)$ for a complexity below its (relational) value, imposing an imposition for $(n = 1)$ complexity.

Meaning of Complexity

Complexity for a relation is observed through objective measurement. It essentially "skews" the curve of a relationship:

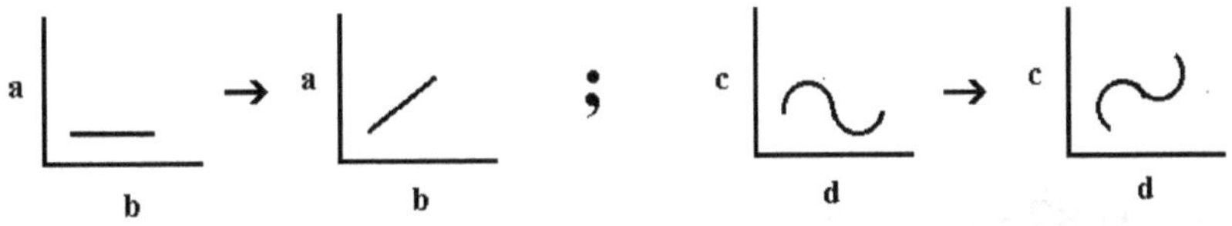

etc., via modification of shapes

This skewing becomes more complicated as the effective number of variables changes during (the expression or expedience of) a relation:

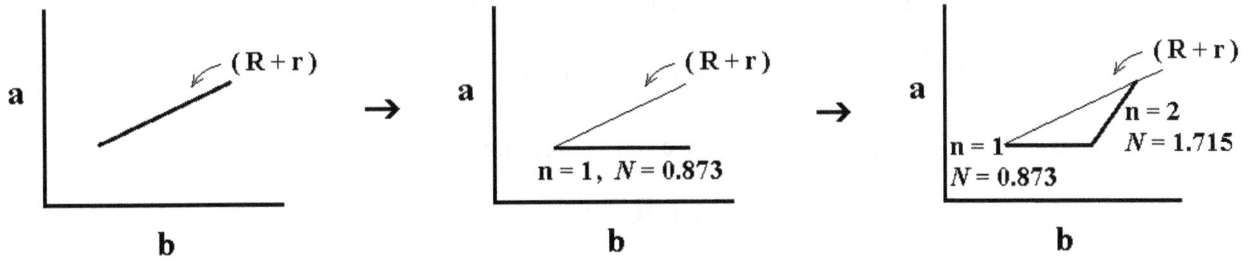

Therefore, for some relation like speed: $s = d / t$, if distance remains constant:

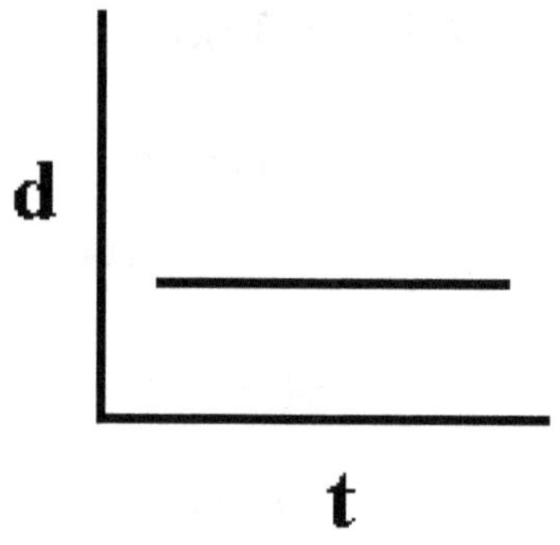

there is no movement.

But in terms of the complexity of this ($n = 1$) relation, the distance actually contracts:

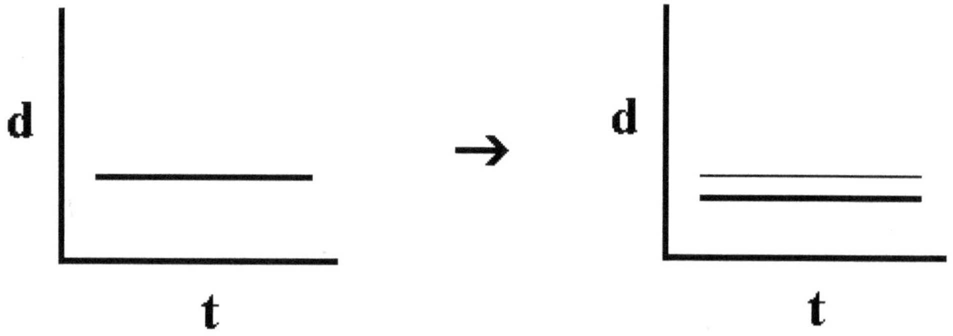

$d \rightarrow 0.873d$, or ($1 / t$) $\rightarrow$ ($0.873 / t$) ; thus, t expands.

With ($n = 2$), d expands or t contracts (of complexity). Possibly, a combination of modification (factorial) of variables occurs, as distributed by the N coefficient, to achieve the relational complexity.

A curious, new operation now presents itself (for a relation):

complexity $== N(R + r) = (R + r)(\Delta l / \Delta r) ==$ Ce (say), or Cp .

Therefore, for any desired complexity we may calculate Δl as:

$\Delta l = Ce\Delta r / (R + r) == Cp\Delta r / (R + r)$, ($\Delta l /. \Delta r) == (1 / r)_{\Delta}$.

Δl can be used to distribute N among the variables:

e.g., if $d / t = (R + r)$, then $\Delta l = (Cp + \Delta r) / d$

Therefore, one can change Δl by increasing or decreasing any/either or both of d and t; thus, one can compensate (i.e. maintain as steady) Δl by manipulating d and t. For example, if we demand a complexity of $\mathcal{C}p = 1$, then $(1/r)_\Delta = t/d$.

The Δl relation seems to decouple $(R + r)$ from an explicit n, and may even allow for a fractional n, based on the complexity (desired). We can symbolize complexity from an explicit n (e.g. as $n = 1$ or 2 for $s = d/t$, d and t being the variables independent) as $\mathcal{C}e$ (the "e" emphasizing **e**xplicitness), and complexity from a merely "relational" n as $\mathcal{C}p$ (the "p" emphasizing com**p**lexity of name). Both $\mathcal{C}e$ and $\mathcal{C}p$ are functionally equivalent. $\mathcal{C}p$, though, is essentially a "variable" (as a) coefficient (for unknown n). So now, by arbitrarily choosing a $\mathcal{C}p$, we are searching for (or calculating) a Δl (between two points of a relation).

For a relation, then, $(1/r)_\Delta$ represents a complexity slope:

e.g., (for) of $s = d/t$, $\Delta l = \mathcal{C}p t \Delta r / d$

$$== (C_n(n+1)t\Delta r / (\Psi^n \bullet d)) \bullet (d/t), (d/t) = (R+r)$$

Thus, $\Delta l / \Delta r == C_n(n+1)/\Psi^n$, $(C_n(n+1)/\Psi^n) \bullet (R+r) = \mathcal{C}p = (L+1)$

Therefore, $(1/r)_\Delta \bullet (R+r) = (L+1)$

We, of course, have specialized or "relational" n values (which can not be actual): e.g., for n where $\Psi^n = 1$ (i.e. $n == 0$),

$$(C_n(n+1)/1) \bullet (R+r) = \mathcal{C}p \text{ ; thus, } C_n = \mathcal{C}p / [(n+1)(R+r)]$$

$$== \mathcal{C}p / (R+r)$$

More generally:

$C_n = \Psi^n \bullet \mathcal{C}p / [(n+1)(R+r)]$ using "relational n"

More specifically:

$C_n = \Psi^n \bullet \mathcal{C}e / [(n+1)(R+r)]$ using "explicit or whole number n"

Purpose of Complexity

Any physical operation is either done (performed) or not done. Even attempts of operations are committed by completion of subsidiary operations (which are successfully performed). This says that, as per our originating circular model, a whole number n (or index) is accomplished (for any physical process). Fractional n (and residual n) conform only to the theoretical (or inspired as predictable) considerations. Physics (i.e., of physical phenomena) is circuitous (i.e. closed or completed circle of events or actions). Theory, however, can remain incomplete(-d), or imagined (or partially accomplished and fractional), which condenses (in general) to conversions of (types of) quantities via division. Therefore, a circuitous and completed route rescues (repairs and recovers from) division.

What we have been calling "complexity" can be evolved formulaically through Ψ definitions, to show that this amounts to a power (consideration) of the C_n combination:

$$\frac{C_n(n+1)}{\ulcorner\!+^{\jmath n}} \equiv (\mathcal{C}e \text{ or } \mathcal{C}p)(R+r)^{-1} \rightarrow (L+l)(R+r)^{-1}$$

$$\frac{C_n}{\ulcorner\!+^{\jmath n}} \hookrightarrow C_n \bullet \left(\left[\sum_{\substack{n=0 \\ \text{relational}}}^{\infty} \right] \frac{n}{C_n} \right)^{-1 \bullet n} \tilde{\equiv} \; C_n^2 / n$$

and $$\frac{C_n^2}{n} \bullet (n+1) \; \tilde{\equiv} \; C_n^2$$

Likewise:

$$C_n \bullet \ulcorner\!+^{\jmath n} \hookrightarrow C_n \bullet \left(\left[\sum_{\substack{n=0 \\ \text{relational}}}^{\infty} \right] \frac{n}{C_n} \right)^{+1 \bullet n} \tilde{\equiv} \; \overset{\text{(formuleic)}}{n}, \quad n \cdot (n+1) \; \tilde{\equiv} \; n^2$$

Rather than calling the two principles "(Ψ) inversions" of each other, let us say they are cognates, such that:

$(C_n / \Psi^n)(C_n \bullet \Psi^n) \rightarrow C_n^{\,2}$, the relative purpose of the complexity.

"Opticals" of Complexity

The power of C_n considered for the complexity is, of course, not 2, but is based on the Ψ^n summation, and is therefore with definition of character (as like a logarithm). We can simply call it δ, with the exponent (summarily) tempered by n. Therefore, the complexity goal (of N) for a relation is $C_n^{\,\delta}$ (== N).

The significance of δ is from the relation:

$$\sqrt[\delta]{C_n^{\delta}} = C_n \text{ , or } \left(C_n^{\delta}\right)^{1/\delta} = C_n$$

In terms of complexity, for a relation (R + r), n is essentially replaced by: n(R + r). This suggests an obvious partition to address. We can provide for an extended δ based on (R)n for the formulation. Thus, of δ_R we create $C_n^{\,\delta R}$, to be further characterized (or analyzed) by $(1 / r)_\Delta$ computations of slope.

(R) $\bullet$ n is restricted to whole (thus possible) numbers, for use of $C_n^{\,\delta R}$:

$$C_n^{\delta_R} \equiv C_n\left(\left(\frac{C_m}{m}\right)^n \cdot R\right)_{m:1\rightarrow n} \cdot (n+1)/(R)$$

$$= C_n\left(\left(\frac{C_m}{m}\right)^n \cdot R(n+1)\right)_{m:1\rightarrow n} / R$$

$$\equiv C_n\left[\sum_{m=1}^{n}\left(\frac{C_m}{m}\right)\right]^n \cdot R(n+1) / R$$

$$C_n^{\delta} \equiv C_n\left[\sum_{m=1}^{n}\left(\frac{C_m}{m}\right)\right]^n \cdot (R+r)(n+1)/(R+r) \cong N$$

We can call, then, terms such as:

$$\left[\sum_{m=1}^{n}\left(\frac{C_m}{m}\right)\right]^{n} \cdot (R+r)(n+1)$$

and (especially)

$$\left[\sum_{m=1}^{n}\left(\frac{C_m}{m}\right)\right]^{n} \cdot R(n+1)$$

"opti(pi)cals" of the complexity, for a relation $(R + r)$, since they provide a means of "observing" the complexity (behavior).

For example, we can examine how

$$\left[\sum_{m=1}^{n}\left(\frac{C_m}{m}\right)\right]^{n} \cdot R(n+1)$$

can be approached by

$$\left[\sum_{m=1}^{n}\left(\frac{C_m}{m}\right)\right]^n \cdot R(m+1)$$

or by

$$\left[\sum_{m=1}^{n}\left(\frac{C_m}{m}\right)\right]^n \cdot {}^{a}R(m+1)$$

where ${}^{a}R$ is an "arbitrary" R (e.g., let ${}^{a}R = m$ or n).

Likewise, the cognate proposition has its opticals:

From $C_n \bullet \Psi^n$ the opticals are

$$\left[\sum_{m=1}^{n}\left(\frac{m}{C_m}\right)\right]^n \cdot (R+r)(n+1) \equiv n^{\delta} \cdot (R+r)$$

and

$$\left[\sum_{m=1}^{n}\left(\frac{m}{C_m}\right)\right]^n \cdot R(n+1) \equiv n^{\delta} \cdot R \; (\equiv n^{\delta_R} \cdot R)$$

The (optical) summation

$$\left[\sum_{m=1}^{n}\right]$$

is only an explicit approximation of

$$\left[\sum_{n=0}^{\infty}\right]$$

Further reduced, and even more explicit, opticals would be

for $C_n \bullet \Psi^n$

for

$C_n \cdot$ ┌+┘^{n}

$$\sum_{m=1}^{n}\left(\frac{m}{C_m}\right) \cdot (R + r)(n + 1)$$

$$\sum_{m=1}^{n}\left(\frac{m}{C_m}\right) \cdot R(n + 1)$$

for C_n / Ψ^n

$$\text{for } \frac{C_n}{\Psi^n} \qquad \sum_{m=1}^{n} \left(\frac{C_m}{m} \right) \cdot (R + r)(n + 1)$$

$$\sum_{m=1}^{n} \left(\frac{C_m}{m} \right) \cdot R(n + 1)$$

where the exponent of Ψ is constrained to be 1 (instead of n).

The most restricted opticals would thereby be:

$$\frac{C_n^{\delta}}{(R+r)^{-1}} \; , \quad C_n \sum_{m=1}^{n} \left(\frac{C_m}{m} \right) \cdot (R + r)(n + 1) \qquad \Big| \qquad C_n \sum_{m=1}^{n} \left(\frac{m}{C_m} \right) \cdot (R + r)(n + 1) \; , \quad \frac{n^{\delta}}{(R+r)^{-1}}$$

$$\frac{C_n^{\delta_R}}{R^{-1}} \; , \quad C_n \sum_{m=1}^{n} \left(\frac{C_m}{m} \right) \cdot R(n + 1) \qquad \Big| \qquad C_n \sum_{m=1}^{n} \left(\frac{m}{C_m} \right) \cdot R(n + 1) \; , \quad \frac{n^{\delta_R}}{R^{-1}}$$

$$(C_n / \Psi^n) \qquad \qquad (C_n \cdot \Psi^n)$$

Note that the opticals are distinct from $(L + 1)$.

The opticals can thus be summarized by:

$$C_n \left[\sum_{m=1}^{n} \frac{C_{(m;n)}}{m} ; \frac{m}{C_{(m;n)}} \right]^{(n;m;1)} \bullet (R+r;R) \bullet (n+1)$$

The 'exponent of Ψ being 1' postulate is useful because it allows for the direct conversion of n (as m) to $n \bullet R$ or $n \bullet (R + r)$, in the optical (i.e. as a multiplicative function of n or n^{-1}). Logistically, then, the least restrictive (formulaically) opticals would derive from the arbitrary incorporation of $(R + r ; R)$ into the Ψ^n summation:

$$C_n \left[\sum_{m=1}^{n} \frac{C_{(m;n)}}{m \cdot (R+r;R)} ; \frac{m \cdot (R+r;R)}{C_{(m;n)}} \right]^{(n;m;1)} \bullet (n+1)$$

where n is replaced by $n \bullet (R + r ; R)$.

The cognates complexity proposal:

$$(C_n / \Psi^n) \bullet (C_n \bullet \Psi^n) \rightarrow C_n^{\,2}$$

suggests that the occurrence of a physical process is the duty of the square of its actuality (or action) as by a combination of its probability to happen (i.e. the probability of a physical occurrence is the square of its action). But we find that the exponent (to C_n) is actually δ.

The number of variables (involved) is found from n^{δ}.

Therefore, we can use "optical" relations to evaluate δ :

$\boldsymbol{N} \bullet (R + r) = (L + 1)$; thus, $\boldsymbol{N} = (L + 1) / (R + r)$ and can be assessed by $(r \text{ vs } 1)$ plots.

Therefore, (we can) set $C_n^{\,\delta} \sim = \boldsymbol{N}$. Thus, $\boldsymbol{N} \sim = C_n^{\,\delta} = C_n \bullet (\text{opitcal}) / (R + r; R)$

$$\log C_n^{\delta} = \log [\, C_n \bullet (\text{ optical }) / (\, R + r; R \,)\,] \text{ ; thus,}$$

$$\delta \log C_n = \log C_n \bullet (\text{ optical }) - \log(\, R + r; R \,) \text{ ; thus,}$$

$$\delta = [\, \log C_n \bullet (\text{ optical }) - \log(\, R + r; R \,)\,] / \log C_n$$

$$= \log C_n \bullet (\text{ optical }) - \log(\, R + r; R \,) - \log C_n$$

$$= \log C_n + \log(\text{ optical }) - \log(\, R + r; R \,) - \log C_n$$

$$= \log(\text{ optical }) - \log(\, R + r; R \,)$$

So, one must fit (or choose) the proper optical to yield the (calculated or) expected $\mathcal{N}$. Thus, observance of δ effectively modifies Ψ (or the base's formation).

The cognates postulate is akin to propositions like "the energy of a wave is proportional to the square of its amplitude," or "the intensity of a light is inversely proportional to the square of its distance from a detector (or screen)," but with the proviso that if some threshold (of energy) is reached then the probability of a physical occurrence is absolute (i.e. the energy can cause some physical disturbance, like an observed (transverse) propagation of wave or detection of intensity or photoelectric effect, etc.).

Probability of an Event via its Combinations of Results

The combination probability for an event occurring seems by definition (thus straightforward): e.g., if $n = 2$, $C_n = 3$: say for actions a, b, a + b .

Therefore, (we) can have a, b each occurring (i.e. a or b), and we can have (a + b) occurring.

Assuming an energy threshold of occurrence is reached, then the probability of a or b or (a + b) occurring is 1/3 :

$$\frac{\mathbf{a}}{\mathbf{a+b+(a+b)}}, \quad \frac{\mathbf{b}}{\mathbf{a+b+(a+b)}}, \quad \frac{\mathbf{a+b}}{\mathbf{a+b+(a+b)}}$$

$$\underbrace{\mathbf{1/3} \qquad \mathbf{1/3}}_{\mathbf{2/3}} \qquad \mathbf{1/3}$$

But this demands that a summed event (i.e. $a + b$) has a lower probability than a single event (i.e. a or b).

Curiously, for $n = 3$ (thus, $C_n = 7$), bi-summed events have the same probability as single events (a, b, c) :

$$\rightarrow \quad \underbrace{\underbrace{a\ ,\ b\ ,\ c}_{3/7}\ ,\ \underbrace{a+b\ ,\ a+c\ ,\ b+c}_{3/7}}_{6/7}\ ,\ \underset{1/7}{a+b+c}$$

But these are probabilities of (combinatorial) choices. For occurrences (out of 3), each single event is $1/3$, each bi-event is $(1/3)(1/3) = 1/9$, each (there is only one) tri-event is $(1/3)^3 = 1/27$.

So, we can multiply choices by occurrences (to adjust their relative occurrences):

single $\rightarrow (1 / 7)(1 / 3) = 1 / 21$

bi $\rightarrow (1 / 7)(1 / 9) = 1 / 63$

tri $\rightarrow (1 / 7)(1 / 27) = 1 / 189$

Then, the probability of occurrences is equal to this (probability) product times the number of events (for each class: single, bi, tri) :

e.g., $(1 / 189) \bullet 7 = (1 / 7)(1 / 27) \bullet 7 = 1 / 27$

Therefore, with the

probability product = (prob. of choices of events)(prob. of occurrences of events)

we have

(prob. of choices of events)(prob. of occurrences of events) • C_n = probability of class events

or

probability product • C_n = probability of class events

Therefore, given a (product) probability, and n, one can calculate (or predict) the number of event occurrences for a class (from C_n).

[e.g., given 1 / 189 and n = 3 → $C_n = C_3 = 7$; thus,

$(1 / 189) \bullet 7 = 1 / 27 = 1 / n^{NEOC} = 1 / 3^{NEOC}$

where NEOC (== # of event occurrences for class) = 3]

The product probability characterizes a class.

Of further interest is determining the number of members in a class. For a given n, the number of members of a class size of 1 is simply n; i.e., n = 3 members are: a, b, c. Thus, the # of members of class size = 1 are 3: (a, b, c).

Empirically observed, the number of members for class sizes larger than 1, with a given n, is:

n! / (class size)! ;

e.g., at n = 2

class size = 1 → 2! / 1! = 2 x 1 / 1 x 1 = 2 members, (a, b)

class size = 2 → 2! / 2! = 2 x 1 / 2 x 1 = 1 member, (ab)

Thus, the total numbers of members = 2 + 1 = 3 members.

[Since 1 x 1 = 1 and 2 x 1 = 2, any n > 2 will lead to a factorial whose value will be greater than n.]

e.g., at n = 3

class size = 1 → 3! / 1! = 3 x 2 x 1 / 1 x 1 = 6 ,

but # members = n; ∴ # members = 3, (a, b, c) → (a, b, c, P_1, P_2, P_3);

One might say 3 members are phantom, or not observable.

class size = 2 → 3! / 2! = 3 x 2 x 1 / 2 x 1 = 3, (ab, ac, bc)

class size = 3 → 3! / 3! = 3 x 2 x 1 / 3 x 2 x 1 = 1, (abc)

Thus, the total number of members = 3 + 3 + 1 = 7 members.

Now, we see the nature of members, as their manners lead to the real relational (or functional) complexities. For, as example, a class with two members in a relationship could be (a + b) or (a)(b), etc.

Therefore, we can consider how a class with one member is (functionally) equivalent to a class with two members: When does a == a + b or a == ab ?

We could say a == a + b when b is absent (rather than zero) and a == ab when b = 1 (or is present multiplicatively). With this method(ology) any great elaboration can be constructed, with combinations within the member associations themselves;

e.g., class of 4 members could be (for: a, b, c, d):

$a + b - bc + c^2 / d$ (as an example) , and various possible (or allowable) combinations of member associations, such as: $b + a - ac + c^2 / d$, $c + a - ab + b^2 / d$, etc.

Thereby, the "guiding combination" (C_n) is made much more complex by the (internal) member (associative) combinations (possibly with their own "sub-guiding combinations," ${}_sC_n$). All of these lead to (distinguishable) complexity coefficients for a relation (which, conceivably, may even become recursive, or resonant). While the coefficients remain ***N*** (of type), particular probabilities for the construction of the coefficients (must) ensue.

Therefore, member associations (manners of association) within classes lead to complexities of their probable states, subordinate to their choices and occurrences (of events).

Using Defined Mathematical Operations to Consider Combinations

A factorial is, of course, a defined (mathematical) operation, rather than a derived one (from mathematics).

In this spirit, we can define another operation, say one based on addition and multiplication arbitrarily applied to suit a purpose. Let us call it an "additorial" operation. It is demonstrated by first using a factorial and then converting to its (equivalent) additorial:

3! / 1! == 3 x 2 x 1 / 1 x 1 → (3 + 2 + 1) / (1 + 1)

Now here, we can define to allow canceling of like terms (as could be done with multiplication's division):

$(3 + 2 + \cancel{1}) / (1 + \cancel{1}) \rightarrow (3 + 2) / 1 = 5 / 1 = 5$

But we can also define that advantageous (to a goal) condensation of terms is allowed:

$(3 + 2 + 1) / (1 + 1) == 6 / 2 = 3$,

or $(3 + 2 + 1) / (1 + 1) \rightarrow (3 + 2 + 1) / 2 == 6 / 2 = 3$

In like manner, using additorials advantageously, we can re-assess the n! / class-size! procedure (using symbol

"$\dot{\downarrow}$"

as an additorial indicator):

n = 2 **class size = 1** $\rightarrow \dfrac{2\dot{\downarrow}}{1\dot{\downarrow}} \equiv \dfrac{2 + \cancel{1}}{1 + \cancel{1}} \rightarrow \dfrac{2}{1}$ **= 2 members**

class size = 2 $\rightarrow \dfrac{2\dot{\downarrow}}{1\dot{\downarrow}} \equiv \dfrac{2 + 1}{1 + 1} \rightarrow$ **1 member**

n = 3 **class size = 1** $\rightarrow \dfrac{3\dot{\downarrow}}{1\dot{\downarrow}} \equiv \dfrac{3 + 2 + 1}{1 + 1} \rightarrow \dfrac{3 + 2 + 1}{2} = \dfrac{6}{2} =$ **3 members**

class size = 2 $\rightarrow \dfrac{3\dot{\downarrow}}{2\dot{\downarrow}} \equiv \dfrac{3 + \cancel{2} + 1}{\cancel{2} + 1} \rightarrow$ **3 members**

class size = 1 $\rightarrow \dfrac{3\dot{\downarrow}}{3\dot{\downarrow}} \equiv \dfrac{3 + 2 + 1}{3 + 2 + 1} =$ **1 member**

Thus, our (use of) additorials de-restrains the procedure (for desired results). Additorials are less restrictive than factorials. They also can be degenerate:

$$\text{e.g.} \quad \frac{3\dot{\downarrow}}{1\dot{\downarrow}} \quad \textbf{can yield 5, can yield 3, etc.}$$

The entire "additorial" operation, with all of its particulars, can be seen (or viewed) as a composite of factorial and addition, where, strategically (or judiciously) an expression or term is first treated as the (or through the) factorial proper, in order to execute any desired divisional cancellations, and then converted to the additional proper, to carry out the additive condensations before any required final division. But all of this is much easily envisaged by simply applying the two additorial definitions espoused earlier and accepting that (divisional) cancellations (as if factors) can be done (from numerator to denominator) across additive expressions:

$$\text{e.g.} \quad \frac{4\dot{\downarrow}}{2\dot{\downarrow}} \equiv \frac{4!}{2!} = \frac{4 \times 3 \times \not{2} \times \not{1}}{\not{2} \times \not{1}} = \frac{4 \times 3}{1} \rightarrow \frac{4+3}{1} = \frac{7}{1} = 7$$

or

$$\frac{4\dot{\downarrow}}{2\dot{\downarrow}} \equiv \frac{4+3+\not{2}+\not{1}}{\not{2}+\not{1}} \rightarrow \frac{4+3}{1} = \frac{7}{1} = 7$$

Of course, one can have "mixed" operations:

e.g. $\frac{4\dot{\downarrow}}{2\,!}$ **, etc.** **They might be projected (proposed) as restricted or limited evaluations.**

$$\therefore \quad \frac{4\dot{\downarrow}}{2\,!} \equiv \frac{4+3+2+1}{2 \times 1} = \frac{10}{2} = 5$$

$$\frac{4\dot{\downarrow}}{2\dot{\downarrow}} \equiv \frac{4+3+\not{2}+\not{1}}{\not{2}+\not{1}} = 7$$

$$\frac{4\,!}{2\,!} \equiv \frac{4 \times 3 \times \not{2} \times \not{1}}{\not{2} \times \not{1}} = 12$$

$$\frac{4\dot{\downarrow}}{2\,!} \equiv \frac{4 \times 3 \times 2 \times 1}{2+1} = \frac{24}{3} = 8$$

While N remains the same (as calculated), several paths towards numbers of members in classes might be invoked (for particular physical problems), not the least of which involving phantom members. This involves expansion of the (n! / class-size!) calculation for the number of members in classes to provide, perhaps (and theoretically) some admixture of additorial(s) with factorial(s). Note, though, that these numbers (of members in classes) must presumably remain whole. A sort of class type probability, or distribution, can be constructed by dividing the # of members of a class by C_n :

members of a class == n! / class-size! ; therefore,

probability of a class ~ ~ # members of a class / C_n

== (n! / class-size!) / C_n = n! / (class-size!C_n)

where the sum of the # of members of all classes == C_n .

Of course, C_n itself is intimately related to n, as it predicts classes and their members and numbers (in a class). So, calculation of the probability of a class is done through an intraoperative assessment, as expected. That is, the probability remains a "pure" and unit-less number.

So, the complexity of a relation is determined by the number of its variables and their probabilistic dispositions (e.g., whether any are treated as constants, etc.).

Complexity of a Relation as an Exponential or Power

But what does a complexity impart to a relation?

N, being a coefficient, imbues a relation with a characteristic power (or exponent), which we can call a complexity power:

$$ax = x^y ,$$

where a == ***N*** , x == relation , y == complexity exponent, and ax == (the) complexity

Thus,

$$\log_x ax = \log_x x^y = y$$

$$y = \log_x ax = \log_x a + \log_x x \text{ , therefore,}$$

$$y = \log_x a + 1$$

Therefore, we can say:

$$c = \log_R \boldsymbol{N} + 1 \qquad (\text{for } \boldsymbol{N}R = R^c)$$

where R == x = the relation, c == complexity exponent (== y).

N and c are alternate manifestations of the same complex*ation*, each relating differently to R. But c is more utilitarian by

1) defining ***N*** (via $\log_R \boldsymbol{N} + 1$) , and

2) being an exponent (i.e. a character of definition).

So, for example,

if $v = d / t$, $n = 2$; say $d = 4.5$ and $t = 5.0$

then $\boldsymbol{N} = 1.7155$ and $\boldsymbol{N}R = 1.5440$ (== L + 1) ;

therefore,

$$NR = R^c , R = 4.5 / 5.0 = 0.9$$

$\mathbf{c = \log_R \mathcal{N} + 1 = \log_{0.9} 1.7155 + 1 == \log_{10} 1.5440 \,/\, \log_{10} 0.9 + 1 = -5.122(5) + 1 = -4.122(5)}$

$\mathbf{\log_R 1.5440 = \log_{10} 1.5440 \,/\, \log_{10} 0.9 = -4.122(7)}$, which is close to **c** since **R** is close to **1** .

$$[\ -4.122(7) == \log_R 1.5440 \,/\, \log_R 0.9 = \log_R 1.5440 \,/\, 1\ ]$$

$$\mathcal{N}\mathrm{R} = 1.5440 == (\ 1.7155\ )(\ 0.9\ )$$

$$\mathrm{R}^c == (\ 0.9\ )\text{-}4.1225 = 1.5440$$

Complexity, then, is a relation brought to a (certain) power. But this imposes a condition, for we have (thus far) been assuming that (for physical processes) any complexity must be positive (since we have taken n to be positive and therefore C_n and Ψ^n also).

What if R is negative, though? The sign of the product $\mathcal{N}$R would be negative, while the sign of R^c would be desultory "+" or "-" (for a negative "R") depending on the odd or even-ness of the c exponent. And what about exponents with fractional parts?

It is of definition to declare that:

(+)(+)	→ +
(+)(-) or (-)(+)	→ −
(-)(-)	→ +

So, we can declare that the fractional part of an exponent takes on the sign of its whole number carrier: e.g., (+ • f) → + , (− • f) → − (for operations).

Yet, assume that (for a physical process) we allow for (some) n to be negative, thereby imparting (possibly) a negative sign to C_n and/or Ψ^n. The situation is (by instruction) analogous to dealing with phantom numbers.

The negative c exponent in our case was due to the R being less than one. But for a negative n we must analyze carefully each derivative part (or application of n):

$$\Psi = \sum_{n=1}^{\infty} \frac{n}{C_n} \ , \quad \Psi^n_m = \sum_{n=1}^{\infty} \frac{1}{C_n} \ , \quad C_n = 1 + n + \sum_{x=1}^{n-2} n^x$$

C_n is a summation that is part power series (thus derivative) as well as discreet (initial) addends (thus integrative). Each part (of distinction) would handle a negative n differently (i.e., characteristically as a sum or exponential rising on n).

Loss of Events, or Negative n Paths

From our model, n represents the (absolute) number of events needed (or used) to complete a circuit (or route of transit). Presumably, negative n would represent a loss of (such) events, during the completion of the circuit. The abstraction, however, suggests that some combination of positive n with negative n (during a physical process) occurs.

If we start on a route ($n = 1$) :

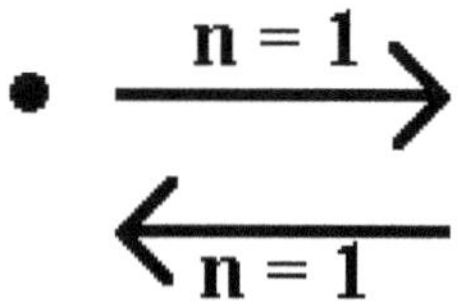

the return route is necessary, to complete the circuit. This return, however, is also $n = 1$ and therefore not negative. The two routes together generate the ($n + 1$) principle necessary for *N*.

So, what constitutes a negative n?

We may postulate that a negative step is "off-route" and that the deviation (from necessary circuit accomplishment) is a matter of judgment, perspective, or contention towards the particular physical problem:

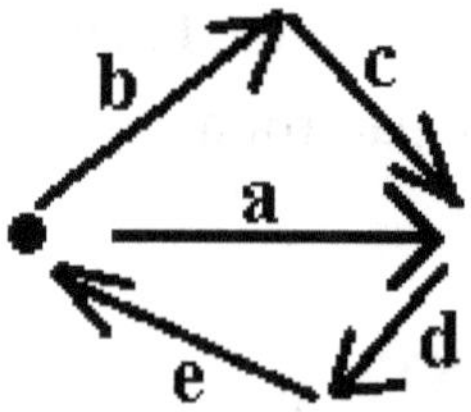

Here, b might be considered a negative step (as a deviation from a), necessitating c-d-e (as recovery positives); or b (as a positive) simply forms part of a route such as b-c-d-e, distinct from a-d-e.

So, from the perspective of the a-d-e transit, step b is negative (with a positive c recovery); while from the perspective of the a-a transit, step d is negative, with a positive step e recovery.

But including negative n presents or suggests a nul/los predilection, since no step is empty (of consequences), and the complexity is dependent on the number of steps rather than their extent (to reach zero or any endpoint).

Therefore, a zero-less shift is suggested, or produced whereby one can go from +1 to −1 without the intervention (or traversal) of a null; e.g., from the a-a perspective:

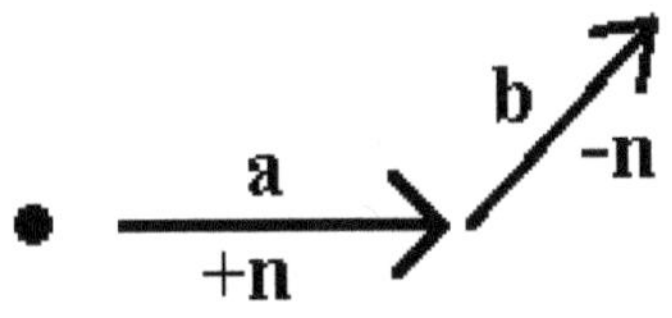

What is the consequence for C_n with such a shift at n = -1 ?

$$C_n = 1 + n + \sum_{x=1}^{n-2} n^x \equiv 1 + (-1) + \sum_{x=1}^{(-1)-2} n^x = \sum_{x=1}^{-3} (-1)^x$$

where the shift is: 1 → -1 → -2 → -3 ; therefore,

$$\therefore \quad \sum_{x=1}^{-3} (-1)^x = (-1)^1 + (-1)^{-1} + (-1)^{-2} + (-1)^{-3}$$

$$= -1 + \frac{1}{-1} + \frac{1}{(-1)^2} + \frac{1}{(-1)^3}$$

$$= -1 - 1 - \frac{1}{1} + \frac{1}{-1}$$

$$= -2 + 1 - 1 = -2$$

For n = ,	$C_n \rightarrow$
-2	$1 + (-2) + (-2)^1 + (-2)^{-1} + (-2)^{-2} + (-2)^{-3 \rightarrow -4} = -3 + (f \text{ sum})$
-3	$1 + (-3) + (-3)^1 + (-3)^{-1} + (-3)^{-2} + (-3)^{-3 \rightarrow -5} = -5 + (f \text{ sum})$
-4	$1 + (-4) + (-4)^1 + (-4)^{-1} + (-4)^{-2} + (-4)^{-3 \rightarrow -6} = -7 + (f \text{ sum})$
-5	$1 + (-5) + (-5)^1 + (-5)^{-1} + (-5)^{-2} + (-5)^{-3 \rightarrow -7} = -9 + (f \text{ sum})$

These values do not show the typical candor of C_n combinations, since they are negative and fractional. Including the null (i.e., anti-shift: +,- $n^0 = 1$) increases C_n by 1. Therefore, only $n = -1$ approaches a conventional (whole number) C_n.

It might seem curious that the negative n series (from $-1 \rightarrow -5$) seems to follow the C_n values: -2, -3, -5, -7, -9 (i.e., an even followed by increasing odds); or, if null is included, then: -1, -2, -4, -6, -8 (i.e., an odd followed by increasing evens); but the C_n values past (for) $n = -1$ are actually fractional.

For example, at $n = -2$, C_n goes to:

$1 - 2 - 2 - ½ + [1 / (-2)^2] - [1 / (-2)^3] + [1 / (-2)^3] = -3 -1/2 + 1/4 - 1/8 + 1/16 \sim= -3.3125$

It's clear that as n extends (to more negative values), the fractional part of C_n (i.e. f sum) gets smaller and smaller and therefore closer to the odd integer value for C_n. While the series is not a limit, thereby allowing all integer values of n to lead to discreet results that are distinctive, it is a (powerful) motor for leading C_n to (or towards) odd values (or towards even values if null or anti-shift is incorporated or employed).

This suggests some (apparent) utility of adding (or combining in some way) both shift (odd) and anti-shift (even) C_n values. Their difference ($+,- n^0 = 1$), of course, indicates the shift (occurring).

The addition would be:

even $$1 + n + n^0 + \left(\sum_{x=1}^{n-2} n^x \right) - n^0$$

odd $$1 + n + \left(\sum_{x=1}^{n-2} n^x \right) - n^0$$

$$2(n+1) + 2\left[\left(\sum_{x=1}^{n-2} n^x \right) - n^0 \right] + n^0$$

$$= 2\left[(n+1) + \left[\left(\sum_{x=1}^{n-2} n^x \right) - n^0 \right] \right] + n^0$$

(n+1) principle (pointing to $(n+1)$); **shift principle** (pointing to the bracketed sum); **null incorporation** (pointing to $+ n^0$)

Since $n^0 == 1$, the result must be odd (from the addition), as would be expected from the addition of an odd and an even. [($+,- n^0 = 1$) refers to ($+,- n)^0 = 1$]

But the zero exponent rule actually requires (or expects) that a base (for the exponent) must be positive, and the negative of n^0 must be applied after the (zero) exponent is applied to the base. Therefore, $-n^0$ [$= -(n)^0$] $= -(n^0) = -(1) = -1$.

We are using (the implication that) $(-n)^0 = 1$ (i.e. that n is negative).

Thus, $(\text{negative } n)^0 = 1$. This is definitional, since the definition of the exponent of a base b is: $b^a / b^c = b^{a-c}$, where $a = c$ for zero exponent. $\therefore$ $b^a / b^c == b^a / b^a == b^c / b^c == 1$.

So, instead of a rate of change (by division), we have an example of 'rite of being', to accommodate a negative n where we allow $n = -z$; thus, $n^0 == (-z)^0 = 1$.

If we don't accept (directionally) a negative n, then the canonical summation

$$\sum_{x=1}^{n-2} n^x$$

yields simply $n^1 = n$, so that $C_n \rightarrow 1 + n + n = 2n + 1$ (a negative, odd number). This is virtually the definition of odd integers.

So, we lead to a curious exposition (on the summations):

The set of negative odd integers is $[C_n]_{n=-1}^{-\infty}$,

The set set of positive odd integers is $-[C_n]_{n=\sim 1}^{-\infty}$,

and the set of even integers is $\{ [C_n]_{n=\sim 1}^{-\infty} , -[C_n]_{n=\sim 1}^{-\infty} \} + 1$.

Then, an n that is negative has some fundamental utility. We can have compounded expressions, such as

$$\left.\begin{array}{l} {}_pC_n \equiv \text{(say)} \; [C_{n=x}]_{+x=-1}^{-n} + C_n \\ {}_p\Psi^n \equiv \Psi^{-n} + \Psi^n \end{array}\right\} \quad \frac{({}_pC_n)}{({}_p\Psi^n)} \equiv \left(\frac{C_n}{\Psi^n} \right)_p$$

which might be useful for demarcating (or displaying) a spectrum of (physical) effects. Note that the compounding uses additions.

${}_P\Psi^{+n}$ can be of various forms; for example, it can be

${}_P\Psi^{+n} == [\Psi]_{+x=-1}^{-n} + \Psi^n == \Psi^{-1} + \Psi^{-2} + \ldots . \Psi^{-n} + \Psi^n$;

or it can be ${}_P\Psi^{-n} == \Psi^{-n} + \Psi^n$.

Likewise, ${}_PC_n$ can be of various forms:

${}_PC_n == [C_{n=x}]_{+x=-1}^{-n} + C_n$, or $(C_x)_{x=-n} + C_n$, where a negative n is accepted (directionally). The forms can be mixed (as specified) for C_n and Ψ^n when computing quantities such as (the) complexity.

Spectrum of Combinations

The notation $[\, C_n \,]_{x=-1}^{-n}$ depicts an array or spectrum of values, which (here) can be (considered) condensed as by a compounding of additions: therefore,

$[\, C_{n=x} \,]_{x=-1}^{-n} == \{\, C_{-1}, C_{-2}, C_{-3}, \ldots.\, C_{-n} \,\} + C_n \rightarrow \{\, C_{-1} + C_{-2} + C_{-3} + \ldots.\, C_n \,\}.$

Or, the notation can be treated to retain the discreet values in the set, to allow for the spectrum (of values) proper. The same is, thus, for $[\, \Psi \,]_{x=-1}^{-n}$ and the $(\, C_n \,/\, \Psi^n \,)_P$ operation.

The conjugation (additions) can be restrained to either or both terms (C_n, Ψ^n), or neither. But note that additions do not alter the character (quality) of each term.

So, we have introduced array operations.

For example: $\{\, C_{-1}, C_{-2}, C_{-3}, \ldots.\, C_{-n} \,\} + C_n$ could mean $\{\, C_{-1}, C_{-2}, C_{-3}, \ldots.\, C_{-n}, C_n \,\}$ (i.e. discreet conjugation);

or it could mean $\{\, C_{-1} + C_{-2} + C_{-3} + \ldots.\, C_{-n} + C_n \,\}$ (i.e. compounding) ;

or, more discreetly, it could mean $\{\, C_{-1} + C_n, C_{-2} + C_n, C_{-3} + C_n, \ldots.\, C_{-n} + C_n \,\}$ (i.e. discreet additioning) ;

or, it could simply stand as (originally) $\{\, C_{-1}, C_{-2}, C_{-3}, \ldots.\, C_{-n} \,\} + C_n$ (i.e. simple conjugation, but interpret(ive)-ly complicated).

We can also allow for a spectrum (of discreetness) on C_n :

${}_sC_n == [\, C_{n=x} \,]_{x=1}^{n} == \{\, C_1, C_2, C_3, \ldots.\, C_n \,\}$

Perhaps $\{\, C_{-1}, C_{-2}, C_{-3}, \ldots.\, C_{-n} \,\} + C_n$ can be interpreted as C_n + either of the discreet $-n$ derived values, which, therefore, results in a sort of (weighting) tare on C_n .

Complications, thus, occur when trying to interpret quality changing operations, such as complexity:

$$[(\{C_{-1}, C_{-2}, C_{-3}, \ldots C_{-n}\} + C_{n}) / (\Psi^{-1}, \Psi^{-2}, \Psi^{-3}, \ldots \Psi^{-n}\} + \Psi^{n})](n+1)$$

since each set can be supposition-ed independently (and therefore possibly differently).

The set or discreet nature of the $-n$ derived terms proposes a sort (or form) of convolution between (or amongst) them. Normally, as we have seen for zero influenced operations, multiplication has precedence over division. But, since we can consider a “numerator-ed” term being principled-ly modified, i.e. $(n+1)$, we can select the denominator as the dominant (or slow) operator of the convolution. Since the division must maintain its quality (or character of units or form) the convolution can be of addition. Here, we have an operation like:

$$\sum_{x=-1}^{-n} \frac{\{C_{-1} + C_{-2} + C_{-3} + \ldots\ldots C_{-n}\}}{\Psi^{J+x}}$$

if we consider only the sets' division (first). Since the summation is addition (of within a convolution), x can go from $-1 \rightarrow -n$ or from $-n \rightarrow -1$ (equivalently); i.e., it is not directionally dependent.

Now, we must consider the addition of this convolution to the formalism of the relation C_{n} / Ψ^{n} such as to satisfy the operation:

$$(\{C_{-1}, C_{-2}, C_{-3}, \ldots C_{-n}\} + C_{n}) / (\Psi^{-1}, \Psi^{-2}, \Psi^{-3}, \ldots \Psi^{-n}\} + \Psi^{n}).$$

The convolution is obviously partitioned among the numerator and the denominator, but by parts of which remain complicated (or obscured of definition).

The most we can say (or insist) is that:

$$(\{C_{-1}, C_{-2}, C_{-3}, \ldots C_{-n}\} + C_{n}) / (\Psi^{-1}, \Psi^{-2}, \Psi^{-3}, \ldots \Psi^{-n}\} + \Psi^{n}) =$$

$$\left[y \cdot \sum_{x=-1}^{-n} \frac{\{C_{-1} + C_{-2} + C_{-3} + \ldots\ldots C_{-n}\}}{\Psi^{J+x}} \right] + \left(\frac{C_{n}}{\Psi^{J^{n}}} \right)$$

when y is a participle dependent on the (C_n / Ψ^n) relation.

In this way, the use of y can yield (to) a tare on (C_n / Ψ^n). The tare adjustment might be empirical if based on (physical) measurement or observation.

The y participle, or convoluting factor (that which validates the supposed convolution), can of course be found explicitly by deigning a complexity for the considered relations:

$$[\,\Psi \bullet [\text{ convolution }] + (\,C_n / \Psi^n\,)\,] \bullet (\,n + 1\,) \bullet R_{(elation)}$$

$$= \text{(relation's) Complexity}$$

Aside from the tarring of the complexity (values), a spectrum of discreet values can be provided for by such terms (of expression) as:

$$\{\,C_{-1} + C_{-2} + C_{-3} + \ldots.\; C_{-n}\,\} \,/\, [\,\Psi^{+x}\,]_{x=-1}^{-n} \quad \text{and} \quad [\,[\,C_x\,]_{x=-1}^{-n} / \Psi^x\,]_{x=-1}^{-n}, \qquad \text{etc.}$$

There are, of course, modified (but coherent) convolutions. Therefore, allowing (or utilizing) negative n (perhaps artificially or by manufacture) provides for a spectrum of results (as desired).

Let us take an example, for a velocity's speed, at $n = 3$ artificially extended to $n = -3$.

At $n = 3$ for $s = d / t$, there must be a phantom variable (we can call "ϕ"). [Note that a phantom variable is unobserved, rather than equal to zero.]

Therefore, $s = \phi(\,d / t\,)$ or $s = \phi + (\,d / t\,)$, etc.; $C_{-1} = -1$, $C_{-2} = -3$, $C_{-3} = -5$.

Using discreet, or spectrum values (sans null):

$$[\,C_x\,]_{x=-1}^{-3} == \{\,-1, -3, -5\,\}$$

Therefore, we have:

$$\left[\left[y \bullet \frac{\{-1,-3,-5\}}{\Psi^{x}} \right]_{x=-1}^{-3} + \frac{7}{\Psi^{3}} \right] \bullet (3+1) \bullet \left(\frac{d}{t}\right) = \text{complexity}$$

$$\equiv \left[\frac{-y\,,\,-3y\,,\,-5y}{\Psi^{x}} \right]_{x=-1}^{-3} \bullet \frac{4d}{t} + \left(\frac{28}{\Psi^{x}}\right) \cdot \left(\frac{d}{t}\right)$$

$$\equiv \left\{ \frac{-y}{\Psi^{-1}} ; \frac{-3y}{\Psi^{-1}} ; \frac{-5y}{\Psi^{-1}} ; \frac{-y}{\Psi^{-2}} ; \frac{-3y}{\Psi^{-2}} ; \frac{-5y}{\Psi^{-2}} ; \frac{-y}{\Psi^{-3}} ; \frac{-3y}{\Psi^{-3}} ; \frac{-5y}{\Psi^{-3}} \right\} \bullet \frac{4d}{t} + \frac{28\,d}{\Psi^{x}t}$$

thus, 10 terms (of the spectrum), or 9 with the concentrating (condensing) addition.

This presents an interesting condition of discreetness, since, if for a relation a particular complexity is constant (of value), then each of the y factors must be different unless the terms are additive. Thus, we can say that discreetness with individual (and distinct) y factors is "dissolute" of manner, while additive distinctness (where a common y factor may be factored out to be solved) is "solute" of manner.

Therefore:

dissolute is

$\{ -y_1 / \Psi^{-1} , -3y_2 / \Psi^{-1} , -5y_3 / \Psi^{-1} ; -y_4 / \Psi^{-2} , -3y_5 / \Psi^{-2} , -5y_6 / \Psi^{-2} ; -y_7 / \Psi^{-3} , -3y_8 / \Psi^{-3} , -5y_9 / \Psi^{-3} \}$ and

solute is

$y\{ -1/\Psi^{-1} + -3/\Psi^{-1} + -5/\Psi^{-1} + -1/\Psi^{-2} + -3/\Psi^{-2} + -5/\Psi^{-2} + -1/\Psi^{-3} + -3/\Psi^{-3} + -5/\Psi^{-3} \}$

For example, (a) dissolute solution (for y_2) :

$[(4d / t) \bullet (-3y_2 / \Psi^{-1}) + (28d / \Psi^3 t)] = \text{Complexity}$; thus,

$(d / t)[(-12y_2 / \Psi^{-1}) + (28 / \Psi^3)] = \text{Complexity}$; thus,

$[(-12y2 / Y-1) + (28 / Y3)] = \text{Complexity} \bullet (t / d)$,

$-12y_2 / \Psi^{-1} = \text{Complexity} \bullet (t / d) - 28 / \Psi^3$; thus,

$y_2 = [\text{Complexity} \bullet (t / d) - 28 / \Psi^3] \bullet (\Psi^{-1} / -12)$

The distinguishing factor (to yield to y_2) appears to be: $\Psi^{-1} / -12 == \Psi^{-1} / 4(-3)$

Therefore, these factors are: $\Psi^{\{-1, -2, -3\}} / 4\{-1, -3, -5\}$== "dissolution factors"

$== \Psi^{\{-1, -2, -3\}} / (n + 1)\{-1, -3, -5\} == \Psi^{\{-1, \ldots. -n\}} / (n + 1)\{ odd_{-1}, \ldots. odd_{-n} \}$

[e.g., $odd_{-3} = -5$, the 3rd negative odd number]

It is interesting, then, to conjecture on complementary dissolution factors that may be possible for "positive discreetness" :

$\Psi^{\{1, \ldots. +n\}} / (n + 1)\{ even_1, \ldots. even_n \}$

and, of course, contrasting dissolution factors (theoretical):

$\Psi^{\{1, \ldots. +n\}} / (n + 1)\{odd_{-1}, \ldots. odd_{-n} \}$, $\Psi^{\{-1, \ldots. -n\}} / (n + 1)\{ even_1, \ldots. even_n \}$

A negative exponent (negative n) would of course defer to a fractional (sub-unitary) value applied from its base. Thus, were negative n accepted, the resultants would be smaller and smaller magnitudes as n (or $|n|$) increased.

If we follow such a directive for C_n, at n = -3, (and sans null), we yield to:

$$\mathbf{C_n = 1 + n + \sum_{x=1}^{n-2} n^x}$$

$$\mathbf{n - 2 \equiv -3 - 2 = -5}$$

$n - 2 == -3 - 2 = -5$; therefore,

$C_{-3} == 1 + (-3) + (-3^1) + (-3^{-1}) + (-3^{-2}) + (-3^{-3}) + (-3^{-4}) + (-3^{-5})$

$= -2 - 3 + (5 \text{ base } -3 \text{ fractions})$

$\sim= -5 + (-0.333 = -3^{-1}) = -5 - (0.333) \sim\sim -5.333$

That is, $C_{-3} == 1 - 3 - 3^1 + (-1 / 3^1) + (-1 / 3^2) + (-1 / 3^3) + (-1 / 3^4) + (-1 / 3^5)$

$\sim= -2 + (-3) + (-0.333) = -5 + (-0.333) = -5.333$

which is like the 3rd negative odd integer plus a (negative) fractional component:

$odd_{-3} + (-0.333)$.

Therefore,

$$C_{(-n)} \equiv \underbrace{\underbrace{odd_{(-n)}}_{1 + n + n^1}}_{(2n+1)} + \left(\text{negative fractional component } \sum_{x=-1}^{n-2} n^x \right)$$

here $x = 1$ of $\sum_{x=1}^{n-2} n^x$

here $x = \{ -1, \ldots\ldots -n \}$

Only when negative $n = -1$ is there no fractional component: $C_{-1} = -1$ (as follows)

So, $C_1 = 1$ and $C_{-1} = -1$

Note, $C_1 = 1$ by definition (of possible combinations). If we use the formulaic route insisted for $n = -1$, then

$$C_{-1} = 1 + (-1) + \sum_{x=1}^{n-2} n^x \equiv 1 - 1 + (-1)^{-1} + (-1)^{-2} + (-1)^{-3}$$

$$-1 + 1 + (-1)$$

$$-1 = C_{-1}$$

But, at $n = 1$, $n - 2 = 1 - 2 = -1$. Thus, $C_1 \rightarrow 3 + 1^{-1} == 4$.

And, if we simply disallow the summation (direction) from +1 to a negative number, then

$C_1 \rightarrow 1 + n = 1 + 1 = 2$

The formulaic

$$\mathbf{C_n = 1 + n + \sum_{x=1}^{n-2} n^x}$$

was designed (defined or prepared) for n >= 2 ; i.e., if n = 1, n – 2 = -1. Therefore, the summation is aborted (the direction +1 → -1 not being valid).

If n = 2, n – 2 = 0 ; thus, the summation is also aborted (directionally).

If n = 3, n – 2 = 1 ; thus, the summation with n = 1 (one term), is carried out, etc.

If accepting negative n, does n = -1 conform to the formulaic (i.e. accepting this negative direction)?

Using +1 as the starting point, and sans null, then -1 is the 2nd point, therefore adoptable to the formulaic. When adopting null (thus: +1, 0, -1 directionally), then -1 is the 3rd point (therefore also comforted by the formulaic, for negative direction). Thus,

(-n) ….. , -2 , -1 , 1 , 2 , ….. (+n) **(sans null)**

for C_n

$$\mathbf{odd_{(-n)} + \sum_{x=-1}^{n-2} n^x} \qquad \mathbf{1 + n + \sum_{x=1}^{n-2} n^x}$$

(functional component)

Possibly, the reference point (here +1) may be adjusted to any (particular) + or – n value, as a null/los shifting proposal.

This poses interesting initial complexity coefficients:

$N_{\Psi n} = (C_n / \Psi^n)(n + 1)$

For n = -1, (n + 1) = -1 + 1 = 0 ; but a **nul/los** shift would make this (factor) equal to 1 :

$$\mathbf{C_{-1} = 1 + (-1) + \sum_{x=-1}^{-3} -1^x = -1 + (-1)^{-1} + (-1)^{-2} + (-1)^{-3} + 1}$$

$$\downarrow \quad \downarrow \quad \downarrow \quad \downarrow$$

$$\mathbf{-1 + (-1) + (+1) + (-1)}$$

$$\mathbf{= -2 + 1 = -1}$$

$\Psi^{-1} \sim === 2.2805^{-1} = 0.43689$; thus, $N_{\Psi\text{-}1}$

$\sim = (-1 / 0.43689)(1) = -2.2905 = -\Psi$ base.

For n = +1 : $C_n = 1$, (n + 1) = 1 + 1 = 2 , $\Psi^1 = 2.2905^1 = 2.2905$

$\therefore N_{\Psi 1} \sim = 2 / 2.2905 = 0.87317 == 2 / -(N\Psi^{-1})$;

$\therefore (N_{\Psi 1})(N_{\Psi\text{-}1}) = -2 == -(n + 1)$

If we apply C_1 strictly to the formulaic for which C_{-1} is subjected, then

$$\mathbf{C_n \equiv 1 + n + \sum_{x=1}^{n-2} n^x \quad \text{sans null}}$$

$$\mathbf{\therefore \; C_1 = 1 + 1 + \sum_{x=1}^{-1} 1^x = 2 + 1^1 + 1^{-1} = 4}$$

(as seen earlier).

Therefore, a span of from C_{-1} to C_1 ($-1 \rightarrow 4$) is reached. This is a (symmetrical) span of (sans null) 4 units: $-1 \rightarrow 1$, $1 \rightarrow 2$, $2 \rightarrow 3$, $3 \rightarrow 4$. [With null it would be a symmetrical span of 6 units, including $-1 \rightarrow 0$ and $0 \rightarrow 1$.]

If the formal (sigma) summation is disqualified because of (negative) direction:
$x = +1 \rightarrow -1$, this leaves us with $C_1 = 2$ and a span of

(units; $C_{-1} \rightarrow C_{+1}$) 2: $-1 \rightarrow 1$, $1 \rightarrow 2$ (still symmetrical).

["Symmetrical" simply implies an even number of units.]

The definition order of operations, for C_n, however, is

$$\mathbf{C_n = n + 1 + \sum_{x=1}^{n-2} n^x}$$

$$\downarrow$$

$$\mathbf{C_1 = 1 \leftarrow n = 1}$$

$$\mathbf{C_2 = 3 \longleftarrow n = 2}$$

$$\mathbf{C_n = 3 + \overset{n + 1 + \cdots\cdots}{\cdots\cdots} \longleftarrow n \geq 2}$$

This emphasizes that C_n is a (compacted) positional (or functionally productive) operation (i.e., it differs at various positions of the n number-line). Thus, the suggestion is that it may change with (the nature of) the number-line.

While the complexity is a power series (essentially) on C_n, its complement: $\Psi^n \bullet C_n$ is essentially an additive series on n.

Primordially, one can argue (based on the primary formuli) that, as

$$\ulcorner\!\!+\!\!\lrcorner^{n} \equiv \left(\sum_{n=1}^{\infty} \frac{n}{C_n} \right)^{n_{\text{defined}}} \equiv \left[\sum_{k=1}^{\infty} \frac{k}{C_k} \right]^{n}$$

then

$$\Psi_{(n=1)}{}^{n} \bullet C_n \sim == (1/C_n)(1+2+3+\ldots.\infty) \bullet C_n = (1+2+3+\ldots.\infty).$$

This, then, is an additive simplicity, that might be used to estimate a determinable n (for a particular relation of measurable values). Again, the nature of n can modify (a simplicity's) meaning. The adjective “simple” is used here to indicate “straight forward” in thought or contemplation, as additive operations are.

The such represents, then, a fundamental distinction between multiplication and addition: multiplication modifies (of value as opposed to division's modification of numeric quality), while addition leaves quantities distinct (or rigidly coherent).

But Ψ^n is of a derived base (Ψ) from additive summations extending indefinitely (i.e. to ∞), while C_n is a definite number of additive summations extending up to n. Therefore, as far as C_n is concerned, a (type of) simplicity can be (inductively) related to any base:

Simplicity $== S = b^n \bullet C_n$, where b is a considered base.

Thus, we can (for example) simplify to $b = 1 : \therefore S_{b=1} = 1^n \bullet C_n = C_n$ for all positive n.

This seems to be a particularly easy (or straight forward) way to access the qualitative value of a (functional) base.

$$\Psi \bullet C_n \sim\sim (1/C_n)(1+2+3+\ldots.\infty) \bullet C_n = (1+2+3+\ldots.\infty)$$

$$\Psi^n \bullet C_n \sim\sim [(1/C_n)(1+2+3+\ldots.\infty)]^n \bullet C_n$$

$$= (1/(C_n)^n) \bullet (1+2+3+\ldots.\infty)^n \bullet C_n$$

$$= (1+2+3+\ldots.\infty)^n \bullet (C_n)^{-n} \bullet C_n^{+1} = (C_n^{1-n}) \bullet (1+2+3+\ldots.\infty)^n$$

$$= (1+2+3+\ldots.\infty)^n \bullet (1/(C_n)^{n-1})$$

Interesting bases (for a simplicity) could be $b = 1, \Psi_{()}, e, n, C_n, i$; simplicity itself is essentially an enhanced, “activated” or aggrandized base.

Of course, the same might be considered for the complexity, with changes of base. But while complementary, division is fundamentally different from multiplication. It is quality altering, as opposed to a comparison of multiplication and addition/subtraction.

Therefore, for complexity, Ψ^n must retain as a base to its exponent. And this is why complexity can be regarded as a power series on C_n. This is even emphasized by multiplying the complexity and simplicity components:

$$(C_n / \Psi^n)(n+1) \bullet (\Psi^n \bullet C_n) = C_n^{\,2} \bullet (n+1)$$

[Since the simplicity is also applied to an action of $(n + 1)$, this factor is only used once in the multiplication (of complexity to simplicity), here.]

One can also consider the term: Ψ^{Cn}, which yields to: $\log\Psi^{Cn} = C_n \log\Psi$.

This proposes a boundary (marker) to the influence (observance) of C_n, for a physical process (i.e. ~0.36 of C_n).

So, C_n can be interpreted as so many $\log s_{(\text{at base } 10)}$ of Ψ : i.e., $\log_\Psi \Psi^{Cn} = C_n$.

Using a primordial or primogenitural analysis, $\Psi^{Cn} \sim (n / C_n)^{Cn} = n^{Cn} / C_n^{\,Cn}$.

Therefore, n^{Cn} is also a term to consider (for trends or inclinations or projections, etc.):

e.g.	$\mathbf{1^1}$	$\mathbf{2^3}$	$\mathbf{3^7}$	$\mathbf{4^{25}}$
	↓	↓	↓	↓
	1	**8**	**2187**	**1125899906842624**

(rising) major ratio : **8 (= 8/1) , 273.375 (= 2187/8) , ~ 514814772218.85**

(falling) minor ratio : **0.125 , ~0.003658 , ~1.9424462 x 10^{-12}** .

But this (primogenitural) analysis poses an interesting contradiction, since (as C_n increases) Ψ^{Cn} must become very large, while (as n increases) the term $(n / C_n)^{Cn}$ must become very small. Therefore, there is a divergence from the primogenitural (i.e. $n = 1$), of effect; and $(n = 1)$ acts as an origin (of initiation).

Again, the value 0 (or zero) does not seem to occur; for with $n = 0 \rightarrow 1$, in both cases Ψ^{Cn} and $(n / C_n)^{Cn}$ increase.

This is because, when $n = 0$ or 1, then $n = C_n$, as far as $(n / C_n)^{Cn}$ is concerned. Ψ^{Cn}, then, increases from 1 to Ψ. So, $(n = 1)$ demarcates both an initiation point and a transition point (divergence of Ψ^{Cn} and $n^{Cn} / C_n^{\ Cn}$).

The nul/los shift here, from $0 \longrightarrow 1$, is analogous to the summation of digits of a number, where the digit zero can be ignored entirely while a nine (or a summation of nine) signals the result is either nine or whatever is left of the (summed) digits: from 1 to 8. Zero never occurs (as a summation of positive numbers), unless the only digits are zeros. Then "1" acts as an initiation point and "9" acts as a transition point (away from 9 unless only 9 is left, for a single digit sum). The base for this sum(mation) is, of course, base 10. So this is an automatic conversion from base 10 to base 9 (i.e. loss of the digit "0"), or from decimal to 'noneral'.

[It is interesting that the base name: base 10, includes just the two digits that contend for (or with) each other at replacement (or for intervention).]

$C_n^{\ Cn}$ is simply a nominal power (series) on C_n. It can be analyzed (or studied for various potentials) as if: $C_{\{1, 2, ... n\}}^{\ C\{1, 2, ... n\}}$, as indeed n^{Cn} can be so adjudged

as if $[\ \{1, 2, ... n\}\ ]^{C\{1, 2, ... n\}}$.

The obvious expansion of the $(\ n / C_n\ )^{Cn}$ term, then, is:

$$(\ \{\ 1, 2, ... n\ \} / C_{\{\ 1, 2, ... n\ \}}\ )^{C\{\ 1, 2, ... n\ \}} \sim \Psi^{C\{\ 1, 2, ... n\ \}} \bullet a$$

where $a ==$ a compensation factor (for the above arguments).

Primordially (i.e. at $n = 1$), $a = 1$ since the representation for Ψ (say, Ψ_P) is

equivalent to $(\ n / C_n\ )$.

Therefore, $\Psi_P == n / C_n$; thus, $a == [\ (\ n / C_n\ ) / (\ n / C_n\ )\ ]^{Cn}$.

But, we can substitute terms to get $[\ (\ n / C_n\ ) / \Psi_P\ ]^{Cn} == [\ n / (\ \Psi_P \bullet C_n\ )\ ]^{Cn}$

and then replace Ψ_P for the proper Ψ (which is of course serially defined) to yield a particularly 'summarial' (as in summarize) condensation of terms derived from C_n :

$$a \sim [\ n / (\ \Psi \bullet C_n\ )\ ]^{Cn} \sim\sim 1 / \Psi \qquad (\text{at } n = 1\ )$$

This illustrates a "rite of being" for the accustomed component of the complexity relation:

$(\ C_n / \Psi^n\ )(\ n + 1\ ) ==$ complexity coefficient

The component $(\ C_n / \Psi^n\ )$ is thus related to $\Psi^{\{\ 1, 2, ... n\ \}}$ when $n == C_n$ for Ψ^n.

Therefore, we may replace for Ψ^{n} :

$$(C_n / (\Psi^{C\{1, 2, ... n\}} \bullet a))(n + 1) == (C_n / (n / C_n)_{\{1, 2, ... n\}}{}^{Cn})(n + 1)$$

$$\sim == (n + 1) / (n / C_n)_{\{1, 2, ... n\}}{}^{Cn - 1} \rightarrow \text{(primordially) } 2, \text{ or } (n + 1) == 1 + 1 = 2$$

[i.e., as action + reaction, or movement and return movement (positionally);

note that at $n = 1$, $C_n \sim = (n / C_n)^{Cn}$]

"Rite of Being" for the Complexity of a Relation

As in analogy to (a calculus') rate of change for a derivative having as its integration the positional summation of these changes, a summation of rites of being can be employed to recover an original (or initiating) relationship.

This particular "rite" seems to hang (depend or be accomplished by) the thin thread (of being) that $n = 1$, but the tether is as strong and durable as if $n = 0$. And the manipulation of the "rite of being" is not merely imitative of the (classical) calculus, since it acts as a tying spoke to an entire family of derivations (of relational method). Therefore, the common thread is not so slender (as to be easily disrupted, nor broken or dissolved). ($n = 1$) must be an important rite for many relationships (to become of proper being).

One can have an accelerated series in C_n, where n becomes C_n of the previous n, etc.

n :	**1**	**2**	**3**	**4**		
C_n :	**1**	**3**	**7**	**25**		
n_{accel} :	**1**	**| C_2 = 3**	**C_3 = 7**	**C_7**	**$C_{(C7)}$**	
or	**(1)**	**| 2**	**C_2 = 3**	**C_3 = 7**	**C_7**	**$C_{(C7)}$**

Note that the series can not begin with ($n = 1$), since $n = 1 = C_1$.

So the series can not advance (always returning to 1), so powerful is ($n = 1$) to the C_n relation grasping. Of course, the corresponding complexity and simplicity coefficients rise also. But they can not (curiously) start or initiate from ($n = 1$). Perhaps the initiations could be backward (i.e., lead towards decreasing n in direction) of progression. Clearly, observations are results of procedures that might be traced backwards to an initiation of effect (or action).

Working backwards, for example, say we have an observed speed ($= S = d / t$),

with $n = 2$, (d, t).

If we assume the observed is a complexity coefficient times an "effective speed" ($== S_e$), then

$NS_e = S$; therefore, ($C_2(2 + 1) / \Psi^2) \bullet S_e = S$;

thus, $S \sim == (3 \bullet 3 / \Psi^2) \bullet S_e \sim = 9S_e / 8.439 = 1.066S_e$.

Here, we can only start (to go backwards) with $n = 2$, since we can not start (going backwards) with $n = 1$. We see that S_e is close to S (observed).

Now, suppose we "explode" (d, t) into variables with two values each: $(d_1, d_2; t_1, t_2)$,

so that $S == (d_1, d_2) / (t_1, t_2) \rightarrow 4$ (possible) speeds of velocities:

$d_1 / t_1, \; d_1 / t_2, \; d_2 / t_1, \; d_2 / t_2$.

Here $n = 4$, $(d_1, d_2; t_1, t_2)$; if we assure (or insist on) an accelerated series (of C_n), going backwards:

$S_4 == (C_{(C3)} \bullet (4 + 1) / \Psi^4) \bullet S_e$, $S_3 == (C_{(C2)} \bullet (4 + 1) / \Psi^4) \bullet S_e$

$S_2 == (C_2 \bullet (4 + 1) / \Psi^4) \bullet S_e$, $S_1 == (C_1 \bullet (4 + 1) / \Psi^4) \bullet S_e$

Therefore, $S == (C_{(7, 3, 2, 1)} \bullet (4 + 1) / \Psi^4) \bullet S_e$. Thus, a spectrum of values occurs.

If we assign ($d_1 / t_1, d_1 / t_2, d_2 / t_1, d_2 / t_2$) to (a, b, c, d), then

$S_{(a, b, c, d)} == (C_{(7, 3, 2, 1)} \bullet (4 + 1) / \Psi^4) \bullet S_{e(a, b, c, d)}$

assuming each S yields to a particular S_e.

If we allow that $S_a < S_b < S_c < S_d$, then we can graph as (say):

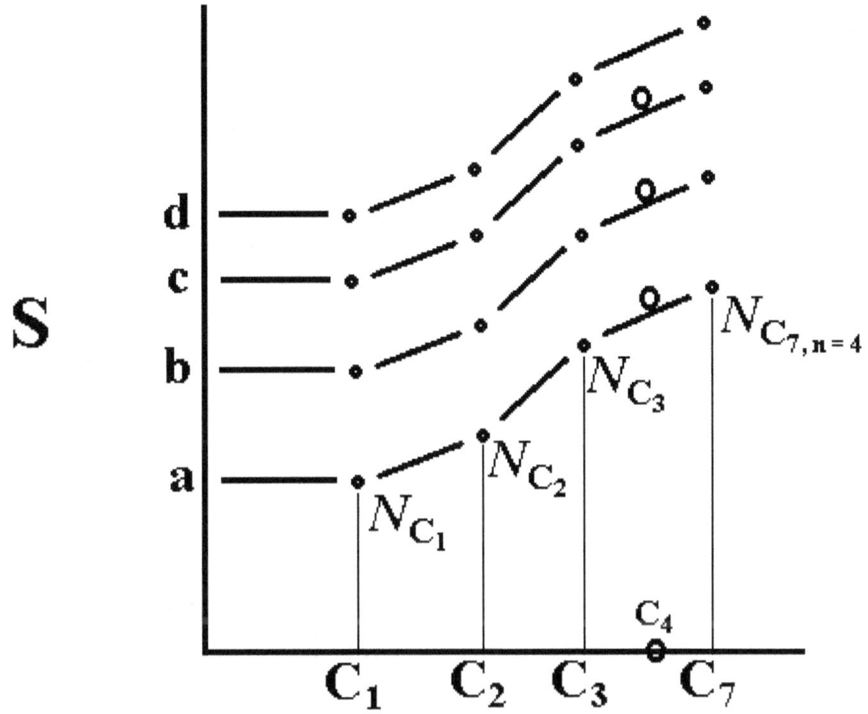

Note that the C_4 values (circles, in graph), i.e. $C_4(4 + 1) / \Psi^4$ coefficient, act as sort of anti-limits to the accelerated (C_n) values; i.e., instead of (values) going towards C_4, they go away from C_4.

The plot trajectories, however, are all of the same shape (form). [The "being" remains the same, for each S.] This is due to coefficient evolutions rather than variable modifications. Alterations (in shape) of the (complexity) trajectories can be accomplished simply by assigning to the (complexity) coefficient a variable: e.g., "y"; $\therefore N == (C_n(n + 1) / \Psi^n) \bullet y$

y would then have to have units based on the relationship considered. (The other components of the coefficient are pure numbers.) In order to make the complexity a pure (unit-less) number, "y" would have to have reciprocal units to the considered (or utilized) relation. So, we can declare that the distinction between y and S_e is that y is a component of the coefficient that renders the complexity unit-less. It also allows for a modification of the complexity trajectory. To maintain a relationship of equality (i.e. an equation), y becomes the reciprocal of the coefficient ($= 1 / N$). These (procedures) make postulates of the complexity.

Complexity Applications

But how might we apply such conjectures? The applications will have "sculptural" characteristics.

For example:

The fastest speed measurable is c (in a vacuum), which insists that no energy (nor matter) or information may travel faster. Superluminal velocities and speeds are of course possible (i.e. by the definition of wave propagation, the phase of the wave may travel faster than the wave period). However, no information can be transferred faster. This is akin to saying, essentially, that no speeds faster than c are perceivable. How, then, can we demonstrate (characteristics) of superluminal transit?

Since energy travels as an electromagnetic wave, it can be assigned to the energy of work:

$$E == \text{work} = \text{force x distance}$$

Distance, like temperature, is simply a physical measurement (or difference between two measurements). But force is a physical result (or resultant of physical processes) which, therefore, is subject to perception. We seek, then, a force so minimal (or small) as to be imperceptible. This must be coupled to a distance so great as to yield (by the work relationship) a measured (or measurable) energy, E.

By the complexity methods, then, we can apply (or "slap" on) the complexity coefficients to the work relationship:

$$E = f \bullet d$$

$$E_{comp} == (C_n (n + 1) / \Psi^n) \bullet (f \bullet d)$$

This can be reinterpreted (or manipulated) as:

$E_{comp} / (n + 1) =$	f / Ψ^n	$\bullet$	$(C_n \bullet d)$
a minimal energy	minimal or imperceptible force		enlarged distance

Here, n itself is the variable to be considered (or found). Formally, however, we request to find the (size of) minimal or imperceptible force as well as the corresponding (or necessary) C_n enlarged distance.

This, of course, can be assessed empirically, by insisting

$E_{comp} / (n + 1) ==$ a minimal energy == 1 quantum of energy (of a particular frequency); i.e., a photon's energy.

Therefore, if we set $E = E_{comp}$, and $E / (n + 1) = h\nu$; thus, $E / h\nu = n + 1$,

whereby $n = N + f$**(raction)** may be obtained.

Therefore, (using f and d for $E = f \bullet d$), the imperceptible force is reached when $n = N + 1$;

thus, imperceptible force == $f / \Psi^{(N+1)}$; necessary distance == $C_{(N+1)} \bullet d$

The Plank constant, h, is an empirical (i.e. determined or discovered by experiment) quantity, as well as c; therefore,

minimal energy made imperceptible ~== $h\nu \sim= E / (N + 1 + 1) = E / (N + 2)$

Setting $E = E_{comp}$, for $E = f \bullet d$, allows one to calculate S_e (or the essential E), with $n = 2$ (variables: f, d) for the relation; thus, $E_{Se} = E / 1.066$

Let us try an (empirical) example:

Say, a force of 1 newton is applied a distance of 1 meter, thereby resulting from an energy of 1 newton-meter.

The transit (motion) is of visible light (photons) of a given frequency, wavelength $\lambda = 600$nm;

thus, $freq^{-1} \sim== 2.00 \times 10^{-15}$ sec ; therefore, freq $\sim= 5 \times 10^{14}$ oscillations/sec

$h \sim= 6.626 \times 10^{-34} J \bullet s$ (with 1 newton-meter / $J_{(oule)}$)

$= 6.626 \times 10^{-34}$ newton-meter $\bullet$ s

Therefore, the energy of one (600nm) photon is $hf_{(req)}$:

$E_p = (6.626 \times 10^{-34} \text{n-m} \bullet s)(5 \times 10^{+14} s^{-1}) = 33.13 \times 10^{-20}$newton-meters

$E_{comp} == E = 1$ newton-meter (done to try to determine a value of n)

$E / (n + 1) = E_p$; $\therefore$ $E / E_p = (n + 1) == 1 / 33.13 \times 10^{-20} \sim= 3.018 \times 10^{+18}$.

[Dropping the fractional part, subtracting 1 and then adding 1 (for superluminal's n) yields essentially the same number for the value of $n == n_{sup}$.]

∴ the (superluminal speed's) imperceptible force

$== 1$ newton-meter / $\Psi^{3.018 \times 10+18}$ meters $\rightarrow C_{nsup} >>> \Psi_{nsup}$;

the C_n enhanced (or enlarged) distance $== C_{(3.018 \times 10+18)} \bullet 1$ meter

But, E_{comp} is actually $(C_n(n + 1) / \Psi^n) \bullet f \bullet d$; and with $n_{sup} \sim= 3.018 \times 10^{18}$,
E_{comp} is obviously much greater than E_p.

With $n == n_{sup}$ (as found here), is E_{comp} the (huge) energy necessary for superluminal speed?

This depends on whether our definition of complexity can have wrought the very extremes of variables, here the minimum f and a corresponding expansion of d.

Functionally, of course, without relevance to superluminal speeds,

E_{comp} is simply $(2 / \Psi^1)$ newton-meters, with the solution that $n = 1$:

$E_{comp} = (C_1(1 + 1) / \Psi^1) \bullet 1$ newton (= f) $\bullet$ 1 meter (= d) $\sim= 0.87$ newton-meters

What is the significance of $C_{nsup} \bullet 1$ meter? Is that the (or an) extent of the universe?

What is the significance of Ψ^{nsup} meters (as for 1 newton-meter / Ψ^{nsup} meters)?
Is this the maximum extent of a linear signal?

Deflection Medium Energy

At least we have means procuring these large values. But such a linear (distance) limit suggests, once the distance is met, the (radiant) energy must change direction (therefore, cause an acceleration). This might be due to some sort of deflection (e.g. reflection or refraction), which implies (existence of) a medium of deflection. But this opens an interesting conduit to answer the question: Why does radiant energy radiate (or travel)? Perhaps it is to keep up with the accelerating C_n / Ψ^n ratio for increasing n, where n is proportional to distance traveled.

This suggests a concomitant expenditure of energy (from a 1 newton force) of:

a sum of ($1 / \Psi^n$, for n values up to n_{sup}) x 1 newton-meter .

For $n = 1 \rightarrow 10$, this comes to (~ ~) :

$(0.87 + 0.12 + 0.08 + 0.036 + 0.016 + 0.007 + 0.003 + 0.001) \bullet 1$ newton-meter

$= 1.133$ n-m, which is more than the input energy by ($1.133 - 1.000 =$) 0.133 n-m .

Therefore, it takes more energy to radiate energy (than the energy radiated itself), which drives the C_n / Ψ^n acceleration. We might call the extra 0.133 newton-meters as the "deflection medium energy," or DME. For our supposed model (of motion), typical geometric constraints, descriptions, and postulates may be applied.

If we say our "universe" has a diameter (in meters) of C_{nsup}, for the shape of a sphere, then its volume is: $(4 / 3)\pi r^3$, where $r == (C_{nsup}$ meters$) / 2$; therefore,

$$V_u == (4 / 3)\pi(C_{nsup} / 2)^3 = (4 / 3)\pi C_{nsup}{}^3 / 8 = (4 / 24)\pi C_{nsup}{}^3 = (1 / 6)\pi C_{nsup}{}^3 .$$

If we say our radiant energy conduces a sphere (of potential direction or orientation of travel) with a radius (i.e. central initiation) of Ψ^{nsup} meters, and these spheres are compacted within the universe, then the volume of each (maximum) radiant energy sphere is $(4 / 3)\pi(\Psi^{nsup})^3$ and the maximum number (of maximum spheres) for the universe is :

$$\mathbf{(1 / 6)\pi C_{nsup}{}^3 / [(4 / 3)\pi(\Psi^{nsup})^3] == (1 / (2 \bullet 4))(C_{nsup} / \Psi^{nsup})^3 = (1 / 8)(C_{nsup} / \Psi^{nsup})^3} .$$

We assume, for light (or energy) transit, refraction between spheres (each sphere surfaced by DME) is only allowed where the sphere surfaces do not abut. For where they do, only reflection (back into the sphere) is allowed, the double DME (coating) preventing passage inter-sphere. Deformation of the interstices (i.e. "melting") due to ill-formed submaximal spheres are conceivable.

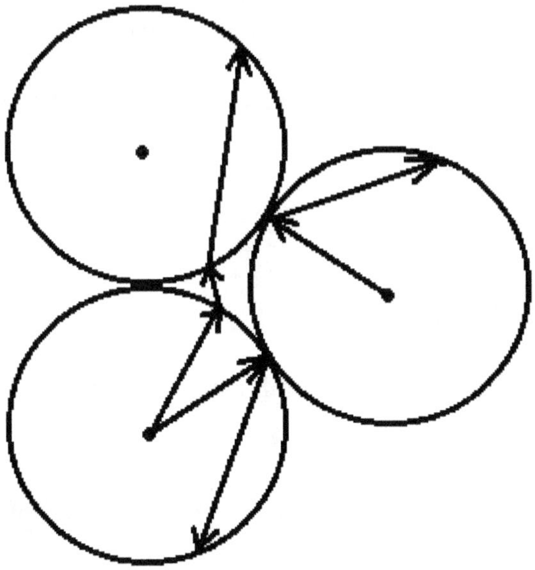

Note that, by going through spheres between spheres, the rays have spent or depleted energies to prevent them finding the centers of visited spheres. [Each sphere is of course theoretical or probabilistic for any particular ray to initiate (or be generated).]

The process could be called: 'Rays chasing (theoretical) chords'. The lengths of the (chased) chords would be less than Ψ^{nsup}. Of course, the "light circles/spheres" are opportunistic of time and circumstance and are ever varying or changing according to opportunity or crisis of need.

The C_{nsup} / Ψ^{nsup} ratio is so characteristic of the complexity that it can be "canonized" as: Complexity Coefficient / n_{sup} where $n_{sup} + 1 \sim\sim n_{sup}$, as complexity coefficient is $C_n(n + 1) / \Psi^n = \boldsymbol{N}$. Therefore, the number of spheres in the universe is around:

$(1 / 8)(\boldsymbol{N} / n_{sup})$ or $(1 / 8)(\boldsymbol{N} / n)_{sup}^{\ 3} == (\boldsymbol{N} / 2n)_{sup}^{\ 3}$.

The term $(\boldsymbol{N} / 2(n + 1))^3$ can be considered as a volumetric principle of (the) complexity. This is, of course, equivalent to $(C_n / 2\Psi^n)^3$ and provides an easy (or straight forward) contemplation of numbers (of components) in a volume, though it is more like a sculpturing of numerical analysis than a direct definition (of entities). It is a curious reduction of the cube of the C_n / Ψ^n ratio: $(1 / 8)(C_n / \Psi^n)^3 == (1 / 2^3) \bullet (C_n / \Psi^n)^3$.

This would tend to treat the ratio as a length, to provide a volume or space (of consideration). This is a revelation, that the ratio might (symbolically) substitute of (particular) physical properties or qualities, provided the proper numerical factors necessary for the (ratio's) elevation from a pure number. (But the actual physical property considered, here length or distance, must be rendered unit-less by the ratio, to keep the ratio unit-less.) Thus are some consequences of "slapping on" the complexity coefficient to a given relation (among variables).

Modes of Complexity

The mode of the C_n / Ψ^n ratio is (geometrically): $C_n^{\ 2} / n == C_n / (n / C_n)$.

Therefore, the mode for the geometric term for the ratio appears to be:

$2^{-1}C_n^{\ 2} / n == C_n / (n / 2^{-1}C_n) = C_n^{\ 2} / 2n$.

Presumably higher modes are available as $m^{-1}C_n^{\ 2} / n$, $m = 2^{(m)-1}$ (here) with $m >= 1$, correspondent with various physical processes (types or procedures). Although, conceivably, m may entertain (apply) any manner of complexity or sophistication to the fundamental mode $(C_n^{\ 2} / n)$.

Thus, we have a

Length directive:	$(1/2^{+1})\bullet(C_n/\Psi^n)^1$	**m = 1 mode** $2^{-1}C_n^{\,2}/n$,
Area directive:	$(1/2^{+2})\bullet(C_n/\Psi^n)^2$	**m = 2 mode** $2^{-2}C_n^{\,2}/n$,
Volume directive:	$(1/2^{+3})\bullet(C_n/\Psi^n)^3$	**m = 3 mode** $2^{-3}C_n^{\,2}/n$,
m dimension: directive	$(1/2^{+m})\bullet(C_n/\Psi^n)^m$	**m = m mode** $2^{-m}C_n^{\,2}/n$

Base 2 appears (here) geometrical (or spacial-ly dimensional), for the factor to the fundamental (of complexity).

Why would (base) 2 be geometrical: for 2 points define a length?

Other bases might apply to (or signal) different physics (i.e. physical phenomena). Of natural interest would be Ψ^m (i.e. base Ψ):

$$(1/\Psi^m)\bullet(C_n/\Psi^n)^m == C_n^{\,m}/\Psi^m\Psi^{mn} = C_n^{\,m}/\Psi^{mn+m} = C_n^{\,m}/\Psi^{m(n+1)} == (C_n/\Psi^{n+1})^m$$

which we might say appears as a modified or enhanced complexity coefficient, compared to $N = (C_n/\Psi^n)\bullet(n+1)$; its mode is $C_n^{\,2}/(\Psi^m\bullet n)$ or, as applied:

$$(C_n^{\,2}/(\Psi\bullet n))^m == (1/\Psi^m)\bullet(C_n^{\,2}/n)^m.$$

What can the applied mode show us (here)?

If we set $(C_n/\Psi^{n+1})^m = (C_n^{\,2}/(\Psi\bullet n))^m$

[an idealization for 1 (as a ratio), which allows $\{ n \mid 0, +\text{-}\infty \}$ as the solution set; although definitionally C_n for $n <= 0$ does not actually occur]

then $C_n/\Psi^{n+1} = C_n^{\,2}/(\Psi\bullet n)$;

$\therefore\ 1/\Psi^{n+1} = C_n/(\Psi\bullet n)$, $C_n = \Psi\bullet n/\Psi^{n+1} = n\Psi^{-n}$

Minor Values of Combinations

Therefore, (with adaptation) C_n becomes (here) minor C_n (or the minor value of C_n), which is a fraction of n (for n >= 1) or a (Ψ based) exponential factor for n <= 0 (or n < 1), [a positive or negative fraction of n].

Presumably, then, ${}_{minor}C_n = C_n - {}_{minor}C_n$, or $C_n == {}_{major}C_n + {}_{minor}C_n$

[Obviously, for $n >= 1$, C_n must be larger than any fraction of n.]

For example, for $n = 3$, $C_n = 7$

$n\Psi^{-n} == 3 \bullet \Psi^{(-3)} \sim= 0.24965 == {}_{minor}C_n$

$C_n / {}_{minor}C_n \sim= 28.04$, ${}_{major}C_n = 7 - 0.24965 = 6.7504$

$C_n / {}_{major}C_n \sim= 1.037$

Because the reduction in n is (Ψ) fractional, even with large negative n the minor C_n (in absolute value or magnitude) is less than C_n (in absolute value or magnitude) eventually.

$C_n / {}_{minor}C_n$	n	C_n	${}_{minor}C_n$	${}_{major}C_n$	$C_n / {}_{major}C_n$
34.41	**4**	**25**	**0.14532**	**24.85468**	**1.006**
28.04	**3**	**7**	**0.24965**	**6.7504**	**1.037**
7.870	**2**	**3**	**0.38121**	**2.61879**	**1.146**
2.291	**1**	**1**	**0.43659**	**0.56241**	**1.775**

Therefore, as $n\uparrow$, the ratio $C_n / {}_{major}C_n$ tends towards 1.000 (i.e., 1), which is expected as ${}_{minor}C_n$ decreases with $n\uparrow$.

$C_n / {}_{minor}C_n$	n	C_n	${}_{minor}C_n$	${}_{major}C_n$	$C_n / {}_{major}C_n$
+0.2271	**-4**	**(-25) -110.1**		**+85.1**	**-0.2938**
+0.1942	**-3**	**(-7) -36.05**		**+29.05**	**-0.2410**
+0.2860	**-2**	**(-3) -10.49**		**+7.49**	**-0.4004**
+0.4366	**-1**	**(-1) -2.2905**		**+1.2905**	**-0.7749**

This indicates that C_n (or n, ${}_{minor}C_n$, ${}_{major}C_n$) is a disparate function, with (while continuous) positive n essentially different from negative n, in behavior. [The $n = 1$ or $C_n = 1$ point forms a (substantially) sharp angle (vertex) for the function's graph, $(n, C_n = 0)$ excluded.]

We can extend the analysis by considering summations of the form:

$\Sigma(1, 2, ... n)\Psi^{-\{1, 2, ... n\}}$, with degrees of term via $\{1, 2, ... n\}$. Then, the degree corresponds to the ${}_{minor}C_n$ based on the summation of terms for $n\Psi^{-n}$ for $\{1, 2, ... n\}$, etc. The element $n\Psi^{-n}$ is, of course, exponential, and should show a characteristic decay (of values). However, it is also (decidedly) integral (of n) and therefore is a step function. Allowing for fractional (non-integral) and negative (non-physical) values, though, the plots reveal finer details of behavior. (This is in part due to the greater expansiveness using negative n.)

Using the previously given data, including negative and positive n from $\{ C_n \mid -25 \ldots +25 \}$, but not including zero point data $(n = 0,\ C_n = 0,\ {}_{minor}C_n = 0)$ to maintain an active consideration of n, we may plot $(n,\ {}_{minor}C_n)$ as an actual function: ${}_{minor}C_n = n\Psi^{-n}$. We can also plot $(C_n,\ {}_{minor}C_n)$ for a comparison of activity.

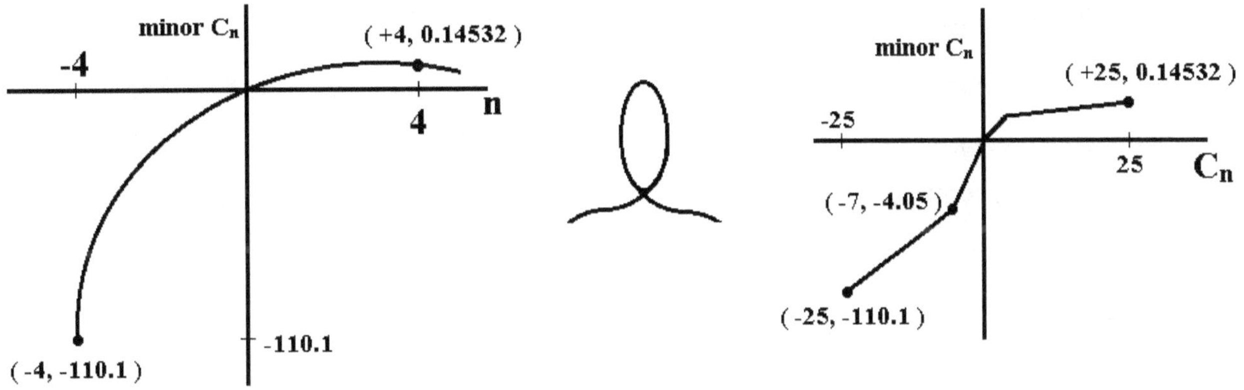

To highlight finer detail, we can restrict the x-axis to the closest negative n or C_n to the positive (i.e. $n,\ C_n = -1$):

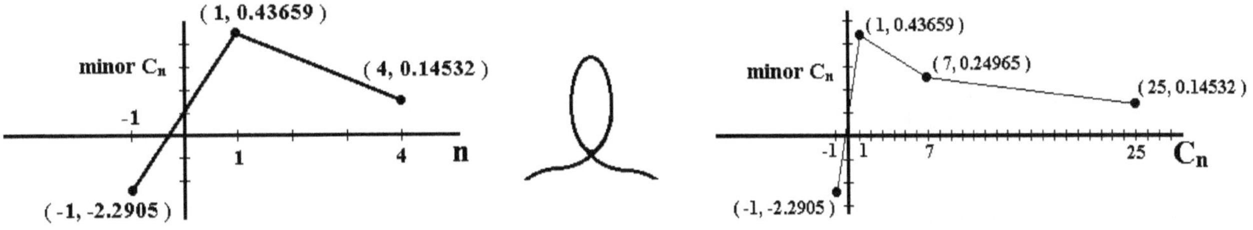

For the latter data restricted plots, inclusion of the zero point data yields a perturb(-erance) bend (i.e. a protuberance).

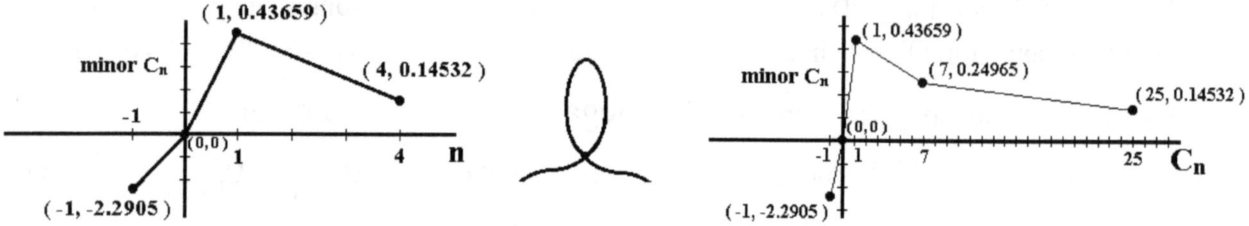

The zero point data suggests yet another change in functional behavior, distinct from both positive n and negative n. This could be a sign (as for negative n) that it does not (realistically) exist (of function for C_n). The symbol "ℓ" (de)marks a process to convert one plot (for n) into another (for C_n), here whereby from $(n, {}_{minor}C_n)$ the n is first used to find C_n, and then the C_n is used to plot $(C_n, {}_{minor}C_n)$ correspondingly (to n). [This is in contrast to a convolution of terms.]

Therefore, $\ell \bullet (n, {}_{minor}C_n) \rightarrow (C_n, {}_{minor}C_n)$

or ${}_{n}\ell_{Cn}(n, {}_{minor}C_n) \rightarrow (C_n, {}_{minor}C_n)$

or $(n, {}_{minor}C_n)\ {}_{n}\ell_{Cn}\ (C_n, {}_{minor}C_n)$ or $({}_{n}\ell_{Cn}, {}_{minor}C_n)$

and of course obvious related notations:

e.g., $({}_{n}\ell_{Cn}, {}_{minor}C_n\ \ell\ {}_{major}C_n) == (C_n, {}_{major}C_n)$, a nearly 45° (line) plot of curve, because $C_n \sim\sim {}_{major}C_n$ in size (value).

And this can be advanced with:

$({}_{n}\ell_{Cn}, \Sigma[{}_{minor}C_n\ \ell\ {}_{major}C_n])$, etc., or $({}_{n}\ell_{Cn}, \Sigma[{}_{minor}C_n]\ \ell\ {}_{major}{}^{\Sigma}C_n)$

where ${}_{major}{}^{\Sigma}C_n = C_n - \Sigma\{1, 2, \ldots. n\}\Psi^{\{1, 2, \ldots. n\}}$.

For example,

$n = 1$, $_{minor}{}^{\Sigma}C_{n} = 1 \bullet \Psi^{-1}$

$n = 2$, $_{minor}{}^{\Sigma}C_{n} = \{1, 2\}\Psi^{-\{1, 2\}} == 1 \bullet \Psi^{-1} + 1 \bullet \Psi^{-2} + 2 \bullet \Psi^{-1} + 2 \bullet \Psi^{-2}$.

$_{minor}{}^{\Sigma}C_{3} == 1 \bullet \Psi^{-1} + 1 \bullet \Psi^{-2} + 1 \bullet \Psi^{-3} + 2 \bullet \Psi^{-1} + 2 \bullet \Psi^{-2} + 2 \bullet \Psi^{-3} + 3 \bullet \Psi^{-1} + 3 \bullet \Psi^{-2} + 3 \bullet \Psi^{-3}$

$_{minor}{}^{\Sigma}C_{4} == 1 \bullet (\Psi^{-1} + \Psi^{-2} + \Psi^{-3} + \Psi^{-4}) + 2 \bullet (\Psi^{-1} + \Psi^{-2} + \Psi^{-3} + \Psi^{-4}) + 3 \bullet (\Psi^{-1} + \Psi^{-2} + \Psi^{-3} + \Psi^{-4})$
$+ 4 \bullet (\Psi^{-1} + \Psi^{-2} + \Psi^{-3} + \Psi^{-4}) == (1 + 2 + 3 + 4 = 10) \bullet (\Psi^{-1} + \Psi^{-2} + \Psi^{-3} + \Psi^{-4})$

$\therefore \Psi^{-1} \sim= 0.43659, \Psi^{-2} \sim= 0.19061, \Psi^{-3} \sim= 0.083216, \Psi^{-4} \sim= 0.0363311$

$(\Psi^{-1} + \Psi^{-2}) \sim= 0.43659 + 0.19061 = 0.62720$

$(\Psi^{-1} + \Psi^{-2} + \Psi^{-3}) \sim= 0.62720 + 0.083216 = 0.710416$

$(\Psi^{-1} + \Psi^{-2} + \Psi^{-3} + \Psi^{-4}) \sim= 0.710416 + 0.0363311 = 0.7467471$

Therefore,

$_{minor}{}^{\Sigma}C_{4} == (0.7467471) \bullet 10 = 7.467471$

$_{minor}{}^{\Sigma}C_{3} == (3 + 2 + 1)(0.710416) = 4.262496$

$_{minor}{}^{\Sigma}C_{2} == (2 + 1)(0.62720) = 1.88160$

$_{minor}{}^{\Sigma}C_{1} == (1)(0.43659) = 0.43659$

Therefore,

n	C_n	$_{minor}{}^{\Sigma}C_n$	$_{major}{}^{\Sigma}C_n$	$C_n / {}_{minor}{}^{\Sigma}C_n$	$C_n / {}_{major}{}^{\Sigma}C_n$
4	**25**	**7.46747**	**17.53253**	**3.34785**	**1.42592**
3	**7**	**4.26250**	**2.73750**	**1.64223**	**2.55708**
2	**3**	**1.88160**	**1.11840**	**1.59439**	**2.68240**
1	**1**	**0.43659**	**0.56341**	**2.29048**	**1.77491**

For $n = 2$, $_{minor}{}^{\Sigma}C_{2} > {}_{major}{}^{\Sigma}C_{2}$ (as a consequence of manner of addition?)

$n = 3$, $_{minor}{}^{\Sigma}C_{3} > {}_{major}{}^{\Sigma}C_{3}$.

Graphs show a divergence between [(n, C_n), ${}_{minor}C_n$ and ${}_{minor}{}^{\Sigma}C_n$];

i.e., ${}_{minor}C_n$ falls as ${}_{minor}{}^{\Sigma}C_n$ rises.

n Σ factor	**And for negative n ?**	**minor$^{\Sigma}C_n$**	**(C_n)**	**n**
-1	**$\Psi^{-(-1)} = \Psi^1 \sim= 2.2905$**	**-2.2905**	**-1**	**-1**
-3	**$(\Psi^1 + \Psi^2) \sim= 2.2905 + 5.2464 = 7.5369$**	**-22.6107**	**-3**	**-2**
-6	**$(\Psi^1 + \Psi^2 + \Psi^3) \sim= 7.5369 + 12.0169 = 19.5538$**	**-117.323**	**-7**	**-3**
-10	**$(\Psi^1 + \Psi^2 + \Psi^3 + \Psi^4) \sim= 19.5538 + 27.5246$**	**-470.784**	**-25**	**-4**

One can plot (n, C_n) vs $({}_{minor}C_n, {}_{minor}{}^{\Sigma}C_n)$ with zero point data, and then ignore any lines leading to $(0, 0)$ to show the determinant nature (features) of the graph. ${}_{minor}C_n$ and ${}_{major}C_n$ are additive since the (basic) mode (of the complexity coefficient) is from an additive (sigma) function.

The basic or fundamental mode (for the coefficient) is:

$C_n / \Psi == C_n / (n / C_n) = C_n^2 / n$, therefore the 1st degree (towards n).

Higher modes are available, of course. For example, the 2nd degree would be:

$C_1 / \Psi^1 + C_2 / \Psi^2 == C^1 / (1 / C_1) + C_2 / (2 / C_2) == (1/1)C_1^2 + (1/2)C_2^2$,

for $n = 2$.

Or, for general n: $(1/(n-1)C_{n-1}^2 + (1/n)C_n^2$.

The applied $(1/\Psi^m)$ mode, then, is : $(1/\Psi^m)[(1/(n-1)C_{n-1}^2 + (1/n)C_n^2]^m$;
therefore, we can set: $C_n / \Psi^{n+1} = [(1/(n-1)C_{n-1}^2 + (1/n)C_n^2](1/\Psi)$,

as before for the 1^{St} degree.

For 3rd degree: $\mathbf{C_n / \Psi^{n+1} = [(1/(n-2)C_{n-2}^2 + (1/(n-1)C_{n-1}^2 + (1/n)C_n^2](1/\Psi)}$.

And for n^{th} degree:

$$C_n / \Psi^{n+1} = [(1/(n-(n-1)))C_{n-(n-1)}^2 +$$
$$\ldots. + (1/(n-2)C_{n-2}^2 + (1/(n-1)C_{n-1}^2 + (1/n)C_n^2](1/\Psi),$$

where $(1/(n-(n-1)))C_{n-(n-1)}^2 = (1/1)C_1^2$.

Or,

$$\frac{C_n}{r^{n+1}} \cdot r^j = \frac{C_n r^{j1}}{r^{n+1}} \equiv C_n r^{j-n} = \sum_{a=1}^{n} \left(\frac{1}{a}\right) C_a^2$$

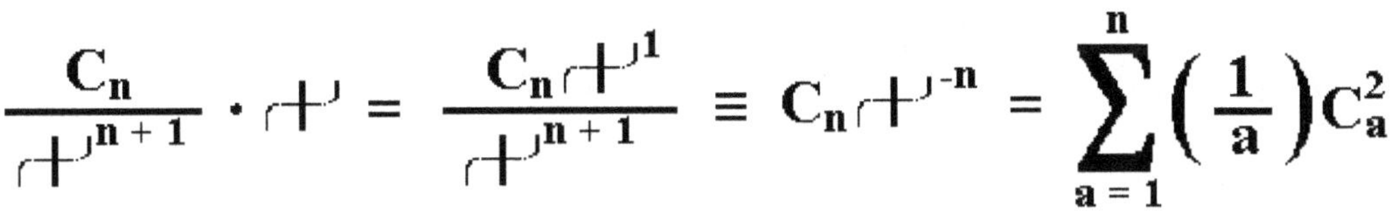

n^{th} degree, for positive n

Note that this seems to occur as an end (i.e. n) to initial (i.e. "1") progression, to establish the degree.

Complexity Sphere Model for a Universe

As regards superluminal possibilities, we consider that the maximum speed obtainable is (for light) energy photons ("particles") in a vacuum; that is, c. Say, then, that there is only one universe, and that this universe contains "everything" that is possible to exist. Now, say that this universe (by its inherent nature) rotates (of a constant regularity, as we could ever observe and measure it). If the universe is a perfect sphere, its absolute center has no motion, or such location would define the most (and only) fundamental particle, as never having motion (there). Then likewise, everywhere else there is motion, and the motion revolves about the center (of motionless-ness). The speed of the motion would have to be maximal for each circumferential position (of a particle) rotating about the center. For if the speed were ever greater, the particle may leave the surface (shell) of the universe, where it was once rotating along the surface. Therefore, each circumferential position has a maximal speed; and the closer to the (universe's) center, the smaller the maximum.

We can divide the universal sphere into (decided) shells of circumference (the outermost being the surface), as well as planes with circumference:

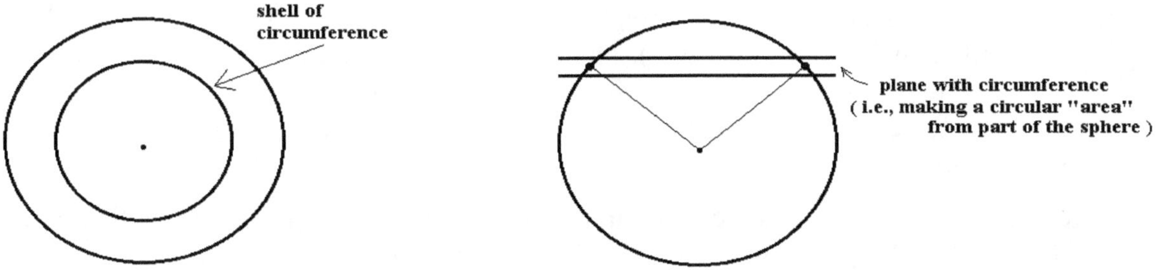

Each shell as well as each plane would have a maximal speed of motion, the plane idealized as a circular area of negligible thickness compared to the size and volume of the (whole) sphere.

Let up propose that all action requires energy, and therefore also all (initiation of) motion. We know that when a mass is affected (impinged) by a force it acquires a constant speed (or velocity, directionally) until affected by another force. Since this speed is movement, it must be induced by some energy. But why is its speed constant? Presumably its motion is maintained (supported) by the sphere's rotation (energy of rotation). Since its (the mass') speed remains the same, and time must increase proportional to distance, then the distance gained must be balanced by a loss of distance due to the rotation, to keep a proportionality with time elapsed.

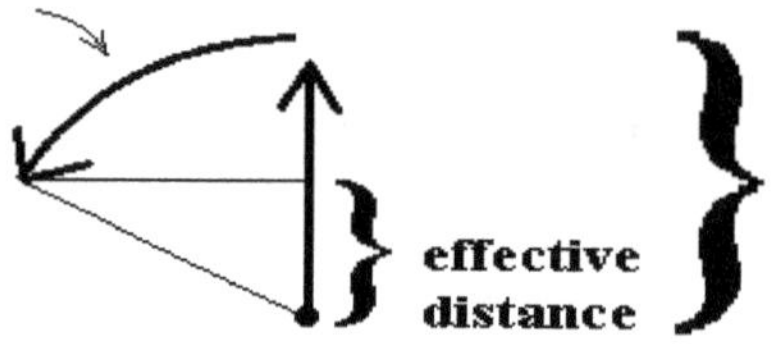

The rotation is of axial vector(s), while the (linear) velocity remains polar. Although what the universe rotates "in" must somehow be itself.

The magnitude of a mass' constant velocity (from an impinging or applied force) can be described by the following relation:

$(s + x) / (t + y) = s / t$, where x is the increase in distance and y is the increase in time.

$\therefore t(s + x) = s(t + y)$, $st + xt = st + ys$; thus, $xt = ys$, $y = x(t / s)$

But this is for the mass' linear velocity, not the angular velocity of the rotating sphere. Using the sphere as a reference frame, the mass is always moving. And changes in its motion (speed) are modifications of this basic motion (or action). So we can postulate that one force, at an instant of time (or a definite time) causes one change in motion (to at most a single velocity for the mass, i.e. that is constant).

But, what is an "instant" in time, as to cause an acceleration that results in a constant (linear) velocity, for a mass? If all of the mass(es) of the sphere absolutely stayed in place (i.e. didn't move) during the (sphere's) rotation, then there would be no method for bringing any masses closer or farther apart (from each other). Therefore, attraction and repulsion are due to forces modifying the motion of masses. Otherwise, free moving (non-fixed) masses would be subjected to a force due to the rotational acceleration of the sphere (be it from angular velocity or acceleration).

Therefore, assuming a mass is movable (in relation to others), it is already in motion (in relation to others). But this motion is in conjunction to other freely moving masses, i.e. due to the same force applied. Although the force may be time varying (but we can approximate it as persistent). Thus, any other (or additional) force applied (to a mass) must conform to the persistence of the original (progenitive) force from the sphere's rotation. So, the constant velocity due to an impinging force on a mass relates to the persistence of the (sphere's original) progenitive force.

This relation applies to the complexity of the (movable or moving) components of a mass.

One consequence of the rotational universe model is that, for any mass within, in general, observations of all other masses suggest they are moving away (from the observing mass), particularly if the observing mass is subjected to impinging forces. The impinging forces tend to randomize the motions (and velocities), but the progenitive force is (more) persistent to divert masses from one another (on average).

Complexity of Energy

Because energy (as, e.g., radiant energy or light) is simplistic (of complexity), it conduces to only one speed of travel, matter (or mass) being (much) more complex and therefore adoptable to more speeds. Yet, as light travels (at a constant and singular speed), with distance its intensity wanes. What, then, is so of mass? Possibly its complexity wanes likewise. The internal energy of a mass must be expended (and lost) throughout its travel and (internal, external) motion. This loss supports the mass' constant velocity (directive). It can continue, of course, until there is no mass left (as there is no longer any energy).

What, then, is the "mass" of radiant energy? This must come down to its inherent (though simplistic) complexity. For even an n of "1" has an inherent residue of complexity (i.e., if it exists, it is complex). So, we can apply our complexity physically (here) as follows:

Energy is of an $n = 1$ relation (or "substance") yielding a complexity coefficient of
$(C_n / \Psi^n)(n + 1) == (C_1 / \Psi^1)(1 + 1) \sim= 0.873$

Note that this is the only (integer n) coefficient of existence (i.e. $n > 0$) which reduces the physical relation (since it is less than one); although using Ψ_m $[== (C_1 / \Psi_m^{\,1})(1 + 1)]$ retains a coefficient > 1.

This $n = 1$ coefficient is for

or

or

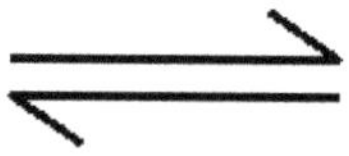

$(1 + 1 \rightarrow 2$, or $n + 1)$ relationship of action.

For $n = 2$ (and higher, of integer), matter (with energy) ensues, and the complexity enhances the relation (the coefficient is > 1). $(2 + 1 \rightarrow 3, n = 2) \rightarrow$

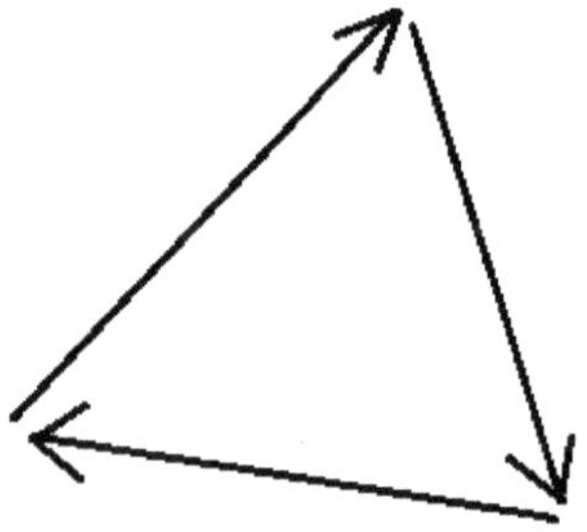

, etc.

of physical relation.

As previously suggested, a speed == s / t (s as a distance, here) would have an $n = 2$ relation (s, t). Presumably, then, a velocity would be an $n = 3$ relation, with direction the third variable: (s, t, d) [d as direction, here]. This might be notated as ${}^{d}S / t$, where ${}^{d}S$ denotes a further (mathematical) relation.

An $(n + 1)$, $n = 3$ relation would be as:

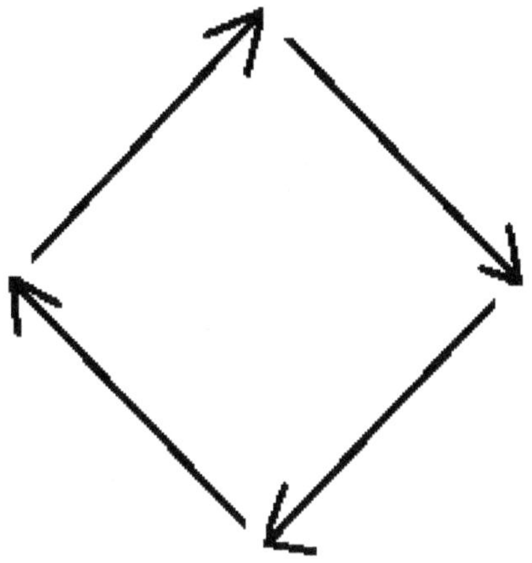

etc.

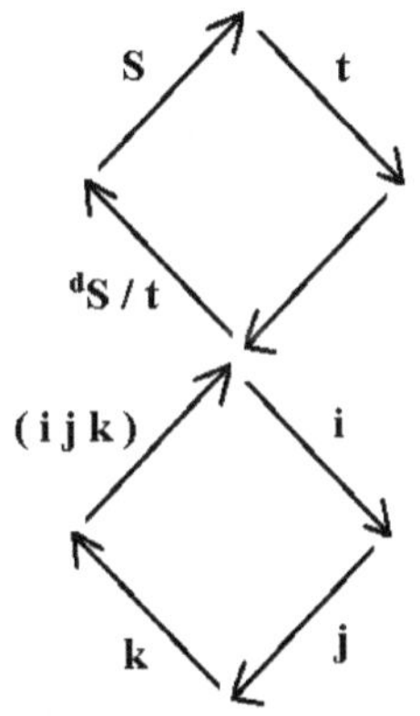

or

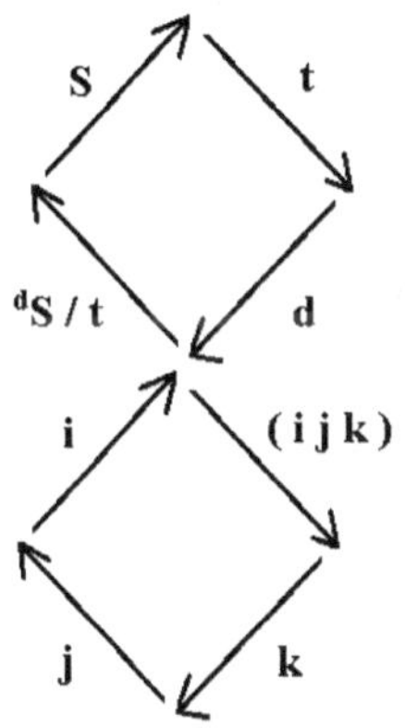

(constructive and contrasting) →

"constractive" (or "determative") [decisive]	**conductive (or explorative)**
historical or passive	**current or active**
∴ **The complexities (for the two loops) add (as if like probabilities).**	**The complexities (for the two loops) multiply (as if like probabilities).**

Though note: the derivation of this third variable (direction) might encounter its own (series of) complexity (for the vector). So we might have "loops" or "rings" of (interconnecting) complexities. They can be called 'associated complexities': e.g.,

For the addition or multiplication of (loop) complexities, they may be (for convenience) normalized (each) by the (total) sum or product (as denominator). This yields the "relative" complexity for each loop of a process.

So we see here, in our example, that the direction of a vector (velocity) can be as important (in terms of complexity) as the relation it modifies (speed) or activates. This suggests loop operations, such as conjugation (addition, contrasting) and integration (multiplication, conducive), as shown or demonstrated above. Separation (subtraction) and excision (disintegrating) logically follow, a means of "clarifying" complexities.

Mathematical Comparison of Complexity Processes

A process can be “resolved” as by the following example:

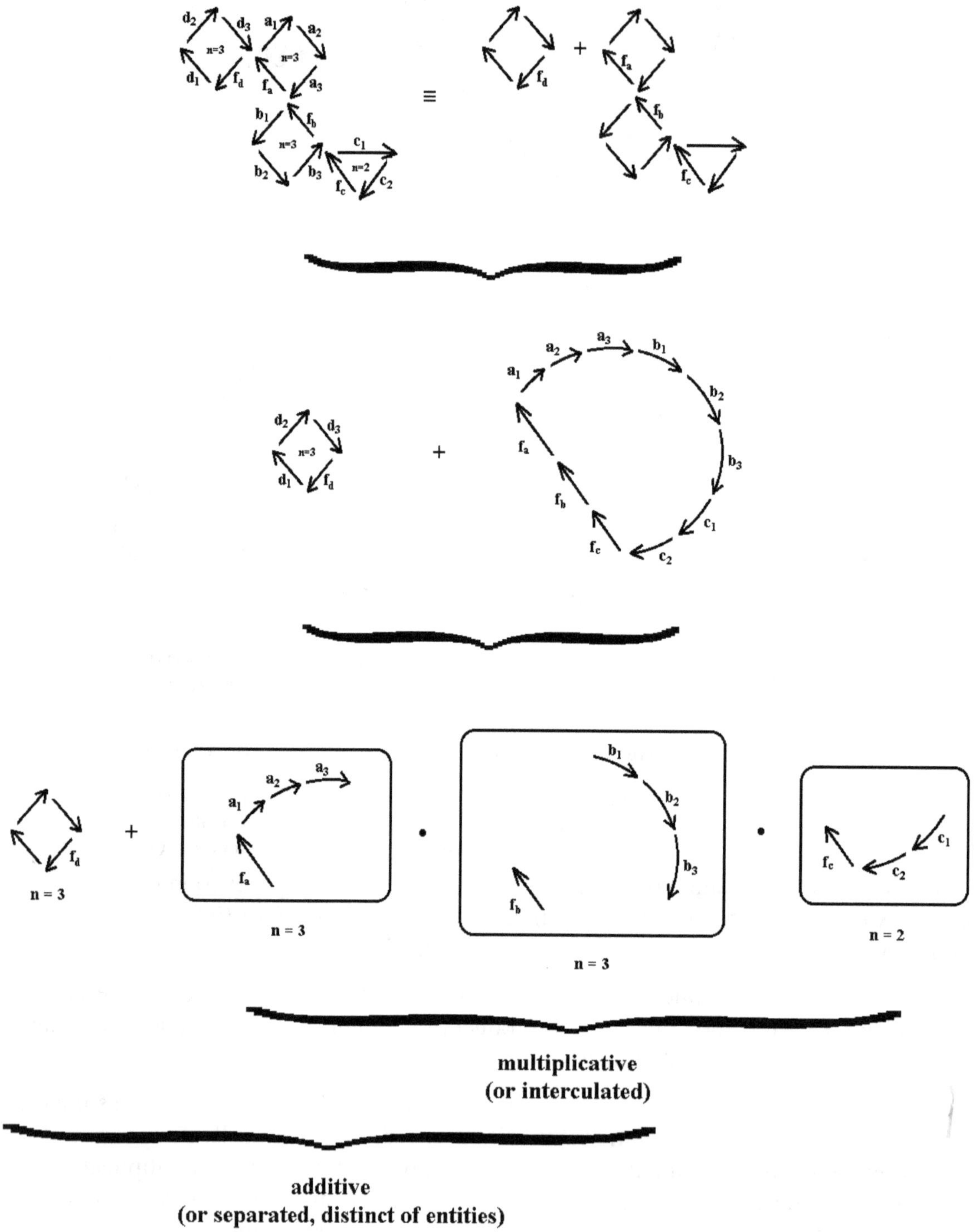

We see demonstrated, then, physical or procedural meanings for processes of addition and multiplication (through their complexity adoptions). This functionally shows the distinction between (2 + 2) and (2 x 2), not via results but by (methods of) action.

Using n = 1 "modules" (i.e. loops), and integers for the variable solutions (i.e. of x), the distinction between (x + x) and (x)(x) or x^2 is easily shown (at n = 1, $N \sim= 0.873$) [attributing each variable to its own complexity → maintains n = 1]:

at x = 2

Complexity of (x + x) is ~ 0.873 + 0.873 = 0.873(2x) == 0.873 • 4 = 3.492

Complexity of (x • x) is ~ (0.873x)(0.873x) = $(0.873)^2 \bullet x^2$ == $(0.873x)^2$ == $(0.873 \bullet 2)^2$ = 3.049

at x = 1

Complexity of (x + x) is ~ (0.873 • 1 + 0.873 • 1) = 2(0.873) = 1.746

Complexity of (x • x) is ~ (0.873 • 1 • 0.873 • 1) = $(0.873)^2$ = 0.7621

Yet, at x = 3

Complexity of (x + x) is ~ (0.873 • 3 + 0.873 • 3) == 2(0.873)(3) = 5.238

Complexity of (x • x) is ~ (0.873)(3)(0.873)(3) == $[(0.873)(3)]^2$ = 6.859

at x = 4

Complexity of (x + x) is ~ (0.873 • 4 + (0.873 • 4) = 2(0.873)(4) = 6.984

Complexity of (x • x) is ~ (0.873)(4)(0.873)(4) == $[(0.873)(4)]^2$ = 12.194

Therefore, $(x = \Psi)_{compl}$ when $[(x + x) = (x \bullet x)]_{compl(exity)}$

[as can be seen by plotting concurrently

x vs Complexity of (x + x)

and x vs Complexity of (x • x)]

That is, 2(0.873)(2.2905) ~ = [(0.873)(2.2905)]2 = 3.99(8 or 9)

or $2N_1\Psi^1 = (N_1\Psi^1)^2 \sim= 4$

or Complexity of (Ψ + Ψ) = Complexity of (Ψ • Ψ)

Complexity of 2Ψ = Complexity of Ψ^2.

$2\Psi \sim= 4.581$, $\Psi^2 \sim= 5.246$

Note that $N_1\Psi^1 == [(C_n(n + 1) / \Psi^n) \bullet \Psi^n]_{n=1} == (C_1 \bullet 2 / \Psi^1) \bullet \Psi^1 = 2C_1 = 2$

Therefore, $2 \bullet N_1\Psi^1 == 2(1) \bullet (2 = 1 + 1) = 2 \bullet 2 = 4$

$(N_1\Psi^1)^2 == [C_1 \bullet (1 + 1) = 1 \bullet 2 = 2]^2 = 4$

Of course, $2 + 2 = 4 = 2 \bullet 2$; therefore, the complexity equivalence [for (x + x) and (x)(x)] is due to the definition of N (where $N_1 \bullet \Psi^1 = 2$).

A plot of x vs Complexity for (x + x) yields a curve with a constant slope of $2N_1$ (thus, a straight line).

A plot of x vs Complexity of (x • x) yields the curve of an exponential (power of 2).

At n = 1 values, the curves intersect at a point where $[(x + x) = (x \bullet x)]_{compl}$, $x = \Psi^1$, and the corresponding complexity = 4 (= 2 + 2 or 2 • 2).

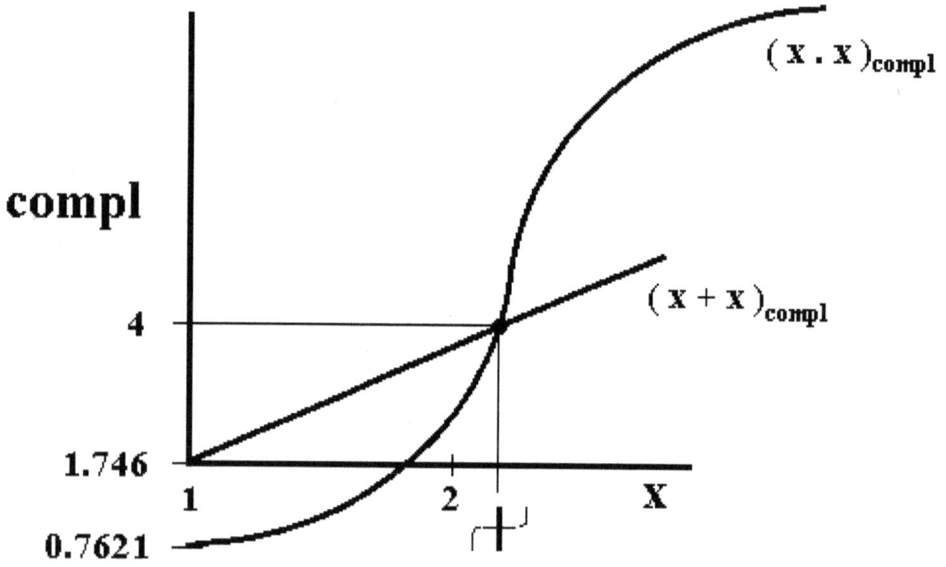

So, this demonstrates a method of deriving Ψ (from two different curves) using the (2 + 2) to (2 • 2) equivalence.

There are two basic types of relationship:

additive, e.g. $x + x$, where the variables remain distinct

multiplicative, e.g. $x \bullet x$, where each (individual) variable is modified of type (so they don't remain distinct)

Here, we have an etymology of (x, y being variables):

relation 1	relation 2	
x	x	
$x + c$	$x \bullet c$	$c == \text{constant}$
$x + x$	$x \bullet x$	(i.e. a power of self)
$x + y$	$x \bullet y == xy$	

The x relationship ($x = x$) is simply identity. The $x + x == x \bullet x$ relationship we have seen as discussed.

For further curve comparisons (in terms of our complexity), we have:

$$x + c = x \bullet c = xc$$

and $$x + y = x \bullet y = xy$$

$x + c = xc$; therefore, $x = xc - c = c(x - 1)$; thus, $c = x / (x - 1)$

Likewise, for $x + y = xy$, $y = x / (x - 1)$

For the ($x + c$), xc relations, n remains 1.

But for the ($x + y$), xy relations, n becomes 2 ; therefore, there is a true complexity distinction between $c = x / (x - 1)$ and $y = x / (x - 1)$.

Here, for example, c must equal 2, while y remains variable (i.e. has a range). Also, y may attain infinity (keeping the equivalence of relations): $x + \infty == x \bullet \infty == \infty$.

At $n = 2$ (and for $n > 1$), each relation (rather than variable) has an assigned $\boldsymbol{N}_n$;

therefore, we compare	$\boldsymbol{N}_2(x + y)$ to $\boldsymbol{N}_2(xy)$	|
as opposed (to)	$\boldsymbol{N}_1 x + \boldsymbol{N}_1 x$ to $\boldsymbol{N}_1^{\,2} x^2$	| definition distinctions
or	$\boldsymbol{N}_1(x + c)$ to $\boldsymbol{N}_1(xc)$	|

But one might also broaden or generalize a relationship, to make the N (and thus n) assignment term specific.

Therefore, for (x + y) to xy, the complexity (assigned) could be $N_1 x + N_1 y$ to $N_2(xy)$.

Consistency of analysis would be determinant of result. For example, for (x + c) to xc we could have $N_1 x$ to $cN_1 x$ (as per calculus), meaning the constant (c) has no (assigned) complexity. This distinguishes term specific from relation specific complexity.

Lack of Complexity Absence

But what does "no variable" complexity specify?

$$\text{Complexity} == (C_n / \Psi^n)(n + 1)$$

If there is no n (i.e. no variable), then n == 0 and we have

$$(C_0 / \Psi^0)(0 + 1) \rightarrow (C_0 / 1) \bullet 1$$

What does C_0 specify?

One might consider it zero, so that the complexity falls to zero.

But one might consider it (for a nul/los analysis) any convenient constant pursuant to the physics. Therefore, we can write:

$$(C_{phys} / \Psi^{phys})(phys + 1) \rightarrow (C_{phys} / 1) \bullet (phys + 1) \rightarrow \text{phys(ical) constant}$$

$$[\text{i.e.}, (C_{phys} / 1)(0 + 1) \rightarrow (C_{phys} / 1) \bullet 1 \rightarrow C_{phys} \rightarrow \text{phys const.}].$$

This assumes that "zero" does not physically exist (i.e. everything varies; all things vary).

Therefore, for to xc, the complexity runs to:

$N_1 x + N_{phys} c$	to	$(N_1 x)(N_{phys} c)$
↓		↓
(physical const.) • c		(physical const.) • c
$\rightarrow N_1 x + c^*$	to	$(N_1 x) c^*$

where c* == c enhanced by the physical constant.

So what is a residual complexity to non-action $(n < 1)$?

It must be a background complexity, and backgrounds depend on the physical process considered. Therefore, a background complexity might be determined from the relation studied. Perhaps a comparison of curves, as we have done, may help.

We know that when the $(x + c)$ and xc curves are equal (at points of equivalence), then $c = x / (x + 1)$. This allows $c = 0$ as a solution, at $x = 0$.

Another solution, as we have seen, is when $c = x$. The point of equivalence (for the two curves) occurs when $x = \Psi$.

Models Avoiding Zero Complexity

So, we have (in terms of complexity):

$$(C_* / \Psi^*)(* + 1) \bullet (x + [c = 0]) = (C_* / \Psi^*)(* + 1) \bullet (x \bullet 0)$$

and $(C_* / \Psi^*)(* + 1) \bullet (x + [c = \Psi]) = (C_* / \Psi^*)(* + 1) \bullet (x\Psi)$.

We can modify these equations with preference for finding C_* by yielding to $\Psi^* \sim\!\longrightarrow \Psi^0 = 1$ and $(* + 1) \sim\!\longrightarrow 1$. Therefore,

$$(C_* / 1)(1) \bullet (x + 0) = (C_* / 1)(1) \bullet (x \bullet 0)$$

$$(C_* / 1)(1) \bullet (x + \Psi) = (C_* / 1)(1) \bullet (x\Psi)$$

This is assuming that the effective n (of non-action) is very small (but positive).

The $(c = 0)$ equation can be simplified to:

$(C_* / 1) \bullet 1 \bullet x = 0$; thus, $xC_* = 0$

Therefore, this simplification (as an equation) can be added to the $(c = \Psi)$ equation:

$xC_* + xC_* + \Psi C_* = x\Psi C_*$, thus, $C_*(2x + \Psi) = x\Psi C_*$.

Therefore, $2x + \Psi = x\Psi$ (i.e. independent of C_*).

Since Ψ itself is a constant (or pure number: ~ ~ 2.2905), then

$2x + 2.2905 \sim = 2.2905x$; therefore,

$2x - 2.2905x = -2.2905$

$x(2.2905 - 2) = 2.2905$

$0.2905x = 2.2905$

$x = 2.2905 / 0.2905 = 7.885$

Now, we have $xC_* = 0 = 7.885C_*$, with $C_* = 0$ as one solution.

We also have $C_* \bullet (x + \Psi) = C_* x\Psi$; therefore,

$C_*(7.885 + 2.2905) = C_*(7.885)(2.2905)$

$10.1758C_* = 18.06C_*$, with $C_* = 0$ as the solution.

Clearly, if C_* is made (positively) very tiny (of value), then each product:

$7.8850C_*$ |

$10.1758C_*$ |→ will also approach $C_* = 0$.

$18.0600C_*$ |

We can consider these products as a solution set and arbitrarily assign the factors (of C_*) to a circle (of the solution set), 3 points defining a circle:

1	7.8850	|	The order of points (i.e. choice
2	10.1758	|	of x coordinate for (x, y))
3	18.060	|	yields unique circle and radius.

This yields (of course) the equation of a circle (of the form $Ax^2 + Ay^2 + Bx + Cy + D = 0$) with a definite radius (r).

Therefore, we can "choose" a C_* (and thus a physical constant) by slightly adjusting $r \longrightarrow r'$, setting the deviation from (solution set) ideality as simply r' / r .

This will yield a new factor for each point (i.e. new y). Setting (for each/any point) this equal to the old product (original, ideal or solution set product) and solving for C_* gives us an " estimate(d) of C^* " (at a deviation of r' / r);

e.g. $7.885C_* =$ (new factor)(chosen C_*)

Solve for C_* ; it will be based on a deviation of r' / r .

Then, the physical constant (accepted) is simply from

$C_* =$ (physical constant) • C == ("solved" C_*) • C , at deviation of r' / r .

The "chosen" C_* should, of course, be close to zero (for a "valid-able" estimate), countenanced by the (accepted) deviation.

We are, effectively, "oscillating, by radius" about the (a) solution set circle.

The "oscillation" is by adjusting the radius away from r (to r'), and the argument is that (physically) the radius is never exactly r (because the circle is for a "zero" solution set, which can not occur).

We might also try to compare term specific complexity to relation specific complexity.

For example, for the relation : $s = x + 2y + 3xy$, we have

term specific	$\boldsymbol{N}_1x + 2\boldsymbol{N}_1y + 3\boldsymbol{N}_2xy$	| definitional
relation specific	$\boldsymbol{N}_2(x + 2y + 3xy)$	| complexities

Therefore, set $\boldsymbol{N}_1x + 2\boldsymbol{N}_1y + 3\boldsymbol{N}_2xy = \boldsymbol{N}_2(x + 2y + 3xy)$

$$\boldsymbol{N}_1(x + 2y) + 3\boldsymbol{N}_2xy = \boldsymbol{N}_2(x + 2y + 3xy)$$

$$\boldsymbol{N}_1(x + 2y) = \boldsymbol{N}_2(x + 2y + 3xy - 3xy) \text{; thus,}$$

$$\boldsymbol{N}_1(x + 2y) = \boldsymbol{N}_2(x + 2y)$$

Since $\boldsymbol{N}_1$ =/= $\boldsymbol{N}_2$, equivalence can only be obtained (via self comparison of the relation, towards complexities, with ' $x + 2y =$ zero ' not considered) by treating ($x + 2y$) as $\boldsymbol{N}_1x + 2\boldsymbol{N}_1y$ or ($x + 2y$) as $\boldsymbol{N}_2x + 2\boldsymbol{N}_2y$ (i.e. adjusting one complexity coefficient to equal the other), and loosing the $\boldsymbol{N}_2xy$ term (altogether) as consistent. [This is just saying that the two "specifics" (term, relational) are not equivalent. Either give ($x + 2y$) the $\boldsymbol{N}_1$ factor or $\boldsymbol{N}_2$ factor.]

The result, here, implies that the relational complexity is more complex than the term complexity by ($\boldsymbol{N}_2 / \boldsymbol{N}_1$)($x + 2y$) , or $\boldsymbol{N}_1(x + 2y)$ • ($\boldsymbol{N}_2 / \boldsymbol{N}_1$) converts term to relational.

That is, $S_{compl} = N_1(N_2 / N_1)(x + 2y) + 3N_2xy$ is used to suggest complexity of $N_2(x + 2y + 3xy)$ when the increase in (total) complexity is $(N_2 / N_1)(x + 2y)$.

We can call $N_1(x + 2y)$ an “impinge-nt” term, because it wants to impinge on (or towards) $N_2(x + 2y)$, to bring the relation $N_1(x + 2y) + 3N_2xy$ to $N_2(x + 2y + 3xy)$.

Therefore, $N_1(x + 2y) \rightarrow N_2(x + 2y)$ is an 'impingent' (as impingement) relation

[i.e., there is a logistic(al) tension, of this means, towards $N_2(x + 2y)$].

So, the assumption is that a relation tends towards the maximum complexity it can attain, or that to restrict complexity requires greater relational effort, similar to the entropy for the doings of a system aiming (always) to increase. The equating of the term and relational specifics was simply to uncover the 'impingent' term(s) explicitly.

Assigning Complexities to Relations via Processing

The two types of relationships, additive and multiplicative, can of course be generalized: e.g., addition and subtraction versus multiplication or any other modification of variable type (e.g. square roots or division, etc.). Complexities (of relationship) may ensue, as in compounding of terms. But the maximum complexity (allowable) is determined by the number of variables (distinct) used; i.e., N_n is a common factor for the relation.

This makes it particularly straight forward to allot a complexity to any given relation (no matter how convoluted). Because of this ease of complexation, one can “process” a relation before assigning a complexity (value): e.g., de-unitation, converting each variable into a pure (unit-less) number by consigning to a base, the most propitious of which could be Ψ :

Say, $S = x + y + 2xy + 3x / y$.

Regardless of the units of x and y , we simply consign each variable to base Ψ (i.e. use as an exponent); therefore, (with ul == 'unit-less' , in the notation)

$$S_{ul} == \Psi^x + \Psi^y + 2\Psi^x\Psi^y + 3\Psi^{x-y} .$$

Now, we can assign a complexity:

$$(S_{ul})_{compl} == N_2(\Psi^x + \Psi^y + 2\Psi^x\Psi^y + 3\Psi^{x-y}) = C_2(2 + 1)(\Psi^{x-2} + \Psi^{y-2} + 2\Psi^{x+y-2} + 3\Psi^{x-y-2}),$$

where $N_2 == (C_2 / \Psi^2)(2 + 1)$.

$\therefore (S_{ul})_{compl} = 3C_2(\Psi^{x-2} + \Psi^{y-2} + 2\Psi^{x+y-2} + 3\Psi^{x-y-2})$, where $C_2 = 2$; thus,

$$(S_{ul})_{compl} = 9(\Psi^{x-2} + \Psi^{y-2} + 2\Psi^{x+y-2} + 3\Psi^{x-y-2}) = 9(\Psi^{x-2} + \Psi^{y-2}) + 18\Psi^{x+y-2} + 27\Psi^{x-y-2} .$$

We can compare a relation's behavior to its "process" behavior using these assigned complexities.

For example, for a speed $S = d / t$, $N_2 \sim= (3 / (2.2905)^2(2 + 1) = 1.7155$

$S_{compl} == N_2(d / t)$, $(S_{ul})_{compl} == N_2\Psi^{d-t}$:

	(a) const. Speed				||	**(an) acceleration**			
d	**3**	**6**	**12**	**24**	||	**1**	**2**	**4**	**16**
t	**1**	**2**	**4**	**8**	||	**3**	**3**	**3**	**3**
s	**3**	**3**	**3**	**3**	||	**0.333**	**0.667**	**1.333**	**5.333**
S_{compl}	**[**	**5.1464**		**]**	||	**0.571**	**1.144**	**2.287**	**9.149**
$(S_{ul})_{compl}$	**9**	**47.2**	**1299.7**	**984635.4**	||	**0.327**	**0.749**	**3.939**	**81937.9**
S_{ul}	**5.246**	**27.524**	**757.604**	**573964.1**	||	**.19061**	**.43659**	**2.2905**	**47763.2**

The S_{ul} and $(S_{ul})_{compl}$ terms are of course exponential. Instead of S_{ul} or $(S_{ul})_{compl,}$ for convenience one can plot: $\log_{\Psi}\Psi^{d-t} = d - t == \log_*$ as **S vs log(** S_{ul} **)** , S_{compl} **vs** $\log_*[(S_{ul})_{compl}]$ and even combine the curves to yield an "area" of complexity (or plot region leading, or lending, towards complexity).

For example: $\log_*(S_{ul}) == d - t$, $\log_*(S_{ul})_{compl} == N_2(d - t)$ by definition $= N_2\log_*(S_{ul})$.

[For formal "area" (or region) graphs, one would plot $(S ; S_{compl})$ vs $(S_{ul} ; S_{ul,\,compl})$ to yield 4 distinct curves that may act as boundaries for a confined (or even open) area.]

The complexity coefficients, of course, simply magnify the behaviors of the relations (shifting the curves to higher values but of the same proportions and slopes), as by being (common) factors. The "processing" makes the relation an exponential (argument or) function (for Ψ).

Pre-processing Relations for Complexities

A form of processing can be "term specific" or variant processing, which is a manner of pre-processing (or partial processing).

For example, using the relation $x + y + 2xy + 3x / y$ one can pre-process in various ways (for specific reasons):

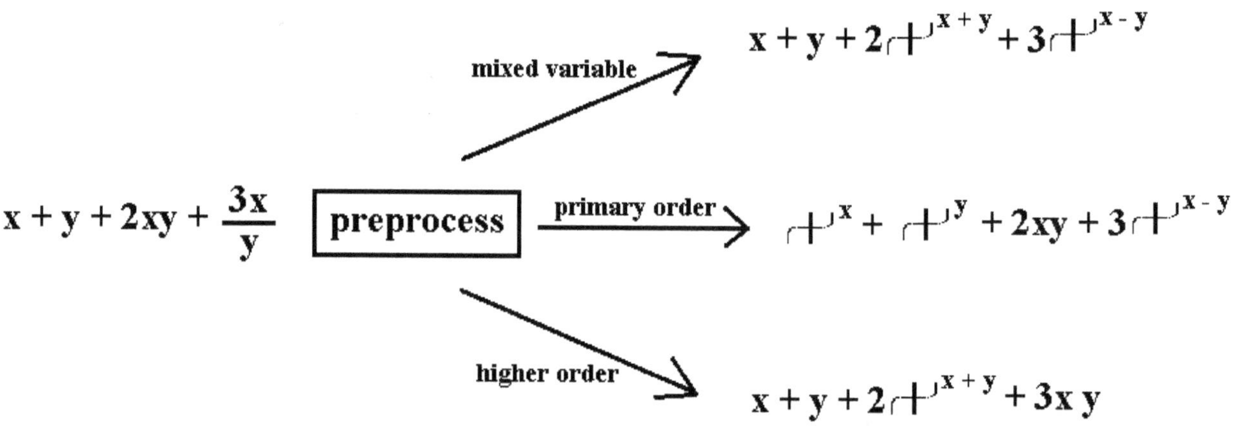

Now, one can naturally (and very reasonably) ask: which pre-processed relation approaches (in value) the complexity of the original relation? since this might reveal the functional relevance (or impetus) of the variables involved (for the physics to be accomplished); i.e., which variables or variable terms are most relevant to the particular (physical) problem at hand (or being considered given certain conditions)? A greater (larger) complexity yields (to) greater "freedom" of solution, or possibilities of solution. The calculated complexity might show (via the pre-processing) what ranges of values (of variables) are most appropriate for a specific problem.

Say, for the above relation (ill-regardless of unitage) we assign $x = 1$, $y = 3$.

Then $\mathbf{x + y + 2xy + 3x / y = 1 + 3 + 2(1 \bullet 3) + 3(1) / 3 = 4 + 6 + 1 = 11}$; $N_2 \sim = \mathbf{1.7155}$.

Therefore, the relation's complexity $\sim = (1.7155)(11) = 18.8705$

Mixed variable pre-processing →

$$\mathbf{x + y + 2\Psi^{x+y} + 3\Psi^{x-y} = 1 + 3 + 2\Psi^{4} + 3\Psi^{-2} = 1 + 3 + (55.05) + (0.5718) = 59.62}$$

Primary order pre-processing →

$$\Psi^{x} + \Psi^{y} + 2xy + 3\Psi^{x-y} = \Psi^{1} + \Psi^{3} + 2 \bullet 1 \bullet 3 + 3\Psi^{-2} = (2.2905) + (12.02) + 6 + (0.5718) = 20.88$$

Higher order pre-processing →

$x + y + 2\Psi^{x+y} + 3x / y = 1 + 3 + (55.05) + 1 = 60.05$

[The term ' $3x / y$ ' is technically reduced order, or lower order, the corresponding Ψ exponent being < 1 (here). But it can formally be considered as higher order, to lead towards another Hop value.]

The (calculated) complexity is clearly closest (in value) to that produced via the Primary order pre-processing, which suggests that terms with the variable to primary (or 1st) order (i.e. x^1, y^1) are most appropriate for solution using the assigned (x, y) values (here).

This provides, in the instance given (above), an interesting formulation for the complexity coefficient in terms of the relationship used, since

$$N_2(x + y + 2xy + 3x / y) == \Psi^x + \Psi^y + 2xy + 3\Psi^{x-y}$$

yields exact values for x and y ; therefore,

$$N_2 = (\Psi^x + \Psi^y + 2xy + 3\Psi^{x-y}) / (x + y + 2xy + 3x / y)$$

or $N_2x - \Psi^x + N_2y - \Psi^y + 2N_2xy + 3N_2x / y = 2xy + 3\Psi^{x-y}$,

or $x(N_2 - 1) + y(N_2 - 1) + 2N_2xy + (3N_2x / y) + x + y = \Psi^x + \Psi^y + 2xy + 3\Psi^{x-y}$.

Therefore,

$x(N_2 - 1) + y(N_2 - 1) + 2N_2xy - 2xy + (3N_2x / y) + x + y = \Psi^x + \Psi^y + 3\Psi^{x-y}$.

Therefore,

$x(N_2 - 1) + y(N_2 - 1) + 2xy(N_2 - 1) + x + y = \Psi^x + \Psi^y + 3\Psi^{x-y} - 3N_2x / y$.

Therefore,

$(N_2 - 1)(x + y + 2xy) + x + y = \Psi^x + \Psi^y + 3\Psi^{x-y} - 3N_2x / y$.

Therefore,

$(N_2 - 1)(x + y + 2xy) = \Psi^x + \Psi^y + 3\Psi^{x-y} - (3N_2x / y) - x - y$

$= \Psi^x - x + \Psi^y - y + 3(\Psi^{x-y} - 3N_2x / y)$

The last equation, on the right side, outlines the terms specifically affected by the 'Primary order' (pre-)processing; i.e. x^1 , y^1, $3x^1 / y^1$ terms.

Therefore,

$$(N_2 - 1) = [(\Psi^x - x)(\Psi^y - y) + 3(\Psi^{x-y} - N_2 x / y)] / (x + y + 2xy)$$

The terms:

$$(\Psi^x - x) / (x + y + 2xy) , (\Psi^y - y) / (x + y + 2xy) , 3(\Psi^{x-y} - (N_2 x / y)) / (x + y + 2xy)$$

can be called "complexing predicates" of the (original) relation. The last term includes all of the term types (of the original relation) as well as a recursion of N (the sum of the predicates yields an expression for N directly).

Using our assigned values of $x = 1$, $y = 3$ for the relation:

$x + y + 2xy = 1 + 3 + 6 = 10$ | relation: $x + y + 2xy + 3x / y$

$\Psi^x - x = 2.2905 - 1 = 1.2905$ | Primary

$\Psi^y - y = 12.0169 - 3 = 9.0169$ | order : $\Psi^x + \Psi^y + 2xy + 3\Psi^{x-y}$

$\Psi^{x-y} = \Psi^{-2} = 0.1906$ | pre-processing

$N_2 x / y = (1.7155)(1) / (3) = 0.5718$; therefore,

$(\Psi^x - x) / (x + y + 2xy) = 1.2905 / 10 = 0.12905$

$(\Psi^y - y) / (x + y + 2xy) = 9.0169 / 10 = 0.90169$

$3(\Psi^{x-y} - N_2 x / y) / (x + y + 2xy) = 3(0.1906 - 0.5718) / 10 = -1.1436 / 10 = -0.11436$

∴ the sum of predicates $== N_2 - 1 = 0.12905 + 0.90169 - 0.11436 = 0.91638$

$0.91638 = N_2 - 1$; thus, $N_2 = 1.91638$ as estimated ($== N_{est}$)

1.91638(18.8705, relation's complexity / 20.88, Primary order pre-processing) = 1.7316 ~ 1.7155

Therefore, the (primary order) pre-processing overestimates the expected (complexity) coefficient by: 1.7316 / 1.7155 = 1.0094 , or just under 1% (== 100.94% - 100% = 0.94%).

The 0.94% discrepancy is presumably due (in part) to the choice of (x, y) values for the enforced equality used.

[(xy) is considered (here) higher order, while (x / y) is not, employing the distinction between multiplication and division, whereas addition and subtraction are considered equivalent (operations) functioning.]

So, the complexity of a relation can be (reasonably) made equivalent to the relation's pre-processing, by converting its variables to exponential arguments of the base Ψ, but in a particular manner.

If we use a relation that is a summation of basic (primary order) variables: $x_1, x_2, x_3, \ldots$

we have

$$\frac{C_n}{\Psi^n}(n+1)\sum_{m=1}^{n} x_m \equiv \sum_{m=1}^{n} \Psi^{x_m}$$

$$\therefore \quad C_n(n+1)\sum_{m=1}^{n} x_m = \Psi^n \sum_{m=1}^{n} \Psi^{x_m} \equiv \sum_{m=1}^{n} \Psi^{x_m + n}$$

For example, for x_1, $n = 1$: $(1)(2)x_1 = \Psi^{x1+1}$

$n = 2$, $(3)(3)(x_1 + x_2) = \Psi^{x2+2} + \Psi^{x2+2} == 9(x_1 + x_2)$

$n = 3$, $(7)(4)(x_1 + x_2 + x_3) = \Psi^{x1+3} + \Psi^{x2+3} + \Psi^{x3+3} == 28(x_1 + x_2 + x_3)$

$n = 4$, $(25)(5)(x_1 + x_2 + x_3 + x_4) = \Psi^{x1+4} + \Psi^{x2+4} + \Psi^{x3+4} + \Psi^{x4+4} == 125(x_1 + x_2 + x_3 + x_4)$,

etc. equivalencies (applied).

The factors $[\, C_n \bullet (n+1) \,]$ series: (2, 9, 28, 125, …) is evident.

If we take our speed, or divisional, relation:

$s = d / t$

and use primary order (pre-)processing, then

$\boldsymbol{N}_2 s == \boldsymbol{N}_2(d / t) = \Psi^d / \Psi^t = \Psi^{d-t}$; therefore, (as for above)

$9(d / t) = \Psi^{d-t+2}$.

Now, if we (arbitrarily) consign $t = 1$, we establish a fundamental unitage (for d / t) of division; therefore,

$9d / 1 = \Psi^{d-t+2} = \Psi^{d+1}$.

Let $d / 1 == {}^*d$ (i.e. with this fundamental unitage for d / t) ; therefore,

$9 \bullet {}^*d = \Psi^{d+1}$; thus,

$\log\Psi(9 \bullet {}^*d) = \log_\Psi \Psi^{d+1} = d + 1$.

But, $d == s$; therefore, $\log_\Psi(9s) = d + 1$ (unit-less, pure numbers).

Therefore, $\log_\Psi 9 + \log_\Psi s = d + 1$

$\log_\Psi s = d + 1 - \log_\Psi 9$

Therefore, changing to base 10:

$\log_{10} s / \log_{10} \Psi = d + 1 - (\log_{10} 9 / \log_{10} \Psi)$; thus,

$\log10s / 0.3599 \sim = d + 1 - (\log_{10} 9 / \log_{10} \Psi) = d + 1 - 2.6512$

Therefore, $\log10s = 0.3599(d + 1 - 2.6512) = 0.3599d - 0.5943$; thus,

$10\hat{}\log_{10} s = s = 10^{(0.3599d - 0.5943)}$.

Note that in this relational formulation s is always positive and substantial (i.e. extant of some "moving" value). Its smallest value is $10^{-0.5943} = 0.2545$, suggesting a "stationary" or base speed (without a distance traveled) of some fundamental at $t = 1$ in unitage (therefore a minimum motion w/o measurable distance?).

Perhaps of more pertinence, for speed with displacement, accepting the pre-processing equivalence as valid, is to allow for $d = 1$ in unitage (as $t = 1$) to yield a fundamental value for speed (or division) as: $s = 10^{[(0.3599 \bullet 1) - 0.5943]} = 0.5829$ (of unitage).

The number values found here, then, are derived ultimately from (only) $\boldsymbol{N}$, Ψ, and n.

But the (speed) pre-processing, unless imaginary or virtual terms are also employed, involves a single term (of two primary order variables), and therefore can be said to undergo a full processing. That, however, requires (from our definitions) the employment of the equivalence:

$$\boldsymbol{N}_2 \mathrm{s} == \boldsymbol{N}_2(\mathrm{d} / \mathrm{t}) = \boldsymbol{N}_2 \Psi^{\mathrm{d}-\mathrm{t}},$$

which only occurs when $\mathrm{d} = \mathrm{t}$ (while restricting units).

The more naturally derived equivalence, then, (freeing up the values of s, d, and t) is:

$$\boldsymbol{N}_2 \mathrm{s} = (\boldsymbol{N})_{\Psi} \Psi^{\mathrm{d}-\mathrm{t}} == {}_{\Psi}\boldsymbol{N} \Psi^{\mathrm{d}-\mathrm{t}},$$

where ${}_{\Psi}\boldsymbol{N}$ is the resulting complexity coefficient (from the equivalence imposed) and an emphasis is on establishing (or finding) the value of the effective n for the relationship.

The value of ${}_{\Psi}\boldsymbol{N}$ can readily be calculated for any s, by (here) allowing $\mathrm{t} = 1$ and (thus)

$\mathrm{d} = \mathrm{s} \bullet \mathrm{t} = \mathrm{s} \bullet 1$ (units still restricted).

For example, let $\mathrm{s} = 10$. Thus, $\mathrm{d} = 10$, $\mathrm{n} = 2$

Therefore, $\boldsymbol{N}_2 10 = {}_{\Psi}\boldsymbol{N} \Psi^{10-1}$; thus,

$(9)(10) = {}_{\Psi}\boldsymbol{N} \Psi^{10-1+2}$; thus,

$90 = {}_{\Psi}\boldsymbol{N} \Psi^{10+1}$; thus,

$${}_{\Psi}\boldsymbol{N} = 90 / \Psi^{11} \sim= 0.0099 == (\mathrm{C}_{\mathrm{x}} / \Psi^{\mathrm{x}})(\mathrm{x} + 1)$$

where x == effective n (or fractional n).

When $\mathrm{n} > 1$ (i.e. $\mathrm{n} = 2$), $\mathrm{C}_{\mathrm{n}} > \Psi^{\mathrm{n}}$; therefore, $\boldsymbol{N}_{\mathrm{n}} > 1$; thus, the effective n (here, x) must be < 1 .

It can be estimated as close to 0.01 (if we assume $\mathrm{n} \sim\sim \mathrm{C}_{\mathrm{n}}$ here) :

$$(0.01 / \Psi^{0.01})(0.01 + 1) \sim= 0.0100 .$$

One can now argue, here, that C_{x} is slightly less than x, in order to bring ${}_{\Psi}\boldsymbol{N}$ down from 0.0100 to $\sim= 0.0099$ (or $90 / \Psi^{1}$) ;

i.e., $(\mathrm{C}_{\mathrm{x}} / \Psi^{0.01})(0.01 + 1) = 90 / \Psi^{1}$.

Therefore, $C_x = (90 \bullet \Psi^{0.01}) / ((1.001) \bullet \Psi^{11}) \sim= 0.00987$ (or 0.00986932).

$\therefore$ when $n = 90 / \Psi^{11} \sim= 0.009886$, then $C_x = 0.00987$ (or 0.00986932).

[We are approximating Ψ as ~ 2.2905, here.]

Thus, the difference: $n - C_x \sim= 0.000022542$.

But, we note that with this method of formulation (or procedure), as s increases (i.e. s↑), ${}_{\Psi}N$ decreases and so does C_x↓, x↓ ; so that essentially $x \sim\sim C_x \sim\sim {}_{\Psi}N == N_x$ (thus is this a characteristic "rite of being" presented, here).

This behavior is characteristic of x (fractional n; $n < 1$ and positive) as an exponent to some base (e.g. similar to Ψ_m, e), and is a function of the exponential descent being indefinitely long without reaching zero for the exponential function's value.

The algebraic architecture in our case of "$(x / \Psi^x)(x + 1) = x$" is that a base brought to a small number, though positive, exponent or power results already to a value of: 1 + a small number. This result simply (or readily) cancels the $(x + 1)$ factor to approximately lend to x as the answer of the expression.

Therefore, the nature of x being a fractional n yields to (approximately) x as the magnitude of the (complexity) coefficient.

Thus, $x \sim\sim N_x$, since the broadness (or broad domain) of x values simply approximates (or comes towards) 1 when a base is brought to x as a power.

This is why we can make such formulaic approximations as, say for $s = d / t$:

$N_2 s \sim\sim {}_{\Psi}N\Psi^{s-t}$, where t is assigned 1 . [Therefore, "magnitudinally" s == d , though unitage is restricted.]

Therefore, $9s \sim\sim {}_{\Psi}N\Psi^{s-t+2} == {}_{\Psi}N\Psi^{s+1}$ (x == effective n is < 1).

As s↑ , ${}_{\Psi}N$↓ (past $\Psi^{s+1} > 9$).

But, to conserve units, since Ψ^{s+1} is a pure (unit-less) number, the unitage of s must be conferred (on)to ${}_{\Psi}N$, distinguishing (functionally) ${}_{\Psi}N$ from N_x and N_n in general.

Therefore, ${}_{\Psi}N == \kappa N_x$, where κ is a unitage conveyor.

But this is dependent on the legitimacy of the processing. Therefore, κ is a processing function (i.e., it is required here). Apparently, κ can carry more than just unitage, but also conversions, which must also be part of the processing function.

The purpose of (forming) the (pre-)processing equivalency is to try to determine the nature of a relation's complexity and its behavior under various (physical) circumstances: e.g., comparison (for our previous example) of $n = 2$ to effective $n \sim\sim 0.01$ suggests that the (Ψ base) pre-processing amplifies the complexity of the ($n = 2$, thus 2) variables by $2 / 0.01 \sim == 200$ times, which might not be too surprising when the variables are elevated to (be) exponents of the (Ψ) base.

Of course, we can have partial-pre-processing, or processing with respect to some variable (as opposed to others), similar to (a requirement of) partial differentiation in calculus.

Using our example of $s = d / t$, $d = 10$, $t = 1$

Full pre-processing:

$$N_2(10 / 1) = {}_{\Psi}N\Psi^{10} / \Psi^{1} = {}_{\Psi}N\Psi^{10-1}\text{ ; therefore,}$$

$$9(10) = {}_{\Psi}N\Psi^{10-1} \bullet \Psi^{2} = {}_{\Psi}N\Psi^{10-1+2} = {}_{\Psi}N\Psi^{11}\text{ ; thus,}$$

$${}_{\Psi}N = 90 / \Psi^{11} \sim= 0.0099$$

d pre-processing:

$$90 = ({}_{\Psi}N\Psi^{10} / 1) \bullet \Psi^{2} = {}_{\Psi}N\Psi^{12}\text{ ; therefore,}$$

$${}_{\Psi}N = 90 / \Psi^{12} \sim= 0.0043$$

t pre-processing:

$$90 = ({}_{\Psi}N \bullet 10 / \Psi^{1}) \bullet \Psi^{2} = {}_{\Psi}N \bullet 10 \bullet \Psi^{2-1} = 10{}_{\Psi}N\Psi^{1}\text{ ; therefore,}$$

$${}_{\Psi}N = 90 / 10\Psi^{1} = 9 / \Psi^{1} \sim= 3.9293$$

The corresponding relational complexities, as distinct from the variation or variable complexities (n / x), is afforded by $N_n / {}_{\Psi}N$ (for pre-processing):

relational	full pre-processing	9 / 0.0099 = 909.0909
numeratoral	d pre-processing	9 / 0.0043 = 2093.0233
denominal	t pre-processing	9 / 3.9293 = 2.2905

Such a distribution of (pre-processed) complexities might be typical of a divisional relation, since the denominal variable often assumes a value of 1. This over-emphasizes the numeratoral variable, while the full pre-processing "averages" the numeratoral to the denominal.

Note that with t kept as ($t = 1$), the singular characteristic of yielding a factor Ψ^{n-1} is applied.

For example, $9 \bullet d = ({}_{\Psi}N\Psi^{d} / \Psi^{1}) \bullet \Psi^{n} == {}_{\Psi}N \bullet \Psi^{d} \bullet \Psi^{n-1}$.

$\therefore$ with (at $n = 2$) $9 == C_{n} \bullet (n + 1)$, we have $C_{n} \bullet d = {}_{\Psi}N \bullet \Psi^{d} \bullet \Psi^{n-1} / (n + 1)$;

thus, $C_{n} \bullet d / {}_{\Psi}N \bullet \Psi^{d} = \Psi^{n-1} / (n + 1)$, which seems independent of d;

i.e., $(C_{n} / {}_{\Psi}N)(d / \Psi^{d}) = \Psi^{n-1} / (n + 1)$ at $t == 1$ (in unitage); therefore, $s == d$ (in magnitude).

This provides a direct determination of ${}_{\Psi}N$, or an estimation of C_{n} for a given ${}_{\Psi}N$, with n and d known (with validity of the pre-processing assumed).

Therefore, using our full pre-process example values:

$\mathbf{(C_{n} / {}_{\Psi}N)(d / \Psi^{d}) = \Psi^{n-1} / (n + 1) ; \{ n = 2, C_{n} = 3, {}_{\Psi}N \sim= 0.0099 \} \rightarrow d = 10}$ **(at t = 1).**

That is, $(3 / 0.0099)(10 / \Psi^{10}) = \Psi^{2-1} / (2 + 1) , \Psi \sim\sim 2.2905$;

$\Psi^{1} / 3 \sim= 0.7635 , (3 / 0.0099)(10 / \Psi^{10}) \sim= 0.7624$

Comparing (d / Ψ^{d}) to ($\Psi^{n-1} / (n + 1)$), the (full) pre-processed relation seems to give d attributes akin to n. Since $t == 1$, d can be substituted with s ($= d / t$) applying unitage conversion defined via κ (kappa); i.e. ${}_{\Psi}N == {}_{\kappa}N_{x}$.

With n never zero (physically), as well as for Ψ (and being always positive), any limiting values (for s) must be caused by N_{x} (in its formation).

Applying Units to Consider Complexities

For calculating s, N_{x} is numerical:

$$s = (\Psi^{n-1} / (n + 1)) \bullet \Psi^{d} \bullet (N_{x} / C_{n}) \quad t == 1;\text{ thus } s == d .$$

For large d, since d is being used as an exponent the right side of the relation is much larger. Yet magnitude of s == 'that of d' ; therefore, with $n >= 1$ and $t = 1$, N_{x} must become exceedingly small to bring the right side (back) down to s. This assumes that n is low in number (for the relation $s = d / t, n = 2$). For a proper s unitage, a $\underline{\kappa}$ would have to be applied to (the) N_x (to make ${}_{\Psi}N$). Here, with $t = 1$ assumed and the value of d as a pure number, the nature of $\underline{\kappa}$ as a conversion factor would be in relation to (the desired type of) s.

For example, say s is in meters / second ; the units of $\underline{\kappa}$ would presumably be something like (meters / feet)(1 / sec) as a conversion from feet / sec to meters / sec . Then, N_x would have to be in feet. This permits N_x to deviate from integer values (as can N_n and ${}_{\Psi}N$). But the presumed "effective n" for N_x also deviates from a (proper) integer (value for) n (allowing N_x to become exceeding small).

Alternatively, $\underline{\kappa}$ can be used as a simple unitage (here of meters / time), allowing N_x to remain "pure" of number.

Logarithmic Considerations to Complexity

With a very large d, computational analysis might be aided (or made more convenient) by the relation to logarithms (of base Ψ):

$s = (\Psi^{n-1} / (n + 1)) \bullet \Psi^{d} \bullet (\underline{\kappa} N_x / C_n)$ $t == 1$; thus $s == d$.

Therefore, $\log_{\Psi} s = (n - 1) + d + \log_{\Psi} \underline{\kappa} + \log_{\Psi} N_x - \log_{\Psi}(n + 1) - \log_{\Psi} C_n$.

Changing to base 10 :

$\log_{10} s / \log_{10} \Psi = (n - 1) + d + (\log_{10} \underline{\kappa} / \log_{10} \Psi) + (\log_{10} N_x / \log_{10} \Psi) - (\log_{10}(n + 1) / \log_{10} \Psi) - (\log_{10} C_n / \log_{10} \Psi)$; therefore,

$[\log_{10} s - \log_{10} \underline{\kappa} - \log_{10} N_x + \log_{10}(n + 1) + \log_{10} C_n] / \log_{10} \Psi = n - 1 + d$,
$\log_{10} \Psi \sim= 0.3599 \sim\sim 0.36$; thus,

$\log_{10} s - \log_{10} \underline{\kappa} - \log_{10} N_x + \log_{10}(n + 1) + \log_{10} C_n \sim= 0.36(n - 1 + d) \sim\sim 0.36d$ for large d and low n. [This is for a pre-processed relation, with $t == 1$; thus $s == d$.]
An equivalency is, thus, established to our earlier formulation:

$N_2 s = N_2(d / t) = \Psi^{d} / \Psi^{t} = \Psi^{d-t}$; therefore,

$9(d / t) = \Psi^{d-t+2}$, where $N_2 == C_2 \bullet (2 + 1) / \Psi^{2}$; therefore,

$\log_{\Psi}9 + \log_{\Psi}s = d + 1$ (assuming $t == 1$);

thus, $\log_{\Psi}s = d + 1 - \log_{\Psi}9$;

thus, $\log_{10}s / \log_{10}\Psi = d + 1 - (\log_{10}9 / \log_{10}\Psi)$

But (here) **$9 == C_2(2 + 1) = C_n(n + 1)$**; therefore, **$\log_{10}9 == \log_{10}C_n + \log_{10}(n + 1)$**

$\therefore (\log_{10}s / \log_{10}\Psi) + (\log_{10}C_n / \log_{10}\Psi) + (\log_{10}(n + 1) / \log_{10}\Psi) = d + 1$.

Thus, **$\log_{10}s + \log_{10}C_n + \log_{10}(n + 1) = (\log_{10}\Psi)(d + 1) \sim= 0.36(d + 1; n = 1 == 2 - 1)$.**

Therefore, the difference between the relations is simply:

$[\log_{10}s + \log_{10}C_n + \log_{10}(n + 1)] - [\log_{10}s - \log_{10}\underline{\kappa} - \log_{10}\mathcal{N}_x + \log_{10}C_n + \log_{10}(n + 1)]$

$== \log_{10}\underline{\kappa} + \log_{10}\mathcal{N}_x == \log_{10\Psi}\mathcal{N}$; i.e., $(\underline{\kappa})\mathcal{N}_x = {}_{\Psi}\mathcal{N}$, the "effective" complexity coefficient for the (presumed) pre-processing.

${}_{\Psi}\mathcal{N}$ was, of course, "assigned" to the (latter) pre-processing relation (to allow for full processing equivalence).

Let's try this with a partial pre-processing (of $s = d / t$) , where the subject of (Ψ) exponentiation is d : $n = 2$, $s = d / t$

$(C_n(n + 1) / \Psi^n) \bullet s = \Psi^d / t$; therefore,

$(C_2(2 + 1) / \Psi^2) \bullet s == 9s / \Psi^2 = \Psi^d / t$; thus,

$9s = (1 / t)\Psi^{d+2}$.

If, here, $t = 1$, then $9s = (1 / 1)\Psi^{d+2} = \Psi^{d+2}$.

Again, here, $9 == C_n(n + 1)$.

Therefore, $\log_{\Psi}9s = \log_{\Psi}C_n + \log_{\Psi}(n + 1) + \log_{\Psi}s$

$\log_{\Psi}\Psi^{d+2} = d + 2$; thus,

$(\log_{10}C_n / \log_{10}\Psi) + (\log_{10}(n + 1) / \log_{10}\Psi) + (\log_{10}s / \log_{10}\Psi) = d + 2$; therefore,

$\log_{10}C_n + \log_{10}(n + 1) + \log_{10}s \sim= 0.36(d + 2)$,

which is (approximately) 0.36 greater than the (full) pre-processing result of $0.36(d + 1)$.

Perhaps more interesting is (in this case) a denominal pre-processing:

$(C_n(n + 1) / \Psi^n) \bullet s = d / \Psi^t$; therefore, $\Psi^{t-n} C_n(n + 1) \bullet s = d$.

With $t = 1, n = 2 : \Psi^{-1} C_n(n + 1) \bullet s = d$; thus $(9 / \Psi) \bullet s = d , 3.9293s = d$.

But, with $s == d$ (in magnitude), as $t = 1$,

then clearly ${}_\Psi\boldsymbol{N}$ must be applied as $(9 / \Psi) \bullet s = {}_\Psi\boldsymbol{N}d$

so that ${}_\Psi\boldsymbol{N} = 3.9293$ (as seen earlier).

We can also logarithmically analyze this:

$(9 / \Psi) \bullet s = d \quad == \quad (9 / \Psi) \bullet s = {}_\Psi\boldsymbol{N}d$

$(\log_\Psi 9 / \log_\Psi \Psi^1) \bullet \log_\Psi s = \log_\Psi({}_\Psi\boldsymbol{N} \bullet d) = \log_\Psi {}_\Psi\boldsymbol{N} + \log_\Psi d$

Therefore, $(\log_\Psi 9 / 1) \bullet \log_\Psi s = \log_\Psi {}_\Psi\boldsymbol{N} + \log_\Psi d$; thus,

$\log_\Psi 9 + \log_\Psi s = \log_\Psi {}_\Psi\boldsymbol{N} + \log_\Psi d$; thus,

$\log_\Psi 9 + \log_\Psi s - \log_\Psi \Psi^1 = \log_\Psi {}_\Psi\boldsymbol{N} + \log_\Psi d$,

$\log_\Psi 9 + \log_\Psi s - 1 = \log_\Psi {}_\Psi\boldsymbol{N} + \log_\Psi d$; therefore,

$\log_\Psi 9 + \log_\Psi s - \log_\Psi {}_\Psi\boldsymbol{N} - \log_\Psi d = 1$; and

$(\log_{10} 9 / \log_{10} \Psi) + (\log_{10} s / \log_{10} \Psi) - (\log_{10} {}_\Psi\boldsymbol{N} / \log_{10} \Psi) - (\log_{10} d / \log_{10} \Psi) = 1$

Therefore, $\log_{10} 9 + \log_{10} s - \log_{10} {}_\Psi\boldsymbol{N} - \log_{10} d = \log_{10} \Psi \sim= 0.36$

Since $s == d$, we have:

$\log_{10} 9 - \log_{10} {}_\Psi\boldsymbol{N} + (\log_{10} s - \log_{10} d) = \log_{10} \Psi \sim= 0.36$,

or $\log_{10} 9 - \log_{10} {}_\Psi\boldsymbol{N} \sim= 0.36$

Therefore, $\log_{10} {}_\Psi\boldsymbol{N} \sim= \log_{10} 9 - 0.36 \sim= 0.59424$; thus (again)

${}_\Psi\boldsymbol{N} \sim= 3.9286 (== 3.9293) == 10^{0.59424}$

We, here, for this denominal pre-processing, see that ${}_\Psi\boldsymbol{N}$ becomes independent of d (and thus also of S).

Simple algebraic manipulation shows that multiplication of the partial pre-processed results (for this case) recovers the original term:

$$(\Psi^{d} / t)(d / \Psi^{t}) = (d / t)(\Psi^{d} / \Psi^{t}) == (d / t)\Psi^{d-t} .$$

Term Summations and Idiotypes

What of summation terms?

Say $P = x + y$; then we have $\Psi^{x} + y$ and $x + \Psi^{y}$; therefore,

$$(\Psi^{x} + y)(x + \Psi^{y}) = x\Psi^{x} + xy + \Psi^{x}\Psi^{y} + y\Psi^{y} == x\Psi^{x} + xy + \Psi^{x+y} + y\Psi^{y} .$$

The original resultant (i.e. $x + y$) appears as an exponential term for (base) Ψ : i.e. Ψ^{x+y} .

But this is a binary pairing, and would not be applicable to a sum of terms greater than two:

$$(a + b + c)(a + b + c) \rightarrow$$

$$\begin{array}{llllllllllll} & a^2 & + & ab & + & ac & & & & & & \\ + & & & ab & & + & b^2 & + & bc & & & \\ + & & & & & ac & & + & bc & + & c^2 & \\ \hline & a^2 & + & 2ab & + & 2ac & + b^2 & + & 2bc & + & c^2 & . \end{array}$$

[i.e., there are no abc products]

Yet, C_n pairing (by its very nature) can afford for sums of terms greater than two:

$$C_3[(a + b + c)(a + b + c)] \rightarrow (a, ab, ac, b, bc, c, abc)(a, b, c, ab, ac, bc, abc)$$

$$\rightarrow \qquad (a^2, a^2b, a^2c, ab, abc, ac, a^2bc)$$

:

: etc.

:

$$(a^2bc, a^2b^2c, a^2bc^2, ab^2c, ab^2c^2, abc^2, a^2b^2c^2)$$

For Ψ pre-processing, we would actually have:

$$C_3[(a + \Psi^{b} + c)(\Psi^{a} + b + c)(a + b + \Psi^{c})]$$

or some such suitable (physically) combination.

The corresponding C_2 pairing (for our example) would be:

$$C_2[\,(\Psi^x, y, y\Psi^x)(x, \Psi^y, x\Psi^y)\,] == x\Psi^x + xy + xy\Psi^x$$
$$+ \quad \Psi^x\Psi^y + y\Psi^y + y\Psi^x\Psi^y$$
$$+ \quad x\Psi^y\Psi^x + yx\Psi^y + xy\Psi^x\Psi^y$$

==

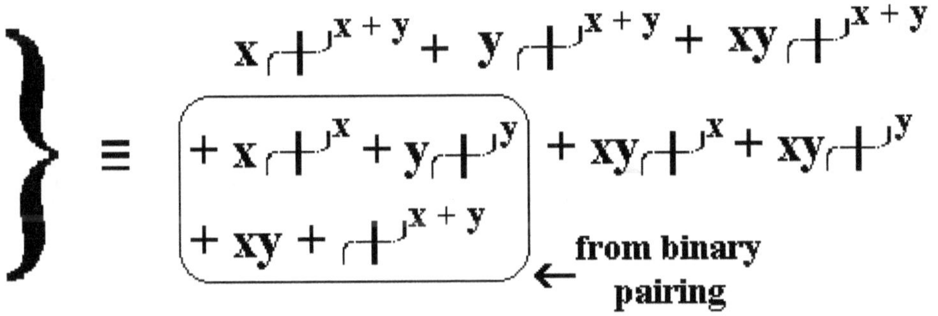

For a single term construct, we have (as before):

$C_2[\,(\Psi^d / t, d / \Psi^t, d\Psi^d / t\Psi^t)\,]$ from partial pre-processing of $s = d / t$, $n = 2$.

Therefore, (with this notation) $C_n[\,(\ ,\ ,\ ,\ \ldots)\,]$ shows C parsed in parentheses.

For a mixed, single term multi-variable with single term variable,

e.g. $Q = (d / t) + m$, thus $n = 3$ (but one term has two variables), we could construct:

$$C_3[\,((d / \Psi^t) + m)((\Psi^d / t) + m)((d / t) + \Psi^m)\,]$$

$\rightarrow$ "term" or quasi C_2 parsing

$$C_3[\,(d / \Psi^t, m, dm / \Psi^t)(\Psi^d / t, m, m\Psi^d / t)(d / t, \Psi^m, d\Psi^m / t)\,]$$

$\rightarrow (d / \Psi^t, m, dm / \Psi^t), (\Psi^d / t, m, m\Psi^d / t), (d / t, \Psi^m, d\Psi^m / t)$

$+ \quad (d / \Psi^t, m, dm / \Psi^t)(\Psi^d / t, m, m\Psi^d / t), (d / \Psi^t, m, dm / \Psi^t)(d / t, \Psi^m, d\Psi^m / t)$

$+ \quad (\Psi^d / t, m, m\Psi^d / t)(d / t, \Psi^m, d\Psi^m / t)$

$+ \quad (d / \Psi^t , m , dm / \Psi^t)(\Psi^d / t , m , m\Psi^d / t)(d / t , \Psi^m , d\Psi^m / t)$

[i.e., 7 "groupings" for C_3],

where, for example,

$(d / \Psi^t , m , dm / \Psi^t) == (d / \Psi^t) + m + (dm / \Psi^t)$, and

$(d / \Psi^t , m , dm / \Psi^t)(\Psi^d / t , m , m\Psi^d / t) == (d\Psi^d / \Psi^t t) + (m\Psi^d / t) +$
$(dm\Psi^d / \Psi^t t) + (dm / \Psi^t) + m^2 + (dm^2 / \Psi^t) + (dm\Psi^d / \Psi^t t) + (m^2\Psi^d / t)$
$+ (dm^2\Psi^d / \Psi^t t)$.

Here, already, the original resultant, $(d / t) + m$, can be "extracted" from, say, terms (in summation):

$$(dm\Psi^d / \Psi^t t) + m^2 == [(d / t)(\Psi^d / \Psi^t) + m]m == [(\Psi^{d-t})(d / t) + m]m$$

These "recoveries" are of course only "images" of the resultant, such as they are found variously modified or enhanced (of factors).

The elements of a "grouping" can, of course, be more than merely variables (or numbers), but also operations on or involving variables and numbers (here multiplication, division, exponentiation of base Ψ). The collectivity of the elements (here) is through addition.

The element summations have a characteristic pattern [**Eigenadditionstrukfur**] based on the C_n construction.

For example,

if we have ' a ' and a modification of: $a \longrightarrow a^*$,

and ' b ' and a modification of: $b \longrightarrow b^*$,

so that our original relation is $(a + b)$

and the pre-processing yields $(a + b^*)$, $(a^* + b)$,

then :

$$C_2[\,(\,a\,,\,b^*\,,\,ab^*\,)(\,a^*\,,\,b\,,\,a^*b\,)\,]$$

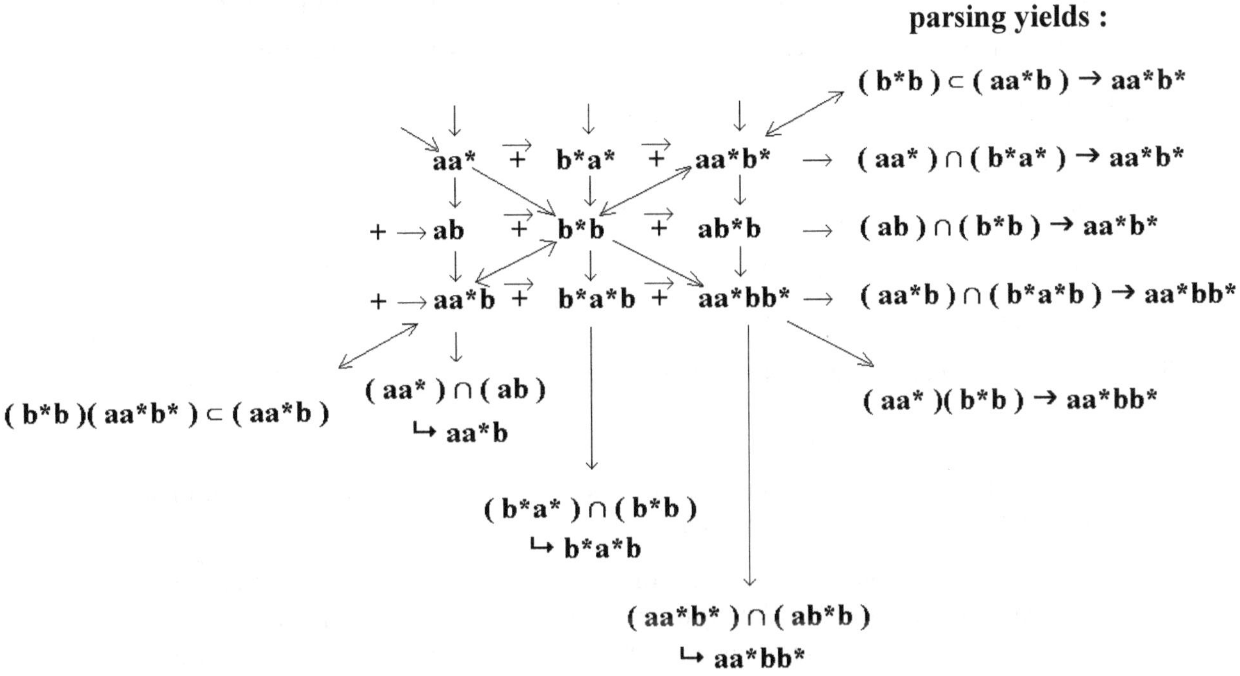

parsing yields :

$(\,b^*b\,)\subset(\,aa^*b\,)\rightarrow aa^*b^*$

$aa^* \;\overrightarrow{+}\; b^*a^* \;\overrightarrow{+}\; aa^*b^* \;\longrightarrow\; (\,aa^*\,)\cap(\,b^*a^*\,)\rightarrow aa^*b^*$

$+\rightarrow ab \;\overrightarrow{+}\; b^*b \;\overrightarrow{+}\; ab^*b \;\longrightarrow\; (\,ab\,)\cap(\,b^*b\,)\rightarrow aa^*b^*$

$+\rightarrow aa^*b \;\overrightarrow{+}\; b^*a^*b \;\overrightarrow{+}\; aa^*bb^* \;\longrightarrow\; (\,aa^*b\,)\cap(\,b^*a^*b\,)\rightarrow aa^*bb^*$

$(\,b^*b\,)(\,aa^*b^*\,)\subset(\,aa^*b\,)$

$(\,aa^*\,)\cap(\,ab\,)$
$\hookrightarrow aa^*b$

$(\,aa^*\,)(\,b^*b\,)\rightarrow aa^*bb^*$

$(\,b^*a^*\,)\cap(\,b^*b\,)$
$\hookrightarrow b^*a^*b$

$(\,aa^*b^*\,)\cap(\,ab^*b\,)$
$\hookrightarrow aa^*bb^*$

where " $\cap$ " indicates an operation of exclusive inclusion

[i.e. including for the final product only a novel factor amongst the factor set: for the factor set $(\,aa^*\,)$ and $(\,ab\,)$, the novel factor (for the final product) is b, therefore $(\,aa^*b\,)$]

and " $\subset$ " indicates an operation of exclusive replacement

[i.e. where a forced exchange (to modified or unmodified element) is made:

$$(\,aa^*b^*\,)\subset(\,b^*b\,)\rightarrow aa^*b \qquad \text{i.e. } b^* \text{ is exchanged for } b$$
$$(\,aa^*b\,)\subset(\,b^*b\,)\rightarrow aa^*b^* \qquad \text{i.e. } b \text{ is exchanged for } b^* \,].$$

Note: one can (an)notate (as)

$$(\,aa^*b^*\,)\subset(\,b^*b\,)=(\,b^*b\,)\subset(\,aa^*b^*\,)\rightarrow(\,aa^*b\,)$$

or $(\,aa^*b\,)\rightarrow(\,aa^*b^*\,)\supset(\,b^*b\,)=(\,b^*b\,)\supset(\,aa^*b^*\,)$

For our $(\,s=d\,/\,t\,)$ examples, variable modification would be exponentiation of base Ψ.

The 'reverse' operations $(\,\cup\,,\,\supset\,)$ are not as sanguine (i.e. are results degenerate) compared to $(\,\cap\,,\,\subset\,)$, perhaps due to the forward nature of the summations. The operations are directional (of procedure) in nature.

The summations develop from a general set (or line) of *idiotypes* of operation:

identical factors			**C_2 factors**			**non-identical factors**			
(a , b , c)(a , b , c)			(a , b* , ab*)(a* , b , a*b)			(a , b , c)(d , e , f)			
a^2	ab	ac	aa*	a*b*	aa*b*	ad	bd	cd	
ab	b^2	bc	ab	bb*	abb*	ae	be	ce	idiotypes
ac	bc	c^2	aa*b	a*bb*	aa*bb*	af	bf	cf	

But each idiotype has its own characteristic **Eigenadditionstrukfur**.

The pattern of generation (across idiotypes) thus proceeds (here) as like:

aa → aa* → ad , etc.

There are other idiotypes between the extremes of identical and non-identical factors, and conceivably transitions between idiotypes.

The mode of operation used here (to accomplish the summation) is through multiplication. But this can be generalized (or liberalized) to any means of operation simply by establishing a distinction between passive elements and active elements (performing the operations):

(a , b , c) Op (a as active element ,

passive elements ↑

operation

When Op == multiplication, then: (a , b , c) Op (a , → (aa , ab , ac)

The connectivity between the resulting elements has been addition thus far (therefore, **Eigenadditionstrukfur**): (aa , ab , ac) → aa + ab + ac

But the (method of) connectivity may also be liberalized for whatever the resulting **Eigenstrukfur** is desired (even if of very complicated connectivity).

Therefore, we can notate more specifically an idiotype to specify (distinctly) both operation and connectivity. For example,

interchangeable

passive **active**

$$\cdot\begin{bmatrix} aa^* & b^*a^* & aa^*b^* \\ ab & b^*b & ab^*b \\ aa^*b & a^*bb^* & aa^*bb^* \end{bmatrix}_+ \leftarrow C_2[\,(a, b^*, ab^*) \cdot (a^*, b, a^*b)\,]$$

active (element)
operation ≡ multiplication

connectivity ≡ addition (∴ summation)

Each element of an idiotype (can) result from operations with other elements. This is evident for elements at the ends (edges) of the idiotype table (as can be indicated by arrows). For internal elements, “angle-d” arrows leading to the element resultant may be indicated:

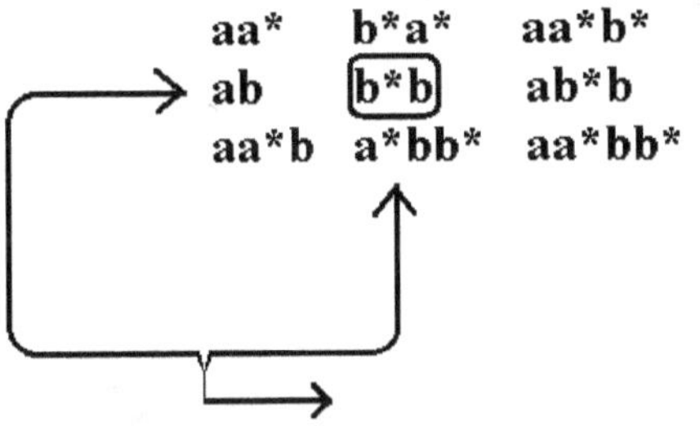

ab ∪ a*bb* → bb* (i.e. exclusion of a, a*)

or

ab ∩ b*a* → bb*

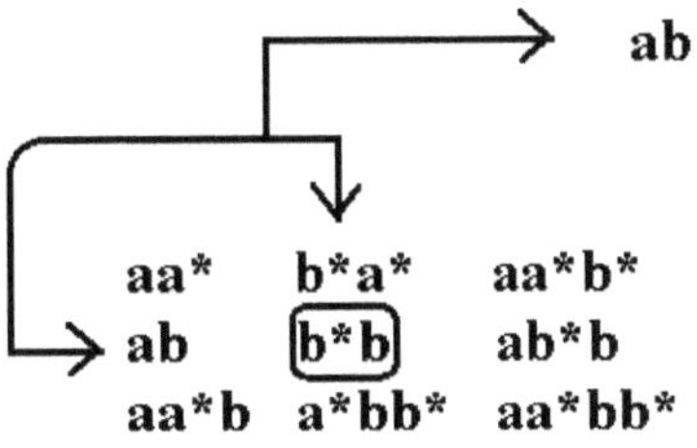

etc.

or, (a) shift of table is allowed: e.g.

aa*b	**aa***	**b*a***	**aa*b***		**aa*b***	**aa***	**b*a***		
abb*	**ab**	**b*b**	**ab*b**	≡ →	**ab*b**	**ab**	**b*b**	→	**abb* ∪ ab → bb***
aa*bb*	**aa*b**	**a*bb***	**aa*bb***		**aa*bb***	**aa*b**	**a*bb***		(i.e. exclusion of ab, a*)

The “value” of the shifted table's idiotype remains unchanged (due to the connectivity that is restricted to its range of elements; no new elements are introduced nor lost).

[Basically, the constituents of an element can be found among operations involving other elements.]

Transitions between idiotypes can of course be accomplished via elements shared or common among the idiotypes:

e.g., (symbolic) construction of the element ab*b U b*b → ab (i.e. exclusive of b*)

from C_2[(a , b* , ab*)(a* , b , a*b)]

can be used for the element ab from [(a , b , c)(a , b , c)] idiotype.

Therefore, one can try (for the idiotype's table):

$$\boxed{}\,(\, ab \,)$$

| a^2 | $ab*b \cup b*b$ | ac |

| ab | b^2 | bc | **as a transition from C_2 to identical factors.**

$_\bullet$| ac | bc | c^2 |$_+$

Introduction of the transition proposes possible further modifications of the table.

Consider, if $ab*b \cup b*b \rightarrow ab$, then (conceivably) $[\, (\, ab*b \cup b*b \,) \, / \, a \,] \rightarrow b$

Therefore, one might have:

$$
\bullet \left[
\begin{array}{ccc}
a^2 & ab*b \cup b*b & ac \\[2ex]
ab & b^2 & \dfrac{(\, ab*b \cup b*b \,) \cdot c}{a} \\[2ex]
ac & bc & c^2
\end{array}
\right]_+
$$

(with arrows indicating transitions, and a "(+)" diagonal)

etc.

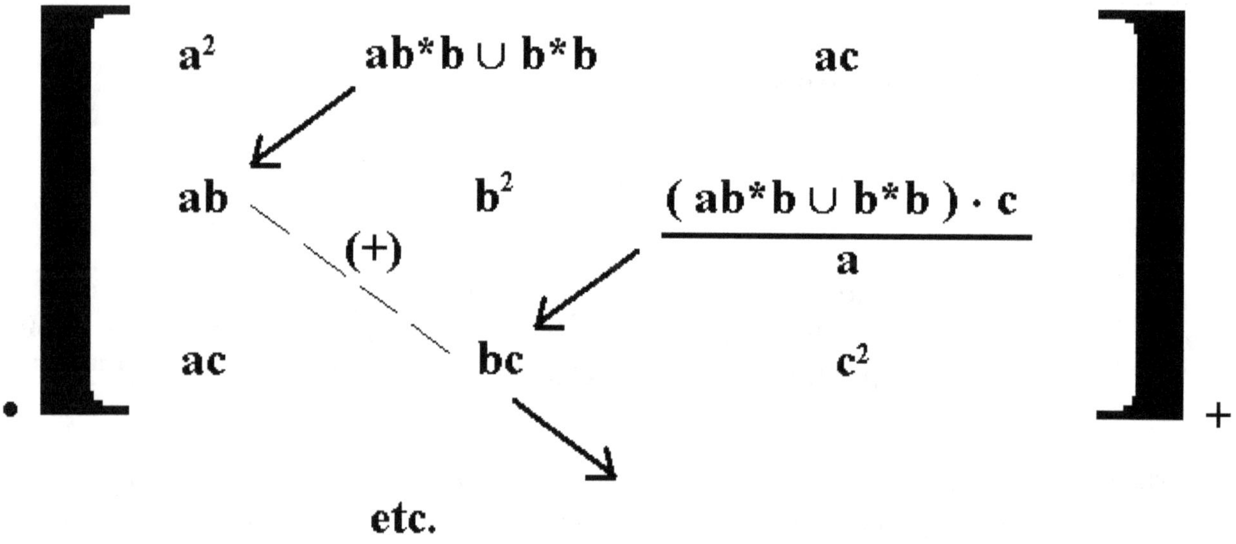

etc. So now we have surrogates for both ab and bc.

[The degenerate nature of the " $\cup$ " operation allows for even further possibilities (of modification).]

Naturally, by this scheme, since $(\, ab + bc \,) = b(\, a + c \,)$, then

$$[\, (\, ab*b \cup b*b \,) \, / \, a \,](\, a + c \,) \rightarrow ab + bc \, ,$$

and so forth more algebraic manipulations (are possible), although the manipulations are only "inspired" algebraically and induced symbolically (for the possible, or passable). Such sums would form part (if desired) of the (modified) idiotype's **Eigenadditionstrukfur**. It almost seems like a transition from identical factors to C_n factors, or a gradual (progressive) mutation of the idiotype's table. But it is essentially a transformation into pre-processed coefficients for a relation.

Now we can present the formal Symbolic Complexity of the relation (a + b), after pre-processing, where C_n is assigned to C_2[(a + b*)(a* + b)] :

Complexity = (C_n / Ψ^n)(n + 1)(relation) ; therefore, with n = 2 ,

(C_n / Ψ^n)(n + 1)(a + b) = ((n + 1) / Ψ^n) • C_n • (a + b)

Therefore,

((2 + 1) / Ψ^2) • C_2[(a + b*)(a* + b)] • (a + b) == (3 / Ψ^2) • .[(a + b) $]_+$ • (a + b)

The 3 / Ψ^2 factor ~ ~ 0.5718 ; therefore, we have

| aa* b*a* aa*b* |

(0.5718) | ab b*b ab*b | (a + b) ==

.| aa*b a*bb* aa*bb* $|_+$

| (a^2a* + aa*b) (aa*b* + a*bb*) (a^2a*b* + aa*bb*) |

(0.5718) | (a^2b + ab^2) (abb* + b^2b*) (a^2bb* + ab^2b*) |

| (a^2a*b + aa*b^2) (aa*bb* + a*b^2b*) (a^2a*bb* + aa*b^2b*) |

The orders of the terms in the summation run 3, 4, 5 (as the C_2 terms run 2, 3, 4).

The lowest order grouping (in the above design) is of 3 order terms: (aa*b* + a*bb*), terms which also occur in C_2[(a + b*)(a* + b)] (as well as terms aa*b and abb*).

Therefore, all of the 3 order terms of C_2[(a + b*)(a* + b)] also occur in the complexity summation. (We might call them “core complexity” terms. They, of course, derive from the C_2 2 order terms.) Only 2nd order terms don't occur for the complexity table.

Note that C_n refers to the number of terms derived:

e.g., for (resultant) a + b , $C_2[\,(a+b)\,] == C_2 = 3$ i.e. (a , b , ab)

Pre-processing:

$C_2[\,(a+b^*)(a^*+b)\,] \rightarrow 3 \bullet 3 = 9$ terms ; thus an action on C_2

Complexity on Pre-processing:

Comp{ $C_2[\,(a+b^*)(a^*+b)\,]$ } , i.e. { $\mathbf{C_2[\,(a+b^*)(a^*+b)\,]}$ } • (**a + b**) ,

$\rightarrow 2 \bullet 3 \bullet 3 = 18$ terms ; thus an enhanced action on C_2 .

Derived Table Terms of Idiotypes

But what of the (derived) terms themselves?

They can be plotted (of numerical value) in table order, to compare the behavior of idiotypes. For example, for the complexity table (as presented earlier) the table order can be 6 columns and 3 rows or, as for the pre-processing table, a grouping of 3 columns and 3 rows (each grouping being a sum itself):

(a^2 +)	(+)	(a^2 +)
($a^2 + b^2$)	(+ b^2)	($a^2 + b^2$)
($a^2 + b^2$)	(+ b^2)	($a^2 + b^2$)

b^2 regions

a^2 regions

Say, we let $a = 2$, $b = 3$, and our variable modification be Ψ exponentiation. Then, for our pre-processed idiotype, the numerical values look as:

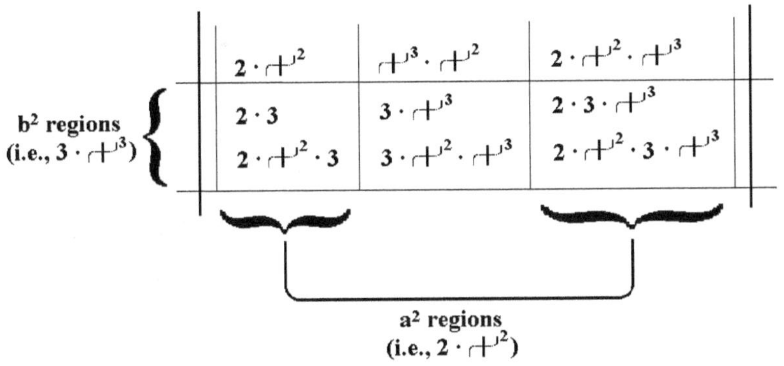

≅

10.49	63.05	126.09
6	36.05	108.15
31.48	189.14	378.27

sum = 948.72

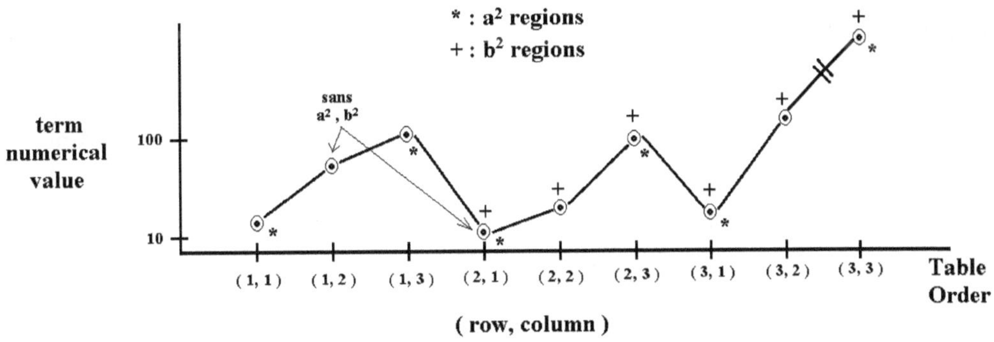

We can repeat the exercise, but giving b a value less than one, e.g. $b = 0.3$ (so that Ψ^b will be more fractional, closer to 1.0) or $b = -3$ (Ψ^b will be entirely fractional), and $b = 0$ (essentially only a , a^* terms).

$$\begin{vmatrix} 2\Psi^2 & \Psi^b\Psi^2 & 2\Psi^2\Psi^b \\ 2 \bullet b & b\Psi^b & 2 \bullet b\Psi^b \\ 2\Psi^2 \bullet b & b\Psi^b\Psi^2 & 2\Psi^2 \bullet b\Psi^b \end{vmatrix}$$

For $b = 0$:

$$\begin{vmatrix} 2\Psi^2 & (1) \bullet \Psi^2 & 2\Psi^2 \bullet (1) \\ 0 & 0 & 0 \\ 0 & 0 & 0 \end{vmatrix} == \begin{vmatrix} 10.49 & 5.25 & 10.46 \\ 0 & 0 & 0 \\ 0 & 0 & 0 \end{vmatrix} \quad \text{sum} = 26.23$$

For b = 0.3 :

$$\begin{vmatrix} 10.49 & (5.246)\Psi^{0.3} & (10.49)\Psi^{0.3} \\ 2(0.3) & (0.3)\Psi^{0.3} & 2 \bullet (0.3)\Psi^{0.3} \\ (10.49)(0.3) & (5.246)(0.3)\Psi^{0.3} & (10.49)(0.3)\Psi^{0.3} \end{vmatrix} = \begin{vmatrix} 10.49 & 6.727 & 13.45 \\ 0.6 & 0.3847 & 0.7694 \\ 3.147 & 2.018 & 4.0353 \end{vmatrix}$$

sum = 41.62

For b = -3 :

$$\begin{vmatrix} 10.49 & (5.246)\Psi^{-3} & (10.49)\Psi^{-3} \\ 2(-3) & -3\Psi^{-3} & 2 \bullet (-3)\Psi^{-3} \\ (10.49)(-3) & (5.246)(-3)\Psi^{-3} & (10.49)(-3)\Psi^{-3} \end{vmatrix} == \begin{vmatrix} 10.49 & 0.4366 & 0.8729 \\ -6 & -0.2496 & -0.4993 \\ -31.47 & -1.310 & -2.619 \end{vmatrix}$$

sum = -30.55

How can we structurally deal with the zeros derived?

For the b = 0 disposition we can collapse its table to: | 10.45 5.246 10.49 | .

Traditionally the effect of zero is suggested by assuming it is actually a limit as some variable goes to zero. In this way a (0 / 0) term (or quantity) can be evaluated as:

(limit as such goes to zero / limit as such goes to zero) = 1 , thereby avoiding the (infinite) consequences of division by zero (as is done for the calculus).

But for our cases we can simply re-define the effects of zero via the influences of unitage; thus, we (can) say "division by zero" means "no division" occurs (physically; a literal interpretation).

Therefore, (x / 1) = x ; but (x / 0) = x where the distinction is through unitage:

(x units a / 1 units b) == x units a / units b ;

while (x units a / 0 units b) == x units a (i.e., no division occurred).

For the summation (i.e. addition) case, the unitage carries (through):

a + b = a units a + b units b

If b = 0 , then a + b = a + 0 = a (etc.)

The structural argument that

x units a / 0 units b == x units a

is as follows:

If we have a fraction (or rationalization) X / a , where X is constant and "a" varies, then from either direction (positive or negative) as a goes towards zero the magnitude (or absolute value) of the fraction increases. Yet, if we plot $(a, X / a)$ it is clear that the (X / a) axis is asymptotic to the curves; i.e. , a can never reach $a = 0$ from (or on) the curves. Therefore, the division X / a does not (and can not) ever reach $X / 0$ (or $a = 0$).

Thus, $X / 0$ does not occur, even though the point $a = 0$ is definite (rather than infinite of location).

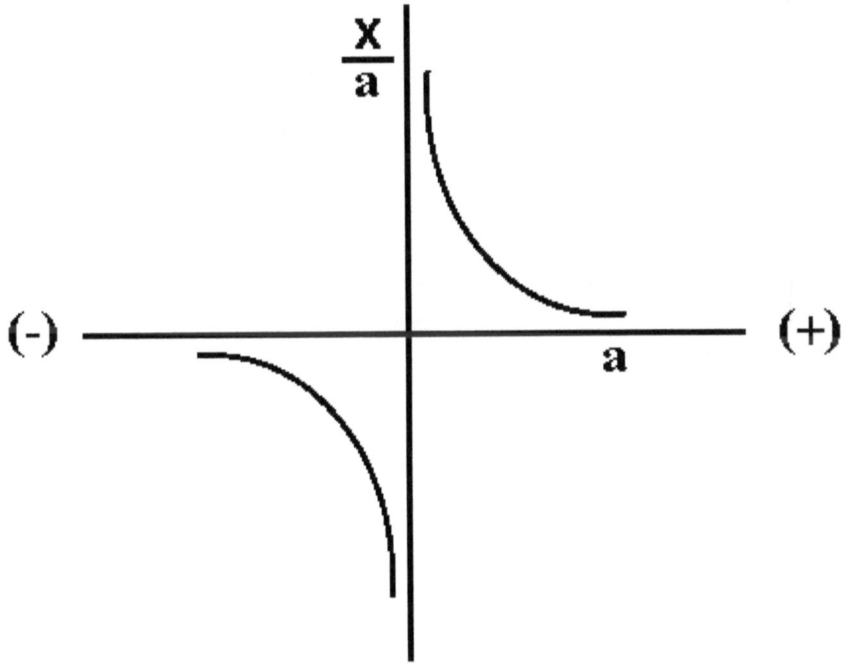

(If X is negative, the plot is of mirror image.)

Since the X units a / 0 units b division can not occur then the ratio is yielded to X units a , rather than being meaningless. On the other hand, if $X = 0$, then the curves condense to the a axis.

Physically, when one has an object (say X units a) and provides an operation on it (say division) but that operation does not (and can not) occur (say division by zero), then one is still left with the object intact (X units a) and unaltered.

Our pre-processed example, for b = (3 , 0 , 0.3 , -3) at a = 2 , plots as follows:

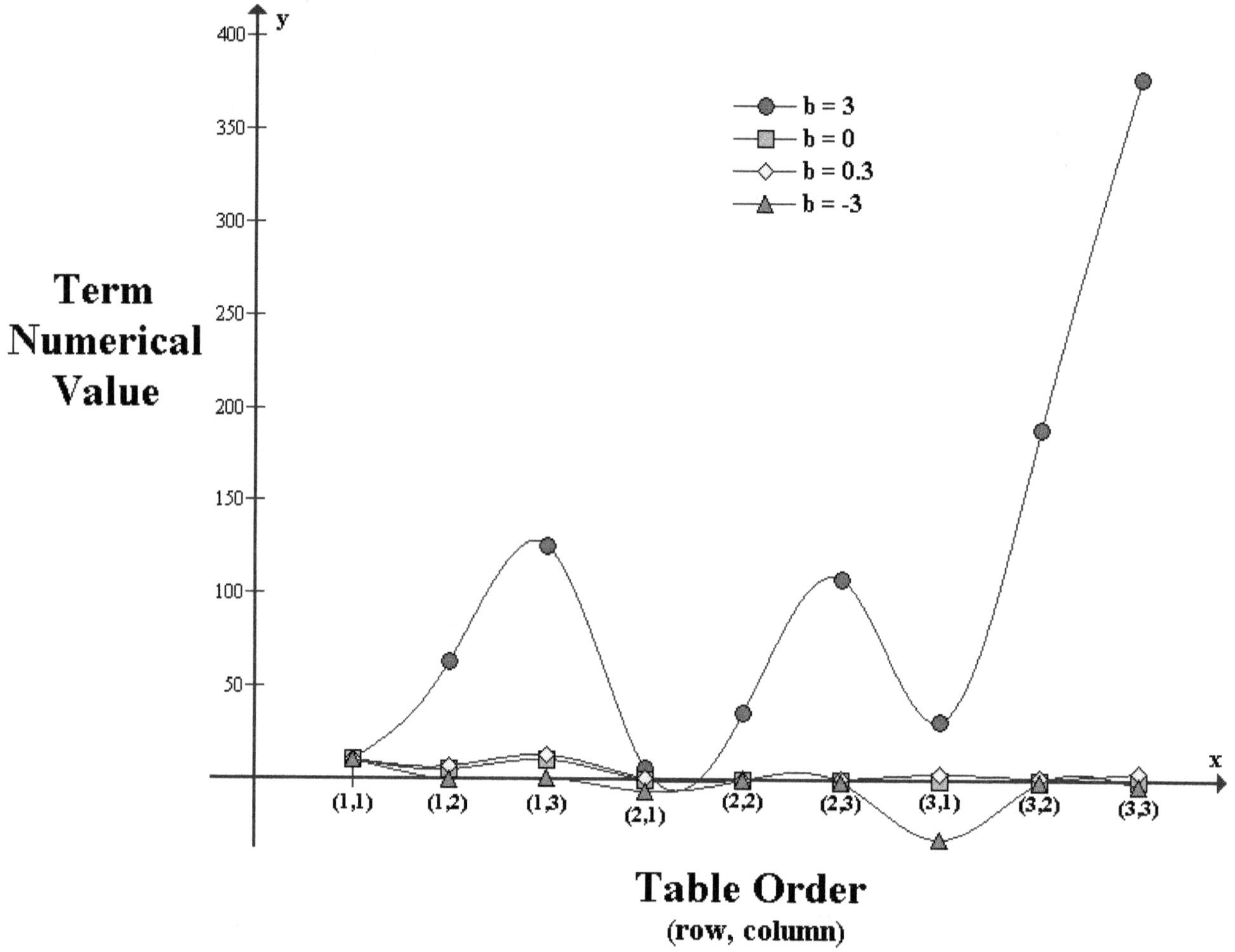

The lower plots expanded (are):

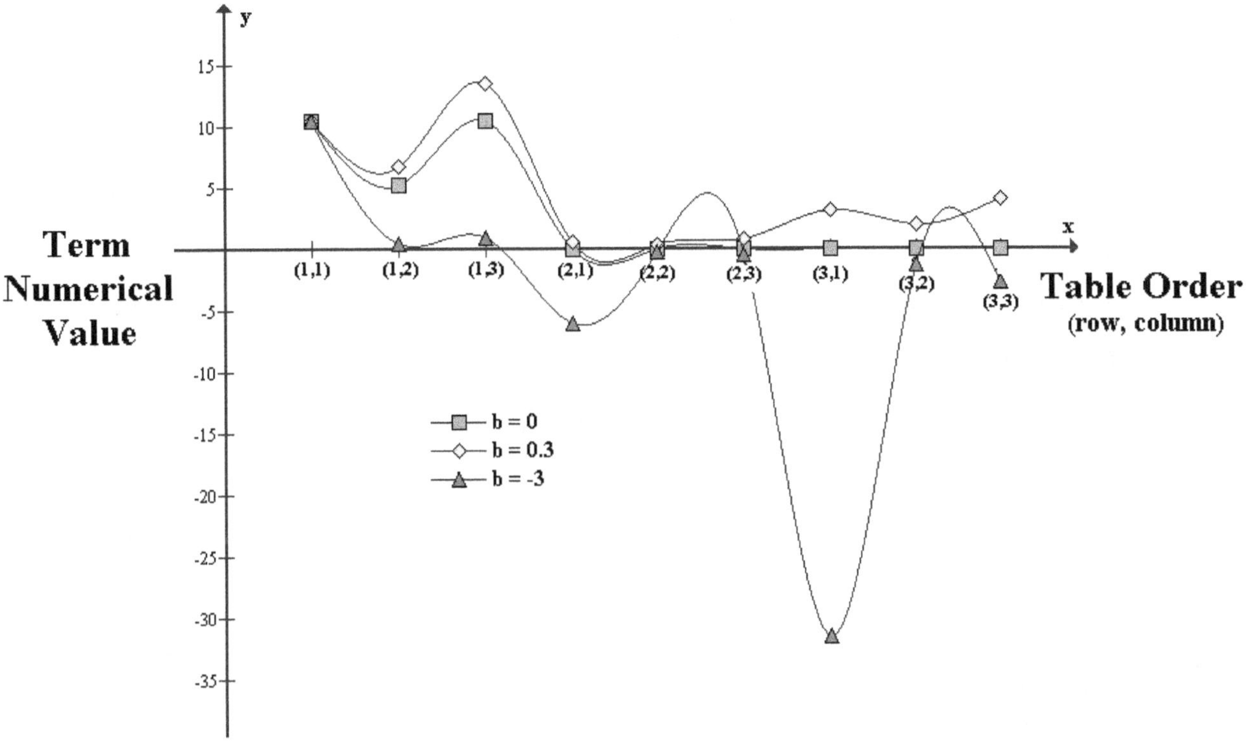

As for an equation, the idiotype (table) is static while the variable (values) are dynamic (of results). The "Ψ" effect is of course exponential of condensation from the lower (values) exponents:

		a*b* terms				
(a = 2) b exponent	**Σ terms**	**a*b***	**aa*b***	**a*bb***	**aa*bb***	**Σ a*b* terms**
3	**948.72**	**63.05**	**126.09**	**189.14**	**378.27**	**756.55**
0.3	**41.62**	**6.727**	**13.45**	**2.018**	**4.0353**	**26.23**
0	**26.23**	**5.246**	**10.49**	**0**	**0**	**15.74**
-3	**-30.35**	**0.4366**	**0.8729**	**-1.310**	**-2.619**	**-2.620**

Due to the manner of the table generation, the (directionally) positive and negative peaks of the plots occur at the internal edges of the row ends: (1 , 3), (2 , 1), (2 , 3), (3, 1) ; but for the lower 3 plots (b = 0.3, 0, -3) peaks (negative direction) also occur at (1 , 2) and (3 , 2) as noticeable, where a peak is defined as a positive or negative change in direction (or progression along the table order or abscissa).

Therefore, the b = 3 plot is distinguished from the others, by points (1, 2) and (3 , 2) not being transitional.

Perhaps the lower exponential plots (being of lower values) are more easily (or readily) disturbed of transitions. This sensitivity would be of a mathematical character (rather than physical), although it may supply a background (level) for physical (measurements of) properties.

Points (1, 2), (3 , 2), and (1, 3) are most directly related to the "Ψ" influence (exponentially), all three being a*b* terms.

[a*b* == Ψ^{a+b} ; thus (they have) most Ψ base influence (or powered base)]

This is especially evident in the b = 3 plot for the very steep rises of (to) points (3 , 2) and (3 , 3), also one term of a*b*. But these are limited terms (to 9 in number), so (3 , 3) is the highest term value reached (for this plot).

Composite and Forced Bases

The terms are idiotype (table) limited, to reach a definite sum. The idiotypic sums result from modes of curve behavior or manifestation, here for points (1 , 1) to (3 , 3). This suggests that any such sum can be so depicted, based on a relation's complexity and pre-processing. It follows, then, that adding up these sums represents a form of summation for (of) the various curves presented for (or from) a particular idiotype. But because of the Ψ exponential pre-processing there is an evident logarithmic relationship between (the exponential) b and the (total) idiotype (table) sum (and an inherent one with the exponential a). That might suggest a composite base to be derived (from the particular relation used in pre-processing).

Thus, $x^3 \sim\sim 948.72$, etc. (i.e. $x^b \sim\sim \Sigma_{(a\,;\,b)}$).

But $x^0 == 1$, not 26.23 ; and $x^{-3} ==$ a positive number (not -30.35);

and $(41.62)^{(1/0.3.)} = 250{,}046.8$ base, not $(948.72)^{(1/3.0)} = 9.826$ base (of **b = 3**) .
So we "force" the 9.826 (as) base (the lower exponent values being treated as residual):

$x^3 = 948.76$	$x^0 == 1$
$x^{0.3} = 1.9848$	$x^{-3} = 0.0010541$

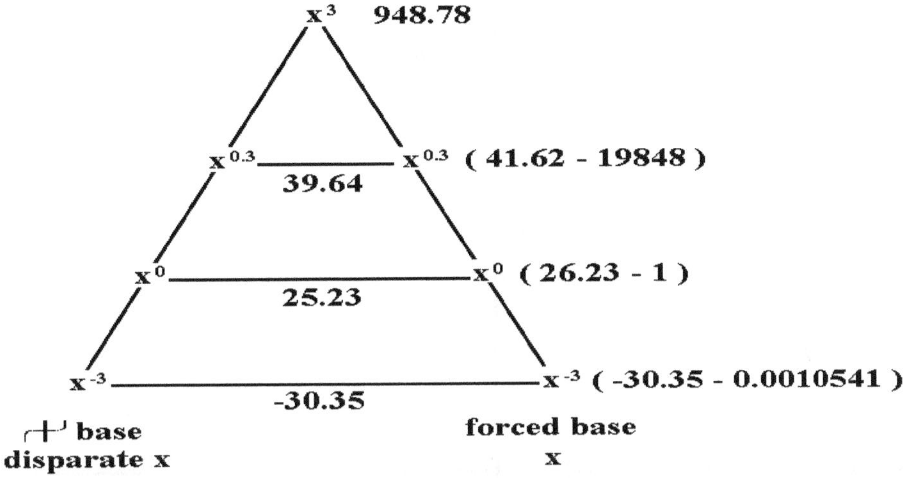

The issue is resolved by making (or taking) a "composite" base; say, using the x^0 example:

$y \bullet x^0 \rightarrow 26.23$; therefore, $y = 26.23$, $x = x$ and can equal forced base. Thus,

$y_{x0} \bullet x^0 = 26.23$, $y_{x0} = (26.23 / 1) = 26.23$

$y_{x-3} \bullet x^{-3} = 30.35$, $y_{x-3} = (-30.35 / +0.0010541)$

$y_{x0.3} \bullet x^{0.3} = 41.62$, $y_{x0.3} = (41.62 / 1.9848)$

$y_{x3} \bullet x^3 = 948.76$, $y_{x3} = (948.76 / 948.76) = 1$

The y factors demonstrate the Ψ base influence. Therefore, the extreme deviations (in y) that occur in the residual (lower exponent) values are a manifestation of the Ψ base influence.

Residual values can be information intensive.

Here, we have $y = (1 ; 20.97 ; 26.23 ; -28{,}792.335)$, in descending exponential order.

We can plot (b, y) points. But since the original relation has contributions from both variables a and b, with a being constant (for our example), it might be more appropriate to plot (a • b, y) points, a being used as a coefficient of b. The effect is to broaden the (b, y) curve:

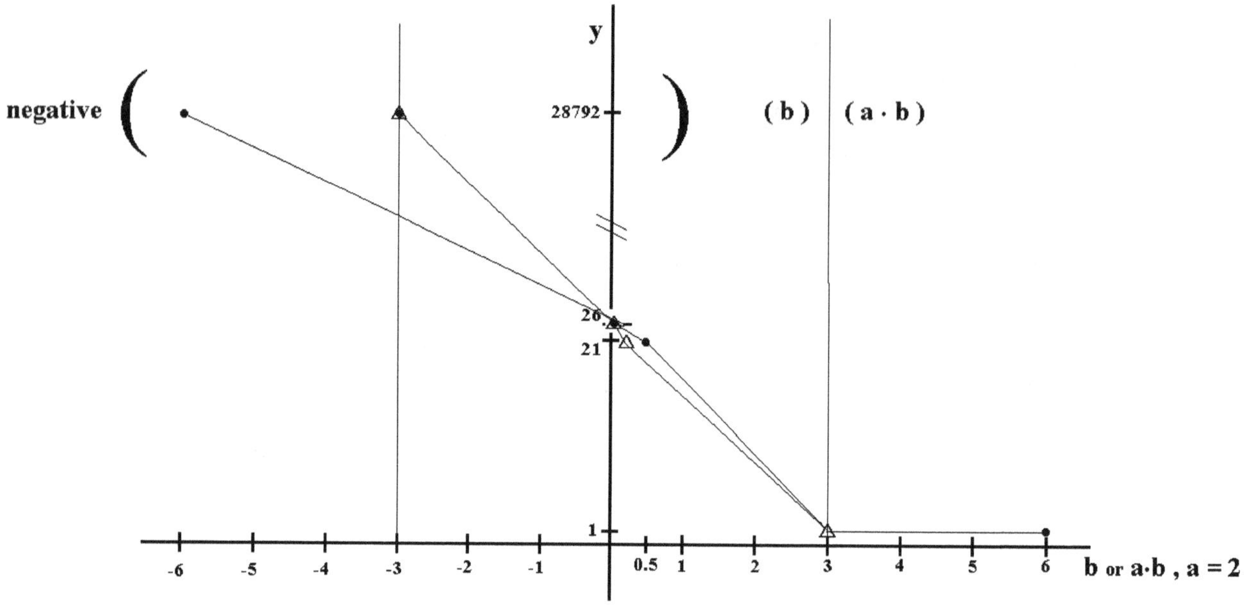

• (a · b , y) : (2 · 3 , 1) (2 · 0.3 , 20.97) (2 · 0 , 26.23) (2 · -3 , 28792)

△ (b , y) : (3 , 1) (0.3 , 20.97) (0 , 26.23) (-3 , 28792)

The disposition of the curve(s) is that of an exponential extinction (similar to remnants from disintegrations due to half-lives). Asymptotically, b reaches towards (+, −)∞ while y tends (with b↑) towards zero (but can never obtain it). This is of course through a forced base (of 9.826). y can never extinguish (itself) down to zero because Σ_{terms} must always be divided by x^b, and x^b is always greater than zero. Therefore, to reach zero, Σ_{terms} must become zero, which can only occur (here) when both a and b are zero in our idiotype (table). But this would be akin to having no (variable) relation at all (i.e., the relation could only be, or result of, a constant value); or, the relation (values) could be made to depend on unconsidered (or not accounted for) variables. [Whenever there is a forced base at a particular exponent, then at that exponent y remains 1 in value. Here, the exponent was b = 3 .]

We can also "force" the Ψ base (somewhat) by "pumping" it towards the 9.826 base:

i.e., **Ψ^x = 9.826 , xlogΨ = log 9.826** ; therefore, **x ~ = log(9.826) / log(2.2905) = 2.757**

Thus, (we) "pump" with $\Psi^{2.757}$:

$x^3 == (\Psi^{2.757})^3 = \Psi^{8.271} = 948.38 (\sim\sim 948.76)$

Therefore,

$$y_{x3} \bullet \Psi^{8.271} \sim= 948.76$$

$$y_{x0.3} \bullet | x^{0.3} = (\Psi^{2.757})^{0.3} = \Psi^{0.8271} | \sim= 41.62$$

$$y_{x0} \bullet | x^{0} = (\Psi^{2.757})^{0} = \Psi^{0} | \sim= 26.23$$

$$y_{x-3} \bullet | x^{-3} = (\Psi^{2.757})^{-3} = \Psi^{-8.271} | \sim= -30.35$$

Therefore, we would (for comparison) plot ($\Psi^{\text{pump exponent}}$, y) ;

{ $\Psi^{\text{pump exponent}}$: 8.271 , 0.8271 , 0 , -8.271 }

The "pumping" acts as a further broadening of the curve(s).

Now, one can study the (curve) effects of "pumping up" the Ψ factor from some (characteristic) number or starting point to the determined $\Psi^{2.757}$ value. A good starting point might be, for example, Ψ^{-3}, so that the x^3 factor condenses down to (Ψ^{-3})^3 or Ψ^{-9} (to study residual effects). Or, one might try $\Psi^{1/3}$, so that ($\Psi^{1/3}$)$^3 \rightarrow \Psi$ (to study the Ψ characteristics for this relation).

A sort of "net" can then be constructed connecting the various x curves via selected Ψ pumps (or nodes). For example, one can plot curves (for or from):

$$\Psi_{xb} \bullet (\Psi^{1/3} , \Psi^{2/3} , \Psi , \Psi^{4/3} , \Psi^{5/3} , \Psi^{6/3} , \Psi^{2.757})^{b}$$

with (b or a • b, y) points plotted.

The cross-hatching (to the curves) can be based on, say, b or a • b values, or equivalent y values, etc., for y value tracking modes.

Because of the b = 0 conjunction, the (above) collection of curves can have (the appearance of) a "bow-tie" effect, with all circuits running through this "tie" knot (dependent on the determined Σ_{terms} for the idiotype). Specific trajectories can then be designed, with the "knot" as a common feature amongst the curves:

Ψ pump	**Ψ_{x3}**	**$\Psi_{x0.3}$**	**Ψ_{x0}**	**Ψ_{x-3}**	**e.g.**
$\Psi^{1/3}$	**414.22**	**38.31**	**26.23**	**-69.52**	$\Psi_{x3} = 948.76/(\Psi^{1/3})^3$
$\Psi^{2/3}$	**180.84**	**35.26**	**26.23**	**-159.23**	
Ψ	**78.95**	**32.46**	**26.23**	**-364.71**	$\Psi_{x0} = 26.23/(\Psi)^0$
$\Psi^{4/3}$	**34.47**	**29.88**	**26.23**	**-835.37**	
$\Psi^{5/3}$	**15.05**	**27.50**	**26.23**	**-1913.42**	$\Psi_{x0.3} = 41.62/(\Psi^{5/3})^{0.3}$
$\Psi^{6/3}$	**6.570**	**25.31**	**26.23**	**-4382.69**	
$\Psi^{7/3}$	**2.868**	**23.30**	**26.23**	**-10038.55**	
$\Psi^{8/3}$	**1.252**	**21.45**	**26.23**	**-22993.29**	$\Psi_{x0} = 26.23/(\Psi^{8/3})^0$
$\Psi^{2.757}$	**1.000**	**20.97**	**26.23**	**-28783.41**	$\Psi_{x-3} = -30.35/(\Psi^{2.757})^{-3}$

b = -3 results in exorbitantly negative y values (in terms of absolute magnitude) compared to the positive y values (in magnitude) from the positive exponents (of b). This suggests a leaning of the "bow-tie" towards favoring the negative y realm (upon employing the forced base).

The increasingly negative exponent, of course, causes division with smaller and smaller (positive) fractional denominators (for y), yielding larger and larger (absolute) magnitudes. The effect is one of extreme distortion through elongation of the (lower) left side of the "bow-tie" (as plotted).

We can try a reductionist analysis, by changing the (primary or originating) base from Ψ to 1, in order to try to discern the relative effect of the Ψ base as distinguished from the relation.

A tenet of nul/los is that when one eliminates a variable its relational value conduces to 1 (rather than zero) within a term. [While within a combination of terms variable elimination simply removes or extinguishes the addition.] Therefore, for example:

ab {sans b} → a(1) == a (with ab unitage) , while

ab + c {sans c} → ab (with ab unitage)

This simply amounts to the eliminated variable being (physically) unavailable (or non-available).

(The base) 1 to any exponent is 1 , and therefore the denominator = 1 for y term, so that the y term is simply Σ_{terms} , allowing the (Ψ) comparison to be:

$$y_{(\Psi)} / y_{(1)} == y_{(\Psi)} / \Sigma_{terms} = [\, \Sigma_{terms} / (\, \Psi^{exponent})^b \,] / \Sigma_{terms} \rightarrow (\, \Psi^{exponent})^b .$$

Therefore, the result(ant) is entirely of the Ψ base function-ed (i.e. independent of the formal base).

For our idiotype table, the reductionist comparison yields:

Ψ pump	y_{x3}	$y_{x0.3}$	y_{x0}	y_{x-3}	
$\Psi^{1/3}$	**0.437**	**0.9205**	**1**	**2.2905**	**← Ψ base**
$\Psi^{2/3}$	**0.191**	**0.8473**	**1**	**5.2464**	
Ψ	**0.083**	**0.7799**	**1**	**12.0169**	
$\Psi^{4/3}$	**0.036**	**0.7178**	**1**	**27.5246**	
$\Psi^{5/3}$	**0.016**	**0.6607**	**1**	**63.0451**	
$\Psi^{6/3}$	**0.007**	**0.6082**	**1**	**144.4048**	
$\Psi^{7/3}$	**0.003**	**0.5598**	**1**	**330.7593**	
$\Psi^{8/3}$	**0.0013**	**0.5153**	**1**	**757.6042**	
$\Psi^{2.757}$	**0.0011**	**0.5039**	**1**	**948.3824**	recovers formal base: 9.826
$\Psi^{9/3}$	**0.0006**	**0.4743**	**1**	**1735.2924**	

This of course presents what amounts to 'standard curves' using the Ψ base.

It is evident that the (Ψ) reductionist curves track oppositely (in curvature or direction) from its comparison relation. Therefore, the relation counters the Ψ based complexity. The relation "resists" the (Ψ) complexity. But the resistance seems to be, in general, residual (of power). At $\Psi^{2.757}$, $(\, \Psi^{exponent})^{-b}$ yields (essentially) $948.3824^{1/3.0}$, i.e. recovering the 9.826 forced base (as expected). [948.3824 ~ ~ 948.72 ; $948.72^{1/3} \rightarrow 9.826$]

Complexity Bases and Mixed Bases

Similar to changing bases through logarithms, other comparisons (aside from reductionist here) can be made by using different comparison bases (from 1). That is, using:

$(\, \Psi^{exponent})^b / (\, [\text{ new base }]^{exponent})^b$; e.g., $(\, \Psi^{exponent})^b / (\, 10^{exponent})^b \rightarrow \Psi / 10$, etc.

The actual rationale would be:

$[\Sigma_{terms} / (\Psi^{exponent})^{b}] / [\Sigma_{terms} / ((\text{new base})^{exponent})^{b}] == ([\text{new base}]^{exponent})^{b} / (\Psi^{exponent})^{b}$
$\rightarrow [\text{new base}] / \Psi$, assuming equal exponents to each base.

$(10 / \Psi)$ would clearly be an amplification. But exponents varying for each base can also be used (i.e. only specific base dependent exponents), as is conceptually profitable.

In fact, the (comparison) term can be totally generalized of exponents for a particular physical process (or consideration):

$[(\text{comparison base})^{a}]^{b} / (\Psi^{c})^{d}$.

Note that unless comparison base $== \Psi$, the exponents (a, b, c, d) are not relational until $a = c$ and/or $b = d$, etc. (e.g., $a = xc$, $b = xd$, with $x ==$ common factor).

This promotes the interesting aspect of a "core" mixed base, e.g. $10 / \Psi$, produced by having $a \bullet b \bullet x = c \bullet d \bullet y$:

$([(\text{comparison base})^{a}]^{b})^{x} / [(\Psi^{c})^{d}]^{y}$,

whereby mixed base $== (\text{comparison base} / \Psi)^{1/(a b x = c d y)}$, i.e. the root.

The term is very reminiscent for our expression of complexity:

$(C_{n} / \Psi^{n})(n + 1)$, if we allow for $C_{n}(n + 1) = nC_{n} + C_{n} == C_{n}^{n}$.

We can reduce the stringency by having $C_{n} \rightarrow C$ for any considered or desired combination:

therefore, $C(n + 1) / \Psi^{n} == C^{n} / \Psi^{n} = (C / \Psi)^{n}$, $C(n + 1) == C^{n}$.

Our mixed base is now C / Ψ, with its exponent being $n == 1 / m$; $C(n + 1) = C^{n}$ holds when $n = 0$ and $C = 1$ as a trivial result (of a non-variable condition).

$C(n + 1) = C^{n}$ does not hold when $C = 2$ and $n = 1$: $2(1 + 1) = 2 \bullet 2 = 4 = 2^{2} =/= 2^{1} == 2^{n}$.

$C(n + 1) / C = C^{n} / C == C^{n-1}$; thus, $n + 1 = C^{n-1}$, $n = C^{n-1} - 1$.

Therefore, for a given n "an appropriate" C can be calculated:

e.g., let $n = 5$

$$5 = C^{(5-1)} - 1$$

$$5 + 1 = C^4$$

$$\log(5 + 1) = 4\log C$$

$$\log(6) / 4 = \log C$$

$$10^{\log(6)/4} = C = 10^{\log C}.$$

[as a check: $10^{(0.778/4)} = 10^{0.1945} = C$; $5 \sim= 1.565^4 - 1 \sim= 4.999$]

Therefore, a combination of (magnitude) 1.565 (or 1.56495) is an appropriate "client" for a 5 variable relation (in terms of the relation's complexity), when using the mixed base:

$C / \Psi \sim= 1.565 / 2.2905 = 0.6832$ (at $n = 5 = 1 / 0.2$; therefore, $m = 0.2$).

Thus, $a \bullet b \bullet x = c \bullet d \bullet y = 0.2$ (if desired amongst exponents).

[The reductionist step would have $b = d$.]

Physical and Estimated Complexities

Of course, there is a substantial difference (in values) between C and C_n (e.g. between 1.565 and C_5). C_n is actual (i.e. physical) numerically, while C is evaluated (estimated).

Therefore, C_n is a real quantity, while C is relational (to methods).

Why, then, is the relational C restricted so far (down) from the physical C_n (in our case from $C_5 = 161$ to $C = 1.565$) ?

The $n = 5$ constriction is $\sim 0.0097C_5 \sim\sim 0.0165 == C_5 / 100$.

$C == 10^{[\log(n+1)/(n-1)]}$.

Therefore, the constriction is: $10^{[\log(n+1)/(n-1)]} / C_n$ of factor.

Note that the constriction goes to ∞ as $n \rightarrow 1$, due to the $(n - 1)$ denominator for the 10's exponent term.

n	**C_n**	**constriction**	**C**	**[log(n + 1)] / (n – 1)**	
1	**1**	**∞**	**∞**	**∞**	
1.5	**cal. C_n = 2.5**	**2.4995**	**6.2488**	**0.7958**	**(informal n)**
2	**3**	**1.000**	**3.000**	**0.4771**	
3	**7**	**0.2857**	**2.000**	**0.3010**	
4	**25**	**0.0684**	**1.710**	**0.2330**	
5	**161**	**0.0097**	**1.5649**	**0.1945**	
6	**1561**	**0.0009454**	**1.4757**	**0.1690**	

[n is formally (here, for C_n) only an integer. The program “rxn_pool” accepts n values with fractional parts, to calculate (or estimate) C_n; at $n = 1$ the constriction and C, etc., are more properly resultant for “undefined value” rather than infinity, since the slightest fractional addition to 1 yields a C_n close to 2 (and far from ∞), as well positive (< 1) yields close to 1 + fractional part and negative (fraction) yields ~ 1 + fractional part or 1 + negative input.]

Starting from $C = 3$ ($C_n = 3$) at $n = 2$, C seems to counter (resist) complexity by decreasing with increasing n (in a manner most opposite of direction and magnitude from C_n).

[Including the informal $n = 1.5$ makes the divergence of C_n from C even more apparent due to the crossover node at (C, $C_n = 3$) more easily demonstrated.]

The limit to C is (somewhat) $10^0 = 1$ (although it is actually $10_{(n=1)}^{-\infty == \log(2/0)}$

or $10^{\log(1/-1)}$ at $n = 0$, assuming n must remain positive).

Therefore, the $n = 1$ point represents a discontinuity,

before n reaches 0 and $10^{\log(1/-1)} == 1$.

Analytically, the n = 1 point is a cusp to the (n, C) curve (also when negative n is included for values). It represents a position where the curve's slope goes to infinity.

(informal) n	**calculated C_n**	**constriction**	**C**	**log(n + 1) / (n – 1)**	
0.5	**1.5**	**0.2963**	**0.4444**	**-0.3522**	The asymptote of
0	**1.0**	**1**	**1**	**0**	a logarithm is zero
-0.5	**0.5**	**3.1748**	**1.5874**	**+0.2007**	from positive side.
-1.0	**0**	**∞**	**∞**	**-∞**	**Calculator**
-1.5	**-0.5**	**NaN**	**NaN**	**NaN**	**Results**

[Logarithms don't take negative arguments.]

[Calculated C_n values come from the "rxn_pool" program.]

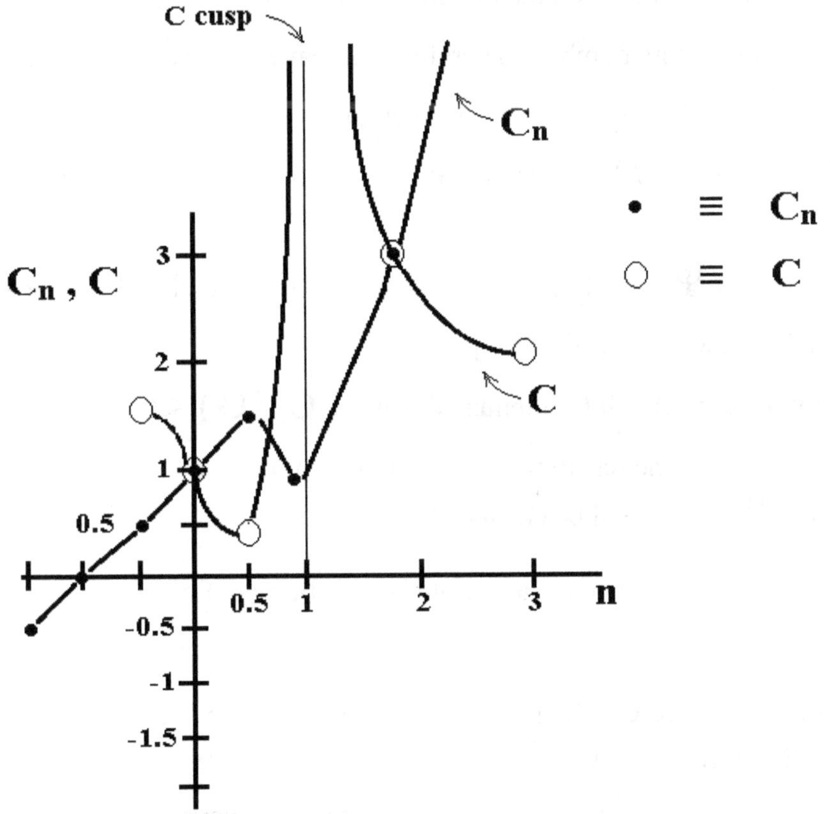

C appears as a "curious" restrictor of complexity (when) by substituting for C_n, curious due to the residual (low C, n) effects (which include infinity). The restriction most potently begins (or starts or initiates) at n = 2 , although some minor attenuation occurs earlier (with informal or fractional and negative n values). An n = 2 is of course a nominal definition for the establishment of a complexity (of variables or defined quantities of consideration). Also of note (to identify) is the conference (equivalence) of C_n and C when $n = 0$, that of having no such quantities for (or towards) complexity.

If we consign C_n to the numerical combination (of n considered quantities or variables) and C to the observed, then it becomes evident that the more quantities (considerable) there are the greater the restriction on their combinations (to be observed). Thus, for example, innumerable molecules collected together form an observable mass (of gas or ether, liquid, or solid). C, then, provides a practical weight on C_n. That weight is the constriction: C / C_n, which is used to devolve to C (from C_n). Then, for any considerable (observable) mass, the constriction is (to its limits) approximately $1 / C_n$ (or C_n^{-1}). This is distinct to the $1 / C_n$ (exact) that occurs (as $1 / 1 = 1$, $n = 0$) during the residual values.

But this is ultimately dependent on the mixed base C_n / Ψ, C_n and C remaining unit-less (to allow C_n / Ψ to stay a pure number). Therefore, we have the base Ψ, and the many bases C_n / Ψ, to peruse. Note that if, for a complexity calculation, you replace Ψ with C_n / Ψ, you are left with Ψ itself. Using C / Ψ to replace Ψ effectively multiplies the resulting Ψ (since $C \rightarrow 1$ with $n\uparrow$).

Therefore, $C_n / (C / \Psi) \rightarrow (C_n / C)\Psi ==$ multiplied Ψ [C_n increases much more than C with $n\uparrow$; in fact, $C\downarrow$ with $n\uparrow$.]

The complexity (resulting) will be attenuated, since $(C_n / C) < C_n$ at first (at low n); but as C quickly goes to $< \Psi$, the resulting complexity (as multiplied by Ψ instead of being divided by Ψ in the (C_n, Ψ) term) will be enhanced.

C, however, is intimately related (or associated) with n, by definition:

$n = C^{n-1} - 1 = -1 + C^{n-1}$.

Therefore, for any assessed C (observed or estimated) we can find an effective n to its relation (as employed). Thus, $n + 1 = C^{n-1}$.

Therefore, for a complexity term (or factor to a relation) we have:

$(C_n / \Psi^n)(n + 1) == (C_n / \Psi^n) \bullet C^{n-1}$, $C_n(n + 1) \sim\sim C^n$ for mixed base C_n / Ψ.

n appears to have two (related but independent) combinatorial terms: (C_n, C) as possible (or expressible). While C_n and Ψ are fixed (absolutely defined) quantities, C is a variable (to be determined of a relation). Therefore, say we assign a (desired) complexity to a relation R :

$$(C_n / \Psi^n)(n + 1)R = C_{(om)px} == (C_n / \Psi^n) \bullet C^{n-1} \bullet R;$$

thus, $C^{n-1} = C_{px} \bullet \Psi^n / (R \bullet C_n)$.

C is choose-able if n is choose-able, allowing C to be an intrinsic combination.

Note, when $n = 1$ (as assigned) this forces:

$C^{1-1} = C^0 = (C_{px} / R)(\Psi / 1) = C_{px} \bullet \Psi / R$; therefore,

$R = C_{px} \bullet \Psi, \Psi = R / C_{px}$

and C_{px} is defined by the relation's value (R) since Ψ is a known; i.e. $C_{px} \sim\sim R / 2.2905$

This, in essence, designates the numerically (or numericological) physical purpose of Ψ.

The variable C factor, then, effectively allows a relation's complexity to be reduced (at $n = 1$) from $(2R / \Psi)$ to (R / Ψ).

When allowing for the mixed base $C_n / \Psi, C_n \rightarrow C$; therefore, there is a clear dichotomy of complexity between C_n and C involving the factor "two" (i.e., there is a cost in complexity while employing C) :

$$(C_n(n + 1) / \Psi^n)R \rightarrow (1(1 + 1) / \Psi^1)R == 2R / \Psi \qquad |$$

$$(C_n / \Psi^n) \bullet C^{n-1}R \rightarrow (1(C^0) / \Psi^1)R == R / \Psi \qquad | n = 1$$

The contrasting factors are, of course, $C_n(n + 1)$ and $C_n \bullet C^{n-1}$; but if $(n + 1) = C^{n-1}$, then they should be the same. Therefore, only when they are the same is the (employed) C valid; and when $n = 1$ they can not be the same:

$n + 1 = C^{n-1}$

$1 + 1 = 2, C^{1-1} = C^0 = 1$

$2 =/= 1$

suggesting a single variable (or quantity) for a relation can not have (C) practical complexity. Away from an exponent of zero, C is allowed to be a variable.

C^0 (as shown earlier) occurs at the "C cusp" position. [That is, C uses a **nul/los** exponent, by necessity.] The C cusp (region or point) is not physically realistic. (Its traversal is accomplished by the totally realistic C_n curve, as compared to a definition for a natural logarithm or similar.)

But what is the (numerical) significance of the factor "two" between C_n and C ?

Aside from the subtractive combination of the n centers

[factors or exponents involving n : $(n + 1) - (n - 1) = 2$],

we can assume that (at $n = 1$) the following prevails:

$(n + 1) / C^{n-1} = 2$; i.e. $(1 + 1) / C^0 = 2 / 1 = 2$

Therefore, a C series (of dependence) on n is established, where only some (particular) values of C satisfy its role as a comparison base (to Ψ), and C can (numerically) take any value in the (its) series. Since both n and C vary, the series is (in a sense) **2-dimensional**. If we elevate C values that can act as comparison bases, the plot becomes **3-dimensional** (or characterized by individual significances). It would be like a "texture" of ranges $(n + 1) / C^{n-1}$, with particular values for C raised [i.e. plotting C vs $(n + 1) / C^{n-1}$].

The 3D plot could be: C vs $(n + 1) / C^{n-1}$ vs n .

From $n + 1 = C^{n-1}$ (for "appropriate" or comparison base C):

at $n = 1.001$, $C = 1.77 \times 10^{301}$ |

at $n = 0.9$, $C = 0.00163$ | calculator results

So we see that "appropriate" C can go from "near" infinity (i.e. essentially $+\infty$) to very small positive values (to zero at $n = -1$); i.e. $0 <=$ "appropriate" $C <= +\infty$ through numerical definition.

When (theoretical rather than physical) $n < -1$, C must become negative (as $n - 1$, then, is always negative) since $(n + 1)$ is always negative; but this is out of the range of logarithmic calculations: since the argument for a logarithm must be positive, so then $(n + 1)$ must be positive.

A short table of $(n + 1) / C^{n-1}$ values (~ **== rounded** ; …. **== repeated digits**) :

↓C	↑ n = 1	2	3	4	5
0.5	**2 (= 1 • 2)**	**6 (= 2 • 3)**	**16 (= 4 • 4)**	**40 (= 8 • 5)**	**96 (= 16 • 6)**
1.0	**2↑**	**3↑**	**4↑**	**5↑**	**6↑**
1.5649	**2**	**1.917055~**	**1.633378~**	**1.304699~**	*1.000472~*
2.0	**2**	**1.5**	*1*	**0.625**	**0.375**
3.0	**2**	*1*	**0.44444~**	**0.185185….**	**0.074074….**
4.0	**2**	**0.75**	**0.25**	**0.078125**	**0.0234375**
5.0	**2**	**0.6**	**0.16**	**0.04**	**0.0096**

[outlined C : "acceptable" C values]

It is very clear that for a given n with its appropriate C (and only at that C) the ratio $[(n + 1) / C^{n-1}]$ equals "1" ; it is also evident that (n = 1) does not have an (any) appropriate C and the ratio always equals "2" in value. We can (here, from this short table) predict that the appropriate C for n = 4 lies between 1.5649 (i.e. that for n = 5) and 2.0 (that for n = 3), to demonstrate an example of table utility. We can predict that for n = 6 , at C = 0.5 the ratio yields:

$((2 \bullet 16) \bullet (6 + 1)) = (32 \bullet 7) = 224$, etc.

[i.e., $2^{n-1} \bullet (n + 1)$ results for the ratio].

$2 == 1 / 0.5$; therefore, the ratio relates to the table as:

$(1 / C)^{n-1} \bullet (n + 1) == (n + 1) / C^{n-1}$, as demanded.

An "acceptable" C is that appropriate for the mixed base C / Ψ , and yields "1" for the ratio.

Therefore, per given n, the acceptable C can be found summarily:

$(n + 1) / C^{n-1} = 1 , n + 1 = C^{n-1}$

$\log(n + 1) = \log C^{n-1} = (n - 1)\log C$

$\log(n + 1) / (n - 1) = \log C ; 10^{[\log(n + 1) / (n - 1)]} = 10^{\log C} = C$

For example, for $n = 2$:

$C = 10^{[\log(2+1)/(2-1)]} = 10^{(\log 3/1)} = 10^{\log 3} = 3$

n	1	2	3	4	5	6
C_{accpt}	$10^{+\infty}$	3	2	1.70998~	1.56508~	1.47577~
C / Ψ	∞	1.310	0.8732	0.7466	0.6833	0.6443

Of the "acceptable" C values listed above (in this limited exposition), the only actual resulting C_n is for $n = 2$; i.e. $C_2 = 3$.

For $n = 3$, it is interesting that, as for $4 / 2 = 2$, $(\log 4) / 2 = \log 2$;
i.e., $C = 10^{\log(3+1)/(3-1)} = 10^{(\log 4)/2}$ $(== 10^{\log 2}) = 2$.

It is also clear that, since "acceptable" C runs down (with increasing n) to its limit of "1" in value, C_2 is the only C_n found appropriate for the comparison base (in the mixed base C / Ψ). Therefore, the (numerical) significance of the number (or factor) "2" is manifest throughout the (or this) analysis. It is the (or a) point of convergence between the C_n and C series.

[If we declare the 0^0 (zero to the zeroth power) $= 1$ (as like $0!$, or zero factorial, equals 1) rather than $0^0 = 0$, then we can suggest $C_n == C_0 = 1$ as another point of convergence (although this would allow $C_1 = C_0$, as like $1! = 0! = 1$).]

These convergences, again, seem to occur during the residual (low n) region of the $(C_n, C; n)$ curves, as can be seen in the previous plotting or graph.

By "acceptable" $C = 2$, the ratio C / Ψ remains < 1 for increasing n, and therefore also powers (> 1) on the ratio. So, how may this (analysis) be utilized in examples?

We have, for preferring a particular complexity (desired) of a relation R:

$(C_n / \Psi^n)(n + 1)R == (C_n / \Psi^n)C^{n-1}R$, with n and thus C_n and Ψ^n fixed and the variable C allowed (as a comparison base).

If we want a relation R to have a given complexity H, then:

$(C_n / \Psi^n)C^{n-1}R = H$; thus $C^{n-1} = H\Psi^n / C_nR$, $\log C^{n-1}$

$= \log(H\Psi^n / C_nR == \log(\text{ all terms fixed })_H$.

Therefore, $(n - 1)\log C = \log (\ )_H$, $\log C = \log(\ )_H / (n - 1)$;

thus, the needed comparison base is: $10^{\log C} = 10^{\log(\ )H / (n - 1)} = C$

Such a calculation forces R to be positive (assuming H, Ψ^n, C_n are positive by default of definition), aside from negative H when R is negative.

Therefore, | R | should be used in $\log(\ )_H$; and the factors $C^{n-1} \bullet R$ can be (effectively) replaced by: $C^{n-1} \bullet R = R_H$.

The Estimated Complexity Ratio

But what is desired (to be uncovered) is the ratio C / Ψ , as evaluated.

Let us, then, take a sample relationship to study the complexity analyses on:

$w = ax^3 + bx^2 + cx$.

To make the summations valid (or proper) we can apply appropriate unitage to the terms and (for convenience) attach them to the coefficients (a, b, c), so that the variables (x^3, x^2, x) remain pure as: $(x^3, x^2, x) == (x, y, z)$ [assuming that the terms of the summations are independent of variable].

Therefore, $w = ax + by + cz$

where, for common unitage u, $a == (u)^{1/3}$, $b == (u)^{1/2}$, $c == (u)^{1/1}$ in terms of unitage and (x, y, z) are pure number variables (i.e. unit-less).

Therefore, $w = u^{1/3}x + u^{1/2}y + u^1z == R$.

The complexity of the relation, with $n = 3$, is defined as:

$(C_3 / \Psi^3)(3 + 1) \bullet R$; therefore, $(7 / \Psi^3) \bullet 4 \bullet R \sim\sim 2.33R$

We set the (common) unitage as $u = 1$;

thus $R == 1^{1/3}x + 1^{1/2}y + 1^1z = x + y + z$ (for unitage).

But the numerical coefficients (while assigned unitage) are independent of unitage.

Therefore, $R = ax + by + cz == a \bullet u^{1/3} + b \bullet u^{1/2}y + c \bullet u^1z$.

We can, of course, change the complexity by changing the values of the variable:

$R_H = C^{n-1} \bullet R$

if we wish to keep the (given) variable values and rather change to a mixed base C / Ψ ,

or $(x, y, z)R \rightarrow (x, y, z)R' == (x', y', z')$, changing variable values.

In either case an effective C can be found:

$C^{n-1} = R_H / R \, , \, [C^{n-1}]_{apparent} = R' / R$

Now, let us apply some values to the relation (arbitrary but correlated to the original proposition):

$(a, b, c) == (3, 2, 1)$ | i.e.

$(x, y, z) == (27, 9, 3)$ | $3x^3 + 2x^2 + 1x^1$

Note the constructed relation to the derivative of the proposition: $dRdx^{-1} == 3x^2 + 2x + 1$.

But the derivative is for a **rate of change** while the former (the proposition) is a **rite of being**.

Therefore, $w = 3x^3 + 2x^2 + 1x^1$

$3x + 2y + 1z = R$

$R = 3 \bullet 27 + 2 \bullet 9 + 3 = 81 + 18 + 3 = 102$

Complexity $\sim= (2.33)(102) = 237.66$

Let's say we want to increase the complexity to 240 :

$240 - 237.66 = 2.34$

therefore, as $240 = (2.33)(102) + 2.34$

$= (2.33)(102) + (2.33)(\phi)$

$= (2.33)(102 + \phi)$

One method would be to increase R by ϕ via adjusting any or some or all of the variables (x, y, z), where $(2.33)(f) = 2.34$; thus, $f = 2.34 / 2.33 \sim= 1.004292$

Therefore, say we increase z by 1.004292 :

$R' = 102 + 1.004292 = 103.004292$

$2.33R' = (2.33)(103.004292) = 240.00000036 \sim= 240.000000$

Now, we can say (or claim) that:

$R' = [\, C^{n-1} \,]_{apparent} \bullet R$

$240 \sim= [\, C^{n-1} \,]_{apparent} \bullet (\, 237.66 \,)$

$[\, C^{n-1} \,]_{apparent} \sim= 240 \,/\, 237.66 = 1.009845998485$

$\log[\, C^{n-1} \,] = (\, n - 1 \,)\log C_{app} = \log(\, 1.009845998485 \,)$; thus,

$\log C_{app} = \log(\, 1.009845998485 \,) \,/\, (\, n - 1 \,)$

$10^{\log Capp} = C_{app} = 10^{\log(\, 1.009845998485 \,) \,/\, (\, n - 1 \,)}$;

with $n = 3$, $C_{app} \sim= 1.004910940574 \sim 1.005$

Another method is to have the variables of the relation as given and determine C (or C^{n-1}) directly (from a desired H): $C^{n-1} = H\Psi^{n} \,/\, (\, C_{n} \bullet R \,)$, as shown earlier.

[Note that H and R_{H} are related, but not the same: $R_{H} = C^{n-1} \bullet R$]

With $H = R_{H} \bullet (\text{ factor term of complexity of R})$, using R_{H} reduces to the first method, but without determining ϕ values for the variables (or needing to).

The corresponding new mixed base, for our example, is: $C \,/\, \Psi \sim= 1.005 \,/\, \Psi \sim\sim \Psi^{-1}$,

and $(\, C \,/\, \Psi \,)^{n} \sim= (\, 1.005 \,/\, \Psi \,)^{3} = 0.0845 \sim\sim \Psi^{-3} \sim 0.0832$;

$(\, C \,/\, \Psi \,)^{3} \bullet R \sim= (\, 1.005 \,/\, \Psi \,)^{3} \bullet (\, 237.66 \,/\, 2.33 \,) = 20.075 \,/\, 2.33$

$(\, C \,/\, \Psi \,)^{3} \bullet R' \sim= (\, 1.005 \,/\, \Psi \,)^{3} \bullet (\, 240 \,/\, 2.33 \,) = 20.273 \,/\, 2.33$

A Complexity "Wavelength"

The $C \,/\, \Psi$ ratio can function as a sort of "frequency" for the relation: there is an obvious structural similarity to $\nu = C_{m} \,/\, \lambda$, where λ is wavelength and C_{m} is the (constant) speed of light in a medium (m).

But for our ratio C is a variable (being the comparison base) and Ψ is the constant.

Thereby, $(\, C \,/\, \Psi \,)^{n}$ yields a frequency of the relation apropos its variables, and: Complexity / $(C / \Psi)^n \rightarrow$ a "wavelength" for the relation whereby the larger the frequency the smaller the wavelength.

[Note, when $(R', R_H) = R$, then $C = 1$ and the freq. $== (\Psi^{-1})^n = \Psi^{-n}$.]

The corresponding (and perhaps more useful) complexity "wavelength" is:

$$\lambda_{cmp} == \text{Complexity} / [(C / \Psi)^n \bullet (R, R', R_H)].$$

This devolves to the complexity factor term itself: $(C_n / \Psi^n)(n + 1)$,

when one allows $C = \Psi$.

Note that the "unitage" of the complexity wavelength is a pure number (i.e. no units, thus unit-less), while that for the relation wavelength is essentially that of the relation's. (So, the complexity λ is as a complexity factor term, while the relation λ is as a relation or its complexity.)

For our example, the complexity wavelength for (or from) R' is:

$$\sim 240 / [(1.005 / \Psi)^3 \bullet (103.004)]$$

$$= 240 / [(0.084471) \bullet (103.004)] = 27.584;$$

therefore, that from R is: $27.584(237.66 / 240) = 27.315$

The (or this) complexity λ for R is relative to that for R' (since the C is calculated for the change of $R \rightarrow R'$).

$240 / 27.584 \sim= 237.66 / 27.315 \sim\sim 8.701$ (full) complexity wavelengths per (total) complexity; i.e., the (total) complexity is a sum of complexity wavelengths, via this construction.

But the actual complexity λ for R (i.e. $C = 1$) is:

$237.66 / [(1 / \Psi)^3 \bullet 102] === 27.9993 \sim\sim 28$ (using $\Psi \sim\sim 2.2905$), yielding

$237.66 / 27.9993 = 8.4881$ (full) wavelengths.

Therefore, extending the complexity from 237.66 to 240 (and R from 102 to R' = 103.004) increases the number of (complexity) wavelengths "spanned" by the (resulting) complexity by $8.701 - 8.488 = 0.213$ (i.e., the complexity λ decreases, from 27.999 to 27.584).

Therefore, increasing the complexity (for a relation) is akin to causing a shorter wavelength and a greater "energy" for the relation; i.e., the frequency (of complexity) for the relation also increases.

Note that the complexity wavelength is fully a quality of complexity term factor, and is thoroughly independent of R, being simply:

$$[C_n(n + 1) \bullet R] / [(1 / \Psi)^n \bullet R] = C_n(n + 1) / (\Psi^n \bullet \Psi^{-n}) = C_n(n + 1) == \tau_{comp} .$$

Thus, **complexity** $\boldsymbol{\lambda \bullet R / [\Psi^n(1 / \Psi)^n \bullet R] = C_n(n + 1) ; \tau_{comp} \bullet R / \Psi^n =}$ **complexity**

[For our example, τ_{comp} is: $\mathbf{7(3 + 1) \bullet R / [\Psi^n \bullet \Psi^{-n} \bullet R] = 7(3 + 1) / 1 = 7 \bullet 4 = 28}$]

We see that τ_{comp} is a function of C_n (and therefore essentially of n itself). As n↑ , τ_{comp} increases tremendously (due to C_n). The delimiter of the complexity, then, is the ratio R / Ψ^n , such that the magnitude of Ψ^n overcomes that of R; i.e., $\tau_{comp} \bullet (R / \Psi^n) =$ complexity .

This configuration allows us to conceptually separate the complexity wavelength from the delimiter, and therefore to distribute the delimitation amongst the terms of the relation R.

For example:

$$\boldsymbol{\tau_{comp} \bullet (R / \Psi^n) == \tau_{comp} \bullet (3x + 2y + 1z) / \Psi^n = \tau_{comp} \bullet [3x / \Psi^n \;+\; 2y / \Psi^n \;+\; z / \Psi^n]}$$

The actual complexity wavelength for a relation presumably must be a positive integer (assuming n is a positive integer). Therefore, manipulation of R can be fixed (or filtered) through the delimitation: let z' ~= z + 1.004 (i.e. z + ϕ) . Thus, we have

$$R' = 3x + 2y + z + \phi == 3x + 2y + z'$$

Therefore, complexity $= \tau_{comp} \bullet [3x / \Psi^n \;+\; 2y / \Psi^n \;+\; (z + f) / \Psi^n]$

and $\tau_{comp} \bullet [3x / \Psi^3 \;+\; 2y / \Psi^3 \;+\; (z + 1.004) / \Psi^3] \sim= 240$

Therefore, we (in this way) keep the proper τ_{comp} for the relation R (and thus also C for the unchanged frequency).

The modification of the complexity (here) for R → R' is, of course, limited to the delimiting term: ϕ / Ψ^n , which is effectively the term added to the original delimiter.

A ϕ can be assigned to any variable considered (i.e. for which there is an "n"). Therefore, our present complexity expression (or formation, formulation) can be more generalized as :

$$\tau_{comp} \bullet [3(x + \phi_x) / \Psi^3 \;+\; 2(y + \phi_y) / \Psi^3 \;+\; z / \Psi^3 \;+\; \phi_z / \Psi^3]$$

so that the relevant contributing delimiting terms (for R → R') are:

$$(3\phi_x / \Psi^3) + (2\phi_y / \Psi^3) + (\phi_z / \Psi^3) == (3\phi_x + 2\phi_y + \phi_z) / \Psi^3 .$$

With n↑, one has C_n↑ and λ_{comp}↑ as well as the complexity increasing. Since λ_{comp} increases, the "energy" for the complexity must decrease (or its energies are dispersed among the rising number of variables considered).

Complexity 'Energisic'

But what is the nature of this complexity energy, for a relation?

From the above formulations, it is distributed among two (relative) parts: λ_{comp} (which is independent of the nature of the relation, with only the number of variables considered being relevant), and the delimiter (which combines the provenance of n with the nature of the variables in the relation). For, with rising n, Ψ^n increases also. [This is accepting that n is a positive integer.] Clearly for the delimiter, as n↑ the delimiter↓ (and therefore the delimitation of λ_{comp} increases).

If we (arbitrarily) make analogies to energy E as E = *F* "a" (i.e., energy = force x "acceleration"), we can assign λ_{comp} as an "acceleration" type (of character or quality) and the delimiter as a force-like quantity; thereby we consider the complexity itself as a sort of energy-sic ('energisic', for the progression of the relation).

Of course, F = ma and W == E = Fd (where d == distance, m == mass). So, the "acceleration" attributed to (or for) λ_{comp} would actually be a term such as: $a \bullet t^2 == d$

(i.e. $a == d / t^2$); i.e., λ_{comp} is a form of distance (analogously, towards energy).

Therefore, $W == \text{work} = E = F \bullet a \bullet t^2 == m \bullet a \bullet a \bullet t^2 = ma^2t^2$; thus,

$W = m(at)^2 = ma \bullet (at^2) = mad == m(d / t)^2 = m(d^2 / t^2)$.

The complexity energisic is such that, while the total complexity of a relation increases (due to the number of variables considered for the relation), the individual energies of each variable decreases combinatorially (to provide, distance-wise, for the resulting increase in λ_{comp} inherent to the "n" used). [Were λ_{comp} to decrease with n↑, the "energies" of the individual variables would have to increase.] Therefore, complexity expends (i.e. uses up) energy; combinatorial (variable) girth requires (relational) 'energy' costs.

The complexity relation, in the form:

$$\text{complexity} = \lambda_{comp} \bullet (R / \Psi^n)$$

presents λ_{comp} as an "n" dependent constant without variables, and R / Ψ^n as a quantity dependent on variables.

Therefore, (by some analogy) considering complexity, as a form of energy, the λ_{comp} term is inertial and can be made to correspond to m mass (of inertia) and the delimiting term R / Ψ^n is of the "acceleration"-type:

$$a \bullet d == (d / t^2) \bullet d = d^2 / t^2 ,$$

which, of course, is a velocity (or speed) squared: $(v)^2 = (d / t)^2$.

Thus, $E = mv^2 == \lambda_{comp} \bullet (R / \Psi^n)$.

[One can also make the argument that λ_{comp} / Ψ^n is the inertial mass and R itself is the corresponding v^2 (dependent on variables); it depends on how one wishes to construct the energisic analogy (i.e. which energisic is most useful).]

To make the (energisic) analogy more general, we can use (or employ) a multiplication of velocities: thus, $E = mv_1v_2 == \lambda_{comp} \bullet (R / \Psi^n)$.

From E's perspective, it itself is (of) a relation and therefore with its own complexity:

Each $v == d / t$, and m (with its inertia) is also a variable.

Therefore, it has 5 variables: $\{ E \mid m , d_1 , d_2 , t_1 , t_2 \}$.

Thus, its $R_E == [\lambda_{comp} \bullet (R / \Psi^n)]_{nR}$; and R_E has its own λ_{comp} and delimiter (terms).

Therefore, $(comp)_E = ((\lambda_{comp})_E \bullet [\lambda_{comp} \bullet R / \Psi^n]_{nR} / (\Psi^n)_{nE})$.

For our example (of R), following the energisic analogy:

$$(comp)_E = (\lambda_{comp})_E \bullet [\lambda_{comp} \bullet R]_{n=3} / \Psi^5 \bullet \Psi^3 == (\lambda_{comp})_E \bullet [\lambda_{comp} \bullet R]_{n=3} / \Psi^8$$

$$== (\lambda_{comp})_{n=5} \bullet [\lambda_{comp} \bullet R]_{n=3} / \Psi^8$$

$$= (\lambda_{comp})_{n=5} \bullet [\lambda_{comp} \bullet (3x + 2y + z)]_{n=3} / \Psi^8$$

$$= (\lambda_{comp})_{n=5} \bullet (\lambda_{comp})_{n=3} \bullet [(3x + 2y + z) / \Psi^8]$$

Thus, the "energy of an energy" is a compilation of complexities.

The energisic resolves to: $C_5(5 + 1) \bullet C_3(3 + 1) \bullet (3x + 2y + z) / \Psi^{(5+3)}$.

If we keep $(C_5, n = 5)$ as a constant (for the energy) and let R be variable, then we have:

EF / n_R	n_R	EF	$E / R = E_R / R = [\lambda_{comp} / \Psi^{nR}]_{nR}$
13.4	1	13.4	0.8732
13.2	2	26.3	1.7155
11.9	3	35.7	2.3301
17.4	4	69.6	4.5414
46.96	5	234.8	15.3224
193.23	6	1159.4	75.6692

[EF = energisic factor for R == $(\lambda_{comp})_{n=5} \bullet (\lambda_{comp})_{nR} / \Psi^{(5+nR)}$]

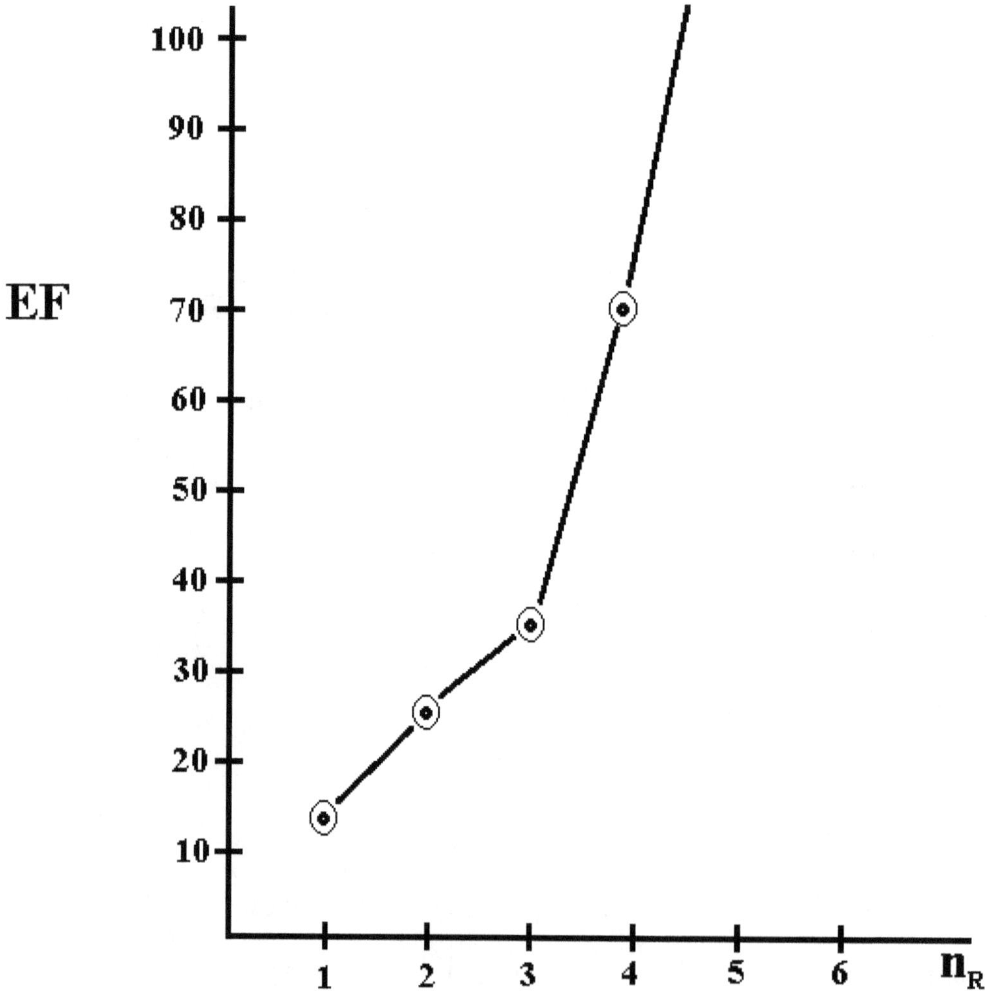

The (n_R, EF energisic) slope quickly rises (of steepness) towards infinity (the vertical or energisic axis); of course, the slope is constant for any particular n_R.

Since $[(\lambda_{comp})_E / \Psi^5]_{n=5}$ essentially acts as a constant, the energisic quality for n_R is established by E_R (i.e. by the component of the energisic with changing n). This follows since the energy is established by E_R (rather than its complexity).

Imparting Complexity to a Relation

So, what does a complexity impart to a relation (or an energy), aside from an evident multiplication?

Perhaps a limit (in rational terms), to a relative ratio, is established.

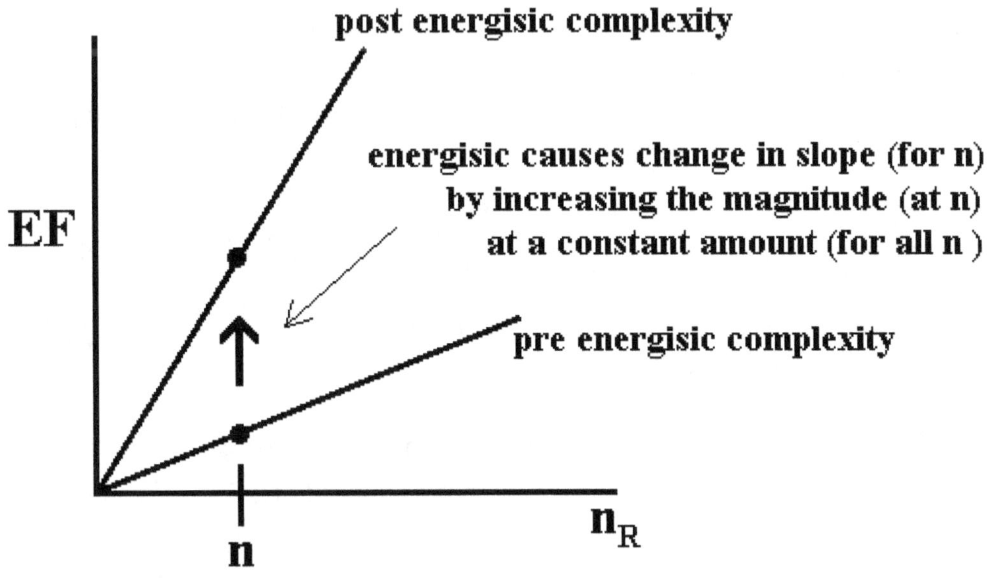

Since EF and n_R are pure (unit-less) numbers, any unitage must derive from R. But how does this relate directly to the EF / n_R ratio? Again, we may employ the coefficients of R to determine (or assign) u as a unitage (to keep the variables unit-less). Then, an appropriate factor may be defined to relate u directly to the EF / n_R ratio (i.e. allowing the ratio to remain unit-less when determining or terming a complexity for the relation).

This is effectively $(EF / n_R) / u == EF \bullet u^{-1} / n_R$ (of employment).

Therefore, the unitage factor is: $(EF / n_R) \bullet u^{-1} == EF / (n_R \bullet u)$.

This also pertains to an energy (E_R):

i.e. $E_R / (n_R \bullet u)$

and indeed any complexity for a relation:

i.e. $(Cpl)_R / (n_R \bullet u)$.

Since the complexity already contains a numerational u, the $(Cpl)_R / (n_R \bullet u)$ term remains unit-less (i.e. sans unit dimensionality).

But, the nature of "u" is compounded, since its utility is in the comparison of comparable units or measures (e.g., meters to feet, etc.). Without such comparisons, "u" can remain (as) "1" (in value).

If there are comparative unitages, say u_a and u_b, then they may be constituted (for a complexity) as by their ratio u_a / u_b to be numeratoritive for R and denominational for the n_R fraction, the resultant term remaining unit-less for the complexity. Yet, as a ratio, this unit-less-ness is assured (or implied). But this presents the curious result that, for a relation employing unitage, the relation without unitage yields to virtually no predictive value upon calculations.

Therefore, we must apply such ratios to an unitage enhanced complexation for the relationship (to try to uncover the ratio): $(Cpl)_R == (C_n (n + 1) / \Psi^n) \bullet u \bullet R$; thus,

unitage factor $== (Cpl)_R / (n_R \bullet u)$, $u = (Cpl)_R / (n_R \bullet$ unitage factor $)$,

where u is (now) a unitage ratio.

This makes the unitage (ratio) inherent of the relation (R). What is more arbitrary, then, is the unitage factor, since it is based on n_R (and not necessarily n). "n_R" is an index that can be "chosen" to apply for R.

The n_R nomenclature, with its "R" subscript, indicates that this "n" is 'applied' to R (rather than constituent of it, the relation). This formulation provides for a small "u" from a large "n_R" application. This is also true for the unitage factor. They (u, unitage factor) are switchable with regards to n_R.

u is fixed by the unit comparison(s) made. n_R may be fixed if it results from a limit for the relation. Otherwise it is arbitrary (to other considerations). Fixing both u and n_R necessarily fixes the unitage factor.

A Complexity Model for Work

But what could limit n_R?

Work = W = m • a • d ; mass and distance can be any number (value). Therefore, acceleration is suspect (here): $a = d / t^2 = (d / t) \bullet (1 / t) == v \bullet (1 / t)$. t is limited by being (only) positive (for physical purposes). Therefore, t is clearly (somehow) the culprit.

Perhaps it (the solution for n_R) is something as simple as comparing the (total) time for each fundamental unit of a mass to reach a given point compared to the time for some physical process (of essentially motion, since work is 'force times distance'). If $t_{work} - t_{mass} < 0$, then (say) the process can not occur and a limit for n_R results.

It would be better to formulate this (inequality) as: $1 + t_{work} - t_{mass} < 1$ (to allow for nul/los considerations). Time (being demonstrated, here) is not allowed to be (or become) zero. Time, as a concept for physics, is compromised and therefore limited by (a) boundary (to remain greater than zero). But this positive (of t) boundary is “closed” and is never open-ended for a (real) process. All processes are definite of extent.

We imagine (for a first or primary model) a mass (of definite shape or dimensions) made up of fundamental (i.e. subject) units [They can be, for this first instance, atoms; but the units can be finer as the model evolves.]. Thus, we pick a position within the mass, not shared of by any of the units, to establish a distance of each unit from this point. Now, during a time (or period) we assess the distance each unit travels to reach this point within the given period, and add all of the (necessary) velocities (as subject or unit speeds) to yield a resultant velocity. The time, of course, is that necessary for some process (physical) involving the mass. Therefore, the resultant is not due to any integration of speeds, but rather of some other type of summation contingent on the physical process considered (or evaluated).

For example, for n (fundamental) units of mass, we have

$$\frac{\sum_{1\rightarrow}^{n} d_{\vec{n}}}{nt}$$

$$\left[\text{i.e. } \sum_{i=1}^{n} d_i \Big/ nt \right]$$

where t is the period of the process, d == distance for each particular unit to the geometrical or spacial position picked. [This position is not coincident with any of the units', to avoid zero d for that unit.]

This is distinct from a summation of the unit velocities:

$$\sum_{1\rightarrow}^{n} \frac{d_{\vec{n}}}{t}$$

$$\left[\text{i.e. } \sum_{i=1}^{n} (d_i / t) \right]$$

The latter summation (rather) is a form of constituent (mass) addition [e.g., if $n = 3$, then the constituent term is $(d_1 + d_2 + d_3) / (t_1 + t_2 + t_3)$, where $t_1 = t_2 = t_3 == t$

(as used; thus, 3t).]

For the summation of velocities, $t == t_{work}$, and of course, discounting vectorial considerations (of direction), the speed (only) increases.

In contrast, the (rational) constituent summation is persistently decreased (of velocity rationalization) by $1 / n$ of factor.

As an example,

let $d_1 = d_2 = d_3 = 1$ distance unit , $t_1 = t_2 = t_3 = 1$ time unit ,

and compare (mass) velocity summation to constituent (mass) summation:

velocity summation	constituent summations
$(d_1 / t_1) + (d_2 / t_2) + (d_3 / t_3) ==$	$(d_1 + d_2 + d_3) / (t_1 + t_2 + t_3) =$
$(1/1) + (1/1) + (1/1) = 3$	$(1 + 1 + 1) / (1 + 1 + 1) = 3/3 = 1$

Here, with $t_{work} = t_{mass}$, the only way t_{mass} can reach t_{work} for their respective fractions (or d / t ratios) is for t_{mass} to decrease to $(1/3)$ its value.

Therefore, the condition $t_{work} - t_{mass} >= 0$ is maintained.

For parity (equivalence) of the velocity and constituent summations, the common constituent t must be made equal to $(1/n)$ of the common velocity t, a further reduction being prohibited (for our model).

A Complexity Model for Motion

Time, by nature of its definition (or usage) being the passage or travel (in motion) of some quality or caliber, is never zero. And when we enlist "$t = 0$" we actually mean the initiation of the 'measurement' of some process.

Therefore, we can ascribe (for a process) a closed formulation (with geometrical analogy, i.e. like as a closed circuit, etc.). Let us, then, by analogy, represent a velocity as that of a closed circle, the circumference of which estimates the extent (size) of velocity. Then, by our model, the radius is $1/n$, and the variation (or deformation) from a (perfect) circle represents changes in velocity and therefore accelerations.

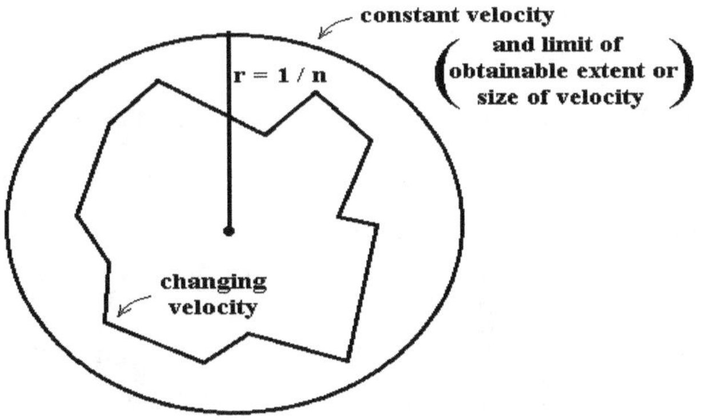

But what of mass-less particles (e.g., photons) which are assumed to have no volume. Then, as (non-self subtractive) operations on infinity, (∞) only yields infinity, the mass-less particle can be anywhere in a volume, and of any volume there are (or can be allowed) an infinity of particles.

Therefore, for such: $r = 1 / n_\infty$, which can only be an approximation if we insist physical reality can not allow of infinite quantities. [Here, a circle of any size is possible since n is (always) n_∞ or infinite.]

By further analogy, rather than a circumference, but a surface area of sphere or even volume of sphere can be used to represent the extent of a (limiting) velocity.

But let us stay (for geometrical analogy) with the circle and its circumference.

If we align circumference (thus, circle size) to velocity, the $v_c == 2\pi r = 2\pi(1 / n)$, with dimensions of v_c being distance / time and (2 and π being pure numbers) dimensions of n necessarily being time / distance.

If, then, we have a limiting velocity for a mass-less particle as the speed of light (say $c \sim= 3 \text{ x } 10^8$ meters / second), then $n_\infty \rightarrow n = 2\pi / c \sim= 2.093 \text{ x } 10^{-8}$ seconds / meter in mass units (for n).

The dimensions of n can be claimed as 'Phantom-Proprietary' units: phantom because they are not real (for n, which is actually a pure number), proprietary because they belong to R (relation) while being coupled to another (or foreign) relation (here the circumference of a circle).

This allows 'time' units to be converted to more tractable measures (i.e. more than merely a passage of quality): e.g.,

(1 / n)t == [1 / (seconds / meter)] • seconds = seconds / (seconds / meter) = meters

for a (common) t, in analogy. For one thing, distance can be subtracted (and made negative, to a direction) but time (properly) can only be extended.

Also, $(1 / n)t$, with $t = 1$ second, represents the radius r (on the circumference side of the equivalence mirror) which when multiplied by 2π yields a circumference of c meters (which on the velocity side of the equivalence mirror is the velocity magnitude of $3 \text{ x } 10^8$ meters / sec). So, the velocity "reflection" has $r = 1 / n$, while the circumference "reflection" allows $r = (1 / n)t$ (i.e. seconds / meter). The "mirror," then, promulgates factors according to the equivalence of (different) relations (i.e. an equivalence transformation).

The accelerations "submissive" to the maximum velocity circle can be represented at specific points by specific velocity circles:

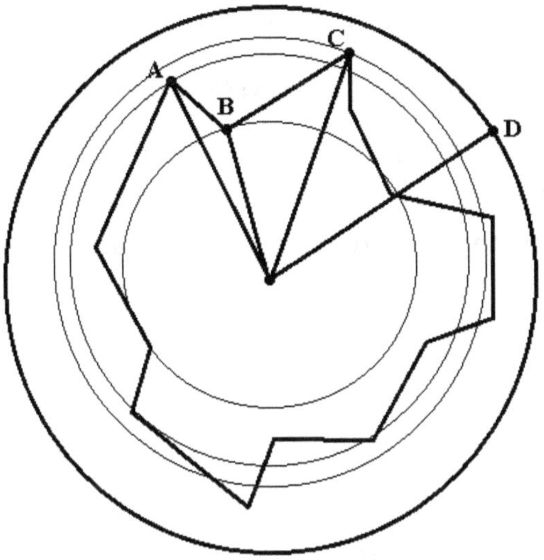

The points establish the various (sub-) radii.

$r_A \rightarrow r_B \rightarrow r_C \rightarrow r_D$
is a trajectory of accelerations (or shifting velocities)

$\therefore$ $(1/n_A) \rightarrow (1/n_B) \rightarrow (1/n_C) \rightarrow (1/n_D)$

Therefore, as n↑ the acceleration↑ (for $r = 1/n$ model) with as $n \longrightarrow n_\infty$ the radius reducing to the center; but as n↓ also the acceleration↑, with a limit (to n) of 2.09×10^{-8} seconds / meter at the maximum velocity (c) of 3.00×10^{8} meters / second (for the max. velocity circle). Thus, as n changes, accelerations are induced.

By similar analogy, we can apply a velocity to the area of a circle (or the surface area or volume of a sphere), keeping $r == 1/n$; therefore,

	R analogy	u	n
$v_c \sim\sim \pi r^2 = \pi(1/n)^2$	**circle area**	(meter • R / second)$^{1/2}$	**1.02×10^{-4}**
$v_c \sim\sim 4\pi r^2 = 4\pi(1/n)^2$	**surface area of sphere**	(meter • R / second)$^{1/2}$	**2.05×10^{-4}**
$v_c \sim\sim (4/3)\pi r^3 = (4/3)\pi(1/n)^3$	**volume of sphere**	(meter • R / second)$^{1/3}$	**2.41×10^{-4}**
$v_c \sim\sim 2\pi r = 2\pi(1/n)$	**circumference of sphere**	[1](meter / second)	**2.09×10^{-8}**

n is in seconds / meter; $v_c \sim\sim 3 \times 10^{8}$ meters / second ,

u == unitage (after R) ; $v_c = u \bullet R$

A more convenient notation for the unitage leading to (or maintaining for) v_c is:

$[\,u_R\,]^{\text{reciprocal exponent}}$ or $(\,u_R\,)^{1/\text{exponent}}$, where u_R == unitage of R and this is applied to the appropriate reciprocal exponent to maintain unitage for v_c;

therefore,

$(u_R)^1$ for circumference : $(meter / sec)^1 = meters / sec$

$(u_R)^{1/2}$ for circle and sphere surface areas : $(meters^2 / sec^2)^{1/2} = meters / sec$

$(u^R)^{1/3}$ for sphere volume : $(meters^3 / sec^3)^{1/3} = meters / sec$

But, of course, these "n"s are graphical. Relationally, each one equals "1" (in quantity). And by (functional) definition, they must (or should) be integers. So again we must consider the significance of fractional n, for a relationship. It appears to result from analogical assignments.

Let us try, then, a more direct comparison of analogy (here):

v == d / t ~ ~	**2π(1 / n)**	
speed	**circumference**	
n = 2	**n = 1 (of some unitage)**	|
	r = 1 / n (i.e. radius)	| let **n == $n_{r(adius)}$** ;

via complexities:

$$(C_n(n + 1) / \Psi^n)(d / t) \sim\sim (C_n(n + 1) / \Psi^n)[2\pi(1 / n_r)] \text{ ; thus,}$$

$$(C_2(3) / \Psi^2)(d / t) \quad \sim\sim (C_1(2) / \Psi^1)[2\pi(1 / n_r)] \quad \text{; thus,}$$

$$n_r \sim\sim (C_1 / C_2)(\Psi^2 / \Psi^1)(t / d) \bullet 2\pi \bullet (2 / 3)$$

Conceptually, then, a motion (speed) is related to a distance (circumference). If there is uncertainty (of measurement) in either, the other 'might' be stated exactly. However, this so called "uncertainty principle" seems somewhat spurious, it being based on measurement (of physical properties), since our sphere of existence (the universe) is of indeterminate (by us) motion (and direction of motion for any totality of compartmentalization or vector sum) and therefore of motions dependent (for results) of it (universal motion).

If we allow, here, that (exactly) $v_c = d / t$, then we can calculate n_r as:

$$(3 / 2)n_r \sim\sim (1 / 3)\Psi^{2-1}(1 / (3 \times 10^8)) \bullet 2\pi \sim= 1.6 \times 10^{-8} \text{ seconds / meter}$$

Therefore, $n_r \sim= 1.1 \times 10^{-8}$ seconds / meter ~ ~ 1×10^{-8} seconds / meter

This is about 1 / 2 times larger than the non-complexed $n_{(r)}$ (making the radius 2 times as large); so this is due entirely to the ratio of the (differing n) complexity factors. Therefore, it is a consequence of (through analogy) converting an n = 2 relation to an n = 1 relation. The conversion "costs" (pays) through a factor of 2 (for the complexity).

So, we have two cases:

uncomplexed

$\mathbf{v_c \sim\sim 2\pi(1 / n_r)\ ,\quad n_r \sim == 2.1 \times 10^{-8}\ sec / meter\ ,\ r_{(radius)} = 4.8 \times 10^{+7}\ meters / sec}$

complexed

$\mathbf{(v_c =)\ d / t \sim\sim 2\pi(1 / n_r) \rightarrow (C_2(2 + 1) / \Psi^2)(d / t) \sim\sim (C_1(1 + 1) / \Psi^1)2\pi(1 / n_r)}$

$\mathbf{n_r \sim == 1.1 \times 10^{-8}\ sec / meter\ ,\ r_{(radius)} = 9.1 \times 10^{+7}\ meters / sec}$

But there are also the original proper (and derivative) relations:

uncomplexed $\mathbf{v_c = d / t \sim = 3 \times 10^8\ meters / second}$

Comp $\mathbf{v_c == (C_2(2 + 1) / \Psi^2) \bullet (d / t) \sim = (1.715)(3 \times 10^8\ meters / sec)}$

$\mathbf{= 5.15 \times 10^8\ meters / sec}$

$\mathbf{C_{(ircumference)} = 2\pi r \sim = 2\pi \bullet (2.1 \times 10^{-8}\ sec / meter)^{-1} \sim = 3.0 \times 10^{+8}\ meters / sec}$

Comp $\mathbf{C = (C_1(1 + 1) / \Psi^1)2\pi r \sim = (0.8732) \bullet 2\pi \bullet (1.1 \times 10^{-8}\ sec / meter)^{-1}}$

$\mathbf{= 4.99 \times 10^8\ meters / sec}$

It is interesting that the C*pl* v_c is slightly greater than the C*pl* C, the circumference being (used as) the analogy for the speed. Only when the complexity factor for C*pl* v_c is less than "1" is the complexity less than 3×10^8 meters / sec , i.e. treating the variables d and t as a term (d / t) of n = 1 , maintaining (by calculation) a maximum circumference "velocity" (as given) for the analogy.

This suggests (for analogies at least) that terms of variables may be reduced in (accepted) n for a relation (when considering possible complexities). This is equivalent, of course, to using one or more variables of a term as constant(s). For (d / t) it is evident (or most convenient) that t is treated as "1" to bring the term to a single (or reduced) value. The speed resulting from an (n = 1) variable (from which 3×10^8 meters / second derives, for some complexity) is:

$$\sim (3 \times 10^8\ \text{meters / sec}) / 0.8732 = 3.44 \times 10^8\ \text{meters / second} .$$

Perhaps a more readily applicable use of n condensation (to lower numbers) via analogy is use of powers of a variable: e.g., instead of circumference employ area or volume.

$v_c \sim\sim \pi r^2$, where r^2 can be viewed as (r • r); thus, n = 2,

but condensing down to n = 1 (as the only essential variable).

$$v_c \sim\sim \pi r^2 == \pi(1/n_r)^2 \;,\; n_r \sim= 1.02 \text{ x } 10^{-4} \text{ sec / meter}$$

[It's clear that as the power of the analogous variable, here r, increases so does n (as seen earlier).]

$^{(r)}\mathbf{n = 1}$ $\quad \mathbf{v_c \sim\sim (C_1(1+1)/\Psi^1) \bullet \pi(1/n_r)^2} \quad \mathbf{n_r = 9.56 \text{ x } 10^{-5} \text{ sec / meter}}$

$^{(r)}\mathbf{n = 2}$ $\quad \mathbf{v_c \sim\sim (C_2(2+1)/\Psi^2) \bullet \pi(1/n_r)^2} \quad \mathbf{n_r = 1.34 \text{ x } 10^{-4} \text{ sec / meter}}$

$^{(r)}\mathbf{n = 1}$ $\quad \mathbf{\mathit{Cpl}\; v_c = (C_2(2+1)/\Psi^2)v_c \sim\sim (C_1(1+1)/\Psi^1 \bullet \pi(1/n_r)^2}$

$\mathbf{n_r = 7.30 \text{ x } 1^{-5} \text{ sec / meter}}$

$^{(r)}\mathbf{n = 2}$ $\quad \mathbf{\mathit{Cpl}\; v_c = (C_2(2+1)/\Psi^2)v_c \sim\sim (C_2(2+1)/\Psi^2) \bullet \pi(1/n_r)^2}$

$\mathbf{\rightarrow v_c \sim\sim \pi(1/n_r)^2} \quad \mathbf{n_r \sim= 1.02 \text{ x } 10^{-4} \text{ sec / meter}}$

Likewise for (spherical) volume:

$v_c \sim\sim (4/3)\pi r^3$; for $r \bullet r \bullet r$, $^{(r)}n = 1, 2, 3$ [i.e.: r^3 ; $r \bullet r^2$; $r^2 \bullet r$; $r \bullet r \bullet r$]

$$^{(r)}\mathbf{n} = \Big]_{1}^{3} \mathbf{i} \,|\, \mathbf{2} \qquad \mathbf{V_c \approx \frac{C_1(i+1)}{\Psi^{i}} \cdot (4/3)\pi(1/n_{r_i})^3}$$

$$v_c \sim\sim (C_i(i+1)/\Psi^i)(4/3)\pi(1/n_{ri})^3$$

$$^{(r)}\mathbf{n} = \Big]_{1}^{3} \mathbf{i} \,|\, \mathbf{2} \qquad \mathbf{\mathit{Cp/}V_c = \frac{C_2(2+1)}{\Psi^{2}} V_c \approx \frac{C_1(i+1)}{\Psi^{i}} \cdot (4/3)\pi(1/n_{r_i})^3}$$

$$\mathbf{\mathit{Cpl}\; v_c = (C_2(2+1)/\Psi^2)v_c \sim\sim (C_i(i+1)/\Psi^i \bullet (4/3)\pi(1/n_{ri})^3} .$$

One notices a trend that as the exponent of r increases, n_r also increases, so that eventually it becomes above fractional (i.e. a whole number) and perhaps even inhabits integer domains.

Therefore, we can intuit (through this pattern) that we can bring n_r to (proper) integer values by further exponentiation of r (or $1 / n_r$) :

$r^1 \rightarrow n_r$ of 10^{-8} |

$r^2 \rightarrow n_r$ of 10^{-4} | $\{ r^1 , r^2 , r^3 ; 10^{-8} , 10^{-4} , 10^{-3} \}$

$r^3 \rightarrow n_r$ of 10^{-3} |

Therefore, we can ask of which r exponent yields n_r of $>= 10^0$ ($>= 1$) .

Since we are using geometrical (closed) curves for circumference (distance), area, and volume using factors of π^1 (in these analogies), one might assume that for r exponentiation > 3, the exponent of π can (of these transitions) rise greater than 1, with steps of 1 (for π) to steps of 3 (for r) : e.g., $\{ r^1 , r^2 , r^3 ; \pi^1 \}$ |

$\{ r^4 , r^5 , r^6 ; \pi^2 \}$ | etc.

Necessary corrections can occur through the numerical factors (NF) that forcibly relate to v_c: e.g.,

$$\{ r^1 , r^2 , r^3 ; \pi ; 2 , 4 , 4 / 3 \}$$

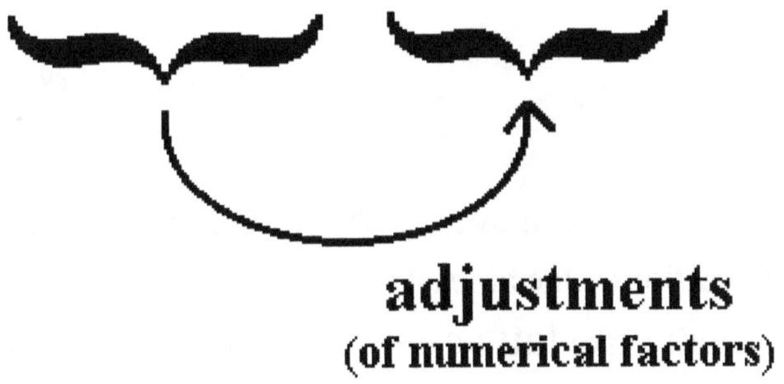

With $n_r = 1$, the exponentiation of r is any number, since $1 / n_r = 1 / 1 = 1$, and (base) 1 to any exponent remains 1. With these precepts at hand, we can at least estimate useful numerical factors for specific (desired) n_r values: e.g., for $n = 2$, assume a practical r exponentiation; let's try the π^2 range: r^4, r^5, r^6 ($r = 1 / n_r = 0.5$) .

Thus,

$v_c \sim\sim (NF)_4 \bullet \pi^2 \bullet (1/2)^\wedge(r)^4$ $(NF)_4 = 4.863 \times 10^8$

$v_c \sim\sim (NF)_5 \bullet \pi^2 \bullet (1/2)^\wedge(r)^5$ $(NF)_5 = 9.727 \times 10^8$

$v_c \sim\sim (NF)_6 \bullet \pi^2 \bullet (1/2)^\wedge(r)^6$ $(NF)_6 = 1.945 \times 10^9$;

i.e., $v_c \sim\sim (NF)_4 \bullet \pi^2 \bullet (1/2)^4$, $(NF)_4 \sim\sim v_c / [(\pi^2) \bullet (1/2)^4] = 4.863 \times 10^8$

Therefore, NF acts as a compensating factor (for selected n_r). When NF is large, r can remain small (as well as n_r); but if r is large (or n_r very small), then NF can remain small. A similar approach can be taken for the π^1 range, to calculate (an) n_r, with the actual (geometrically) correct NFs known (i.e., the spirit of the analysis can be similar; although the calculations are as originally made, without reference to π^1 range).

But concerning the π^1 range, n_r is restricted to a very small number (i.e. fractional). Yet, as an "n" quantity it should be defined as (at least) an integer. This (paradox) seems equivalent to the contraction of time for one coordinate system with (greater) motion compared to another (each having a defined or uniform velocity). The errors (of motion) appear to be related to constraining spacial coordinates (which can be positive or negative or resting in change) to temporal coordinates which (by definition of time) can only be positive. The constraining (here) is therefore improper, and a new (or different) type of temporal coordinate (quality and quantity) must be created and employed, one which also has negative and resting changes.

"Tau"-time and Coordinate System (CS) Drawings

Therefore, (x, y, z) does not "move" or change with t , making (x, y, z, t) an aberration (of description and notation). One at least needs something like (x, y, z; t) or (x, y, z | t) , or utilizing a new quantity, like tau : $\tau \rightarrow (x, y, z, \tau)$. The "tau"-time must probably be one of analogy (in quality).

We can construct our coordinate system as (x, y, z, τ) . There is no natural (i.e. physical) reason to force the coordinate axes to be orthogonal to each other, since this is a simplification to aid in the mathematical analyses (with $\sin(\pi / 2)$ defined as equal to 0 , and $\cos(\pi / 2)$ defined as equal to 1). We shall use a more appropriate simplification to generalize the angles between each axis to be equal. This, however, insists (informs or enforces) that the tau-time axis be "analogous" to one of the other axes, and therefore coincident to one of the other axes, and with the same (distance) unitage.

Thus, our coordinate system looks as (for ∠ xy = ∠ xz = ∠ yz) :

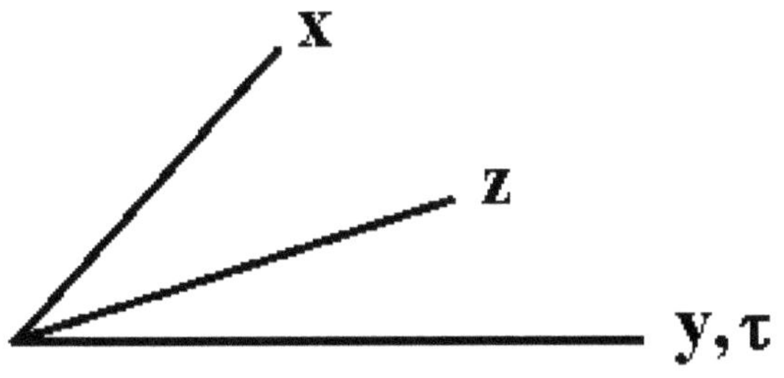

with, say, tau-axis coincident with the y-axis (an equilateral triangle being formed by the points x, y, z).

Now, we can visualize an often used demonstration of coordinate system comparison of motion. Say in CS_1 (a coordinate system) it is considered (of itself) stationary, and we "drop" an object from a height to a floor. The object "moves" with a given time τ ; but actually, the CS_1 "moves" along tau-axis:

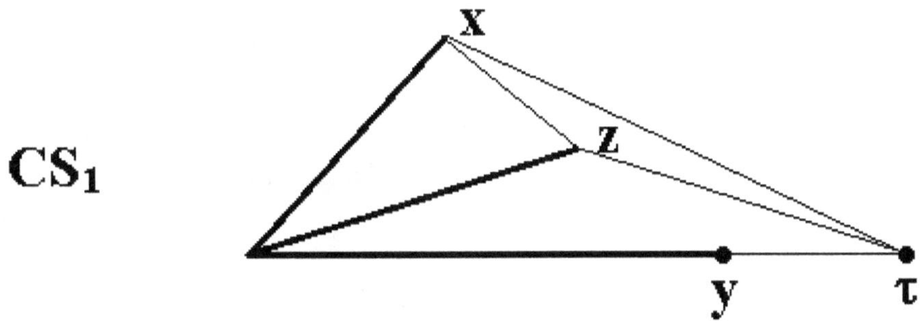

the points (of) x and z making a plane with τ of a particular orientation and extent.

We next compare CS_1 to CS_2 which observes CS_1 externally and discerns that it is moving (along y):

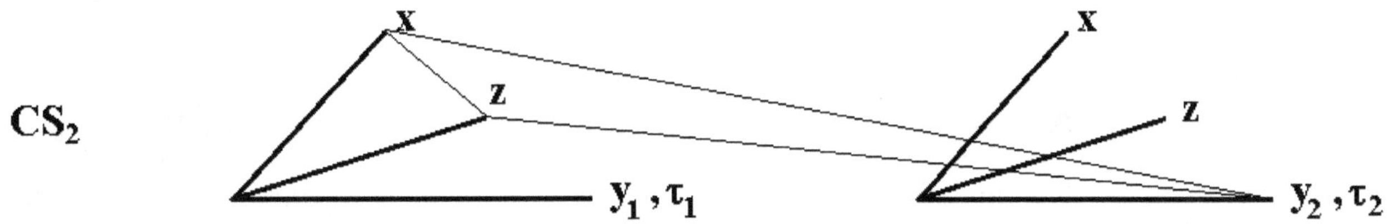

The same plane (of: x, z, τ_2) is formed as for CS_1, with no "contraction" of time (in tau) applied for CS_1 when measured from CS_2.

Note that the generalization of (equal) axes angles also applies to the orthogonal case (with the same condition or restriction for the tau-axis); although it seems suspiciously fortuitous that the electric and magnetic planes of an electromagnetic wave are (found or measured to be) orthogonal to each other.

So, tau-time is as of distance, and presumably energy (dependent on tau-time) is as of motion. (The 'motion' might be just as particular.)

But let us simplify further the CS experiments: say, in CS_1, an object is at rest (x, y_1, z) and remains at rest throughout the observations of CS_2. Therefore, (x, y_1, z) remains, as τ changes from τ_1 to τ_2.

We can draw (depict) this as follows:

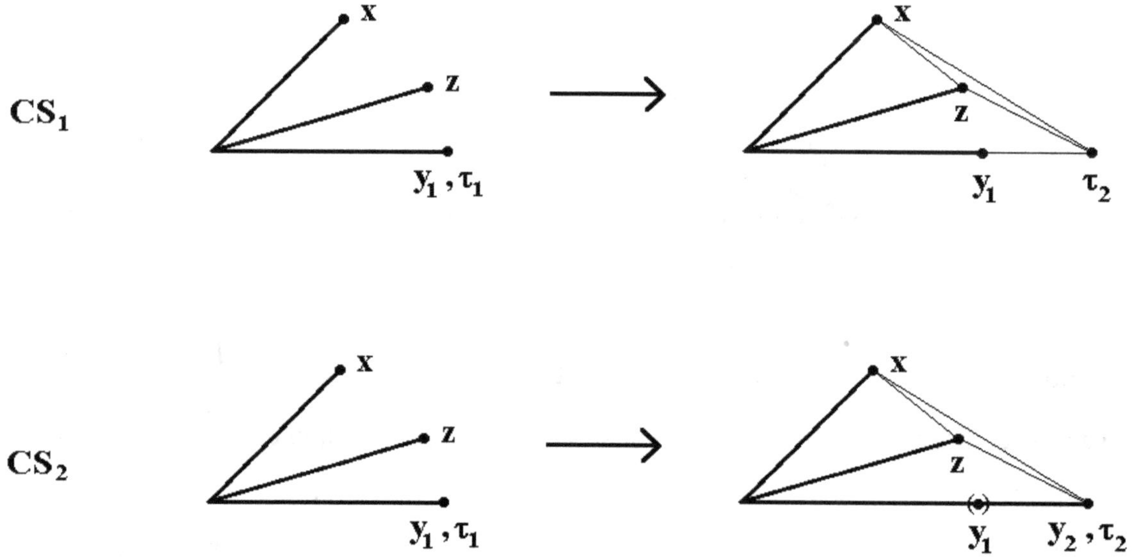

In both cases the (x, z, τ_2) plane results (as the same), but CS_2 perceives a change in y (from $y_1 \rightarrow y_2$) if the direction of motion observed by CS_2 is along the y-axis. Again, there is still no contraction of tau-time perceived by CS_2 compared to CS_1. From CS_2's perspective, the result is the same as for the previous experiment:

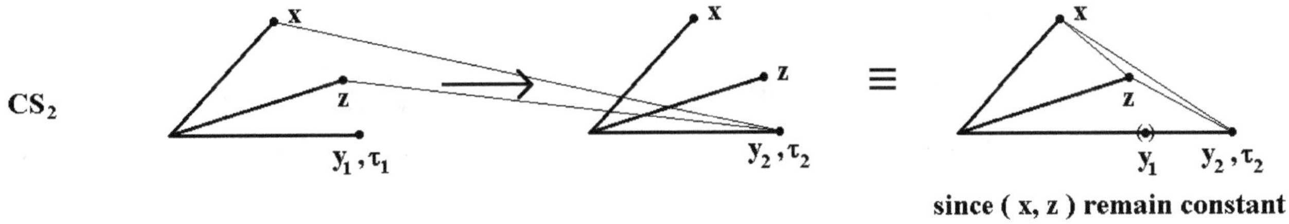

except that the consequences may be different (e.g., in the first case the drop-page: $y_1 \rightarrow y_2$ can or might damage the object, while the object stays at rest in the second case). Events are distinguished by consequences.

The most number of axes that can have (all) equal angles between them is three. This delimits all coordinate systems of measurement, and is independent of "dimensionality" or other (geometric) descriptions. Measurement itself is restricted to such physics.

Line drawings can also be used to depict these (two) conditions, of a more traditional character limited to the events encouraged:

e.g., for case 1

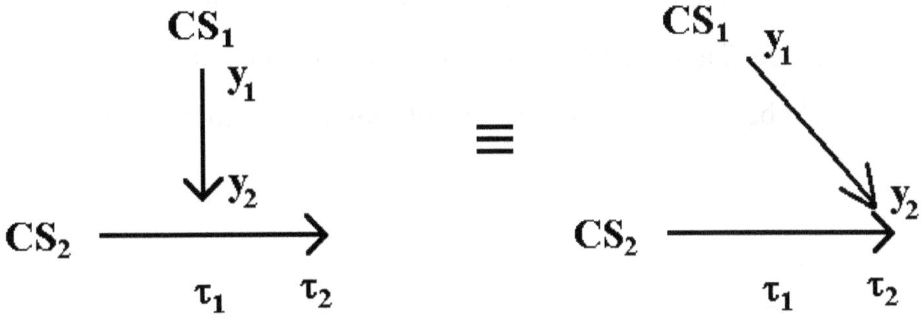

One sees that the distance ($y_2 - y_1$) for CS_1 might be different from that ($y_2 - y_1$) of CS_2's measurement of CS_1.

for case 2

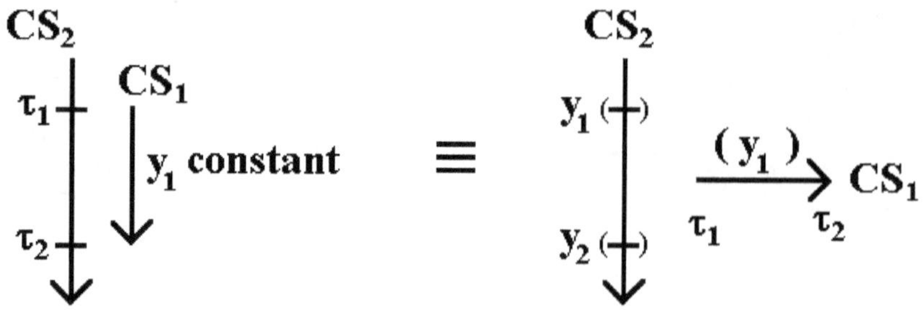

where for CS$_2$ (y_2 - y_1) is the duration of the measurement of CS$_1$

One can tell that these line drawings tend to complicate interpretations compared to the more efficient and general(ized) equal angle CS diagrams. The observation of the line drawings fail to exhibit the equivalence (of results) for the two cases. The equal angle CS diagrams, however, make it more evident that CS_2 is a reference system in which CS_1 is contained. Therefore, CS_2 shows what actually occurs for CS_1 (or at least the results are more clearly understandable and simpler to interpret). That is to say, for the line drawings one has to imagine that CS_1 is in motion (as measured or noticed by CS_2), the relay allowing the drawings to be adjusted to equivalent (==) results.

But such interpretations can be (or seem) confusing. The object in CS_1 for case 2 actually does "fall" a distance of $(y_2 - y_1)$, even though it is considered (for CS_1) at rest. The "fall" might be considered discreet in (tau-)time, depending on how one wants to conduct the analysis:

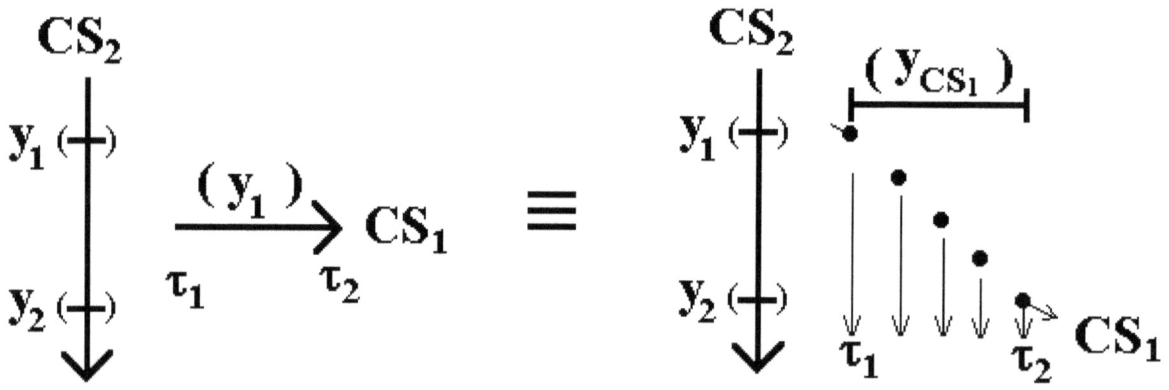

But the interpretation remains somewhat (complicated or) abstract (with these drawings). The equal angle CS drawings (on the other hand) offer a particular manner of coordinate transitions via altering the equal angle of one CS from another when both have a common origin, as well as by rotating CS at the common origin and translations (with rotations) between CS, etc.

If we draw two CSs with a common origin but each having an equal angle (to its 3 axes):

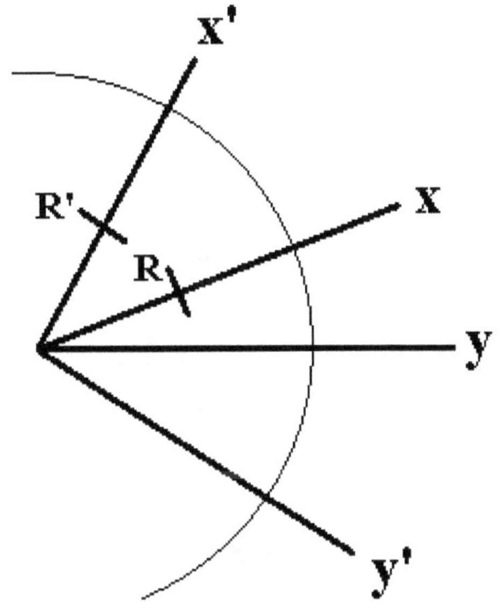

we see that the two CSs have common radii distances (from the common origin);

say, for the x-axes: R' = R .

Yet, consider the perspective of $CS_{(x,\,y)}$ when viewing $CS_{(x,\,y)'}$:

From $CS_{(x,\,y)}$, R' < R

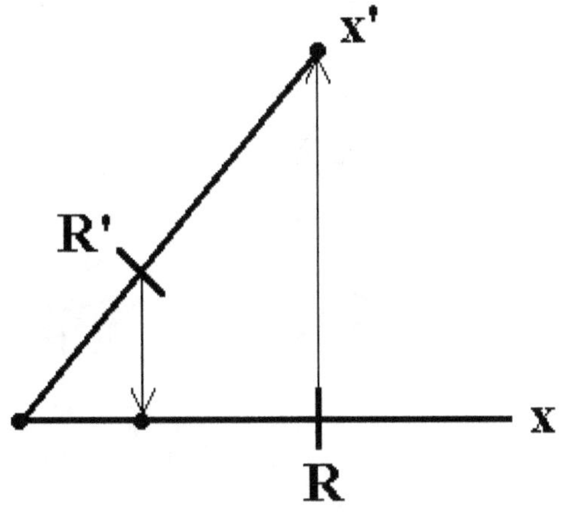

$CS_{(x, y)}$ sees only a projection of R' onto R (or x-axis), and its projection of R onto x'-axis would be larger than R.

Therefore, $CS_{(x, y)}$ and $CS_{(x, y)'}$ represent equivalences of coordinates (with differing equal angles per CS) which yield different coordinates from each CS perspective. Clearly, these perspectives are in error (from reality) when the differences in CS equal angles are not taken into consideration for adjustment of results. If knowledge of each CS equal angle is obscure, then so must be each CS perspective.

This being the case, where is the (most utilizable or opportune) reference CS? It must be external to the circle whose rays are depicted; or it may be the circumference itself (as of measurement of the radii internal and thus of the circle). This, of course, says that a circumference (of a circle) is only noticed externally. It is an interesting notion, then, that a circle's circumference is external of the circle (i.e., is a measurement of the circle).

Obviously, if this external-ity can not be reached (approached or met), then CS perspectives can be in error. One can only assume circumferences and judge (radii) accordingly. Other complications include coincident axes from different CSs of equal angles the same or different, and of course inclusion of the z-axis for a volumetric sphere. Since the limit of equal angles for a geometric figure is 3, there are no further complications (of perception). Perspectives are contained. This implies that all perspectives are in some error since we are ultimately confined to (or are within) a system of observations.

The only appropriate recourse is to provide for inner circles (i.e. sub-circumferences, volumetric surfaces and shapes).

Perspectives and Combinations

In terms of perspectives, then, they run as the C_n for considerations:

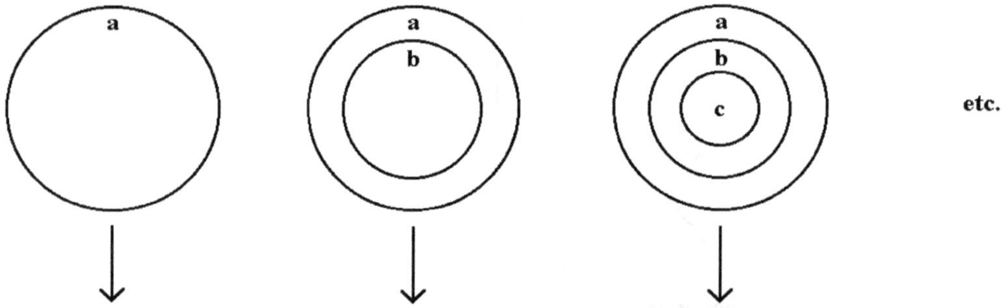

C_n	**1 : a**	**3 : a, b, ab**	**7 : a, b, c, ab, ac, bc, abc**
n	**1**	**2**	**3**

Each perspective is correct for its group numbered by the C_n and distributed as shown: e.g.,

for C_2 , a is correct for a |

b is correct for b | ab is correct for a and b

Therefore, for each grouping (under C_n) error-prone perspectives can be enlisted of analysis: e.g., (again) for C_2

a yields errors (of perspective) for b, ab

b yields errors (of perspective) for a, ab

[One can replace the term "error" to "having a relationship to" as a cause (of action or measurement).]

How can we assess the errors in perspective?

Using C_2

additive errors : **2 • (1 + 1/2) = 3** [e.g., a yields a full error for b and 1/2 error for ab]

a (+ or) b | (1/2)(ab)
b or a

multiplicative errors : 2 • (1) (1 / 2) = 1 !!

a + b | (1/2)(ab)
b or a

Using C_3

The error (counting) is per group here.

additive : **3 • (2 + 1 + 2(1/2) + (1/3)) = 13**

(a + b + c) a, b, c ab ab abc
ac ac
bc bc

e.g. **a (b,c + bc + (ab, ac) + (abc))** the error (counting) being for only a-axis correct (as signaled against error)

multiplicative : (3)(2)(1)(1)(1/3) = 2

But these (calculations) are only partial errors, because they don't consider all of the groups of a set; and having, say, a compared to abc yield an error (per group) of 1 / 3 seems more like an 'error-not-made' for a; and one can very well suggest that if a group is self-consistent (containing) of members (e.g., abc), then any member of the group (e.g., a) is correct (in consideration).

Therefore, let us use more stringent conditions for announcing (or recognizing) errors of perspective, each letter representing an axis (i.e. a different partiality of analysis than used above):

1) if a letter considers a group containing the letter, there is no error

2) all letters of a group show correct perspective of each member

3) all groups must be included for error calculations

$C_2 = 3$ members and groups:

	a ,	b ,	ab
a errors	–	1	–
b errors	1	–	–
ab errors	–	–	–

Therefore, additive errors : 2 , multiplicative errors : (1)(1) = 1

$C_3 = 7$ members and groups:

	a	b	c	ac	ab	bc	abc	
a errors	–	1	1	–	–	1	–	
b errors	1	–	1	1	–	–	–	
abc errors	–	–	–	–	–	–	–	
ac errors	–	1	–	–	½	½	–	
ab errors	–	–	1	½	–	½	–	
bc errors	1	–	–	½	½	–	–	
c errors	1	1	–	–	1	–	–	
sums	3	+ 3	+3	+ 2	+ 2	+ 2		= 15

Therefore, additive errors : 15 ,

multiplicative errors : $(1^9)(1/2)^2(1/2)^2(1/2)^2 = (1)(1/2)^6 = 0.015625$

If we replace the (a, b, c) member names with the traditional (x, y, z) names (for axes), and then include the tau-time (τ) name (as before) coincident with the y-axis, then (for error counting, of perspective) we can simply double each group that includes a "b" member and say that y is correlated to τ. But if τ is a truly independent coordinate, then the number of unique members rises from 3 to 4 (x, y, z, τ) , and C_n increases from $C_3 \rightarrow C_4$ (= 25) groups. Four equal angle but unique axes, however, is not (geometrically) possible to have.

We may annotate errors as like:

a : { b , c , bc } , read : a considers groups (b, c, bc) for errors and finds them to be so in perspective.

Therefore, a { b , c , bc } → 3 (errors) == a : { a , b , c , ac , ab , bc , abc } → 3

Replacing (x, y, z) for (a, b, c) , and correlating y with τ for perspective (of axes), we have (of errors):

$x : \{ y , z , yz \} \rightarrow 3$ | $xz : \{ y , xy , yz \} \rightarrow 2$

$y : \{ x , z , xz \} \rightarrow 3$ | $xy : \{ z , xz , yz \} \rightarrow 2$

$z : \{ x , y , xy \} \rightarrow 3$ | $yz : \{ x , xz , xy \} \rightarrow 2$

$x : \{ \tau , z , \tau z \} \rightarrow 3$ | $xz : \{ \tau , x\tau , \tau z \} \rightarrow 2$

$\tau : \{ x , z , xz \} \rightarrow 3$ | $x\tau : \{ z , xz , \tau z \} \rightarrow 2$

$z : \{ x , \tau , x\tau \} \rightarrow 3$ | $\tau z : \{ x , xz , x\tau \} \rightarrow 2$

We can further concentrate the annotation by creating the function

$$\tilde{\mathbf{y}}$$

connotative of y and τ as its own mini-CS:

$$\tilde{y}\,(\,y,\tau\,):\{\,y,\tau\,\}\rightarrow$$

no (possible) error(s) in perspective.

$\tilde{y}$ then, represents axis coincidence (for our model). This can, of course, be applicable to any axis (and for as many times as desired) of coincidence:

e.g.,

$$\tilde{x}\,,\tilde{y}\,,\tilde{z}\,;\tilde{y}_{(1,2,3)}\,,$$

read as $(y, \tau_1, \tau_2, \tau_3)$ etc. ;

$$\tilde{y}_{1,2,3\,(1,2,3)}$$

is read as $(y_1, y_2, y_3, \tau_1, \tau_2, \tau_3)$ etc.

The "mini-CS" is of course linear (in scale), although its numerical values need not be. The coincidence is (only) linear. It is to be maintained, however, that as stated previously, all axes use the same units. Then, for coincident axes, a unit of one (distance) on one axis can be compared to several units (of distance) on another axis. In fact, one unit (of, say, effort) for one coincident axis can correspond to infinite units (of effort) for another coincident axis (of the same mini-CS). But that is math rather than physics.

For a 3-axis CS (and therefore any CS of equal angles), a plane can be established demarcating an equilateral triangle where each vertex is a point the same distance (from an origin) on each axis. Each angle of the $\triangle$ is equal (being $180^\circ / 3 = 60^\circ$). If one (or any) of the axes is coincident, then the coincident factor (e.g., τ) of unit distance (i.e. "1") represents a deviation from the equilateral $\triangle$ (dimensions):

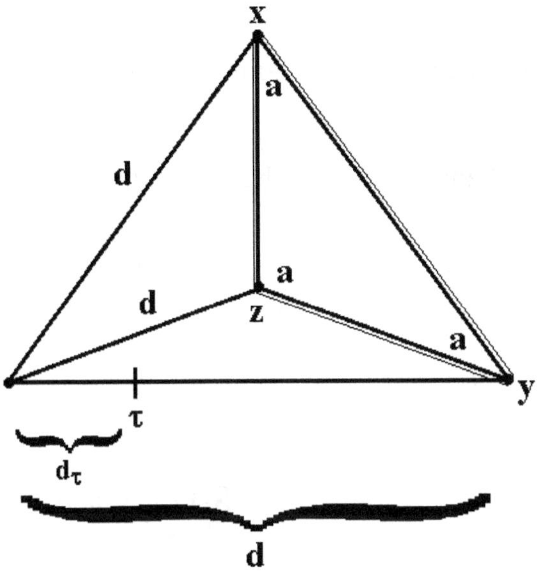

angles ≡ "a"

distances ≡ d , d_τ

We call the (coincidental) factor, τ , as a factor because (essentially): d • τ ≡ d_τ

where d_τ is a fraction of d (i.e. $\tau < 1$, although d_τ can be "unitized" as "1" relative to d).

Therefore, $\tau = d_\tau / d == 1 / d$.

But this belies just how long (or large) d_τ should be in relation to d (if d_τ is of unit length).

With a CS of four axes (two coincident), n = 4 ; therefore, we might try Ψ^{C4}:

$\Psi^{C4} == \Psi^{25} \sim= 1 \times 10^{9} \sim== d$; therefore,

$\tau = 1 / d \sim= 1 \times 10^{-9}$ (d_τ = "1")

[If Ψ^{1} allows for the occurrence of each of the 25 groups (from $C_4 = 25$), then as a set all of the groups occurring is represented by $(\Psi^{1})^{C4} == (\Psi^{1})^{25} = \Psi^{25}$.]

If we instead employ Ψ_m (~ ~ 1.5231), the corresponding $d_{(m)} \sim== \sim 3.7 \times 10^{4} = \Psi_m^{\ 25}$, $\tau_{(m)} = 2.7 \times 10^{-5}$. Unlike Ψ, the base Ψ_m, while derivative of C_n, is not (directly) related to n through its formative (series) summation.

We can also try the base e , to yield the corresponding values:

$d_{(e)} \sim== \sim 7.2 \times 10^{10} = e^{25}$, $\tau_{(e)} \sim= 1.4 \times 10^{-11}$.

e is not related to C_n nor n (in its formative summation), as "n" is defined here (in a purposefully structural sense).

Accepting Ψ as the most appropriate base considered, we can extemporize on the nature of the (our) CS axes. It is characterized as 4 axes with 2 (of the axes being) coincident. Therefore, it has the character of a 3-CS group of axes as well as a CS of 4 axes.

If we average the d per axis, it is both of natures d / 4 and d / 3 , and the true nature lies (perhaps) between these two limits.

First, we can distribute axis properties (to concentrate on the coincident axis, which is of our interest) to be (as estimable) ½ of 3-axis CS and ½ of 4-axis CS ; thus,

$$d((1/4)\bullet(1/2)+(1/3)\bullet(1/2))$$

This yields a factor (for d, for the coincident axis) of:

$$(1/4)(1/2)+(1/3)(1/2)=(1/8)+(1/6)=(6+8)/(8)(6)= 14/48=7/27\sim=0.2917$$

[We do not, however, know the actual partitioning of qualities (of influence) between the two axes of the coincident axis, and merely assume (or select) that it is 1:1 , or ½ for each.]

We are, of course, here redistributing these properties to each axis of the CS, but based on the peculiarities of the coincident axis. This effectively collapses each d (by a common factor) among the CS axes. Using this "equal partitioning" method (as method "a"), the d for the coincident axis comes to: $\Psi^{25}\bullet(7/24)\sim=\sim 2.905\times 10^{8}=d_a$.

For method "b" we might assume that the coincident axis has (only) a weighted influence (for the CS), in favor of the non-coincident axes (or axes before the introduction of the "extra" tau-time axis (i.e. 3 normal axes and one τ axis); thus,

$$d((1/4)\bullet(1/4)+(1/3)\bullet(3/4))\rightarrow d_b\sim=\sim 3.112\times 10^{8}.$$

The average of d_a and $d_b \rightarrow d_{(a,\, b)avg} \sim=\sim 3.009\times 10^{8}$.

One can, of course, also try explicit occurrences of axes (3 normal and 1 $\tau \rightarrow$ 2 normal and 1 coincident):

$$d((1/4)\bullet(1/3)+(1/3)\bullet(2/3))\rightarrow d_c\sim=\sim 3.043\times 10^{8},$$

with $d_{(a,\, c)avg} \sim=\sim 2.974\times 10^{8}$,

and the average of $d_{(a,\, b)avg}$ and $d_{(a,\, c)avg}$ being $d_{\{(a,\, b)avg,\ (a,\, c)avg\}avg} \sim=\sim 2.992\times 10^{8}$.

We are obviously approximating the established speed of light (in a vacuum) in meters / second ($c \sim= 2.998\times 10^{8}$ m/s) by these methods, should our d be so customize-ably map-able to those units (by mere resonance of occurrence). For each method, d_τ = "1" is retained.

The speed of light given is based on its traversal through a vacuum, which is idealistic (physics) since a true vacuum is not obtainable. Such a system would have to be contained by walls which, by definition, obliterate a vacuum. Therefore, the measurement itself is at best an estimate (rather than a limit) for a situation not physically obtainable. It is also somewhat circular (of argument), since the definition (for the extent) of a meter is based (itself) on the speed of light in a vacuum (to derive the corresponding length of path). This is due (allowed) to using "c" as a "known" constant. "Constants" determined by measurements are peculiar: when the measurement changes, so must the constant.

The maximum magnitude of the (tau-based) speed is dependent on the C_n group-age (i.e. number of C_n groups) and therefore on the CS system (i.e. number of coincident axes). But what does it mean to have 1, 2, or 3 tau-(time) coincident axes (for a CS)? What is the physical relevance (of the such)?

Clearly it is apparent that (for a 3-axis CS) only one tau-time (coincident) axis is necessary or needed to describe (coordinate) motion(s). Including (or inducing) one or two more greatly explodes (to much higher values) the magnitudes of d, and the resultant speeds, etc. This is due to the great rise in C_n with n (and therefore Ψ^{Cn}).

$C_5 = 161$, $C_6 = 1561$; therefore, we are dealing with Ψ^{C5} and Ψ^{C6} for C_5 with two coincident + 1 normal (→ 5 axes, 2 coincident) and C_6 with three coincident, thus all coincident (→ 6 axes, 3 coincident), respectively.

We might assume then, physically, that each tau axis is independent and can represent qualities that are distinct (i.e. other than just "time"). [d_τ remains restricted to "1" in size.]

Using method "c" (explicit occurrence of axes), we have: for n = 5

$$\mathbf{d_5((2 / 5)(2 / 3) + (2 / 3)(1 / 3)) \sim= \Psi^{161}((4 / 15) + (2 / 9)) \sim= 5.267 \times 10^{56}},$$

drawn (of the 3 axes) 2 coincident axes and 1 normal axis → 5 axes

(== (1 + 1) + (1 + 1) + 1)

[Tau axes are (share) 2 out of 5 of tau character and 2 out of 3 of normal character; 2 / 3 of the (3) axes are coincident, 1 / 3 is normal.]

For n = 6 , 6 axes (== (1 + 1) + (1 + 1) + (1 + 1)):

$$d_6((3 / 6)(3 / 3) + (3 / 3)(0 / 3)) == \Psi^{1561} \bullet (9 / 18) >>> 5.267 \times 10^{56}.$$

$(1/2)\Psi^{1561}$ is of course an astronomical number:

$$\log_{10}((1/2)\Psi^{1561}) == \log(1/2) + 1561 \bullet \log\Psi \sim= 561.550$$

$$\log_{e}((1/2)\Psi^{1561}) == (\log(1/2)/\log(e) + 1561(\log\Psi/\log(e)) ==$$
$$\ln(1/2) + 1561 \bullet \ln\Psi \sim= 1293.017$$

$$\log_{\Psi}((1/2)\Psi^{1561}) == ((\log(1/2)/\log(\Psi)) + \log_{\Psi}\Psi^{1561}$$

$$== (\log(1/2)/\log(\Psi)) + 1561 \sim= 1560.164$$

Making Physical Analogies and Representations

But what could numbers of such magnitude(s) physically represent?

Well, the numerical (arithmetic) factor for the $n = 5$ result looks suspiciously similar to the established Bohr radius (5.291×10^{-11} meters), which is also measured as a distance (meters). But the ten's factor (10^{-11}) is way off (from the $n = 5$ result: 10^{+56}). Even though the arithmetic factors might be numerical coincidence, it can be useful to see how coincident axes are manipulable for physical qualities (in this instance).

First, we note that as 3 non-coincident ($n = 3$) axes allow for one quantity (e.g., distance) of quality, then $n = 4$ (1 coincident axis) allows for two qualities (e.g., distance, time) and $n = 5$ (2 coincident axes) allows for 3 qualities (distance, time, and other).

Now, using 5.267×10^{56} as our target value (method c), with 5.267 as our (rounded) arithmetic factor and 10^{+56} as our tens factor, we may peruse (or dally) through a list of (established) physical constants (similarly rounded off of arithmetic factor) for any reasonable combinations (towards the target).

We start with the most obvious (or evident), adopting a speed as for $n = 4$.

We choose, therefore, the Bohr radius (B_r) $\sim= 5.292 \times 10^{-11}$m and the Plank time ($P_t$) $\sim= 5.391 \times 10^{-44}$seconds :

$$B_r / P_t == 5.292 \times 10^{-11}\text{m} / 5.391 \times 10^{-44}\text{s} \sim= 9.816 \times 10^{+32} \text{ m/s}$$

discounting the observed phenomenon (or measurement) that the fastest speed obtainable for a transfer of energy is the (constant) c in a vacuum.

Our B_r / P_t result is promising in leading towards a magnitude (in tens) of 10^{+56}, but we are allowed one other unit to extend $10^{+32} \rightarrow 10^{+56}$, which means (or indicates) a factor of

$\sim 10^{+24}$; and searching we find Avogardro's Number (A_{no}) $\sim = 6.022 \times 10^{23}\ mol^{-1}$ which uses a single unit (mol^{-1}). Presumably, if a usable constant (of favorable magnitude) is found that has multiple units (e.g., Boltzman constant $\sim = 1.38 \times 10^{-23} J \bullet k^{-1}$), we can combine those units into one (inseparable) unit for furthering the analysis.

So, we have: (Br / Pt) • A_{no} $\sim = 5.912 \times 10^{+56}$ meters/(seconds • mole) .

This looks very suggestive and illuminating, being at the proper tens factor and having a somewhat close arithmetic factor.

But its arithmetic factor is around: 5.912 / 5.267 = 1.1225 larger than reasonably desired, or almost 10 / 9 larger than needed (i.e. 10 / 8.909 or 5.267 / 5.912 ~ = 0.89087).

Therefore, we apply the given diminishment to arrive at the desired term ("target" quantity) that can be applicable to our n = 5 CS with two coincident axes:

$$(B_r / P_t) \bullet (A_{no} / 1.1925) \sim == (8.909 / 10) B_r \bullet A_{no} / P_t$$

$$\sim\sim (9 / 10) B_r \bullet A_{no} / P_t \sim == \sim 5.267 \times 10^{+56}\ m/(s \bullet mol) ,$$

which suggests a mole (amount) of Br / Pt speed diminished by a factor of ~ (9 / 10) .

The quantity term: $\sim (10 / 9)^{-1} B_r \bullet A_{no} / P_t$ is "map-able" to our n = 5 CS .

B_r represents the (classical) distance of an electron from the nucleus of a hydrogen atom (i.e., which has only one electron), therefore it is the orbital radius of the electron. P_t is the time it takes light (in a vacuum) to travel the distance of one Plank length ($\sim 1.6162 \times 10^{-35}$m). It (for P_t) is a length at which (or within which) quantum gravitational effects are assumed to become significant.

A_{no} (as Avogardro's Constant) is the number of units (entities) per mole of any thing or substance.

We have then, for our n = 5 CS, two coincident axes with tau-time and (now) tau-mole utilized (each case with d_τ = "1" maintained). Each unit (of distance or tau) is of the same extent (on the various axes). This causes the (distance, mole) axis to be (presented at least initially) much longer than the (distance, second) axis. The non-coincident axis is of any (initial) length project-able. Because of the disparity of lengths for the two coincident axes, depending on the (plotted or proposed) length of the non-coincident axis the mapped "space" can be made to look very much as a line projection:

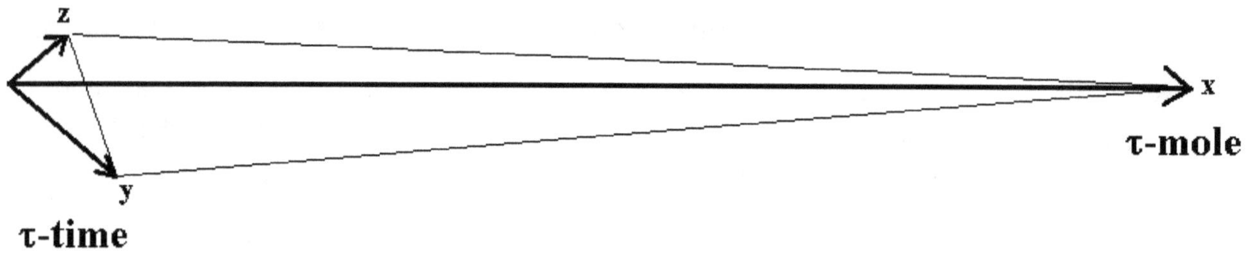

(x, y, z) "space" seems more like a line projection, depending on the (relative) length of z (as plotted).

Therefore, the yx plane can be very large of magnitude and extent and yet (be) extremely thin; i.e., a short change in tau-time can lead to a great change in tau-mole, and one can think of mole equivalents to time (as regards to distance).

But what does (or can) distance / (time • mole) refer to physically?

Keeping d_t = "1" for both of the coincident axes, it means (physically) that for one second of a mole (of some collection of identical particles or idealistic quantities), a distance (of travel) of (collectively or additive-ly) $5.912 \times 10^{+56}$ meters is achieved, each individual particle achieving a distance of $9.816 \times 10^{+32}$ meters.

But there is an apparent contradiction (in distance) of (5.267 / 5.912 =) 0.89 (of the 5.912×10^{56} meters), based on Ψ method "c" for the distance (of axis) d. This is consistent for a n = 5 CS plot(ting). But this still comes down to (essentially) 0.89 times $9.816 \times 10^{+32}$ m/s speed per "particle" involved (of a mole), which is much greater than the established speed of light in a vacuum. Perhaps there is a synergistic or multiplicative (mathematical) effect (for the effective speeds) which can (be used to) maintain the resultant distances achievable. It would tend to imply that the actual electron radius (from the nucleus) is not linear but rather curved or disparate of straight lengths (collectively), assuming the speed of light is the maximum speed measurable.

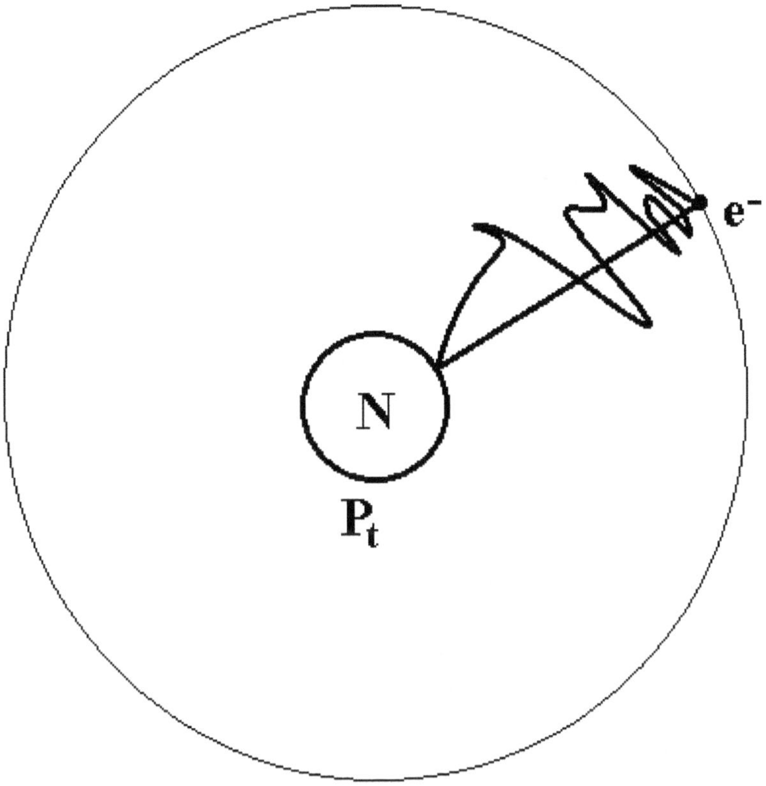

This inspires (the conception of)

$$[\ 0.89 \bullet 9.816 \times 10^{32}\text{m/s}\ ] \ /\ 3.043 \times 10^{8}\text{m/s}$$

simultaneous incidences of virtual electrons (at virtual positions) engaging in transits (at or to [simultaneous – 1] points of curvature) to result in the measured e- radius at any considered position of orbit (about the nucleus or proton), the summation of the (total) distances traversed yielding to our

$(B_r / P_t) \bullet (A_{no} / 1.1225)(m / (s \bullet mol))$ value with d_t = "1" always maintained.

The "position" of the electron (in its orbit) is arbitrarily unknown (i.e. does not inhibit the virtuality of the individual transits involved for its conclusion of orbital location, when this is based on some measurement).

But how do the "virtual electrons" know where to initiate their transits (they know when as being for d_t)?

Presumably the transits can initiate anywhere within (i.e. below) the (real) electron's orbital path, with the accumulative sum of paths ultimately reaching the orbit (location or region) with the proper (Bohr) radius. Any such combination of paths is, then, acceptable. (The paths are, of course, three dimensional in their totality.)

The maximum distance covered by the paths (in total) is, then, estimated to be $5.267 \times 10^{+56}$ meters in one second per mole of paths, or $9.816 \times 10^{+32}$ (times 0.89) meters in one second per (any) successful path (to reach the proper orbit).

From the point of singularity (i.e. the proton) to the (e-) orbit the maximum distance traversed (by the virtual electrons simultaneously) is 9.816×10^{32} meters per (total) path per second. But any circuitous path can reach this maximum and (presumably) most are considerably shorter (initiating anywhere within the *proton*) or by any manner or route-age. There are, then, at most 2.871×10^{24} virtual electrons (i.e. simultaneous incidences) involve-able (in the process of establishing the electron to its Bohr radius orbit) during one second of (d_t) time. This is a more plausible way of expressing the phenomenon, since 10^{+32} meters of travel within an atomic dimension(of volume) seems quite incredulous.

Since the electrons used here are virtual, this amounts to a mathematical manipulation in favor of the Ψ base, similar to inclusion of the (0.89) factor employed.

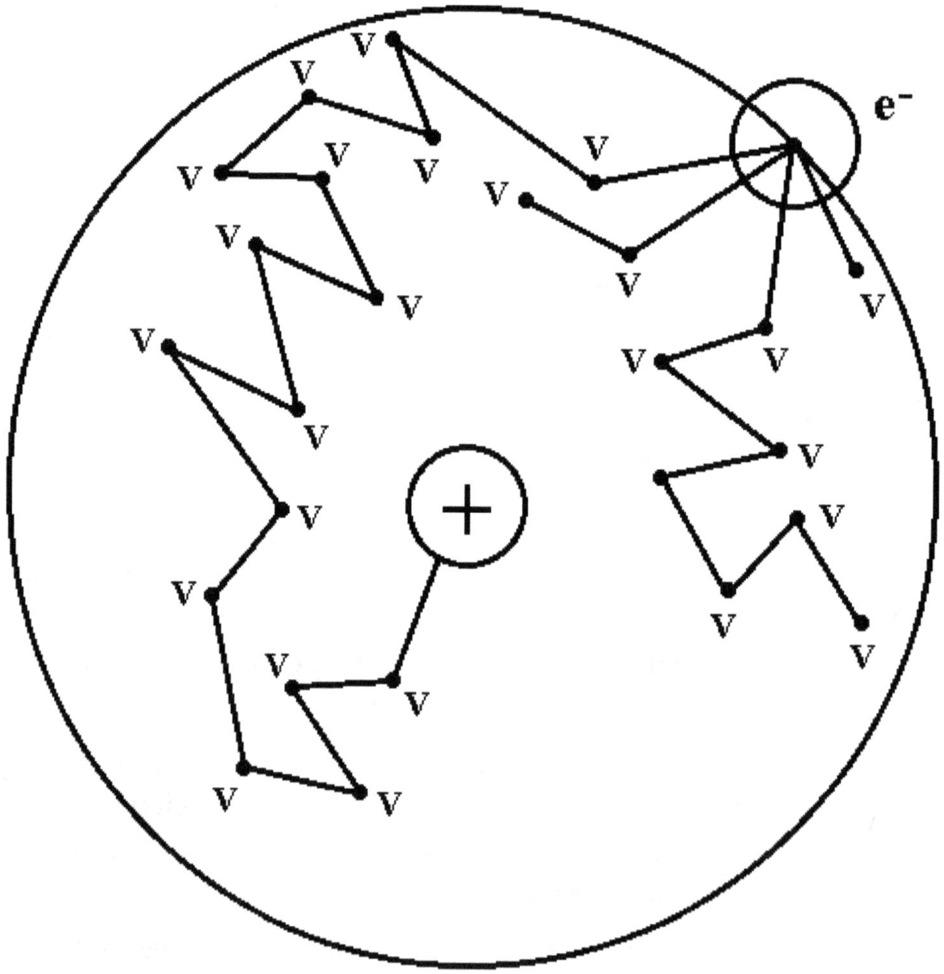

An (hydrogen) electron establishing its orbit (atomic) does so within a minute fraction of a second (for an allowed radius to be established or achieved). So our process is a measure of (hypothetical) virtual e- rates of involvement towards that goal. The virtual electrons are, of course, mass-less and only define positions (of transit). And one can have from one to the simultaneous incidences numbered of them leading up to the e- position at its proper atomic orbit (with Bohr radius).

Therefore, any distance leading up to the orbital distance of position can be accommodated by a virtual electron (within an incidence length of simultaneity). For our curious model, then we can expand the "hydrogen atom" to an electron's orbital radius of $5.267 \times 10^{+56}$ meters (per mole, at d_t = "1"), or we can divide the Bohr radius by that number to get the extent (in fraction of distance) of each simultaneous incidence (for the virtual electrons), as exercises in mathematical physics (or theoretical conjecturing). This would be essentially (for a unit d_t second) a "confined" P_t :

5.292×10^{-11} meters / [(0.89)$9.816 \times 10^{+32}$ meters] ~ = 6.058×10^{-44} (fraction for a second d_t), i.e. slightly larger than P_t (for a second).

A mole of such endeavoring (consecutively simultaneously of electrons) would take a time of:

$(6.058 \times 10^{-44} s)(6.022 \times 10^{+23} mole^{-1}) = 3.648 \times 10^{-22}$ sec/mole .

"Consecutively simultaneously" means simultaneously per (by virtual electrons) atom but consecutively by atom per mole (in this model); i.e., the entire mole acts as (if) one large atom. But this is, of course, conceptual or theoretical (idealistic) physics, here compliant upon adoption of the Ψ base (for the mathematics involved).

The maximum amount of time (per) mole, for the "electron" to reach its (proper) orbit would then be 3.648×10^{-22} seconds. Alternatively, if each virtual electron (even within the entire mole) simultaneously obtained (its real electron's) proper orbital (Bohr) radius, then the minimum amount of time necessary for the phenomena (collectively) would be

6.058×10^{-44} seconds.

The mole (magnitude) extent of the (this) orbit-age is due to the employment of the Ψ base exponentials. With some correction (i.e. 0.89) the orbit-age requires A_{no} of (for) the Ψ exponential's size, with A_{no} "fitted" to the (large) Ψ value. So the exponential-ized (properly) Ψ base can be used to approximate some measured (experimentally) physical quantities, suggesting combinatorial relationships or derivations (to those quantities). But what is the tolerance of correctness for these (Ψ base inspired) values (e.g., is a ~ 0.89 factor "correction" acceptable)?

We can try to find (for our instance) the significance of another notable base, such as "e" whereby the most serviceable exponents may be found: thus,

$$e^{x1} \sim= 5.267 \text{ x } 10^{56},$$

or more properly (for our custom-ed example of $B_r \bullet A_{no} / P_t$)

$$e^{x2} \sim= 5.912 \text{ x } 10^{56} \text{m}/(\text{s} \bullet \text{mol}).$$

A third examination (for base e significance, here) can be to employ the method "c" exponential to it; i.e. $e^{161}((4/15)+(2/9)) = y_{n=5}$.

Therefore, we are studying: $e^{x1} \leftarrow {}_{(e)}y_{n=5} \rightarrow e^{x2}$ in comparison to ${}_{(\Psi)}y_{n=5}$.

$$x_1 = \ln(e^{x1}) = \ln(5.267 \text{ x } 10^{56}) = 130.6062 \; (= 0.9991x_2)$$

$$x_2 = \ln(e^{x2}) = \ln(5.912 \text{ x } 10^{56}) = 130.7217 \; (= x_1 / 0.9991)$$

${}_{\Psi}y_{n=5} < {}_{e}y_{n=5}$ since $\Psi < e$; ${}_{e}y_{n=5} \sim= e^{161} \bullet (0.489) = 4.08 \text{ x } 10^{69}$.

But, if we select $x_{1,2} \sim 131$, then $e^{131} \bullet (0.489) \sim= 3.818 \text{ x } 10^{56}$; i.e., a corresponding (e) base exponent is smaller for base e than base Ψ, in this analysis.

Therefore, the equivalent (of method) exponent to e, for the appropriate Ψ exponent of 161, is x_3 : $e^{x3} \bullet (0.459) == (B_r / P_t)A_{no} = 5.912 \text{ x } 10^{56} \text{ m} / (\text{s} \bullet \text{mol})$

$$\ln(e^{x3}) = x_3 = \ln(B_r \bullet A_{no} / (P_t \bullet (0.459))) = 131.500(4545)$$

Therefore, a tolerable error (for using the Ψ base) would be $\sim\sim 131.5 / 161 = 0.8168$, and (thus) the ~ 0.89 "correction" is within this tolerance (i.e. is better, or closer, to "1.0," than 0.8168) when compared to base e, since 161 is the proper combinatorial number (for our method), at $n = 5$ (i.e., $C_5 = 161$).

Therefore, at least compared to using base e (for the method) a "correction" of ~ 0.89 (after or under base Ψ) is reasonable. [For example, when comparing (results) to employment of the Ψ_m base, where $\Psi_m < \Psi$, a ~ 0.89 correction factor is not reasonable (acceptable) since for both Ψ_m and Ψ the C_n at $n = 5$ is $C_5 = 161$; so the ratio of the exponents remains 1:1 (i.e. 161 / 161), which demands an exactitude of factor (i.e. the "correction" factor = 1).

Still, such comparisons do not assert the correctness or accuracy of the results (i.e. the physics involved). The physical interpretation of the results is somewhat clearer, essentially a mole's summation of speeds as:

$$(m / s)(1 / mole) \rightarrow [\Sigma(m / s)](1 / mole) == \Sigma_{n=1 \rightarrow Ano}(m / s)_n \text{ per mole} .$$

A summation of speeds suggests a potential to utilize them, for some care-taking or annotative purpose. But the usage must (or should) be summational (or collective), rather than of individual speeds. A collective sum of speeds seems statistical rather than physical (i.e. descriptive), such as for the hypothetical agglomeration of a huge atom "of hydrogen type," each electron reaching its (model) appropriate Bohr radius. (One might well consider the virtual protons that must be involved here.) Therein leads to our alternate model that, at $d_\tau = 1$, the sum of the possible Bohr radii (positioned) is $5.267 \times 10^{+56}$ meters, or $(0.89)(5.912 \times 10^{+56}$ meters $)$, for a hydrogen atom. This would (in some way) limit the (positional) spacial disposition for the location of the (real) electron.

With a Bohr radius of $Br = 5.292 \times 10^{-11}m$ the spherical surface area of the hydrogen atom is:

$$SphAr == 4\pi r^2 = 4\pi(5.292 \times 10^{-11}m)^2 = 3.519 \times 10^{-20}m^2 .$$

This would be divided into $5.267 \times 10^{+56}$ surface positions (for the electron):

$$3.519 \times 10^{-20}m^2 / 5.267 \times 10^{+56} = 6.682 \times 10^{-77}m^2 .$$

The radius of each such surface area would be:

$$6.682 \times 10^{-77}m^2 = 4\pi r^2 \text{ ; thus, } r = (6.682 \times 10^{-77}m^2)^{1/2} \sim 2.306 \times 10^{-39}m ,$$

allowing a volume for the electron of (in available space per position):

$$VolSph = (4 / 3)\pi r^3 = (4 / 3)\pi(2.306 \times 10^{-39}m)^3 ==$$

$$(4 / 3)\pi(2.306 \times 10^{-30}nm)^3 = 1.226 \times 10^{-89}(nm)^3 .$$

But, from this (alternate) model, the number of Bohr radii allowed would be

$5.267 \times 10^{+56}m / 5.292 \times 10^{-11}m = 9.953 \times 10^{66}$; therefore, so many surface positions (as this).

Thus, the area of such a position would be: $3.519 \times 10^{-20}m^2 / 9.953 \times 10^{66}$

$= 3.536 \times 10^{-87}m^2$, with a (circular) radius of

$r = (3.536 \times 10^{-87}m^2 / 4\pi)^{1/2} = 1.677 \times 10^{-44}m$, and a resulting (allowed electron) volume of:

$$VolSph = (4 / 3)\pi r^3 = (4 / 3)\pi(1.677 \times 10^{-35}nm)^3 = 1.976 \times 10^{-104}nm^3 ,$$

which is $\sim 1.976 \times 10^{-104} / 1.276 \times 10^{-89} = 1.6117 \times 10^{-15}$ times smaller than the earlier volume (i.e., one could put $\sim 6.205 \times 10^{14}$ spherical electrons in the earlier or larger volume).

This is, of course, contrary to the election of viewing the electron as (modifyingly) a "cloud" the volume of which covers the (hydrogen) atom, which is unsatisfactory when considering it as a (moving) particle. As a "point-charge" it would have no spacial extent. The volumes presented here are (necessarily) between these extremes. But the probabilities of location (for an electron) within the (hydrogen) atom seem limited (or adjustable) from the Ψ based derivations (of radii, volumes, etc.).

Axial Mechanics

It is notable that since the time axis (d_τ) can only occur after the three spacial (distance) axes (x, y, z), then a spacial component (of volume) must adopt a time component (in order to be measurable), other than for just an idealistic description (as for length and area); [i.e., measurement takes time.]. Therefore, to notice (all) matter and energies (or qualities of a phenomenon) requires temporal enlistments (of labor), meaning: things must (actually) be noticed in 3 dimensions (at least), with utilization of time (d_τ). For more mathematical, inductive, and idealistic analyses, however, d_τ can maintain any axis, be it 1st, 2nd, 3rd, or whatever, as well as can, of course, other (axial) variables. Coincidences to (only, at most) 3 physical (drawn) axes are still observed (as legitimate).

The axial mechanics can be generalized as follows. We have a quantity Q or Qua which is subject to a (or several) disposition dis:

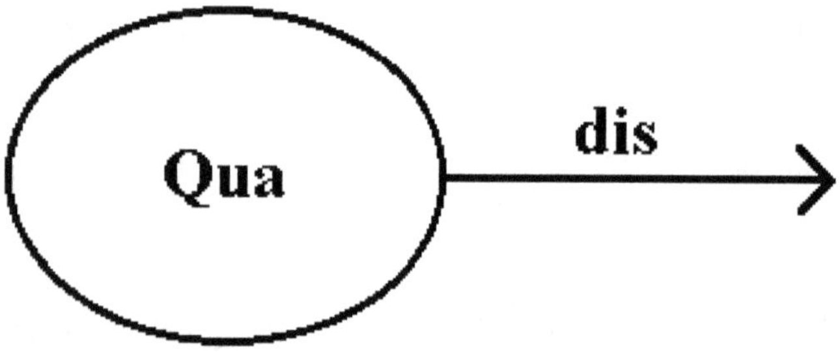

The relation between the two (quantities) is operational as: Q (operation) dis

If, for example, we use the quality mass (as Mas), then

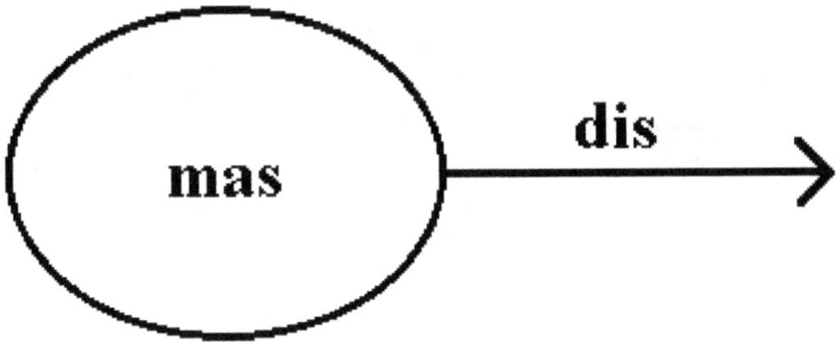

where we can have the disposition as distance, d :

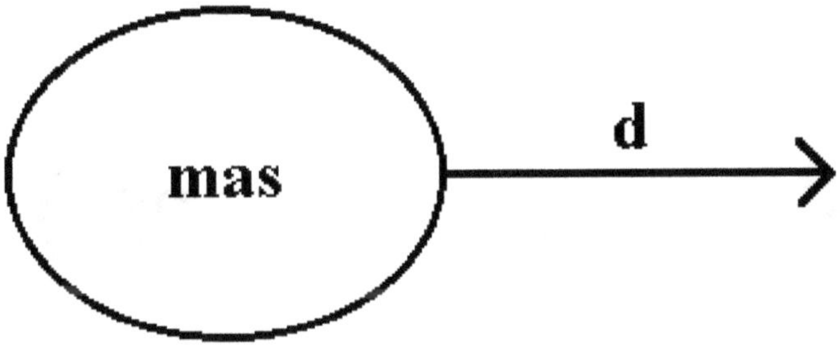

This describes a mass at a certain distance from some arbitrary origin.

Obviously, any analysis that yields to d may be chosen. Therefore, we can have 1, 2, and 3 dimensional axes:

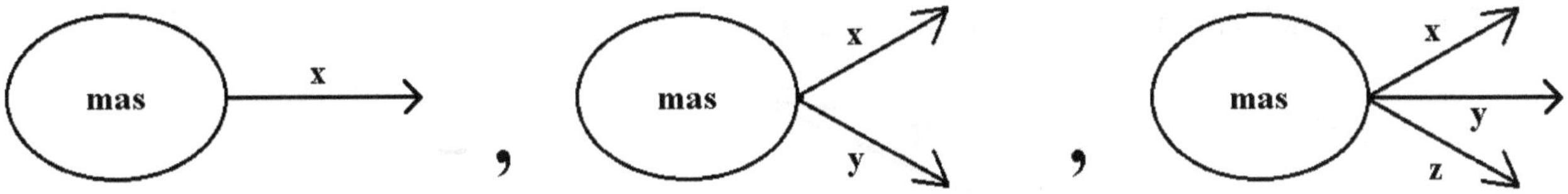

where, somehow, (x) or (x, y) or (x, y, z) defines (for Mas) a distance, and operationally the notation (or graph) represents:

mass (operation) distance.

We may choose, here, the operation to be multiplicative; therefore, (Mas) • d .

Now, coincident axes can be appertained; e.g.

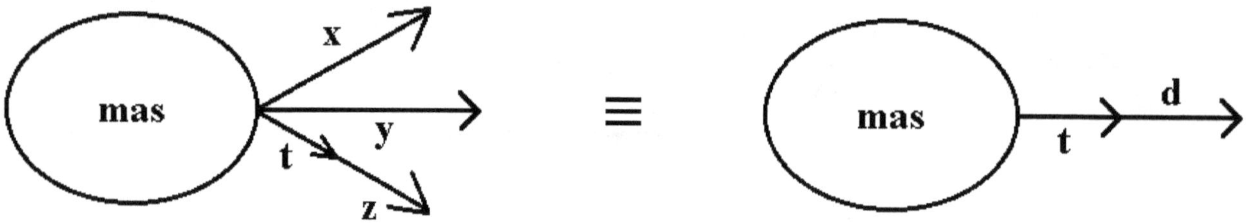

which clearly represents a velocity times mass (as t == time). Thus, the axial relationships are per or divisional,

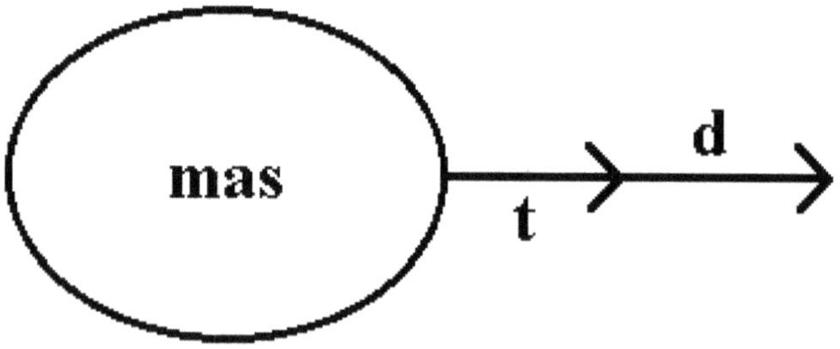

$$\equiv \; (\text{mas})\,\frac{d}{t}$$

thus defining a momentum:

(let Mas == m) ; therefore, m • (d / t) == mv , with v == d / t = a velocity ;

mv == m(d / t) = m • d • (1 / t) or m_{quality} • $(d \cdot 1 / t)_{\text{axial dispositions}}$ = p_{momentum} .

Further coincidences (of t) may be applied, as:

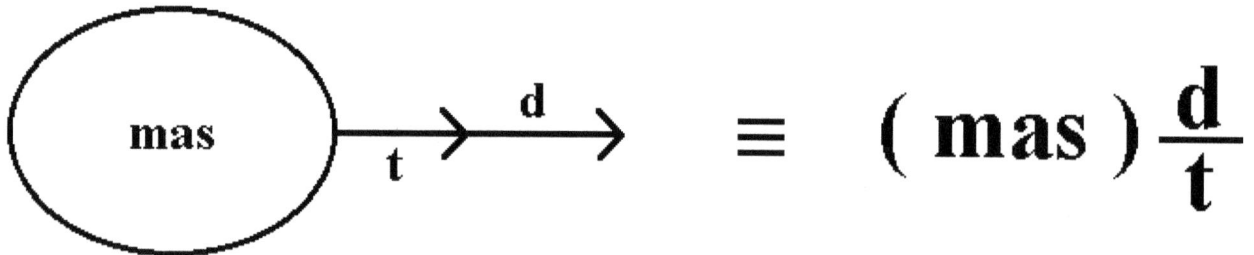

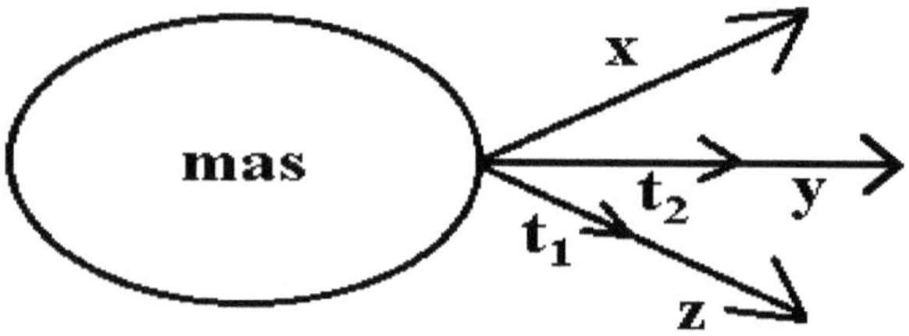

$== m \cdot [\, d \cdot (1 / t_1) \,] / t_2 = m \cdot d \cdot 1 / t_1 t_2 \sim = m(d / t^2)$;

thus defining an acceleration as $a == d / t^2$;

therefore, ma results in a force, as force = ma = f .

We can produce the third (temporal) coincidence with:

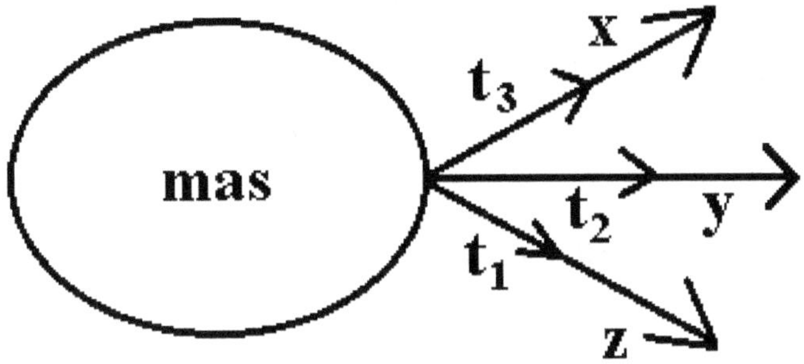

$$== m \bullet \{ [d \bullet (1 / t_1)] / t_2 \} / t_3 = m \bullet d \bullet 1 / (t_1 t_2 t_3)$$

One might call (or label) $m \bullet d \bullet 1 / t_1 t_2 t_3 \sim = m \bullet d / t^3$ (as) an "activity" or activation of mass (i.e. changing acceleration of m).

But t^3 can represent a quantity analogous to d for (x, y, z) ;

let's call it T ; therefore, $m \bullet d / T = A_{(ctivity)}$.

Note that due to the "per" or divisional nature of the axial relations, the temporal (axial) influences follow a series of differentiations (as like from or similar to the calculus): using mass as a constant and dis as a constant

$$\mathbf{d(m \bullet dis \bullet t^0) == md(dis \bullet t^0) == m \bullet dis \bullet d(t^0) = m \bullet dis / [t^{0-1}]^{-1} = m \bullet dis / t}$$

[Here we abrogate d(dis) = 0 and say (define) multiplication by zero does not occur (or exist); for combinatorial analysis this is no more an aberration than 0! = 1! = 1 .]

For the momentum

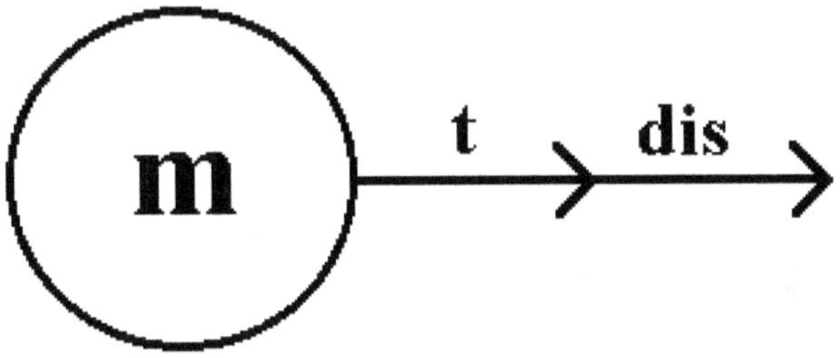

formally the graph is automatically $m \bullet dis / t = p$ (by virtue of the "per" axial nature for t); and, for the force

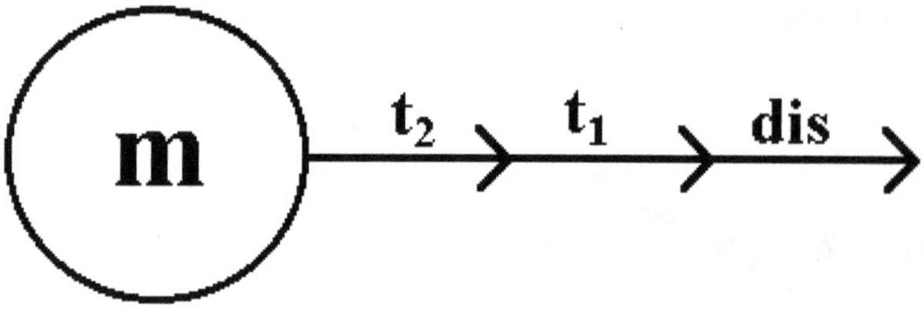

where $m \bullet dis / t == m \bullet dis \bullet t^{-1}$, then $d(m \bullet dis \bullet t^{-1}) == m \bullet dis \bullet d(t^{-1}) = m \bullet dis \bullet (-1) / [t^{-1-1}]-1 = -m \bullet dis / t^2$;

and, for the Activity

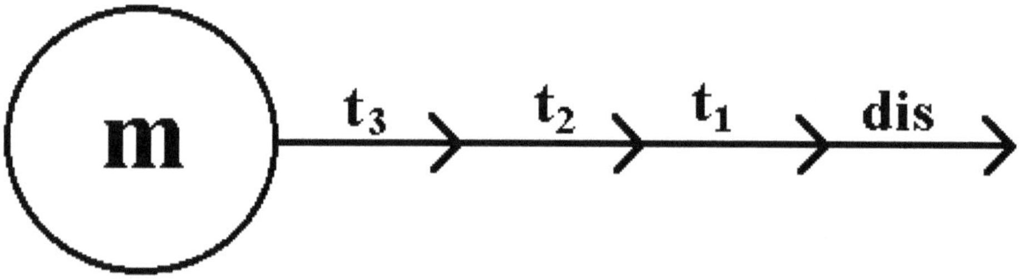

where $-m \bullet dis / t^2 == -m \bullet dis \bullet t^{-2}$,

$\mathbf{d(-m \bullet dis \bullet t^{-2}) == m \bullet dis \bullet d(t^{-2}) = +2m \bullet dis \bullet t^{-2-1} = 2m \bullet dis \bullet t^{-3} = 2m \bullet dis / t^3}$.

This seems to imply that the momentum and the activity are applied for the mass (i.e. are positive), while the force is applied against (the inertia of) the mass (i.e. is negative). Clearly the axial relations must be resolved before operation on the quality (here mass).

The combinatorial conditions (here) depend on the (axial) resolution of dis:

e.g.

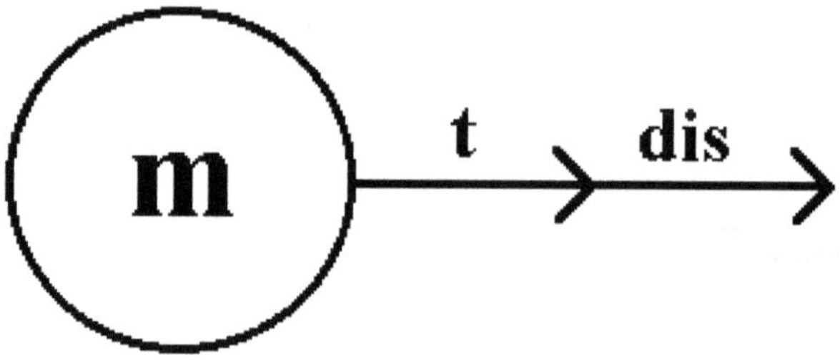

has (for m) $C_n = C_2 = 3$ conditions → m(t), m(dis), m(t, dis), while

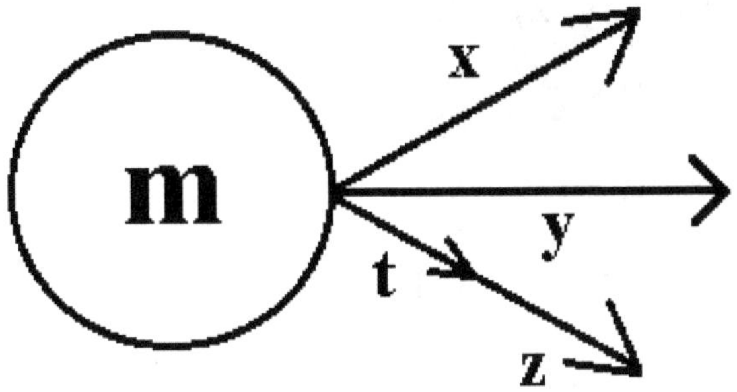

has $C_4 = 25$ conditions (t global is over x, y, z; therefore, it can be on any axis) as interpreted; or, if t is specific (limited) to z, then there are $C_{(3+1+1)} = C_5 = 161$ conditions. They are all locations (of mass) and momentums (of mass).

The "per" nature of the axes clearly involve (axial) coincidences. But what of the discreet axes (themselves) as to how they relate to one another?

We know geometrically that $d == (x^2 + y^2 + z^2)^{1/2}$; therefore, the separate axes (eventually) relate to each other as involved in some operation that includes (or encompasses) a summation. Here, the apparent (axial) summation (rule) is:

$$d^2 = [(x^2 + y^2 + z^2)^{1/2}]^2 = x^2 + y^2 + z^2 .$$

But this is not suggested at all by the (axial) drawing, so it is obscure (of result or resultant) and dependent on geometrical knowledge. The resultant, however, is Cartesian (i.e., the equal angle between axes is $\pi / 2$). The axial equal angles plot can be analyzed as like Cartesian (or cartographically), with P a point on the y-axis:

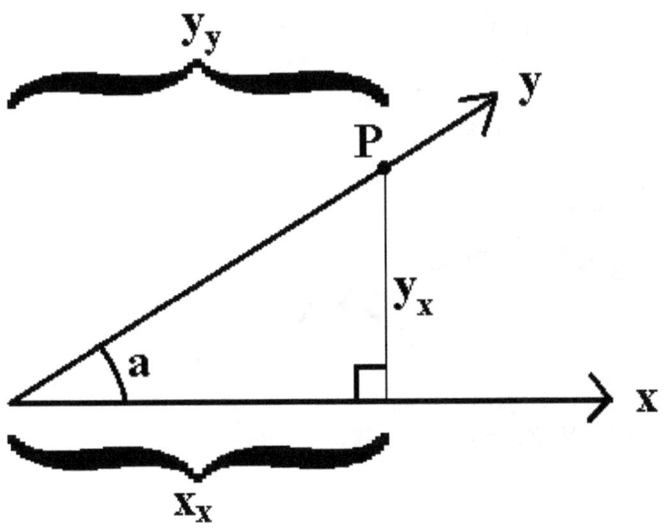

equal angle (between x, y, z) = a

y_x == projection of x on y

x_x == projection of x on x (== x)

y_y == projection of y on y (== y)

Thus, $\sin(a) = y_x / y_y == y_x / y$ |

$\cos(a) = x_x / y_y == x / y$ | $\tan(a) = y_x / x_x == y_x / x$

dist. of P from origin = $(x_x^{\,2} + y_x^{\,2})^{1/2} == (x^2 + y_x^{\,2})^{1/2}$.

With P (being) a point off either axis:

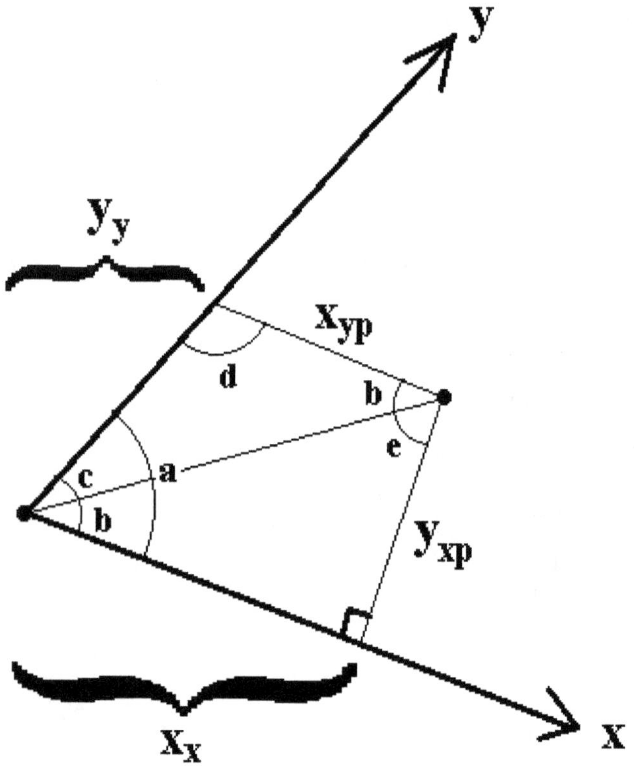

equal angle (between x, y, z) = a

$\angle a = \angle b + \angle c$
$\angle c = \angle a - \angle b$

$\angle d = 180° - \angle c - \angle b$
$\quad = 180° - (\angle a - \angle b) - \angle b$
$\quad = 180° - \angle a \; ; \; 180° = \angle c + \angle b + \angle d$

$\angle e = 180° - 90° - \angle b = 90° - \angle b$

$\sin(b) = y_{xP} / (\text{dist. from origin to P}) == y_{xP} / (y_{xP}^2 + x^2)^{1/2}$

$\cos(b) = x_x / (\text{dist. from origin to P}) == x / (y_{xP}^2 + x^2)^{1/2}$

$\tan(b) = y_{xP} / x_x == y_{xP} / x$

(x_{yP}, y_{xP}) are coaxial (to the horizontal or vertical).

Now we can analyze the same with an axial equal angles disposition:

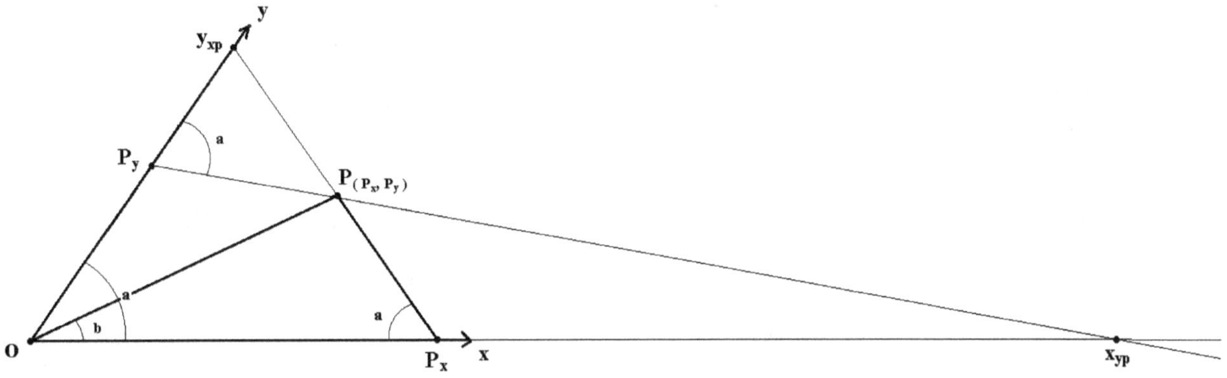

**We can construct new analogous (relevant)
trigonometric functions.**

$$\operatorname{Sin} \left[\frac{a}{a} \right] = 1 \qquad \operatorname{Sin} \left[\frac{0}{a} \right] = 0 \qquad \begin{array}{l} \text{read :} \\ \text{Sin for } \square \end{array}$$

$$\operatorname{Cos} \left[\frac{a}{a} \right] = 0 \qquad \operatorname{Cos} \left[\frac{0}{a} \right] = 1 \qquad \text{Cos for } \square$$

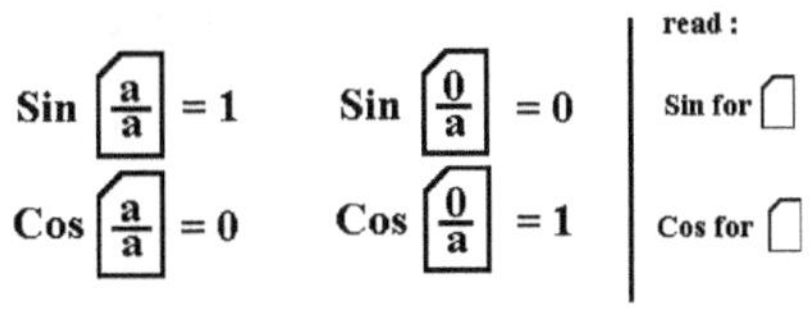

But let us try other periodic functions not thoroughly dependent on 2π ; e.g., (P_x, P_y) vs (x_{yP}, y_{xP}), therefore utilizing P_x / x_{Py} and P_y / y_{Px} .

This is a more **nul/los** approach since both x_{Py} and y_{Px} lead to infinities (of distance) as axial elements (i.e. defining axes).

Since angles $\angle a$ are equal angles (for the axial description), the P_x / x_{Py} and P_y / y_{Px} ratios depend directly on angle $\angle b$ (whether $b = a$ or not).

Note that $P_x \rightarrow 0$ when $x_{yP} \rightarrow 0$ and $P_y \rightarrow 0$ when $y_{xP} \rightarrow 0$; therefore, one is left with ratios of $0 / 0$, which can be interpreted as meaningless, or 'origin-al', or simply equal to "1" for functional usage.

Geometrical Axial Diagrams

Of course, along the $\overline{OP}$ ray (for all points therein), angle $\angle b$ remains constant. Thus, the ratios must stay constant, to the limit of reaching (point) O .

Another ratio to be considered is that of the areas of the two triangles:

$$\text{Area of } \triangle(P - P_y - y_{xP}) / \text{Area of } \triangle(P - P_x - x_{yP}).$$

If one draws the diagram with line $\overline{P_{\bar{y}}}\overline{P}$ properly parallel to the x axis (therefore to line $\overline{OP}$), one finds a fascinating construction where $\triangle(P_y - y_{xP} - P)$ is a similar $\triangle$ to $\triangle(O - y - P_x)$, and $\triangle(O - P - P_x)$ forms a supplement to $\triangle(P - P_x - x_{yP})$. All angles for (these) triangles are known given angle $\angle b$ and equal $\angle a$, seemingly or apparently with $\overline{OP} = \overline{P\, x_{\overline{yP}}}$ (in size or extent) and $\overline{P_{\bar{y}}}\, \overline{y_{\overline{xP}}} = \overline{y_{\overline{xP}}}\, \overline{P}$;

in fact, $\triangle$s $(O - y_{xP} - P_x)$ and $(P_y - y_{xP} - P)$ are equilateral.

Therefore, $\overline{P_{\bar{y}}}\, \overline{P} = \overline{P\, y_{\overline{xP}}} = \overline{y_{\overline{xP}}}\, \overline{P_{\bar{y}}}$.

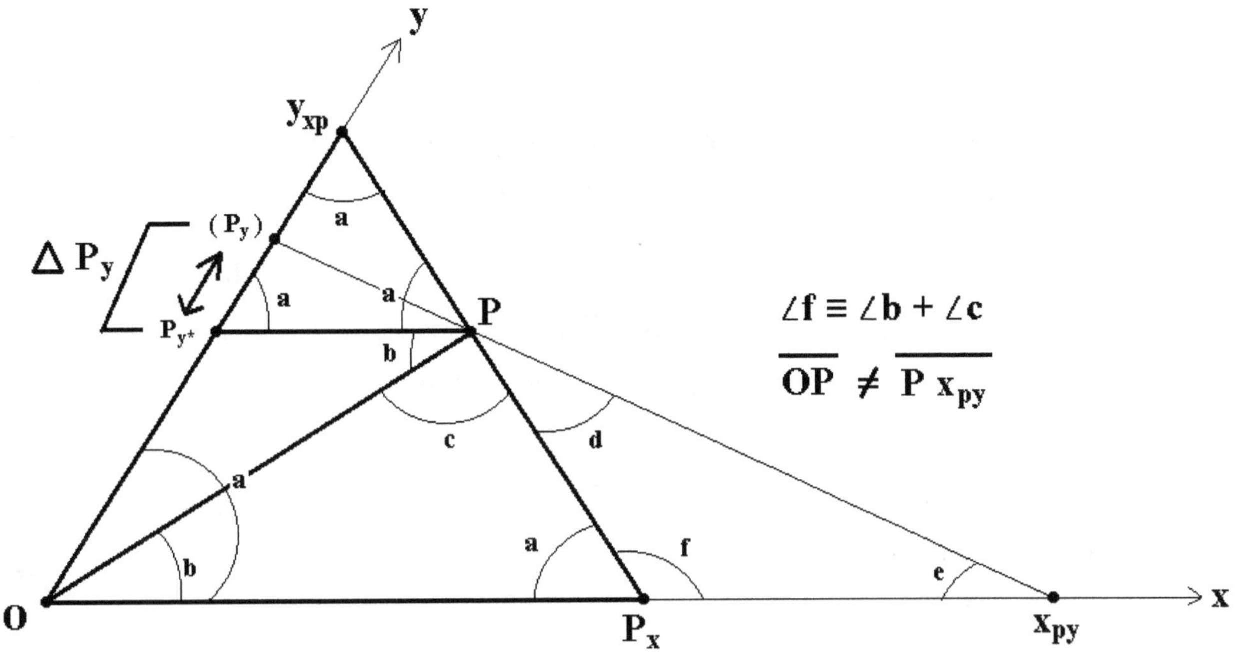

We might label the P_y on a line || to the x-axis as $P_y^{\,x}$ (or $P_y^{\,\|x}$).

But this is contingent on the equal angles being 60° (for equilateral △s).

Without homage to parallel lines, and for a more general equal angle, we can study the determinants of the (geometrical) construction:

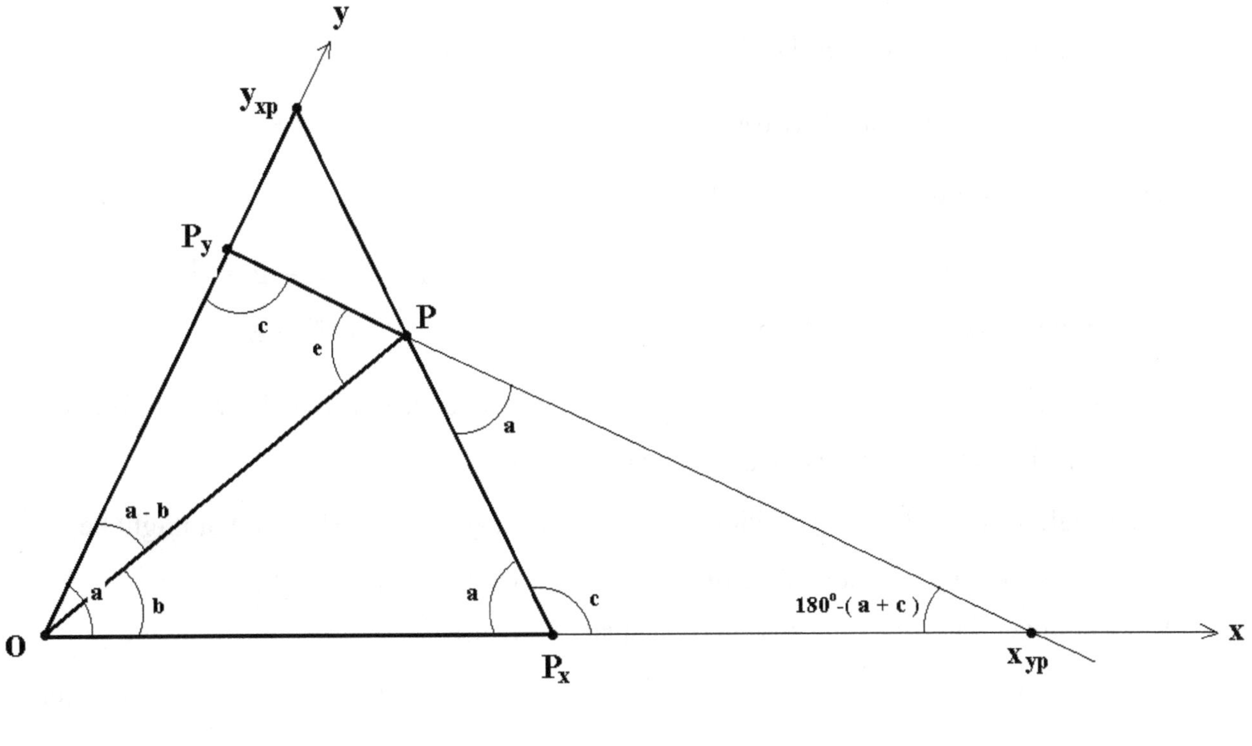

y_{xP} is found from the extension of $\overline{P_x P}$ to y-axis. x_{yP} is found from extension of $\overline{P_y P}$ to x-axis. Both P_x and P_y are found from $\angle a$ angles using point P to the respective x and y-axes; i.e., angle $\angle O - P_x - P = a$, angle $\angle y_{xP} - P_y - P = a$.

Why is angle $\angle(y_{xP} - P_y - P)$ the supplement of $| a - 180° |$, rather than angle $\angle(O - P_y - P) = a$? We assume the lines $\overline{P_x P}$ and $\overline{P_y P}$ take the smaller of the supplementary angles for angle $\angle a$, when contacting their respective axes (x or y). If this is so (or proper or correct), then the other angle (along both the x and y axes) is c as drawn.

An isosceles $\triangle$ is thus established for $\triangle(O - y_{xP} - P_x)$; therefore, $\overline{O\,y_{xP}} = \overline{y_{xP} P_x}$, because the $\triangle$ has two angles $\angle a$.

For $\triangle(O - P_y - x_{yP})$ the angle at x_{yP} is $180° - (a + c)$;

therefore, the angle $\angle$ at $P_x - P - x_{yP}$ is $180° - [c + (180° - (a + c))] = a$,

which is also the angle $\angle$ at $P_y - P - y_{xP}$.

Therefore, $\overline{P_y\,y_{xP}} = \overline{y_{xP}\,P}$ since $\triangle(P_y - y_{xP} - P)$ has two a angles.

Likewise, $\overline{O\,y_{xP}} = \overline{y_{xP}\,P_x}$ since $\triangle(O - y_{xP} - P_x)$ has two a angles.

The angle at y_{xP} is, therefore, $180° - 2a$, for the (2) isosceles triangles.

Thus, we have a distinction between x_{yP}, angle at which is $180° - (a + c)$,

and y_{xP}, angle at which is $180° - 2a$.

Angle $\angle\, b == 180° - a - d$; therefore, angle at $y_{xP} == 180° - a - a + d = b - a + d$; and $(a + e + d) = 180°$. Finally, angle $\angle\, y_{xP} - P - x_{yP} = 180° - a$.

The angle $\angle$ at y_{xP} and the $y_{xP} - P - x_{yP}$ angle are independent of b and can therefore be used as (each) a basis for b (or the deviation from $a \longrightarrow b$).

Since angle $\angle\, y_{xP} - P - x_{yP}$ positions point P at the angle ($= 180° - a$), it might be preferable (for the distinction of the relevance of P), while the angle at y_{xP} is more demonstrative of the graphing system (i.e. equal angle) used.

The basis angle (for P; i.e. angle $\angle y_{xP} - P - x_{yP}$) can serve as a denominational determinant;

therefore, $180° - a$ serves as a denominator.

A numerator-nal determinant must be dependent on b (for its value). A suitable angle might be the angle at x_{yP} ($= 180° - (a + c)$), since angle $\angle c$ is indirectly dependent on b; of course, angle $\angle b$ can be used itself.

However, the x_{yP} angle can be used to demonstrate the periodicity of the (a, b) angular relationship, since when either P is on the x-axis or the y-axis the angle c goes (arguably) to the y_{xP} angle (i.e. in becoming this angle).

So, we have an angular relation such as: $[\ 180° - (a + c)\ ] / (\ 180° - a\ ) =< 1$. Yet, the limiting value for c is the angle at y_{xP} (rather than zero).

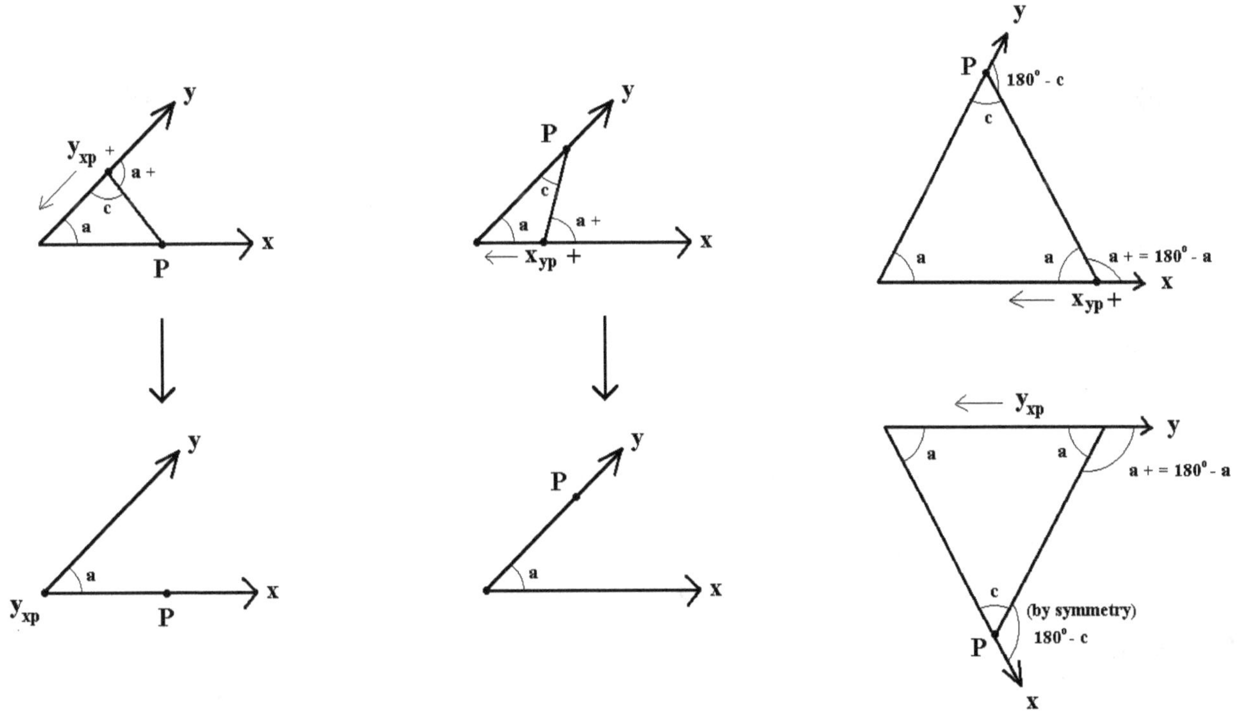

Therefore, we have limits of $(\ x_{yP}, y_{xP}\ ) \rightarrow a$ |

Therefore, $180° - (a + c) \rightarrow a$, $180o - 2a \rightarrow a$ | in terms of angular function

which decrees that $c \rightarrow a$ |

So, we have three (angular) functions established:

$(c \rightarrow a) / (180^o - a) , (y_{xP} \rightarrow a) / (180^o - a) , (x_{yP} \rightarrow a) / (180^o - a)$;
since they each limit to a they are (functionally) equivalent, but of differing trajectories.

Symbolic (Axial) Representations

But the value " 180° " acts very much as a constant within these formuli (of function). Therefore, the relations can be (functionally) reduced in terms as:

$c / -a , (-2a / -a) = 2 , -(a + c) / -a$,

and with $c \rightarrow a$ we have apparent limits $\{ -1 , 2 , 2 \}$. Or, we can maintain the constant as 180° == (symbolically) (I), such as to allow:

$c / ((I) - a) , ((I) - 2a) / -a , ((I) - (a + c)) / -a$ (idiosyncratic of function type, i.e. demonstratively restricted for a type), with $(c \rightarrow a)$ limits $\{ +1$ (if $(I) = 180^o$, $a = 90^o$), $((I) / -a) + 2 , ((I) / -a) + 2 \}$.

We see that, at $(I) = 180^o$, the limits are reset to (adopt) $\{ 1 , 0 , 0 \}$ for the type of (angular) functions involved.

So, the generalized functions are (at $c \rightarrow a$) : $a / ((I) - a) , ((I) - 2a) / -a , ((I) - (a + a)) / -a$, which consists of the types: $a / ((I) - a)$ and $((I) - 2a) / -a$, where any functionally advantageous constant can be used for (I) ; since we deal with 3 equal angles in our angular system (of coordination of points P), we can simply let $(I) = 3a$, yielding for our functions (angular): $a / (3a - a) = a / 2a = 1 / 2$, and $(3a - 2a) / -a = a / -a = -1$.

With denominational correction (i.e. $-a \rightarrow (I) - a$), we have: $a / (3a - a) = a / 2a = 1 / 2 , (3a - 2a) / (3a - a) = a / 2a = 1 / 2$ (i.e. "1 / 2" for both functions).

Note that although $(I) - 2a$ "looks" constant, the condition for 2a (or two "a"s in a triangle) is a determinant for c, and is actually $(I) - (a + c)$, the location for c varying (in the $\triangle$).

Basically (or argumentatively),

when $c \rightarrow 0$, $(3a - (a + 0) / (3a - a) == 2a / 2a = 1$;

when $a \rightarrow 0$, $(3a - 2a) / (3a - a) == a / 2a = 0 / (2 \bullet 0) == 0 / 0 \rightarrow$ "1" ;

when $c \rightarrow a$, $(3a - (a + a)) / (3a - a) == a / 2a = 1 / 2$.

But $(a \rightarrow 0 , c \rightarrow 0)$ each essentially represents (non-angular) lines (axes) on which point P can be positioned. And, when $a \rightarrow 0$ (for one function)

$(3a - (a + c)) / (3a - a) == -c / 0 \rightarrow \sim\infty$.

The other (corresponding) function, $(3a - 2a) / (3a - a)$, might be argued to lead to $+\infty$ (rather than $0 / 0 \rightarrow$ "1").

We can also consider when $c \ll a$ and when $c \gg a$, for functional (angular) trends.

When $c \ll a$, $(3a - (a + c)) / (3a - a) \rightarrow \sim < 2a / 2a \sim\sim 1$ (i.e. close to "1" or "1-").

When $c \gg a$, this forces (by $\triangle$ practicality) that $c \sim < 2a$ for the function $(3a - (a + c)) / (3a - a)$ (to remain positive). This leads to $(2a - c) / 2a \sim\sim 1 - (1-)$, where "1-" $\sim < 1$.

c (angle) is only independent of angle b when angle $\angle b$ does not occur (i.e., P is on the x or y axis), or when angle $\angle b =$ angle $\angle a$ as such.

But how does angle $\angle b$ relate directly to the three functions (angular) we have established:

$$\{ c , \angle \text{ at } y_{xP} , \angle \text{ at } x_{yP} \} / ((I) - a) ?$$

$a == (I) - c$; thus, $(I) - ((I) - c) = c$; therefore, $c / c =$ "1" .

Simple angular $\triangle$ relations suggest the direct relationship to $\angle b$:

e.g., $c == (I) - (a - b) - e$; thus, $((I) - (a - b + e) / c = 1$.

Also, $c == (I) - a$, and $a == (I) - (b + d)$; thus, $((I) - [(I) - (b + d)] / c = 1 == (b + d) / c , c = b + d$.

Other functions follow accordingly:

e.g., angle $\angle$at $y_{xP} / c == ((I) - 2a) / c = ((I) -2((I) - c)) / c = (2c - (I)) / c$,

$\mathbf{(2c - (I)) / c = (2((I) - a) - (I)) / c = ((I) - 2[(I) - (b + d)]) / c = (2(b + d) - (I)) / c}$;

with $c = b + d$, $(2(b + d) - (I)) / c = (2(b + d) / c) - ((I) / c) == (2(b + d) / (b + d)) - ((I) / c) = 2 - ((I) / c) , 2 - ((I) / c) == 2 - ((I) / (b + d))$.

Yet, angle $\angle$ at $\mathbf{x_{yP} / c == ((I) - (a + c)) / c = ((I) - (a - a + (I))) / c = ((I) - (I)) / c = 0 / c}$.

But this occurs when $c \rightarrow a$, as for the other function. This might pertain to when $3a = (I)$ (with $c \rightarrow a$), although the (exercised) $\triangle$ is not established since there is no angle $\angle$ at x_{yP}.

Therefore, if (I) changes then so must (or can) a change.

A varying a suggests a (changing) motion of P, which in turn implies employment of an energy (a force promoting the change in motion). We can track (or follow) the movement(s) of P by the projections: (P_x, x_{yP}) and (y_{xP}, P_y); the point P is of course the intersection of lines $\overline{P_x x_{yP}}$

and $\overline{y}_{\overline{xP}}\ \overline{P}_{\overline{y}}$. The angle $\angle a$ determines the location of $P_{x,y}$ (as a line from point P is projected), and the line $\overline{P}_{\overline{x,y}}\ \overline{P}$ defines the locations of the points x_{yP} and y_{xP} as complementary projections (to the corresponding or "opposing[-ly faced"] axes). Therefore, the slopes of these lines are determined both by a and the position of P (amongst the axes).

The $\overline{P}_{\overline{x}}\ \overline{y}_{\overline{xP}}$ and $\overline{P}_{\overline{y}}\ \overline{x}_{\overline{yP}}$ lines, which together define the location of P, have employable characteristics: the ratios P_x / y_{xP}, P_y / x_{yP} lead to infinities when P is on an axis;

if $(P_x / y_{xP} , P_y / x_{yP}) > 1$, then the (line's) slope is positive, and if it is < 1 then the (line's) slope is negative.

Since all physical processes are only real, physics (per se, as opposed to mathematics) is properly restricted (or limited) to the (special) regions of positive axial coordinates, all matter and energy being (and resulting as) positive.

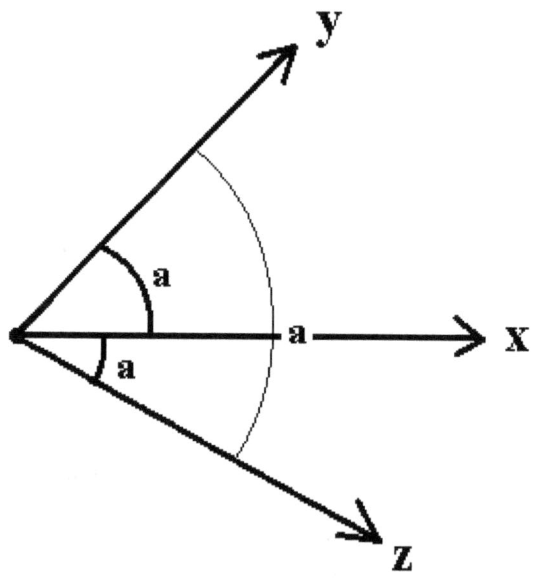

It is then postulated that all "apparent" imaginary or negative (as well as positive) processes must occur within this positive region of the (collective) axes.

Because of the crossing (or intersection) of the lines (at P), the product:

$$(P_x / y_{xP})(P_y / x_{yP}) == (P_x / x_{yP})(P_y / y_{xP})$$

is characteristic for the P location. Due to the opposing slopes of the lines ($\overline{P}_{\overline{x}}\ \overline{y}_{\overline{xP}},\ \overline{P}_{\overline{y}}\ \overline{x}_{\overline{yP}}$), the product lies between the ratios P_x / y_{xP} and P_y / x_{yP} (i.e. between these extremes of value), because one ratio is less than 1 and the other is greater than 1 (unless P is on an axis).

But what is the (coordinate) nature of these points $(P, P_x, P_y, x_{yP}, y_{xP})$ in relation to a and b?

We can define the y coordinate (component) of P as: $_yP == (b / a)P_y$; likewise, the x coordinate of P is: $_xP == ((a - b) / a)P_x$, and similarly in 3 dimensions with the appropriate redundancies and demarcations of angles between axes and planes.

If we assign the angle between the ray P to the y axis as c, then $_cP == (c / a)P_c$; and likewise, if we assign the angle between the ray P to the x axis as d, then $_dP == (d / a)P_d$, (c, d) being (sub-)angles between the z axis and the y, x axes respectively.

Obviously, in terms of redundancies (here): $_cP == {}_zP$ and $_dP == {}_zP$; thus, c = d , a = a .

This is due to the (c, d) angles being made (from the ray to P) to the xy plane (which is common to the x, y-axes).

Nature of Coordinate Points to Physical Processes

If we look at the process more mechanically:

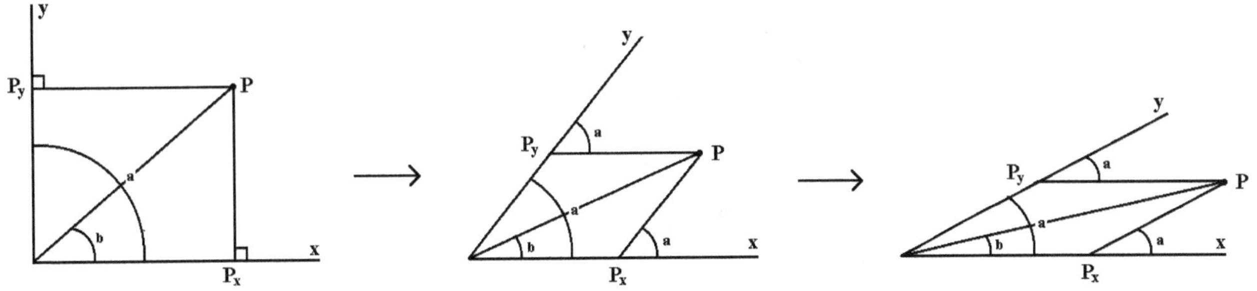

We note that the apparent ray to P increases (in distance) with decreasing a, the limit being the infinities of the (x, y)-axes. But the lines (from P) to (P_x, P_y) increase accordingly, and to the same limits. Presumably the (P_x, P_y) values remain constant until the onset of infinity (or they then become hidden). P may also be hidden within the infinities. But all regions of physical interest are still restricted to the areas bounded by the (various) axes. (x_{yP}, y_{xP}) respond correspondingly, but by projection (manner) they (must) become infinities. Therefore, a collection of points "hidden" by infinities might suddenly be de-collasped by (abruptly) adjusting a (to increase). But each "a" represents a complete physics (of coordinate system).

But if the physics (amongst the "a"s) are equivalent, we notice that the distances (between points) amongst the "a"s must change (or differ), assuming a common unitage for all the coordinate systems (of the axes), unless all points are buried to axes under infinities (i.e., the distances between points remain constant). The angular functions developed previously are independent of points (and thus distances); but they might be used to derive point (locations and) distances if these remain consistent among the "a"s (as given or established).

This would depend on the varying distances of $\overline{P_{x}}\,\overline{P}$ and $\overline{P_{y}}\,\overline{P}$ along the ray to P as angles a and b are established (to be constant). If the points are consistent (i.e. the same) among different "a"s, then the $\overline{P_{x}}\,\overline{P}$, $\overline{P_{y}}\,\overline{P}$ distances must change among the "a"s. Mechanically, though, (as for a machine) these distances would not have to change (i.e. can be made not to change). A function of relative angles to relative lengths might apply (for definitions).

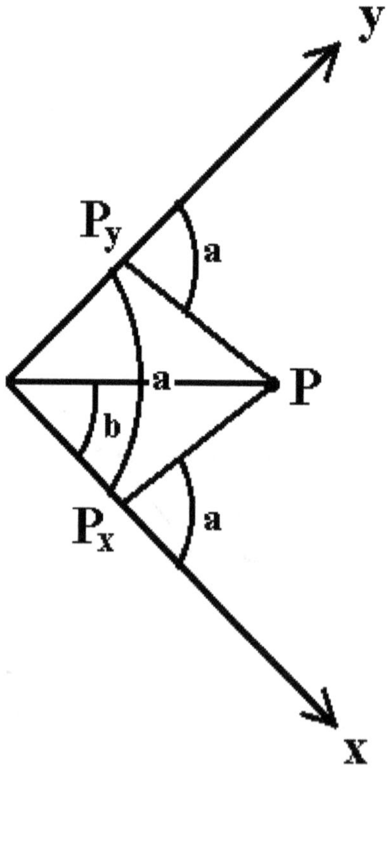

Therefore (algebraically):

$a / b == (\overline{P_x}\overline{P} + \overline{P_y}\overline{P}) / \overline{P_x}\overline{P}$, $a = b \bullet ((\overline{P_x}\overline{P} + \overline{P_y}\overline{P}) / \overline{P_x}\overline{P}) = b \bullet (1 + (P_y P / P_x P))$, $a - b = b \bullet \overline{P_y}\overline{P} / \overline{P_x}\overline{P}$.

The $(\overline{P_y}\overline{P} / \overline{P_x}\overline{P})$ ratio would tend (for this model) to act as (like) a "slope" for the b angle. If $(\overline{P_y}\overline{P} / \overline{P_x}\overline{P})$ remains constant (such as can occur if the $\overline{P_y}\overline{P}$ and $\overline{P_x}\overline{P}$ distances are constant, or if the ratio remains proportionally constant), then $b \bullet (\overline{P_y}\overline{P} / \overline{P_x}\overline{P})$ yields (to) the a complement (i.e. $a - b$). When (for example) the ratio $(\overline{P_y}\overline{P} / \overline{P_x}\overline{P}) < 1$, then $b > (a - b)$, and visa versa.

But, if the a angle is to be maintained (at P_x, P_y), then, as b changes so must P_x and P_y (in terms of distance). If the unitage (of distance) remains the same throughout the changes in b, then (specific) P_x and P_y must then change (per P). This (also) assumes that the distance from P to the origin (of measurement) remains constant. This means that, with changing b, either P is variable or (P_x, P_y) are variable (or possibly, with relaxations, all three).

What might a variable "b," then, signify? This of course allows for the (or a) movement of P.

But what then, so, for a variable "a"? It might indicate "complementary" (or comparative) physics for a variable (e.g., P in different physical realms or influences). This would relate to different movements of P in various (physical) environments (in contrast to variable b).

If equal(ized) units are being employed, this might even be used to extend the (motion-al) range of P:

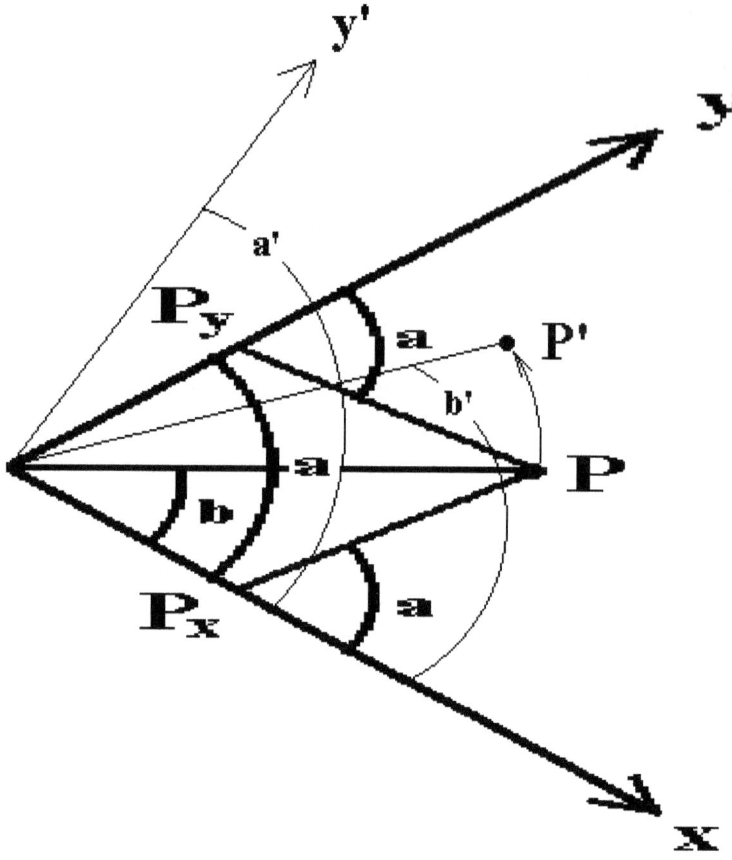

{ a → a' , b → b' } → { P → P' } [e.g., x ~ ~ dist.; y, y' ~ ~ time]

P and P' have distinct physics in their respective movements (e.g., perhaps one is a velocity and the other an acceleration).

But what does a declaration of "distinct physics" mean?

Both (a, a') are unlimited (related-ly) in their (x; y, y') directions. Therefore, the associated areas are so unlimited (in like manner). Each such (a, a') "realm" is a complete physical description (of events). But each (a, a') realm performs a distinct (or different) physical function upon (for this example) $P^{(')}$ (or other variable of interest). So, here we have two velocity realms where changing from one to the other produces an acceleration.

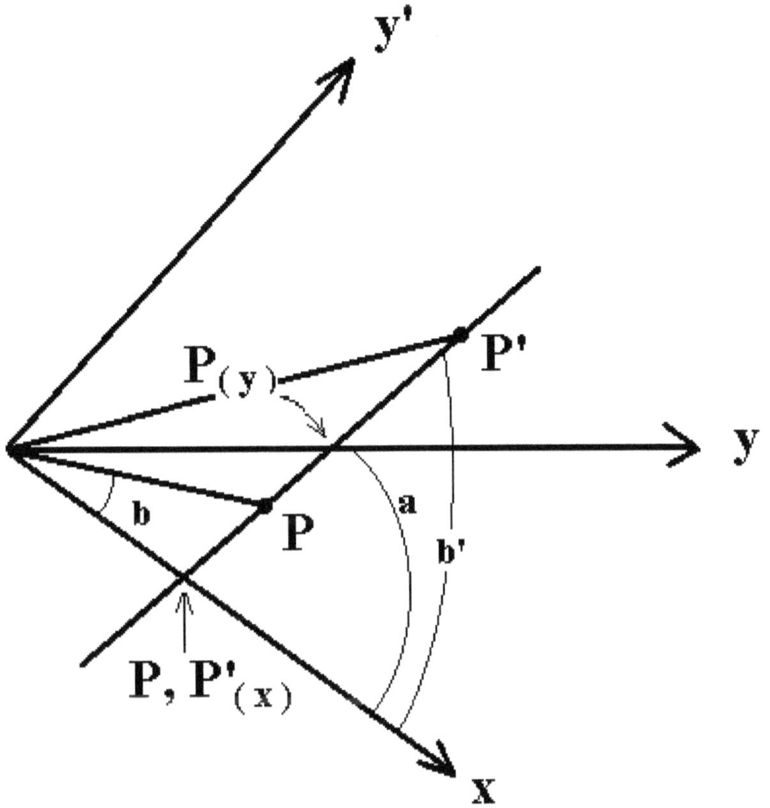

Let $v == \nu^{-1}$ where $\nu_P = P_{(y)} / P_{(x)}$, $\nu_{P'} = P_{(y')} / P'_{(x)}$.

$P_{(x)} = P'_{(x)} == P, P'_{(x)}$ [Note: $P_{(x)}$ does not = P_x, etc. (P_x is a or a' specific.)]

We can symbolically call the acceleration: $a = [(P_{(y)} \rightarrow P_{(y')}) / P_x]^{-1}$. Note that ($P_{(y)} \rightarrow P_{(y')}$) is not the same as $\triangle$y, since $\triangle$y is a value while $P_{(y)} \rightarrow P_{(y')}$ is a process.

[When using the differential calculus, with y == t , we could say

$a = dv / dt == (dx / dt) / dt = dx / (dt)^2 == dx / d^{(2)}t^2$.]

(y → y') occurs in a like manner [but with different route(s)] as does (P → P'), and (realm-)similarly (a → a'). Therefore, we might estimate that $P_{(y')} = (a' / a)P_{(y)}$ for proportional relationships.

Multiple Origins and "Differential" Triangles

But what should the unit-age nature of our (a, a', b, b') angles be? How should we define a (or our) "degree" for these angles?

The angle range must be complete for each realm (i.e. unitary) and yet properly comparative amongst realms. With our (x; y, y') example (coincident x axes), while $a_{R(ealm)} = a'_{R(ealm)}$ == "1" , clearly $a' > a$; or (perhaps, for notations) $\{ (a, a')_R = 1 \mid a' > a \}$.

If $a_R = a'_R$ = "1" and $a' > a$, then $a' = a + (a' - a)$ still; yet, degree (or angular extent)-wise, a is a fraction of a' (and is therefore an angular component of a').

But with regards to relevance for physical phenomena, we find that a' is a component of a (or more properly that they, a and a', are components of each other), since we now come to adopt the usage of multiple origins:

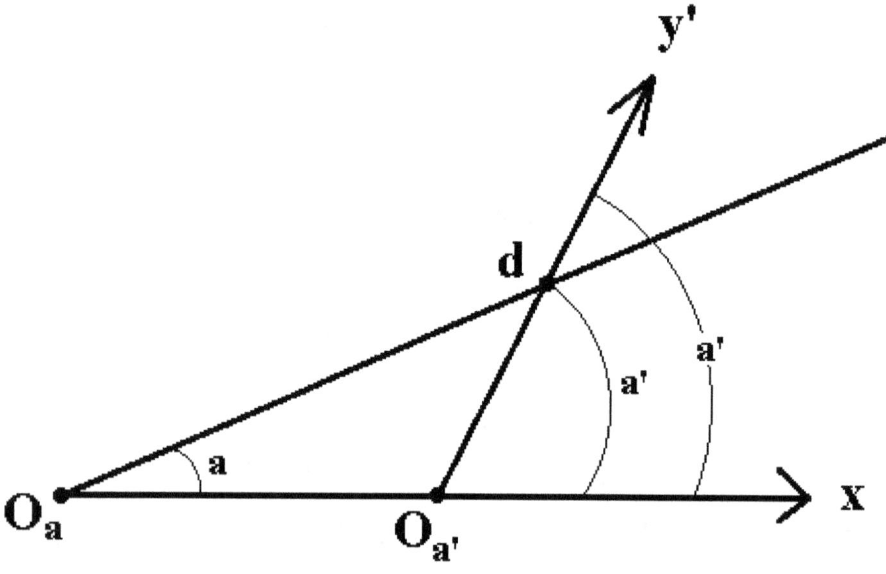

This yields to the determination of (inter-realm-) "differential" triangles [e.g., $O_a - O_{a'} - d$] (or prisms, for more than 2 axes). The extent (area) of the ($O_a - O_{a'} - d$) differential triangle depends on the location of d, the (diagrammatic) location (position) of (point) coincidence for the (y, y') axes.

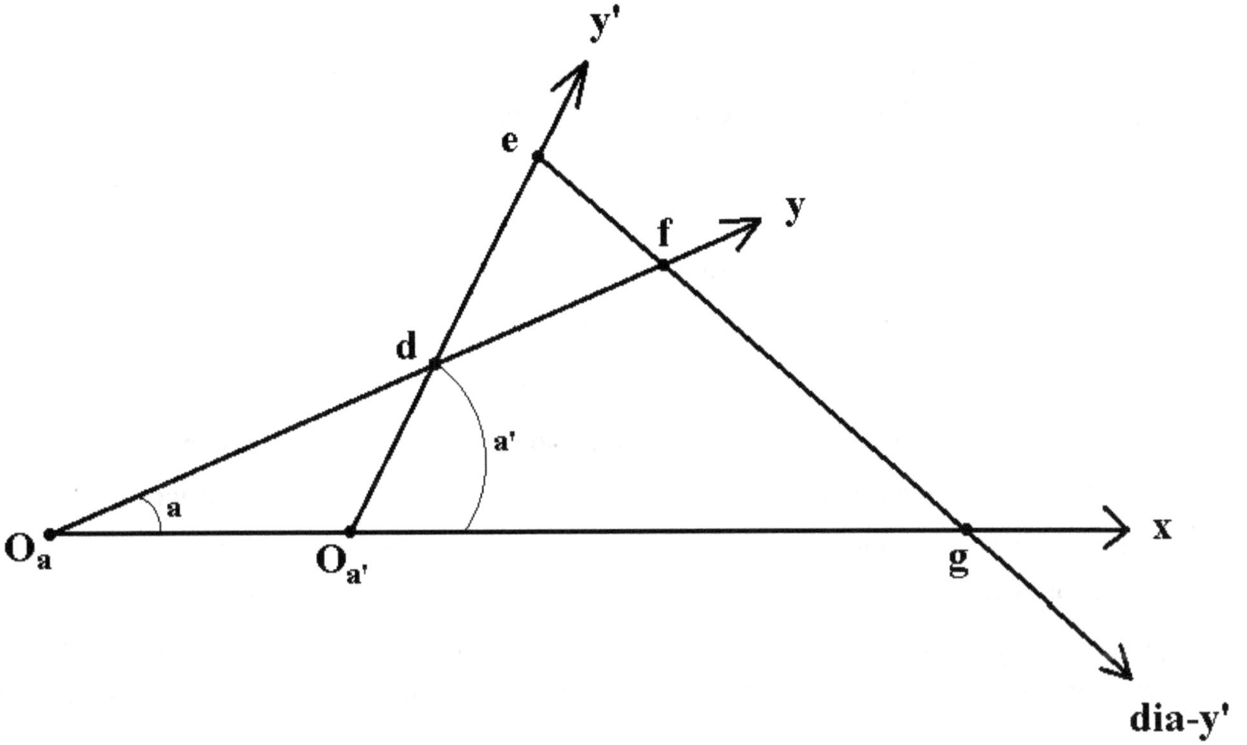

Differential △s : (O_a – $O_{a'}$ – d), (d – e – f), ($O_{a'}$ – e – g), (O_a – f – g) ; they are differential because each involve y and (y' or dia-y') as sides, or x and (y' or dia-y').

The physics (motion, etc.) might be "defined" by changes in the differential △s. This can be extended (of analysis) by using (or allowing for) different units for the various axes (x, y, y', dia-y'). (Physically relevant) Relationships between or among the differential △s may also be found (or investigated). These such comparisons are possible due to the angular components of the corresponding (regular) axes but run in the opposite direction (therefore, the normal Cartesian axes are both regular and dia- in nature). The "dia-" indication does not connote "anti-," but rather related-ly different (and therefore not the same as). The "dia-" axes share the "multiple origins" characteristics of the regular axes. This suggests a certain "nul/los" manner of approach.

Plots can be (made) more generalized by utilizing differing a angles (i.e. specific between axes, at some point of origin or initiation of measurement) and varying trajectories of axes (i.e. allowance for non-linear axes).

A Physics Example to Delineate

But, going away from the more "crystallographic" description of such plots, and back to a common a between axes, we can delineate an example of the physics to deploy of them.

We have (as empirically determined):

$E_{(nergy)} =$	**$mc^2 =$**	**$(1/2)mv^2 =$**	**mgh**
	rest or inertial of a mass (relational-ly)	**kinetic energy (of motion)**	**potential energy (under gravitational influences)**

Therefore, we may employ three axes (one for each energy type) with the commonality of m(ass) allowing for a common a (or angle) between the axes (for each pair of axes).

Now, we may ask: Is this E a different description of m (for each energy type), or does E represent the (maximum obtainable) amount of energy that may be extracted from a particular (and common) m?

We may subsume this question by wondering whether m is simply (the resultant, physically, of) energy not traveling at the speed of light (c), inherently invalidating the insistence that all (radiant) energy (must) travel at this speed.

As dictated (from these relations):

$$c^2 == (d/t)_c^{\,2},\ (1/2)_v^{\,2} == (1/2)(d/t)_v^{\,2},\ gh == (d/t^2)_g \bullet h = d_g \bullet h/t_g^{\,2} \sim\sim (d/t)^2$$

where $d^2 \sim\sim d \bullet h$.

Therefore, an appropriate unit of measurement for each axis could be essentially $(d/t)^2$, meaning v^2.

The problem now is to determine a common "a" which somehow captures a maximum value of v^2 per axis (i.e. c^2, if c is a maximum velocity or speed).

With c (or c^2) as a determinant (for this imperative), a can be assessed as follows:

let the $(1/2)mv^2$ and mgh axes be co-linear but opposite in direction, and with a common origin.

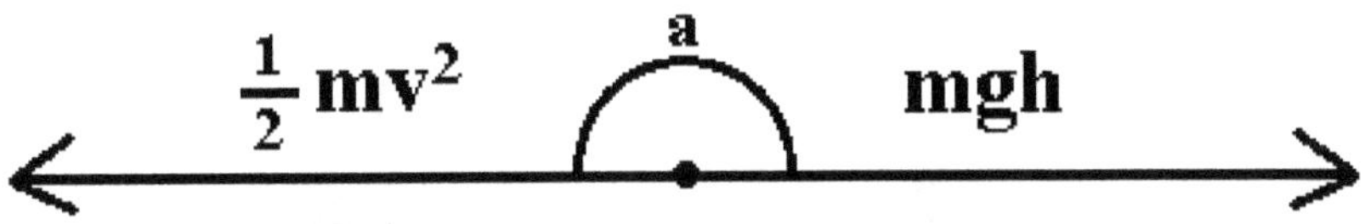

Thus, a is the angle between them (conventionally 180°).

Now, for the mc^2 axis, we demarcate a radius about the previous axes:

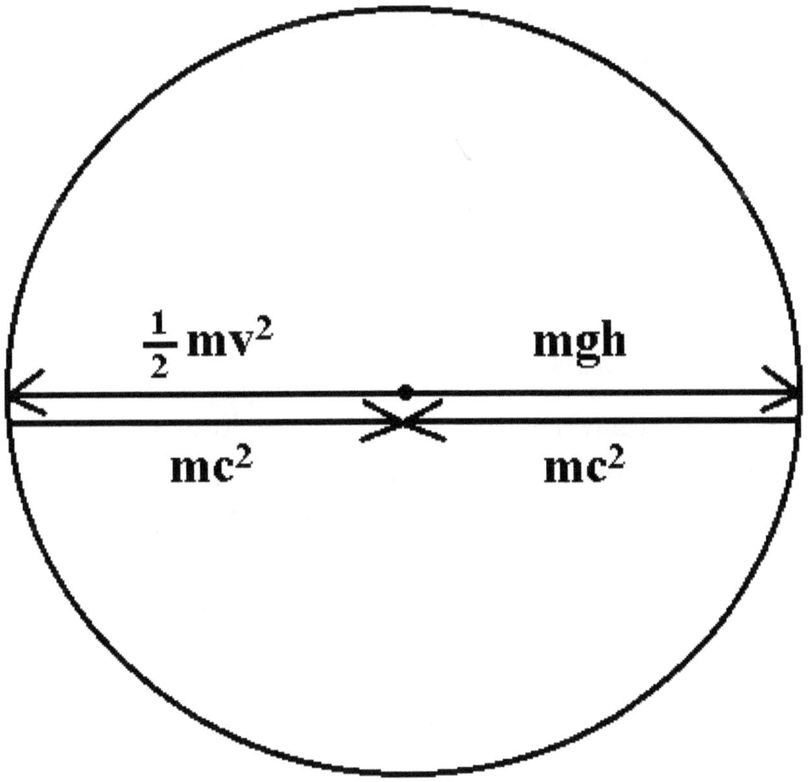

This radius is (of necessity) opposite of direction for each of the previous axes and therefore (subject) of the common a desired.

Employment of this radius (mc^2) also annotates maximum values (for the $(1 / 2)mv^2$ and mgh axes) as specific (or unique) points from those allotted to the mc^2 "circle" (or circumference).

Therefore, the plot has a disposition as:

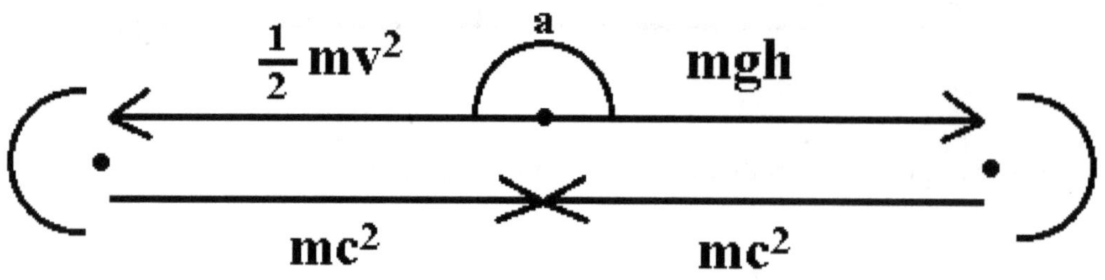

which is dramatically altered (or different) from the Cartesian system.

In this model, all of the axes are co-linear (in relevance).

Note that the radii restrict the extent of the circumference; therefore application of the "c" circle automatically restricts the (unbounded) linear extent(s) of the $(1/2)mv^2$ and mgh axes.

A Description of Gravity

Empirical description of Gravity

Let's say the strength of a gravitational attraction between two masses (M and m) is proportional to Mm (i.e. M•m or M times m).

It is as like the two masses are on separate inclines falling towards a (lower) point of intersection (as a ***cause*** of gravity):

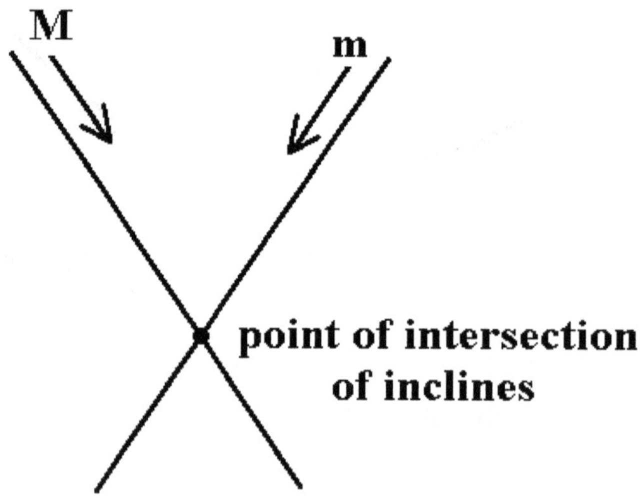

But competing against this gravitational strength is the inclination (or tendency) of one mass to move towards the other.

Say the (this) tendency of M to move towards m is: m / M

Say the (this) tendency of m to move towards M is: M / m

$\therefore$ (gravitational strength)(inclination) $\rightarrow$ gravity (as observed)

The observed gravity of M (in the Mm system) is:

$$(Mm)(m / M) = m^2$$

The observed gravity of m (in the Mm system) is:

$$(Mm)(M / m) = M^2$$

$\therefore$ if $M >> m$, the tendency of m to move towards M is $(M / m) \rightarrow$ very high;

the tendency of M to move towards m is $(m / M) \rightarrow$ very low.

If $M = m$, the observed gravity is:

$$(Mm)[(m / M) = (M / m) == (M / M) == (m / m) = 1] = Mm$$

$\therefore$ the range of gravitational attraction (in the Mm system) is:

$$m^2 \rightarrow Mm \rightarrow M^2$$

The speed of the attraction is the same for M and m, since it is only related to the distance between M and m, which decreases (the same for each) with time, measured at any interval of this time. Since distance is (always) decreasing while time is increasing (as the time remaining, for movements, decreases), an acceleration develops (the distance traveled, for M or m, increasing).

Pictorially, for $M >> m$, if we describe (gravitational) movement as (being) away from the horizontal, then as $m \rightarrow 0$ (compared to M) the inclination (tendency) angle goes $\rightarrow 180^{\circ}$ or π :

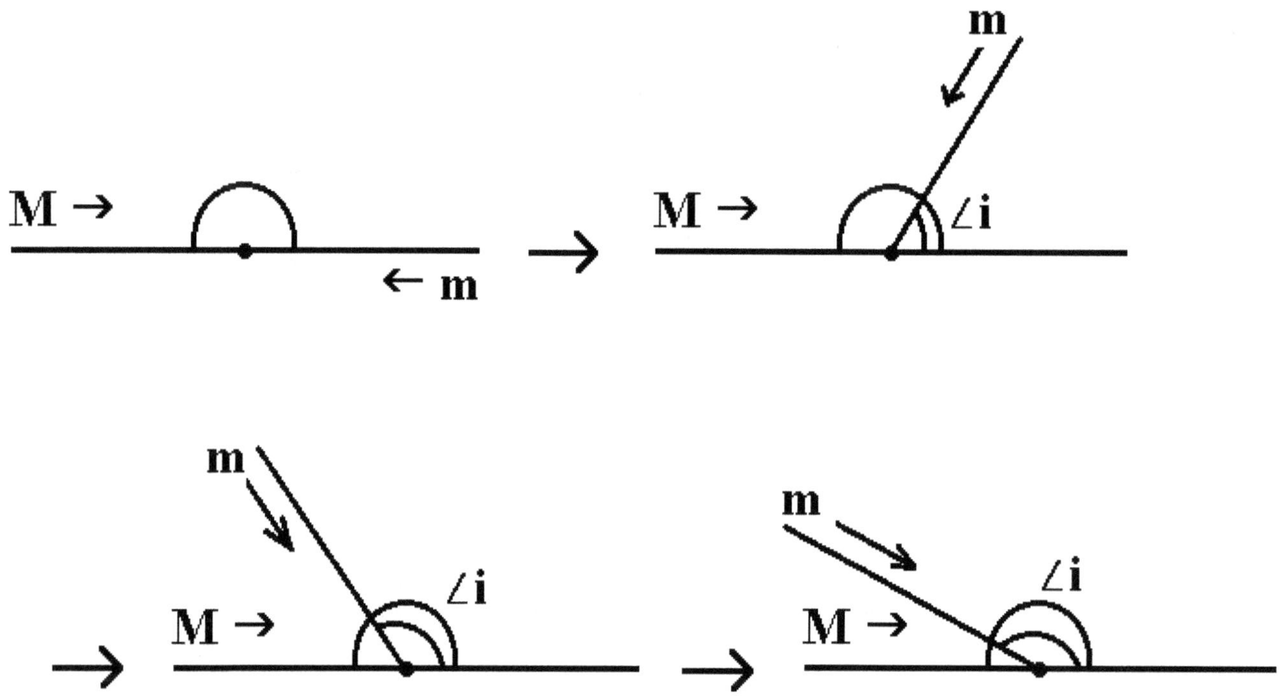

i.e., from the smaller (m) mass' perspective (compared to M) :

$$\text{inclination } \angle == \angle i_{m} = [\,M / (M + m)\,]\pi \quad \text{goes to } \pi \text{ as } m \rightarrow 0$$

From the larger (M) mass' perspective:

$$\angle i_{M} = [\,m / (M + m)\,]\pi \quad \text{goes to } 0 \text{ as } m \rightarrow 0$$

If $M = m$, then:

$$\angle \mathbf{i}_{M=m} = [\,(\mathbf{M} \text{ or } \mathbf{m}) / (\mathbf{M} + \mathbf{m}) == \mathbf{M} / 2\mathbf{M} == \mathbf{m} / 2\mathbf{m}\,]\pi = (\tfrac{1}{2})\pi = \pi / 2 = 90^{\circ} :$$

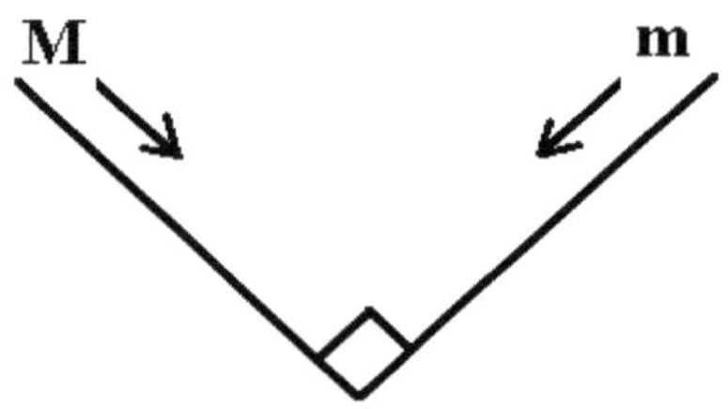

If M = M but m = 0 (e.g. a photon or mass-less quantity), then there is no movement (of the M incline) of the inclination for M; all movement is given to m.

There is a distinction between inclination angle (observation of gravity) movement ($\angle i$) and movement (cause of gravity) towards the point of intersection of the inclines (i.e., movement of masses toward each other). They are different categories of movement. If m = 0 (compared to M), then the strength is: (Mm = 0) , while the inclinations are:

$(M / m) = (M / 0) == \infty$ from the m perspective, and

$(m / M) = (0 / M) == 0$ from the M perspective.

Here we see that either ∞ must take precedence over zero for:

$$(M / m)(M / 0) == \infty = (0)(\infty)$$

or that some gravitational constant (or variable) must be applied for the definitions:

If δ is the gravitational factor (constant or variable), then for the strength it is (of form) δ^2 :

i.e. $(Mm + \delta^2)$,

with δ itself being characteristically small (or minute).

$\therefore$ observed gravities are of the natures:

$$(Mm + \delta^2)[M / (m + \delta)] \;,\; (Mm + \delta^2)[m / (M + \delta)]$$

and we have (proposed) the relations (pictorially):

$$\angle i_m = [M / (M + m)]\pi = (Mm + \delta^2)[M / (m + \delta)] \rightarrow \pi \quad \text{as } m \rightarrow 0$$

$$\angle i_M = [M / (M + m)]\pi = (Mm + \delta^2)[m / (M + \delta)] \rightarrow 0 \quad \text{as } m \rightarrow 0$$

For m = 0 (compared to M), as $m \rightarrow 0$ for $\angle i_m$ we have:

$$\angle i_m = \pi = (0 + \delta^2)[M / (0 + \delta)] = \delta^2 M / \delta = \delta M$$

$\therefore \delta = \pi / M$ in units of mass^2 / radians , to leave δ in units of mass

($\therefore$ the larger is M, the smaller is δ).

For $M = m$, we have:

$$\angle i_{M=m} = \pi / 2 = (M^2 + \delta^2)[M / (M + \delta)] (M^3 + \delta^2 M) / (M + \delta)$$

$$== (m^3 + \delta^2 m) / (m + \delta)$$

or

$$(\pi / 2)\delta - \delta^2 M = M^3 - (\pi / 2) M \text{, etc. (replace } M \text{ with } m)$$

$$\delta[(\pi / 2) - \delta M] = M[M^2 - (\pi / 2)]$$

$$\delta = M \frac{\left(M^2 - \frac{\pi}{2}\right)}{\left(\frac{\pi}{2} - \delta M\right)} ,$$

$$\frac{\delta}{M} = \frac{\left(M^2 - \frac{\pi}{2}\right)}{\left(\frac{\pi}{2} - \delta M\right)}$$

[We are not dealing here with the force of gravity, rather gravity itself: the attraction between masses and *why* it occurs.]

The Nature of Mass as a Goal of Gravity

Now, why are (two) masses (physically) compelled to meet at a conjunction (intersection) of their (respective) inclines, the common point of intersection being a goal?

One corollary assumes that a mass is (always?) in motion along an incline (as adopted by the physical circumstances). ∴ all masses are in motion, and that motion is directed (presumably by other masses).

Implausibles : part 2

Of course, all energy is in motion (or it would not have a speed, constant in any given medium).When an energy (that is uniform) deviates from its constant (and maximum) speed, in a medium, it becomes a mass (that is not uniform), as another corollary (of motion).

Obviously (or apparently), a point of (successful) conjunction is one of (enforced) uniformity. Thus, this is where the masses may retake their "clothing" or substance as (pure) energy, and maximal speed. ∴ Another corollary (of motion) is that all "substances" desire or tend to take their maximal speed given the condition of their (internal) energies, for in this way they may regain (or re-attain) their energetic identities. That is, the non-uniform nature of their components (must eventually) dissipates into the uniform energies of the constituents. This is apparently the physical goal of gravity, since the (substance of components) dissipation requires to be at (or located at) points of conjunction (of inclines).

Then so, (radiant or directional) energies must take a maximum speed in order to allow for the dissipation of masses (into like energies). It is simply a compunction of like wishing to combine with like, or the "need" for like to dissolve in like. The corollary here is that a (special) collection of non-uniform energies (i.e., energies running in different directions) yields a mass, the greater the confusion in (energy) directions (non-uniformity) the denser or more gravimetrical the mass. A non-uniform distribution of energy is characteristic of a mass radiating energy spherically, i.e. in all (allowable) directions.

We might then (and may) conclude that: All measurement requires mass (at the very least, the mass of the measurer), and this mass is ever changing (since the measurer, being alive, is dynamic of change with time).

Since (to date) the mathematics of physics do not take into account of this (measurer) mass, obviously there must be some uncertainty in the results obtained. But all physical results are (by definition of being physical) definite and decidedly not uncertain.

The lack of taking into account (measurer, or any other) mass is all too pervasive in (current) science as "sloppy" physics. This is the fundamental error of Quantum Mechanics, which distributes the (assumed) uncertainties through (alleged factual) probabilities. In fact, they are simply different obtainable results (physical). And some (results) must ***not*** be obtainable (by physical means), as opposed to having extremely low probabilities. Then what of the implausible about our theories and predictions, constructions and concerns? Science proposes, technology demonstrates, research uncovers; all the while logic and fact wrestle with each other to provide an understanding of commonly experienced phenomena.

Appendices

1. Employment of (arbitrary) Coefficient Associations

Let C_n == m(ass) , (n + 1) == d(istance) ,

$(\Psi_{m(inor)})^n$ == m • d , $\Psi^{n\wedge n}$ == t(ime)

$\therefore \log \Psi_\Psi{}^{n\wedge n} = n^2 = \log_\Psi t$ and $n = (\log_\Psi t)^{1/2}$

[If n = 11 and $\Psi \sim\sim 2.2905$, then $\Psi^{n\wedge n} = t = \Psi^{121}$ $\therefore$ $t \sim\sim= 3.5609 \times 10^{+43}$]

$\therefore (\log_\Psi t)^{1/2} + 1 = d$, $[(\log_\Psi t)^{1/2} + 1] / \Psi^{n\wedge n} = d / t$

$\rightarrow (n + 1) / \Psi^{n\wedge n} == (11 + 1) / \Psi^{n\wedge n}$

$\therefore 12 / 3.5609\text{x}10^{+43} = 3.3698876\text{x}10^{-43} \sim\sim= 3.3700\text{x}10^{-43}$

$(\Psi_m)^n = m \bullet [(\log_\Psi t)^{1/2} + 1] == C_n \bullet [(\log_\Psi t)^{1/2} + 1]$

$[(\Psi_m)^n / C_n] - 1 = (\log_\Psi t)^{1/2}$, $([(\Psi_m)^n / C_n] - 1)^2 = \log_\Psi t$

$\Psi\wedge(\log_\Psi t) = t = \Psi\wedge([(\Psi_m)^n / C_n] - 1)^2 \sim\sim \Psi$ if $(\Psi_m)^n << C_n$

What, then, is the apparent n when $t = \Psi$?

$n^2 = \log_{10} 2.2905 / \log_{10} 2.2905 == \log_{10} t / \log_{10} \Psi = 1$, since $t = \Psi$

$\therefore n = (1^2)^{1/2} = 1^1 = 1$, since $1 = 1^2$

$\therefore$ if n = 11 , $n^2 \bullet \log_{10} \Psi = \log_{10} t = (121)(\log_{10} 2.2905) \sim= 43.552$

$\therefore t \sim= 10\wedge(\log_{10} t) = 10^{43.552} = (1 \text{ x } 10^{43})(10^{0.552}) = 3.564 \text{ x } 10^{43}$

If using Ψ_m as the primary base ($\sim = 1.5231$), then, at $n = 11$:

$$n^2 = \log_{10} t / \log_{10} 1.5231 \text{ , } (121)(\log_{10} 1.5231) = \log_{10} t \sim = 22.1101$$

$$\therefore t == t_m \sim = 10^{22.1101} = 1 \times 10^{22.1101}$$

$$= 1 \times 10^{(22 + 0.1101)}$$

$$= 1 \times 10^{22} \times 10^{0.1101}$$

$$= 1.2885 \times 10^{22}$$

$$(t / t_m)_{n = 11} == (t_\Psi / t_{\Psi m})_{n = 11} \sim = 3.564 \times 10^{43} / 1.289 \times 10^{22} = 2.765 \times 10^{21}$$

Then, at what (apparent) n , using base Ψ , is $t == t_m = 1.2885 \times 10^{22}$?

$$n^2 = \log_{10} t / \log_{10} \Psi = \log_{10} 1.2885 \times 10^{22} / \log_{10} 2.2905 \sim = 61.4288$$

$$\therefore n \sim = (61.4288)^{1/2} = 7.8377 \quad (== 7.837652)$$

If

$$C_{11}(11 + 1) / (\Psi_m)^{11} \sim \sim c \text{ in meters / second} \sim = 3 \times 10^8 \text{ m/s}$$

then,

with $C_n \sim \sim$ mass , $(n + 1) \sim \sim$ distance (of meters) ,

$(\Psi_m)^n$ is in units of mass • (time in seconds)

$\Psi^n ==$ functionally to $(\Psi_m)^n \bullet \Sigma n_s == (\Psi_m)^n \bullet$ distance

$$C_{11}(11 + 1) / (\Psi)^{11} \sim = 3.42 \times 10^6 \text{ (m/s)(1/m)} == 1/s$$

$\therefore$ a frequency

$$\text{(m/s) / (1/s)} = m == (3.04 \times 10^8 \text{ m/s) / (} 3.42 \times 10^6 \text{ s}^{-1}) \sim = 88.9 \text{ meters}$$

(a wavelength?)

At $n = 7.84$,

$(7.84)(11) = 86.24$ (meters)

$(88.9 / 11 \sim= 8.08)$

88.9 meters / $3.42 \times 10^{6}\ s^{-1} \sim= 2.6 \times 10^{-5}$ meters • seconds == 26 μ •s

$\therefore$ if $(\Psi_{m})^{n}$ == mass • seconds , then

$(\Psi_{m})^{n} == [\Sigma(1/mass)]^{meters}$ = mass • seconds

or

$[\Sigma(1/mass)]^{dist}$ = mass • time

With mass normalized to "1" , this comes to:

(a sum of reciprocal masses $)^{dist}$ = 1 mass • time

$\therefore$ time = (a sum of reciprocal masses $)^{dist}$ / mass

== 1 / $mass^{(dist + 1)}$ in unitage = $[mass^{(dist + 1)}]^{-1} = mass^{-(dist + 1)}$

$\therefore$ time == $C_{n}^{-(n+1)}$

If, also, time == Ψ^n^2 (coefficient to unitage), then

Ψ^$n^2 = [\Sigma(n / C_{n})]$^$n^2 == n^3 / (C_{n})$^$n^2 = 1 / C_{n}(n + 1)$

$\therefore n^3 \bullet C_{n}$^$n^2 = C_{n}$^$n^2 / C_{n}(n + 1) = C_{n}$^$[n^2 - (n + 1)]$

This is only true (or possible) when $n = 1$ [or $n = 0$, if $0^0 = 1$ and $1 / (1 / 0) = 0$] :

$(C_{n})^{13} / (C_{n})^{12} = (C_{n})^{1} / (C_{n})^{1} = 1 / (C_{n})^{(1+1)} = 1 / (C_{n})^{2} = 1 / 1^{2} = 1 / 1 = 1$

2. Base Coefficient Comparisons

$\Psi_m \sim= 1.5231$ $\qquad$ $e \sim= 2.7183$

$\Psi \sim= 2.2905$ $\qquad$ $\pi \sim= 3.1416$

	$\Psi_m^{e^n}$	Ψ^{e^n}	$\Psi_m^{\Psi^n}$	$\Psi^{\Psi_m^n}$
1	3.1384	9.5148	2.6214	3.5335
2	22.3980	456.2053	9.0921	6.8388
3	4680.2817	16964389.58	156.9717	18.6963
4	9482009004	4.4880×10^{19}	107035.678	86.4990
5	1.3189×10^{27}	2.6313×10^{53}	3.3123×10^{11}	891.7875

$$e = \left(\sum_{n=1}^{\infty} \frac{1}{n!} \right) + 1 = \sum_{n=0}^{\infty} \frac{1}{n!} \qquad (0! = 1)$$

$$\text{For } \sum_{n=1}^{\infty} \frac{1}{n} \ , \quad \int_{n=x_1}^{x_2} \frac{dx}{x} = \ln|x_2| - \ln|x_1|$$

$$\Psi_m = \sum_{n=1}^{\infty} \frac{1}{C_n} \qquad \Psi = \sum_{n=1}^{\infty} \frac{n}{C_n}$$

$C_n\uparrow$ faster than $n!\uparrow$ (with $n\uparrow$) : $\Sigma(1/C_n) \rightarrow 1.5231$, while $\Sigma(1/n!) \rightarrow 1.7183$

If $y^e = \pi$, then $y = (y^e)^{1/e} = (\pi)^{1/e} = 1.5237 \sim= 1.5231 = \Psi_m$

3. Possible Photon Mass (from Ψ_{mSp})

A photon of energy (e.g., light) is *defined* as being mass-less, though maintaining properties of force, speed, and displacement (i.e., from travel). Since the discovery that, contrary to prior assumptions, neutrinos are **not** mass-less, experiments have been made to determine if any mass can be assigned to photons and to find possible lower limits to such values.

If the

$$\Psi_{mSp} = \frac{C_n(n+1)d}{\Psi_m^{n}\, t} \equiv \frac{C_{11}(12)}{\Psi_m^{11}} \bullet \frac{1 \text{ meter}}{1 \text{ second}}$$

derived value is the upper limit for the speed of light ($C_{\Psi m}$), and the (measurable) speed of light (c) is less than this, then the discrepancy ($C_{\Psi m} - c$) might be due to a still (as of yet) not measurable (i.e. "***un***-measurable") mass of a photon.

$$C_{\Psi m} - c \sim= (3.04162423 \times 10^8 - 2.99792458 \times 10^8)\text{m/s}$$

$$= (3.04162423 - 2.99792458) \bullet 10^8$$

$$= 0.04369965 \times 10^8 \text{ m/s} = 4.369965 \times 10^6 \text{ m/s}$$

A photon of wavelength (λ) 0.6 μm has an energy of 2.0663 eV or 3.31×10^{-19} joules .

Assuming the photon has mass and is kinetic (i.e., it is moving):

$$\text{K.E.} = \tfrac{1}{2} mv^2 \therefore m = 2(\text{K.E.}) / v^2 == 2(\text{K.E.}) / (\Delta c)^2$$

With mass (m) in kilograms and velocity (v) in meters/second ,

for K.E. in joules ,

$$\therefore m_{\text{photon}, \lambda = 0.6\ \mu m} = 2(3.31 \times 10^{-19} \text{ joules}) / (4.369965 \times 10^6 \text{ m/s})$$

$$= (\ 1.514886275 \times 10^{-25}\ /\ 4.369965 \times 10^{6}\ )\ \text{kg}$$

corresponding (using $E = mc^2$), for a rest mass, to an energy of :

$$m_{\text{photon}\atop \lambda = 0.6\ \mu\text{m}} = \mathbf{3.466586746 \times 10^{-32}\ kg}$$

$$\xrightarrow{E = mc^2} \mathbf{3.115612791 \times 10^{-15}\ joules}$$

$$\left(\begin{array}{c} \sim < 4\%\ \text{of electron mass: } m_e \cong 9.10938 \times 10^{-31}\ \text{kg} \\ \therefore \rightarrow \sim 26\ \text{photons / electron} \end{array}\right)$$

$$\equiv \mathbf{19{,}449.51876\ eV}$$

$$\frac{2.0663}{19{,}449.51876} \rightarrow \mathbf{1.06239 \times 10^{-4}}\ \textbf{attenuation (in measurement)}$$

4. Mathematical Bases for (chemical, or other) Reactions

Chemical reactions (rxns.) have activation energies that lead to an Activation Complex (AC), an enzyme or catalyst being useful for lowering the AC energy.

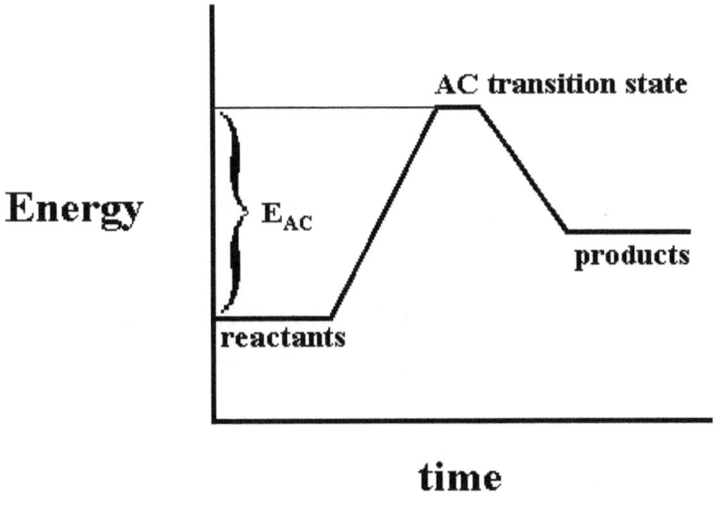

Therefore, it might be useful to use an integer math system based on the 3 (relative) distinctive states of a reaction: reactants, AC, product.

A base 3 system (having digits 0, 1, 2) can be called "trinary" :

reactants	AC	product(s)
0	1	2

with the AC condition leading to either products or back to reactants (∴ "maybe," for rxn.).

We can call the 3 digits a "tryte" : (0, 1, 2) or, for binary type system compatibility, (0, ψ, 1) ;

∴ e.g. 1 0 1 == $1 \times 3^2 + 0 \times 3^1 + 1 \times 3^0$

$6 + 0 + 1 = 7$

Hexidecimal (**digits:** 0, 1, 2, 3, 4, 5, 6, 7, 8, 9, a, b, c, d, e, f) is useful in conjunction with binary because of the easy conversions between the systems:

digits

(0 → f) hexidecimal 16

— = 8 bits per byte (== e.g.: 0 0 0 0 1 0 0 0)

(0 , 1) binary 2

== $(0000) \times 16^1 + (1000) \times 16^0$

In a similar manner, a base 9 ("nonary") system might be useful with the trinary (base 3) :

digits

(0 →8) nonary 9

— = 3 "bits" (or 'tits') per tryte ; e.g.: (1 1 1) ==

(0, 1, 2) trinary 3

$(1 \times 3^2)9^2 + (1 \times 3^1)9^1 + (1 \times 3^0)9^0$

$729 + 27 + 1 = 757$

tryte interpretation (for nonary)

[max: (2 2 2) == 1458 + 54 + 2 = 1514]

As $(1\ 1\ 1\ 1)_{binary} = 15$ == f in hexidecimal, the highest hexidecimal digit, then $(0\ 2\ 2)_{trinary}$ or $(2\ 2)_{trinary} = 8$, which equals the highest digit in nonary.

As like two 4-bits of binary corresponds to an 8-bit byte (for easy conversion between the two systems), two 2-"bit"s of trinary can correspond to a 4-"bit" Quyte, for easy conversion between these two systems: base 3 trinary and base 9 nonary.

How, then, can we "evolve" a sort of reaction using the trinary-nonary Quyte relationship?

Say, we start with a two digit nonary: 74

1) we convert each digit to (two digit) trinary

$$7_{nonary} == 2 \times 3^1 + 1 \times 3^0 = 21$$

$$4_{nonary} == 1 \times 3^1 + 1 \times 3^0 = 11$$

2) so now, we have a Quyte of : 2 1 1 1

or $(2\ 1) \times 9^1 + (1\ 1) \times 9^0 == [7 \times 9^1] + [4 \times 9^0] = 74_{nonary}$ or $63 + 4 = 67_{decimal}$

$$2\ 1\ 1\ 1 == [2 \times 3^3] + [1 \times 3^2] + [1 \times 3^1] + [1 \times 3^0]$$

$$= 2 \times 27 + 9 + 3 + 1 = 67_{decimal}$$

The trinary digit 2 yields towards products (or proceeding);

the trinary digit 1 yields towards 'maybe or maybe not' proceeding;

the trinary digit 0 yields to not proceeding (going back to reactants).

One could assign each (Quyte) digit to a weight of occurrence;

i.e., the 3^3 digit has weight $(3^3) / (3^3 + 3^2 + 3^1 + 3^0) = 27 / 40$

Likewise, 3^2 digit has weight : $3^2 / 40 = 9 / 40$

3^1 digit has weight : $3^1 / 40 = 3 / 40$

3^0 digit has weight : $3^0 / 40 = 1 / 40$

We can assign "Procession" weights to the trinary digits:

$2 \rightarrow 1$; $1 \rightarrow$ random (from $0.0 \rightarrow 1.0$) ; $0 \rightarrow 0$

Absolute Procession == $2\,2\,2\,2 \rightarrow 1 + 1 + 1 + 1 = 4$

Absolute return to reactants == $0\,0\,0\,0 \rightarrow 0 + 0 + 0 + 0 = 0$

Therefore, the Averaged Procession Weight for (2 1 1 1) would be:

2	1	1	1
1 / 4	ran / 4	ran / 4	ran / 4

= (1 / 4) + (random / 4) + (random / 4) + (random / 4)

And this can be convoluted to digit weights:

(1 / 4)(27 / 40) + (random / 4)(9 / 40) + (random / 4)(3 / 40) + (random / 4)(1 / 40)

where, if each of the randoms = 1 ,

the **maximum** (convoluted) **weight == (1 / 4)(0.675 + 0.225 + 0.075 + 0.025) = (1 / 4)(1) = 0.25**

Another convolution of weights (giving each digit equal weight) might be:

[(1 / 4) + (ran / 4) + (ran / 4) + (ran / 4)](1) = 1 at maximum

But what measure of the final (convoluted) weight defines procession (to products)?

That would depend on the reaction considered and its (estimated) spontaneousness (expenditure or release of energy, endothermic or exothermic).

Say, arbitrarily, for example, the FCW for rxn. procession must be > 0.6 , and the given digit randoms (for "1" Quyte digits) are as follows:

2	1	1	1
1 / 4	0.8 / 4	0.7 / 4	0.2 / 4

∴ the Procession Weight is: (1 / 4) + (0.8 / 4) + (0.7 / 4) + (0.2 / 4) = 2.7 / 4 ,

and the final weight is (2.7 / 4)(1) ~= 0.675 :

0.675 > 0.6 , so the reaction proceeds to products.

This is simply equivalent to: (1 + 0.8 + 0.7 + 0.2) (0.25) ;

∴ The method is: (sum of unaveraged "Procession" weights)(0.25) for the Quyte, or more simply: the sum of averaged "Procession" weights.

The digits with convoluted weights would be:

(3^3 procession weight / 4)(27 / 40) + (3^2 pw / 4)(9 / 40) + (3^1 pw / 4)(3 / 40) + (3^0 / 4)(1 / 40)

∴ for Quyte (2 1 1 1), with the given random factors, it would be:

$$(1/4)0.675 + (0.8/4)0.225 + (0.7/4)0.075 + (0.2/4)0.025$$

$$= 0.16875 + 0.045 + 0.013125 + 0.00125 = 0.228125$$

But a maximum, i.e., for Quyte (2 2 2 2), would be:

$$(0.25)(0.675) + (0.25)(0.225) + (0.25)(0.075) + (0.25)(0.025)$$

$$= 0.16875 + 0.05625 + 0.01875 + 0.00625 = 0.25$$

Therefore, we must divide by the maximum to get the true (corrected) Final Convoluted Weight.

For our (2 1 1 1) Quyte with its allotted randoms it is:

$$0.228125 / 0.25 = 0.9125 > 0.6,$$

showing that the 3^3 digit greatly outweighs the others, which is appropriate since it represents the bulk of the number.

∴ the corrected Final (convoluted) weight can be just:

$(3^3 \text{ pw})(27/40) + (3^2 \text{ pw})(9/40) + (3^1 \text{ pw})(3/40) + (3^0 \text{ pw})(1/40)$,
where the maximum → 1.0

But, what can a Quyte indicate of a procession (to products)?

If we have n reactants, then the number of results (possible) is C_n, and the number of products (in type) is $C_n - n$. Therefore, if there are more than ($C_n - n$) results (in type), this means that some reactants (up to n in type) have resulted.

For example, for $n = 3$ reactants the # of resultant types (possible) is $C_3 = 7$.

Putting this in Quyte (trinary) form:

$$(0 \quad 0 \quad 1 \quad 0) \rightarrow (0 \quad 0 \quad 2 \quad 1)$$

$$1 \times 3^1 \qquad\qquad 2 \times 3^1 + 1 \times 3^0$$

reactants products

or, in nonary

$$(0 \quad 0) \times 9^1 \qquad (0 \quad 0) \times 9^1$$

$$\rightarrow$$

$$+ \quad (1 \quad 0) \times 9^0 \qquad (2 \quad 1) \times 9^0$$

$$== \quad 3 \times 1 \rightarrow 7 \times 1$$

$$3 \rightarrow 7$$

nonary: 0 3 → 0 7

The # of product types (other than reactants) possible are:

$C_n - n == C3 - 3 = 7 - 3 = 4 \rightarrow (0 0 1 1)_{Quyte}$

	(0 0 2 1)	rxn. products or results
−	(0 0 1 1)	product types
	(0 0 1 0)	reactants

Here, the "1" digit does not indicate "maybe," but is just a digit for a simple (ordinary) Quyte number.

(0 0 1 0)	→	(0 0 0 1) |	
reactants	+	(0 0 0 1) |	
	+	(0 0 0 1) |	product types (i.e., one of each type)
	+	(0 0 0 1) |	
	+	(0 0 1 0)	reactants
		(0 0 2 1)	results

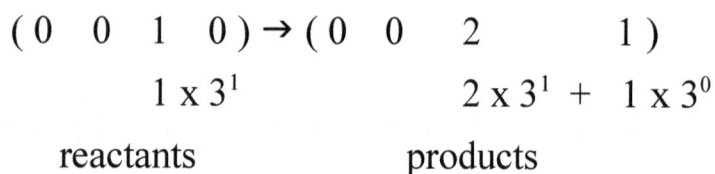

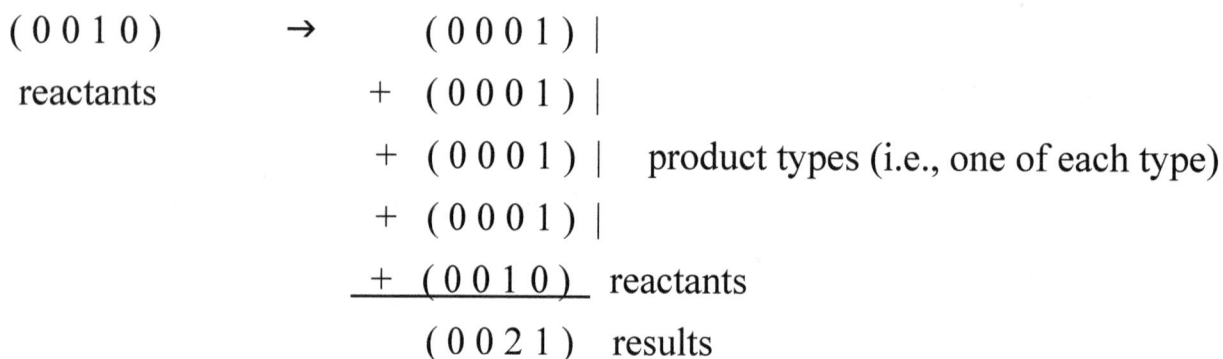

∴ per product type (from its perspective), we have:

$$(0010) \rightarrow (0010) + (0001)$$

reactants reactants product type

$(0010) \rightarrow (0010)$ is evident.

But how does

$$(0010) \rightarrow (0001)$$

occur?

Now we can employ the maybe "1" as (say) : $\psi = 1$ or 0

$$\therefore (0010) \rightarrow (000\psi) \rightarrow (0001) \text{ or } (0000)$$

or

$$\begin{aligned} (0010) \rightarrow\ & (000\psi) \quad | \\ + & (000\psi) \quad | \\ + & (000\psi) \quad | \text{ product Quytes} \\ + & (000\psi) \quad | \\ + & (0010) \text{ reactant Quyte} \end{aligned}$$

or

$$\mathbf{(0010) \rightarrow (0011)(000\psi) + (0010)}$$

$$\mathbf{\equiv (0011)\left[(000\psi) \begin{matrix} \nearrow (0001) \\ \searrow (0000) \end{matrix}\right] + (0010)}$$

product Quyte or
$(0011)_{trinary} \equiv 4$

$(0011)(000\psi) = (00\psi\psi)$, or $(11)(0\psi) = (\psi\psi)$

Therefore, we have:

(0 0 1 0)	→	(0 0 ψ ψ)	+	(0 0 1 0)
reactants		product types		reactants
n		$(C_n - n)$		n

Obviously, in the case of all product types, the product Quyte has been multiplied by (0 0 0 ψ), for its manufacturing (or conception).

Therefore, taking our previous example of (2 1 1 1) as a product Quyte, the equivalent Quyte for product conception of proceeding to be formed would be:

$$(2\ 1\ 1\ 1)(0\ 0\ 0\ \psi) = (2\psi\ \ \psi\ \ \psi\ \ \psi)$$

with $0.675 > 0.6$ → procession to products occurring (based on the final convoluted procession weight).

$\therefore (2\ 1\ 1\ 1) == 74_{nonary}$ or $67_{(decimal)}$ products forming.

This combinatorial approach to reactions (i.e., using C_n) can also be used for conversions:

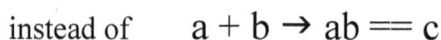

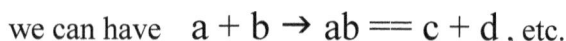

instead of $a + b \rightarrow ab == c$

we can have $a + b \rightarrow ab == c + d$, etc.

("ab" being treated as a transition state form).

Note that operationally, in terms of the mathematics, since the reactants always result in the product pool, the "reaction" can occur indefinitely:

reactants → (combinatorial) products + reactants

↑______________________________↓

Thus, the process is recursive, and time spans (i.e., cycles) can be applied until an arbitrary limit is reached where (it is agreed upon that in a practical sense) no more reactants reside in the pool (i.e., they have been used up or exhausted). The final convoluted procession weight might be used to show the diminishing of the reactant amounts until this limit is reached (the reactant amounts will always remain positive, unless a negative weight is introduced).

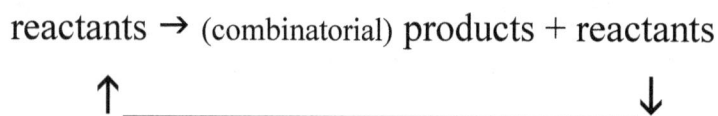

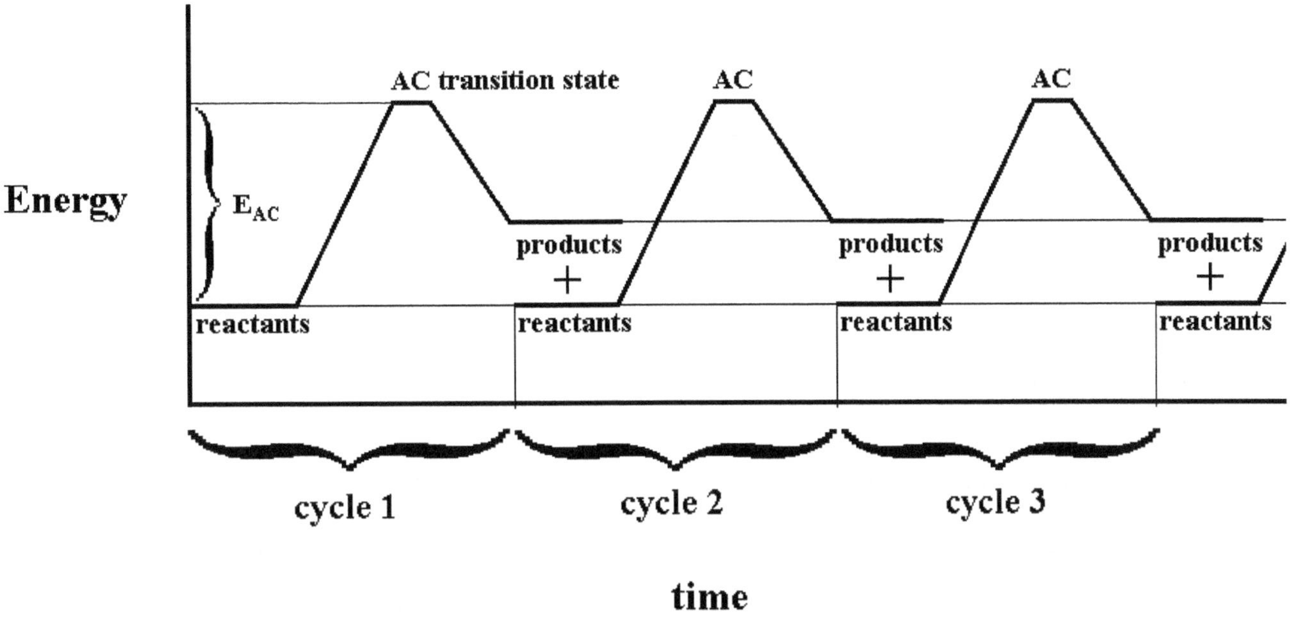

Application of the Quyte digit randoms (rans), as shown earlier, can be formulated as:

(3^3 pw = ran)(27 / 40) + (3^2 pw = ran)(9 / 40) + (3^1 pw = ran)(3 / 40) + (3^0 pw = ran)(1 / 40)

where the maximum = 1.0 (each ran == 1.0), to yield the **F**inal **C**onvoluted Procession **W**eight for the (2 1 1 1) example:

∴ essentially: the FCW for (2ψ ψ ψ ψ) is

(ran)(0.675) + (ran)(0.225) + (ran)(0.075) + (ran)(0.025)

We can combine this with reactant (amount) relationships, to determine and observe the manner of reactant usage (or reduction) towards formation of product types:

e.g., using reactants a, b, and c, we can declare that to form product bc requires 2 times as much b as c, etc. A sort of Reactant Relationship Matrix can then be set up.

Let's take the $C_3 = 7$ case again, but apply some (here, arbitrary) conditions for product formation.

The product Quyte is (0 0 ψ ψ) :

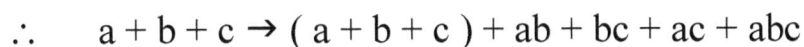

∴ a + b + c → (a + b + c) + ab + bc + ac + abc

Some reactant relationships conditions for product formation can be:

a, b, c start with equal amounts

ab requires equal amounts of a and b

bc requires twice as much b as c

ac requires 5 times as much c as a

abc requires equal amounts of a and b but 3 times as much c as a or b

Other conditions can be:

- the FCW for product formation must be at least 0.675 , meaning that below this products don't form and reactant amounts don't change (although a production cycle, time-wise, occurs).

- Quyte digit randoms (ran) occur for **each** cycle;

- if reactant c falls below a critical value, then production ends; the rxn. stops (i.e., last cycle).

The calculated FCW here is : $[\ (\text{ran})(\ 3^1 \text{ digit })(\ 3 / 4\ ) + (\text{ran})(\ 3^0 \text{ digit })(\ 1 / 4\ )\ ]$, since the # of products = 4 == $(\ 0\ 0\ 1\ 1\ )_{\text{trinary}}$. The # of products = $C_n - n = 7 - 3 = 4$.

For the Reactant Relationships Matrix, we have (assuming a cycle's calculated CW = 0.675) :

ab case	[(0.675)a	+ (0.675)b		](1 / 4)
bc case	[(0.675 / 2)b	+ (0.675)c		](1 / 4)
ac case	[(0.675)a	+ (0.675 / 5)c		](1 / 4)
abc case	[(0.675)a	+ (0.675)b	+ (0.675 / 3)c	](1 / 4)

or

	ab	bc	ac	abc
a	1	0	1	1
b	1	2	0	1
c	0	1	5	3

The table is read: a relates to ab as 1, b relates to bc as 2, c relates to bc as 1, etc.

When the cycle's calculated FCW >= the desired FCW, then the products are made and the cycles continue until the amount of reactant c falls below a critical value.

If the cycle's calculated CW < desired FCW, then production pauses (reactant amounts don't change) to wait for the next cycle.

The resulting reactant amounts per cycle are (given the current calculated FCW):

$$\mathbf{a}_{\text{of current cycle}} =$$
$$\mathbf{a}_{\text{of previous cycle}} \bullet (1 - [(FCW/4)_{\text{ab case}} + (FCW/4)_{\text{ac case}} + (FCW/4)_{\text{abc case}}])$$

$$\mathbf{b}_{\text{of current cycle}} =$$
$$\mathbf{b}_{\text{of previous cycle}} \bullet (1 - [(FCW/4)_{\text{ab case}} + (FCW/(2 \bullet 4))_{\text{bc case}} + (FCW/4)_{\text{abc case}}])$$

$$\mathbf{c}_{\text{of current cycle}} =$$
$$\mathbf{c}_{\text{of previous cycle}} \bullet (1 - [(FCW/4)_{\text{bc case}} + (FCW/(5 \bullet 4))_{\text{ac case}} + (FCW/(3 \bullet 4)_{\text{abc case}}])$$

Corresponding product production within a cycle would be as:

$$\mathbf{ab} = [\, \mathbf{a}_{\text{of previous cycle}} \bullet (FCW/4)_{\text{ab case}}] + [\, \mathbf{b}_{\text{of previous cycle}} \bullet (FCW/4)_{\text{ab case}}]$$
$$\mathbf{bc} = [\, \mathbf{b}_{\text{of previous cycle}} \bullet (FCW/(2 \bullet 4))_{\text{bc case}}] + [\, \mathbf{c}_{\text{of previous cycle}} \bullet (FCW/4)_{\text{bc case}}]$$
$$\mathbf{ac} = [\, \mathbf{a}_{\text{of previous cycle}} \bullet (FCW/4)_{\text{ac case}}] + [\, \mathbf{c}_{\text{of previous cycle}} \bullet (FCW/(5 \bullet 4))_{\text{ac case}}]$$
$$\mathbf{abc} = [\, \mathbf{a}_{\text{of previous cycle}} \bullet (FCW/4)_{\text{abc case}}] + [\, \mathbf{b}_{\text{of previous cycle}} \bullet (FCW/4)_{\text{abc case}}] + [\, \mathbf{c}_{\text{of previous cycle}} \bullet (FCW/(3 \bullet 4))_{\text{abc case}}]$$

Typical results might look like (under program **Rxn_Procession**):

n = 3 **$C_n = 7$**

reactants

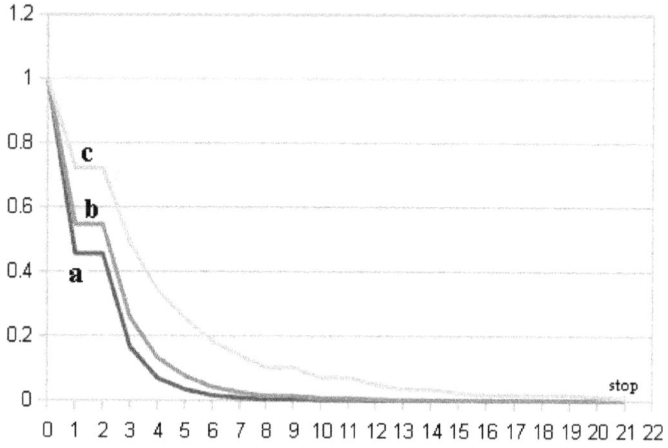

cycles

Reaction Relationship Matrix

	ab	bc	ac	abc
a	1	0	1	1
b	1	2	0	1
c	0	1	5	3

limiting FCW = 0.60 **limiting reactant c = 0.01**

product Quyte

(0 0 1 1)

randomizing Quyte

(0 0 0 ψ)

Another run using the same parameters

(with products and cycle FCW values included)

Reaction Relationship Matrix

	ab	bc	ac	abc
a	1	0	1	1
b	1	2	0	1
c	0	1	5	3

reactants **products**

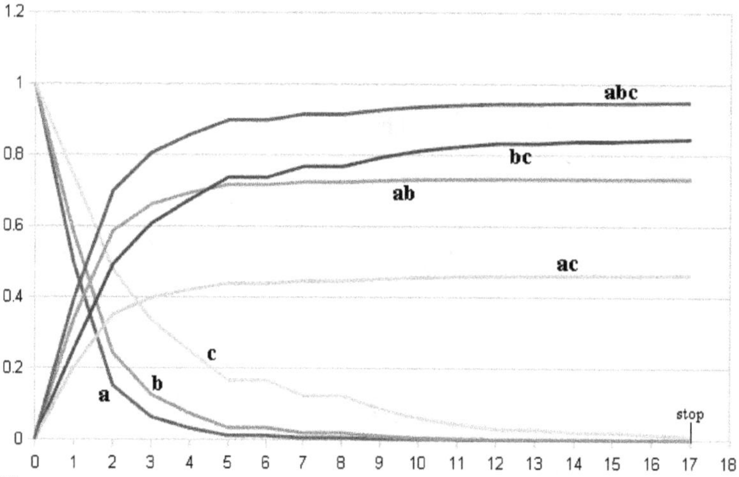

cycle FCW : *.68 .93 .78 .68 .88 .52 .68 .15 .78 .80 .75 .83 .30 .78 .48 .83 .85*

cycles

```
 n  = 3
Cn  = 7

product number = 4     --->    Limiting CW = 0.60   ;   limit on reactant c = 0.01
FCW = 0.68    Cycle # 1    :   a = 0.49375    b = 0.57813    c = 0.74125
FCW = 0.93    Cycle # 2    :   a = 0.15121    b = 0.24390    c = 0.47842
FCW = 0.78    Cycle # 3    :   a = 0.06332    b = 0.12576    c = 0.33629
FCW = 0.68    Cycle # 4    :   a = 0.03126    b = 0.07270    c = 0.24927
FCW = 0.88    Cycle # 5    :   a = 0.01075    b = 0.03294    c = 0.16566
FCW = 0.52    Cycle # 6    :   a = 0.01075    b = 0.03294    c = 0.16566
FCW = 0.68    Cycle # 7    :   a = 0.00531    b = 0.01905    c = 0.12280
FCW = 0.15    Cycle # 8    :   a = 0.00531    b = 0.01905    c = 0.12280
FCW = 0.78    Cycle # 9    :   a = 0.00222    b = 0.00982    c = 0.08632
FCW = 0.80    Cycle # 10   :   a = 0.00089    b = 0.00491    c = 0.05985
FCW = 0.75    Cycle # 11   :   a = 0.00039    b = 0.00261    c = 0.04264
FCW = 0.83    Cycle # 12   :   a = 0.00015    b = 0.00126    c = 0.02916
FCW = 0.30    Cycle # 13   :   a = 0.00015    b = 0.00126    c = 0.02916
FCW = 0.78    Cycle # 14   :   a = 0.00006    b = 0.00065    c = 0.02049
FCW = 0.48    Cycle # 15   :   a = 0.00006    b = 0.00065    c = 0.02049
FCW = 0.83    Cycle # 16   :   a = 0.00002    b = 0.00032    c = 0.01401
FCW = 0.85    Cycle # 17   :   a = 0.00001    b = 0.00015    c = 0.00945
____________________________________________________________________________________
FCW = 0.68    Cycle # 1    :   ab = 0.33750   bc = 0.25313   ac = 0.20250   abc = 0.39375
FCW = 0.93    Cycle # 2    :   ab = 0.58537   bc = 0.49138   ac = 0.35096   abc = 0.69876
FCW = 0.78    Cycle # 3    :   ab = 0.66192   bc = 0.60771   ac = 0.39880   abc = 0.80621
FCW = 0.68    Cycle # 4    :   ab = 0.69383   bc = 0.67506   ac = 0.42083   abc = 0.85703
FCW = 0.88    Cycle # 5    :   ab = 0.71657   bc = 0.73754   ac = 0.43858   abc = 0.89795
FCW = 0.52    Cycle # 6    :   ab = 0.71657   bc = 0.73754   ac = 0.43858   abc = 0.89795
FCW = 0.68    Cycle # 7    :   ab = 0.72395   bc = 0.76828   ac = 0.44598   abc = 0.91464
FCW = 0.15    Cycle # 8    :   ab = 0.72395   bc = 0.76828   ac = 0.44598   abc = 0.91464
FCW = 0.78    Cycle # 9    :   ab = 0.72866   bc = 0.79392   ac = 0.45177   abc = 0.92729
FCW = 0.80    Cycle # 10   :   ab = 0.73107   bc = 0.81216   ac = 0.45567   abc = 0.93545
FCW = 0.75    Cycle # 11   :   ab = 0.73216   bc = 0.82384   ac = 0.45808   abc = 0.94028
FCW = 0.83    Cycle # 12   :   ab = 0.73278   bc = 0.83291   ac = 0.45992   abc = 0.94383
FCW = 0.30    Cycle # 13   :   ab = 0.73278   bc = 0.83291   ac = 0.45992   abc = 0.94383
FCW = 0.78    Cycle # 14   :   ab = 0.73305   bc = 0.83868   ac = 0.46107   abc = 0.94599
FCW = 0.48    Cycle # 15   :   ab = 0.73305   bc = 0.83868   ac = 0.46107   abc = 0.94599
FCW = 0.83    Cycle # 16   :   ab = 0.73320   bc = 0.84297   ac = 0.46193   abc = 0.94754
FCW = 0.85    Cycle # 17   :   ab = 0.73327   bc = 0.84598   ac = 0.46253   abc = 0.94861
```

This behavior simply results from multiplying our product Quyte (0 0 1 1) by the randomizing "maybe" Quyte (0 0 0 ψ), causing a dilation in the number of cycles needed for product creation.

This also suggests and shows that 'time' is not a physical (or real) property (quality or procedure) but is, rather, indicative of the endurance of a process in measurement.

Time is a matter of measurement: when (and where) there is no measurement, there is no time.

5. Peptide Folding example

An example here is given to application of the manner of **peptide folding** based on the disruption of **water molecules** around and between amino acid residues, as described in the text, with the addition of employing $\log C_n$ (in base Ψ, instead of $\ln C_n$, as **n** = # of waters involved for each modification of residue **phi** and **psi** angles) and with the use of '**tryptophane normilization**' : where the tryptophane residue is given the largest **relative hydrophobicity factor** (based on the number of carbon atoms).

The peptide chosen is **(-G-X-P -G-X-P-)**$_4$, the 24 residues acting as a possible internal sequence from a protein, The **X** residues are chosen at random, and this sequence in general is suggested for triple-helix tertiary structure conformations.

Here, the **X** amino acids (randomly picked by computer software) are : **F, C, R, F, C, H, T, H** .

The peptide starts out (program **prot_seqer**) as linear (i.e., **-G-X-P-**), with typical proline bends (where the proline side-chain forms an inter-chain ring with its the main-chain). Thus, all **phi** and **psi** angles are -180.0°, except for proline's with **phi** = -60.2° and **psi** = -178.1° . Each side-chain is of standard conformation off the tetrahedral c-alpha carbon (and remains so after the **phi** and **psi** angle adjustments due to the folding). No (e.g., energy based) geometric optimizations (to relieve steric hindrances) nor any other modifications are applied to the initial (i.e., computational: water molecule influenced) folded result.

The folded peptide (program **p_folding**, with the proline **phi** angle kept within a range of -40° to -80°) presents itself with the cysteine side-chains close enough to possibly condense for a cystine (S-S) linkage. The folding of the main-chain upon itself (without steric distortions) is somewhat prominent. Surface rendering (via solvent excluded volumes) of the peptide shows a deep 'pocket' or cavity, involving the two histadine residues. Upon energy minimization (to adjust side-chain dispositions) it is found that the two histadine side-chains and the arginine side-chain act as sort of a gate or opening boundary to this pocket, formed and anchored by the threonine side-chain at the bottom. Thus, the cavity might theoretically serve as an anionic trap or container.

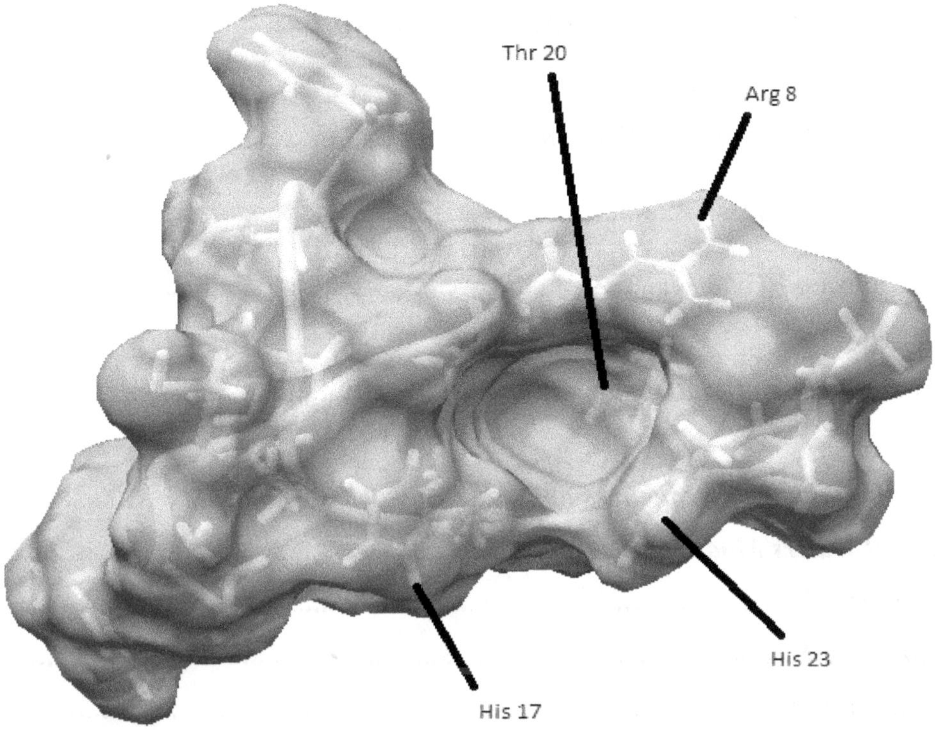

Of course this is all hypothetical, based on a calculative algorithm (rather than a refinement). The sequence may possibly adopt such a conformation. It is striking that these structural details can be derived or defined, designed or discovered, strictly from the amino acid sequence. This technique might be useful to demonstrate the formation of possible (or conceptual) local 'micro'-structures that can be (selectively) embedded within a protein as determined (or guided) by its amino acid sequence alone, a virtual sculpting of conceptual protein structure.

Headings Index

Appendices

n	C_n	$(┌┼┘)^n_m$	$(┌┼┘)^n$	e^n
1	1	1.5231	2.2905	2.7183
2	3	2.3198	5.2464	7.3892
3	7	3.5333	12.0169	20.0859
4	25	5.3816	27.5246	54.5996
5	161	8.1968	63.0451	148.4181

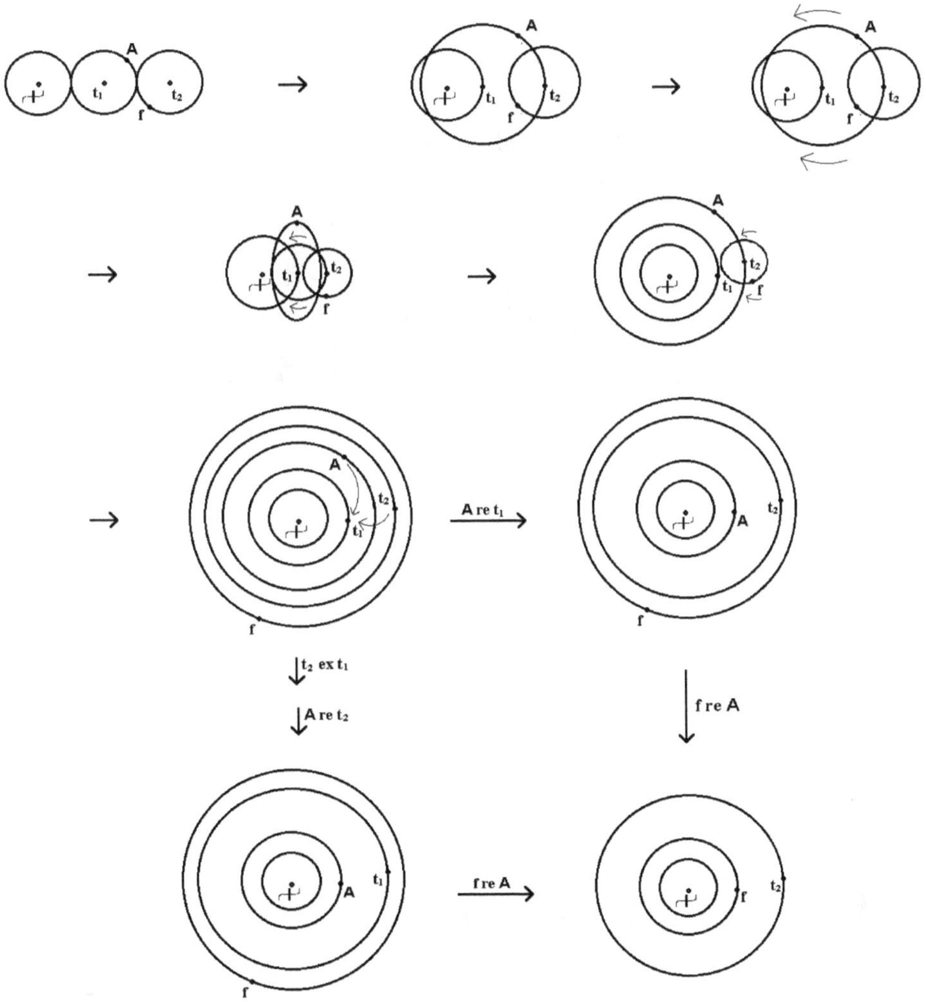

find at www.lulu.com :

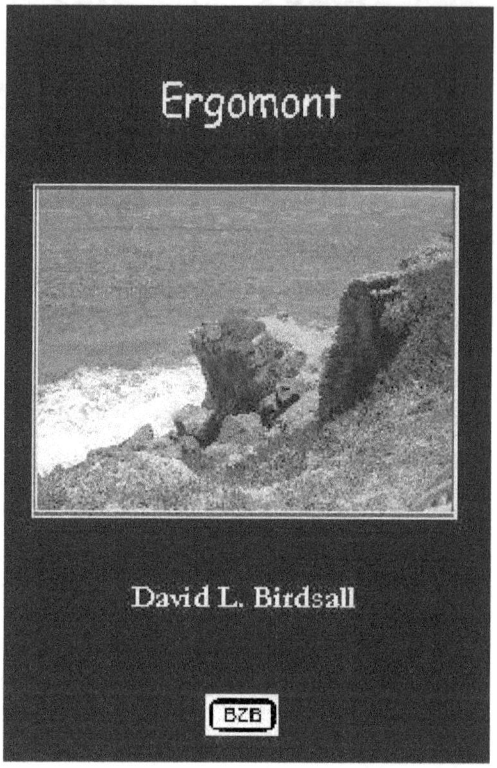

plays:

an Approach to Prescience

Zimmerich

A Pound's Profit

Peccant Pecus

Impasse of a Predicament of Fortitude

Intrepid Trepidations

The Moon Past Noon

also:

treatise on null/los